RANDOM DATA

RANDOM DATA

ANALYSIS AND MEASUREMENT PROCEDURES

Second Edition
(Revised and Expanded)

JULIUS S. BENDAT

Mathematical Consultant in Random Data Analysis

ALLAN G. PIERSOL

Senior Scientist, Astron Research and Engineering

A Wiley-Interscience Publication

JOHN WILEY & SONS

New York · Chichester · Brisbane · Toronto · Singapore

Library of Congress Cataloging in Publication Data:

Bendat, Julius S.
    Random data.

    "A Wiley-Interscience publication."
    Bibliography: p.
    Includes index.
    1. Engineering—Statistical methods. 2. Stochastic
processes. 3. Engineering—Data processing.
I. Piersol, Allan G. II. Title.

TA340.B43 1985   620'.0042   85-17996
ISBN 0-471-04000-2

Printed in the United States of America

10 9 8 7 6 5 4 3

PREFACE

This book is an extensively revised and expanded edition of our 1971 Wiley book by the same title, which in turn was a major revision of *Measurement and Analysis of Random Data*, first published in 1966. This current edition continues the process of improving and updating the appropriate procedures for random data analysis and measurements to reflect recent changes in model formulations, statistical error evaluations, data collection procedures, and computational algorithms. The first four chapters of the book include background material similar to that in the 1971 edition on basic descriptions of data, properties of linear systems, fundamentals of probability, and statistical principles. Chapter 5 presents a comprehensive discussion of stationary random processes which has been expanded on the basic matters and includes further topics such as derivative processes. Chapters 6 and 7 develop and detail input/output relations for single- and multiple-input problems with emphasis on modern analysis procedures that have evolved since the publication of the 1971 edition. Similarly, Chapters 8 and 9 formulate expressions for the statistical errors in various parameter estimates, including advanced estimates for which error formulas were not available in 1971. Chapters 10 and 11 cover updated discussions of data acquisition and digital data analysis procedures with added material on the processing of multiple-input/output data using efficient, iterative algorithms. Chapter 12 presents a completely revised and greatly expanded treatment of nonstationary data analysis, with emphasis on the different spectral descriptions of nonstationary data and the formulation of input/output relations for nonstationary inputs and/or time varying linear systems. Chapter 13 contains new material on Hilbert transform techniques and applications not discussed at all in the previous editions. The reader will also find Appendices A and B to be quite useful. Appendix A lists various statistical tables for convenient reference, and Appendix B gives concise definitions to help standardize many quantities in random data analysis.

Like the 1971 edition, the current book has been written primarily to provide a convenient reference for practicing engineers and scientists. It has further been designed to serve more fully as an analytical companion to the authors' applications oriented book, *Engineering Applications of Correlation and Spectral Analysis* (Wiley-Interscience, New York, 1980). The secondary intent of providing a specialized textbook for students is facilitated by problem sets at the end of each chapter. The reader is assumed to have a working

v

knowledge of calculus and transform methods of applied mathematics. A basic knowledge of system response characteristics, probability theory, and statistics also would be helpful, although these subjects are reviewed in the initial chapters of the book.

We wish to acknowledge the contributions to this book by many colleagues and associates, but in particular, George Richman for his helpful review of portions of the manuscript. We are also grateful to those government agencies and industrial companies that have supported our work, and to the University of California, Los Angeles, the Continuing Education Institute, and other organizations that have sponsored our presentation of short courses on this subject. Our final thanks go to Bette Clark, Ingrid Salazar, and Phyllis Parris for their careful help in preparing the manuscript.

JULIUS S. BENDAT
ALLAN G. PIERSOL

Los Angeles, California
January 1986

CONTENTS

GLOSSARY OF SYMBOLS

a, b	Sample regression coefficients, Arbitrary constants
A	Amplitude, Number of reverse arrangements
$b[\ ]$	Bias error of [ ]
B	Cyclical frequency bandwidth
c	Mechanical damping coefficient, Arbitrary constant
C	Electrical capacitance
C_{xy}	Covariance
$C_{xx}(\tau)$	Autocovariance function
$C_{xy}(\tau)$	Cross-covariance function
$C_{xy}(f)$	Coincident spectral density function (one-sided)
$e(t)$	Potential difference
$E[\ ]$	Expected value of [ ]
f	Cyclical frequency
Δf	Bandwidth resolution (Hz)
$\mathscr{F}[\ ]$	Fourier transform of [ ]
$G_{xx}(f)$	Autospectral density function (one-sided)
$G_{xy}(f)$	Cross-spectral density function (one-sided)
$G_{yy \cdot x}(f)$	Conditioned autospectral density function (one-sided)
$G_{x_i y \cdot x_j}(f)$	Conditioned cross-spectral density function (one-sided)
$\mathscr{G}(f)$	"Energy" spectral density function
$h(\tau)$	Unit impulse response function
$H(f)$	Frequency response function
$|H(f)|$	System gain factor
$\mathscr{H}[\ ]$	Hilbert transform of [ ]
i	Index
$i(t)$	Current
$\mathrm{Im}[\ ]$	Imaginary part of [ ]
j	$\sqrt{-1}$, Index
k	Mechanical spring constant, Index

K	Trace wave number, Number of class intervals
L	Electrical inductance, Length
$L(f)$	Frequency response function for conditioned inputs
m	Mechanical mass, Maximum number of lag values
m_f	Modulation index
n	Degrees-of-freedom, Index
N	Sample size
$p(x)$	Probability density function
$p(x, y)$	Joint probability density function
$P(x)$	Probability distribution function
$P(x, y)$	Joint probability distribution function
Prob[ ]	Probability that [ ]
q	Number of inputs, Number of sample records
$q(t)$	Electrical charge
$Q_{xy}(f)$	Quadrature spectral density function (one-sided)
r	Number of runs, Number of outputs
r_{xy}	Sample correlation coefficient
R	Electrical resistance
$R_{xx}(\tau)$	Autocorrelation function
$R_{xy}(\tau)$	Cross-correlation function
$R(t_1, t_2)$	Double time correlation function
$\mathscr{R}(\tau, t)$	Alternate double time correlation function
Re[ ]	Real part of [ ]
s	Sample standard deviation
s^2	Sample variance
s_{xy}	Sample covariance
s.d.[ ]	Standard deviation of [ ]
$S_{xx}(f)$	Autospectral density function (two-sided)
$S_{xy}(f)$	Cross-spectral density function (two-sided)
$S_{yy \cdot x}(f)$	Conditioned autospectral density function (two-sided)
$S_{x_i y \cdot x_j}(f)$	Conditioned cross-spectral density function (two-sided)
$\mathscr{S}(f)$	"Energy" spectral density function (two-sided)
$S(f_1, f_2)$	Double frequency spectral density function (two-sided)
$\mathscr{S}(f, g)$	Alternate double frequency spectral density function (two-sided)
S/N	Signal to noise ratio
t	Time variable, Student t variable
Δt	Sampling interval
T	Record length, period
T_r	Total record length
u_n	Raw data values
$u(t), v(t)$	Time dependent variables
Var [ ]	Variance of [ ]
W	Amplitude window width
$\mathscr{W}(f, t)$	Frequency-time spectral density function for both positive and negative frequencies (two-sided)

$x(t)$, $y(t)$	Time dependent variables
$\bar{x}$	Sample mean value of x
X	Amplitude of sinusoidal $x(t)$
$X(f)$	Fourier transform of $x(t)$
$X(f, T)$	Fourier transform of $x(t)$ over record length T
z	Standardized normal variable
$\lvert [\ ] \rvert$	Absolute value of [ ]
$[\hat{\ }]$	Estimate of [ ]
α	A small probability, Level of significance, Dummy variable
β	Probability of a Type II error, Dummy variable
$\gamma_{xy}^2(f)$	Ordinary coherence function
$\gamma_{y:x}^2(f)$	Multiple coherence function
$\gamma_{x_i y \cdot x_j}^2(f)$	Partial coherence function
$\delta(\ )$	Delta function
Δ	Small increment
ε	Normalized error
ζ	Mechanical damping ratio
θ	Phase angle
$\theta_{xy}(f)$	Argument of $G_{xy}(f)$
μ	Mean value
ρ	Correlation coefficient
$\rho(\tau)$	Correlation coefficient function
σ	Standard deviation
σ^2	Variance
τ	Time displacement
$\phi(f)$	Phase factor
ϕ	Arbitrary statistical parameter
χ^2	Statistical chi-square variable
ψ	Root mean square value
ψ^2	Mean square value

RANDOM DATA

CHAPTER 1

BASIC DESCRIPTIONS
AND PROPERTIES

This first chapter gives basic descriptions and properties of deterministic data and random data to provide a physical understanding for later material in this book. Simple classification ideas are used to explain differences between stationary random data, ergodic random data and nonstationary random data. Fundamental statistical functions are defined by words alone for analyzing the amplitude, time and frequency domain properties of single stationary random records and pairs of stationary random records. An introduction is presented on various types of input/output problems solved in this book, as well as necessary error analysis criteria to design experiments and evaluate measurements.

1.1 DETERMINISTIC VERSUS RANDOM DATA

Any observed data representing a physical phenomenon can be broadly classified as being either deterministic or nondeterministic. Deterministic data are those that can be described by an explicit mathematical relationship. For example, consider a rigid body that is suspended from a fixed foundation by a linear spring, as shown in Figure 1.1. Let m be the mass of the body (assumed to be inelastic) and k be the spring constant of the spring (assumed to be massless). Suppose the body is displaced from its position of equilibrium by a distance X, and released at time $t = 0$. From either basic laws of mechanics or repeated observations, it can be established that the following relationship will apply:

$$x(t) = X \cos \sqrt{\frac{k}{m}} \, t \qquad t \geq 0 \qquad (1.1)$$

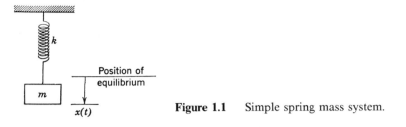

Figure 1.1   Simple spring mass system.

Equation (1.1) defines the exact location of the body at any instant of time in the future. Hence the physical data representing the motion of the mass are deterministic.

There are many physical phenomena in practice that produce data that can be represented with reasonable accuracy by explicit mathematical relationships. For example, the motion of a satellite in orbit about the earth, the potential across a condenser as it discharges through a resistor, the vibration response of an unbalanced rotating machine, and the temperature of water as heat is applied are all basically deterministic. However, there are many other physical phenomena that produce data that are not deterministic. For example, the height of waves in a confused sea, the acoustic pressures generated by air rushing through a pipe, and the electrical output of a noise generator represent data that cannot be described by explicit mathematical relationships. There is no way to predict an exact value at a future instant of time. These data are random in character and must be described in terms of probability statements and statistical averages rather than by explicit equations.

The classification of various physical data as being either deterministic or random might be debated in many cases. For example, it might be argued that no physical data in practice can be truly deterministic since there is always a possibility that some unforeseen event in the future might influence the phenomenon producing the data in a manner that was not originally considered. On the other hand, it might be argued that no physical data are truly random, since exact mathematical descriptions might be possible if a sufficient knowledge of the basic mechanisms of the phenomenon producing the data were available. In practical terms, the decision of whether physical data are deterministic or random is usually based on the ability to reproduce the data by controlled experiments. If an experiment producing specific data of interest can be repeated many times with identical results (within the limits of experimental error), then the data can generally be considered deterministic. If an experiment cannot be designed that will produce identical results when the experiment is repeated, then the data must usually be considered random in nature.

Various special classifications of deterministic and random data will now be discussed. Note that the classifications are selected from an analysis viewpoint and do not necessarily represent the most suitable classifications from other possible viewpoints. Further note that physical data are usually thought of as being functions of time and will be discussed in such terms for convenience. Any other variable, however, can replace time, as required.

1.2 CLASSIFICATIONS OF DETERMINISTIC DATA

Data representing deterministic phenomena can be categorized as being either periodic or nonperiodic. Periodic data can be further categorized as being either sinusoidal or complex periodic. Nonperiodic data can be further categorized as being either "almost-periodic" or transient. These various classifications of deterministic data are schematically illustrated in Figure 1.2. Of course, any combination of these forms may also occur. For purposes of review, each of these types of deterministic data, along with physical examples, will be briefly discussed.

1.2.1 *Sinusoidal Periodic Data*

Sinusoidal data are those types of periodic data that can be defined mathematically by a time-varying function of the form

$$x(t) = X \sin(2\pi f_0 t + \theta) \tag{1.2}$$

where

X = amplitude
f_0 = cyclical frequency in cycles per unit time
θ = initial phase angle with respect to the time origin in radians
$x(t)$ = instantaneous value at time t

The sinusoidal time history described by Equation (1.2) is usually referred to as a sine wave. When analyzing sinusoidal data in practice, the phase angle θ is often ignored. For this case,

$$x(t) = X \sin 2\pi f_0 t \tag{1.3}$$

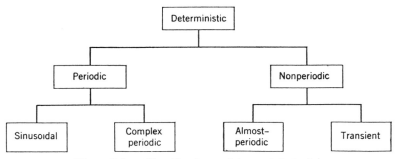

Figure 1.2   Classifications of deterministic data.

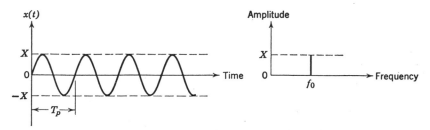

Figure 1.3   Time history and spectrum of sinusoidal data.

Equation (1.3) can be pictured by a time history plot or by an amplitude-frequency plot (frequency spectrum), as illustrated in Figure 1.3.

The time interval required for one full fluctuation or cycle of sinusoidal data is called the period T_p. The number of cycles per unit time is called the frequency f_0. The frequency and period are related by

$$T_p = \frac{1}{f_0} \tag{1.4}$$

Note that the frequency spectrum in Figure 1.3 is composed of an amplitude component at a specific frequency, as opposed to a continuous plot of amplitude versus frequency. Such spectra are called *discrete spectra* or *line spectra*.

There are many examples of physical phenomena that produce approximately sinusoidal data in practice. The voltage output of an electrical alternator is one example; the vibratory motion of an unbalanced rotating weight is another. Sinusoidal data represent one of the simplest forms of time-varying data from the analysis viewpoint.

1.2.2   Complex Periodic Data

Complex periodic data are those types of periodic data that can be defined mathematically by a time-varying function whose waveform exactly repeats itself at regular intervals such that

$$x(t) = x(t \pm nT_p) \qquad n = 1, 2, 3, \ldots \tag{1.5}$$

As for sinusoidal data, the time interval required for one full fluctuation is called the *period* T_p. The number of cycles per unit time is called the *fundamental frequency* f_1. A special case for complex periodic data is clearly sinusoidal data, where $f_1 = f_0$.

With few exceptions in practice, complex periodic data may be expanded into a Fourier series according to the following formula:

$$x(t) = \frac{a_0}{2} + \sum_{n=1}^{\infty} (a_n \cos 2\pi n f_1 t + b_n \sin 2\pi n f_1 t) \qquad (1.6)$$

where

$$f_1 = \frac{1}{T_p}$$

$$a_n = \frac{2}{T_p} \int_0^{T_p} x(t) \cos 2\pi n f_1 t \, dt \qquad n = 0, 1, 2, \ldots$$

$$b_n = \frac{2}{T_p} \int_0^{T_p} x(t) \sin 2\pi n f_1 t \, dt \qquad n = 1, 2, 3, \ldots$$

An alternative way to express the Fourier series for complex periodic data is

$$x(t) = X_0 + \sum_{n=1}^{\infty} X_n \cos(2\pi n f_1 t - \theta_n) \qquad (1.7)$$

where

$$X_0 = a_0/2$$

$$X_n = \sqrt{a_n^2 + b_n^2} \qquad n = 1, 2, 3, \ldots$$

$$\theta_n = \tan^{-1}(b_n/a_n) \qquad n = 1, 2, 3, \ldots$$

In words, Equation (1.7) says that complex periodic data consist of a static component X_0 and an infinite number of sinusoidal components called harmonics, which have amplitudes X_n and phases θ_n. The frequencies of the harmonic components are all integral multiples of f_1.

When analyzing periodic data in practice, the phase angles θ_n are often ignored. For this case, Equation (1.7) can be characterized by a discrete spectrum, as illustrated in Figure 1.4. Sometimes, complex periodic data will include only a few components. In other cases, the fundamental component may be absent. For example, suppose a periodic time history is formed by mixing three sine waves, which have frequencies of 60, 75, and 100 Hz. The highest common divisor is 5 Hz, so the period of the resulting periodic data is $T_p = 0.2$ sec. Hence when expanded into a Fourier series, all values of X_n are zero except for $n = 12$, $n = 15$, and $n = 20$.

Physical phenomena that produce complex periodic data are far more common than those that produce simple sinusoidal data. In fact, the classifica-

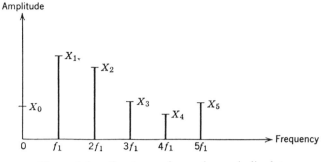

Figure 1.4   Spectrum of complex periodic data.

tion of data as being sinusoidal is often only an approximation for data that are actually complex. For example, the voltage output from an electrical alternator may actually display, under careful inspection, some small contributions at higher harmonic frequencies. In other cases, intense harmonic components may be present in periodic physical data. For example, the vibration response of a multicylinder reciprocating engine will usually display considerable harmonic content.

1.2.3   *Almost-Periodic Data*

In Section 1.2.2, it is noted that periodic data can generally be reduced to a series of sine waves with commensurately related frequencies. Conversely, the data formed by summing two or more commensurately related sine waves will be periodic. However, the data formed by summing two or more sine waves with arbitrary frequencies generally will not be periodic. Specifically, the sum of two or more sine waves will be periodic only when the ratios of all possible pairs of frequencies form rational numbers. This indicates that a fundamental period exists, which will satisfy the requirements of Equation (1.5). Hence

$$x(t) = X_1 \sin(2t + \theta_1) + X_2 \sin(3t + \theta_2) + X_3 \sin(7t + \theta_3)$$

is periodic since $\frac{2}{3}$, $\frac{2}{7}$, and $\frac{3}{7}$ are rational numbers (the fundamental period is $T_p = 1$). On the other hand,

$$x(t) = X_1 \sin(2t + \theta_1) + X_2 \sin(3t + \theta_2) + X_3 \sin(\sqrt{50}\, t + \theta_3)$$

is not periodic since $2/\sqrt{50}$ and $3/\sqrt{50}$ are not rational numbers (the fundamental period is infinitely long). The resulting time history in this case will have an almost-periodic character, but the requirements of Equation (1.5) will not be satisfied for any finite value of T_p.

Based on these discussions, almost-periodic data are those types of nonperiodic data that can be defined mathematically by a time-varying function of the

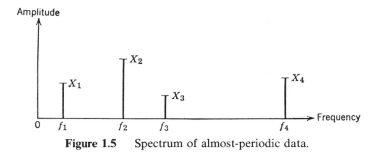

Figure 1.5   Spectrum of almost-periodic data.

form

$$x(t) = \sum_{n=1}^{\infty} X_n \sin(2\pi f_n t + \theta_n) \tag{1.8}$$

where $f_n/f_m \neq$ rational number in all cases. Physical phenomena producing almost-periodic data frequently occur in practice when the effects of two or more unrelated periodic phenomena are mixed. A good example is the vibration response in a multiple engine propeller airplane when the engines are out of synchronization.

An important property of almost-periodic data is as follows. If the phase angles θ_n are ignored, Equation (1.8) can be characterized by a discrete frequency spectrum similar to that for complex periodic data. The only difference is that the frequencies of the components are not related by rational numbers, as illustrated in Figure 1.5.

1.2.4   *Transient Nonperiodic Data*

Transient data are defined as all nonperiodic data other than the almost-periodic data discussed in Section 1.2.3. In other words, transient data include all data not previously discussed that can be described by some suitable time-varying function. Three simple examples of transient data are given in Figure 1.6.

Physical phenomena that produce transient data are numerous and diverse. For example, the data in Figure 1.6(*a*) could represent the temperature of water in a kettle (relative to room temperature) after the flame is turned off. The data in Figure 1.6(*b*) might represent the free vibration of a damped mechanical system after an excitation force is removed. The data in Figure 1.6(*c*) could represent the stress in an end-loaded cable that breaks at time *c*.

An important characteristic of transient data, as opposed to periodic and almost-periodic data, is that a discrete spectral representation is not possible. A *continuous* spectral representation for transient data can be obtained in

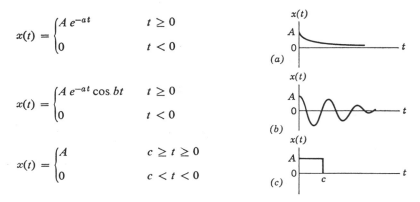

$$x(t) = \begin{cases} A\,e^{-at} & t \geq 0 \\ 0 & t < 0 \end{cases}$$

$$x(t) = \begin{cases} A\,e^{-at}\cos bt & t \geq 0 \\ 0 & t < 0 \end{cases}$$

$$x(t) = \begin{cases} A & c \geq t \geq 0 \\ 0 & c < t < 0 \end{cases}$$

Figure 1.6   Illustrations of transient data.

most cases, however, from a Fourier transform given by

$$X(f) = \int_{-\infty}^{\infty} x(t)e^{-j2\pi ft}\,dt \tag{1.9}$$

The Fourier transform $X(f)$ is generally a complex number that can be expressed in complex polar notation as

$$X(f) = |X(f)|e^{-j\theta(f)}$$

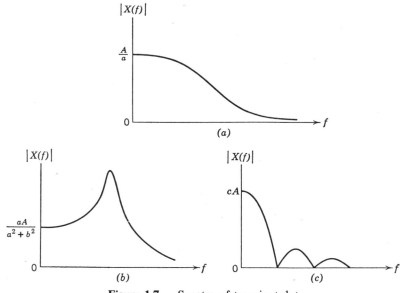

Figure 1.7   Spectra of transient data.

Here, $|X(f)|$ is the magnitude of $X(f)$ and $\theta(f)$ is the argument. In terms of the magnitude $|X(f)|$, continuous spectra of the three transient time histories in Figure 1.6 are as presented in Figure 1.7. Modern procedures for the digital computation of Fourier series and finite Fourier transforms are detailed in Chapter 11.

1.3   CLASSIFICATIONS OF RANDOM DATA

As discussed earlier, data representing a random physical phenomenon cannot be described by an explicit mathematical relationship, because each observation of the phenomenon will be unique. In other words, any given observation will represent only one of many possible results that might have occurred. For example, assume the output voltage from a thermal noise generator is recorded as a function of time. A specific voltage time history record will be obtained, as shown in Figure 1.8. If a second thermal noise generator of identical construction and assembly is operated simultaneously, however, a different voltage time history record would result. In fact, every thermal noise generator that might be constructed would produce a different voltage time history record, as illustrated in Figure 1.8. Hence the voltage time history for any one generator is merely one example of an infinitely large number of time histories that might have occurred.

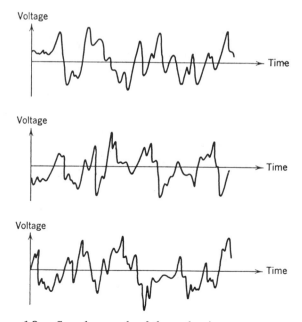

Figure 1.8   Sample records of thermal noise generator outputs.

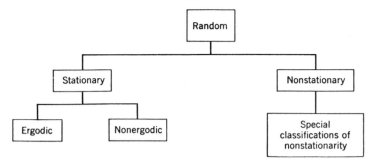

Figure 1.9 Classifications of random data.

A single time history representing a random phenomenon is called a *sample function* (or a *sample record* when observed over a finite time interval). The collection of all possible sample functions that the random phenomenon might have produced is called a *random process* or a *stochastic process*. Hence a sample record of data for a random physical phenomenon may be thought of as one physical realization of a random process.

Random processes may be categorized as being either stationary or nonstationary. Stationary random processes may be further categorized as being either ergodic or nonergodic. Nonstationary random processes may be further categorized in terms of specific types of nonstationary properties. These various classifications of random processes are schematically illustrated in Figure 1.9. The meaning and physical significance of these various types of random processes will now be discussed in broad terms. More analytical definitions and developments are presented in Chapters 5 and 12.

1.3.1 *Stationary Random Data*

When a physical phenomenon is considered in terms of a random process, the properties of the phenomenon can hypothetically be described at any instant of time by computing average values over the collection of sample functions that describe the random process. For example, consider the collection of sample functions (also called the *ensemble*) that forms the random process illustrated in Figure 1.10. The *mean value* (first moment) of the random process at some time t_1 can be computed by taking the instantaneous value of each sample function of the ensemble at time t_1, summing the values, and dividing by the number of sample functions. In a similar manner, a correlation (joint moment) between the values of the random process at two different times (called the *autocorrelation function*) can be computed by taking the ensemble average of the product of instantaneous values at two times, t_1 and $t_1 + \tau$. That is, for the random process $\{x(t)\}$, where the symbol $\{\ \}$ is used to denote an ensemble of sample functions, the mean value $\mu_x(t_1)$ and the

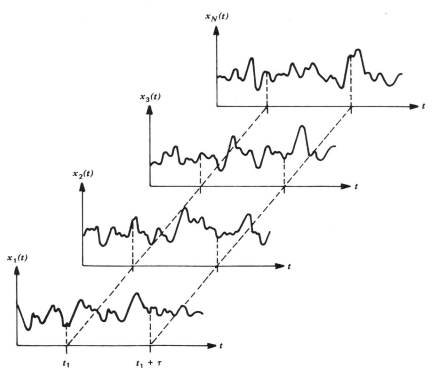

Figure 1.10   Ensemble of time-history records defining a random process.

autocorrelation function $R_{xx}(t_1, t_1 + \tau)$ are given by

$$\mu_x(t_1) = \lim_{N \to \infty} \frac{1}{N} \sum_{k=1}^{N} x_k(t_1) \tag{1.10a}$$

$$R_{xx}(t_1, t_1 + \tau) = \lim_{N \to \infty} \frac{1}{N} \sum_{k=1}^{N} x_k(t_1) x_k(t_1 + \tau) \tag{1.10b}$$

where the final summation assumes each sample function is equally likely.

For the general case where $\mu_x(t_1)$ and $R_{xx}(t_1, t_1 + \tau)$ defined in Equation (1.10) vary as time t_1 varies, the random process $\{x(t)\}$ is said to be *nonstationary*. For the special case where $\mu_x(t_1)$ and $R_{xx}(t_1, t_1 + \tau)$ do not vary as time t_1 varies, the random process $\{x(t)\}$ is said to be *weakly stationary* or stationary in the wide sense. For weakly stationary random processes, the mean value is a constant and the autocorrelation function is dependent only on the time displacement τ. That is, $\mu_x(t_1) = \mu_x$ and $R_{xx}(t_1, t_1 + \tau) = R_{xx}(\tau)$.

An infinite collection of higher-order moments and joint moments of the random process $\{x(t)\}$ could also be computed to establish a complete family of probability distribution functions describing the process. For the special case where all possible moments and joint moments are time invariant, the random process $\{x(t)\}$ is said to be *strongly stationary* or stationary in the strict sense. For many practical applications, verification of weak stationarity will justify an assumption of strong stationarity.

1.3.2   *Ergodic Random Data*

In Section 1.3.1, it is noted how the properties of a random process can be determined by computing ensemble averages at specific instants of time. In most cases, however, it is also possible to describe the properties of a stationary random process by computing time averages over specific sample functions in the ensemble. For example, consider the kth sample function of the random process illustrated in Figure 1.10. The mean value $\mu_x(k)$ and the autocorrelation function $R_{xx}(\tau, k)$ of the kth sample function are given by

$$\mu_x(k) = \lim_{T \to \infty} \frac{1}{T} \int_0^T x_k(t) \, dt \qquad (1.11a)$$

$$R_{xx}(\tau, k) = \lim_{T \to \infty} \frac{1}{T} \int_0^T x_k(t) x_k(t + \tau) \, dt \qquad (1.11b)$$

If the random process $\{x(t)\}$ is stationary, and $\mu_x(k)$ and $R_{xx}(\tau, k)$ defined in Equation (1.11) do not differ when computed over different sample functions, the random process is said to be *ergodic*. For ergodic random processes, the time-averaged mean value and autocorrelation function (as well as all other time-averaged properties) are equal to the corresponding ensemble averaged values. That is, $\mu_x(k) = \mu_x$ and $R_{xx}(\tau, k) = R_{xx}(\tau)$. Note that only stationary random processes can be ergodic.

Ergodic random processes are clearly an important class of random processes since all properties of ergodic random processes can be determined by performing time averages over a single sample function. Fortunately, in practice, random data representing stationary physical phenomena are generally ergodic. It is for this reason that the properties of stationary random phenomena can be measured properly, in most cases, from a single observed time history record. A full development of the properties of ergodic random processes is presented in Chapter 5.

1.3.3   *Nonstationary Random Data*

Nonstationary random processes include all random processes that do not meet the requirements for stationarity defined in Section 1.3.1. Unless further restrictions are imposed, the properties of a nonstationary random process are generally time-varying functions that can be determined only by performing instantaneous averages over the ensemble of sample functions forming the process. In practice, it is often not feasible to obtain a sufficient number of sample records to permit the accurate measurement of properties by ensemble averaging. That fact has tended to impede the development of practical techniques for measuring and analyzing nonstationary random data.

In many cases, the nonstationary random data produced by actual physical phenomena can be classified into special categories of nonstationarity that simplify the measurement and analysis problem. For example, some types of random data might be described by a nonstationary random process $\{x(t)\}$, where each sample function is given by $x(t) = a(t)u(t)$. Here, $u(t)$ is a sample function from a stationary random process $\{u(t)\}$ and $a(t)$ is a deterministic multiplication factor. In other words, the data might be represented by a nonstationary random process consisting of sample functions with a common deterministic time trend. If nonstationary random data fit a specific model of this type, ensemble averaging is not always needed to describe the data. The various desired properties can sometimes be estimated from a single sample record, as is true for ergodic stationary data. These matters are developed in detail in Chapter 12.

1.3.4   *Stationary Sample Records*

The concept of stationarity, as defined and discussed in Section 1.3.1, relates to the ensemble averaged properties of a random process. In practice, however, data in the form of individual time history records of a random phenomenon are frequently referred to as being stationary or nonstationary. A slightly different interpretation of stationarity is involved here. When a single time history record is referred to as being stationary, it is generally meant that the properties computed over short time intervals do not vary significantly from one interval to the next. The word *significantly* is used here to mean that observed variations are greater than would be expected due to normal statistical sampling variations.

To help clarify this point, consider a single sample record $x_k(t)$ obtained from the kth sample function of a random process $\{x(t)\}$. Assume a mean value and autocorrelation function are obtained by time averaging over a short interval T with a starting time of t_1 as follows:

$$\mu_x(t_1, k) = \frac{1}{T} \int_{t_1}^{t_1 + T} x_k(t) \, dt \tag{1.12a}$$

$$R_{xx}(t_1, t_1 + \tau, k) = \frac{1}{T} \int_{t_1}^{t_1 + T} x_k(t) x_k(t + \tau) \, dt \tag{1.12b}$$

For the general case where the sample properties defined in Equation (1.12) vary significantly as the starting time t_1 varies, the individual sample record is said to be nonstationary. For the special case where the sample properties defined in Equation (1.12) do not vary significantly as the starting time t_1 varies, the sample record is said to be stationary. Note that a sample record obtained from an ergodic random process will be stationary. Furthermore, sample records from most physically interesting nonstationary random processes will be nonstationary. Hence if an ergodic assumption is justified (as it is for most actual stationary physical phenomena), verification of stationarity for a single sample record will effectively justify an assumption of stationarity and ergodicity for the random process from which the sample record is obtained. Tests for stationarity of individual sample records are discussed in Chapters 4 and 10.

1.4 ANALYSIS OF RANDOM DATA

The analysis of random data involves different considerations from the deterministic data discussed in Section 1.2. In particular, since no explicit mathematical equation can be written for the time histories produced by a random phenomenon, statistical procedures must be used to define the descriptive properties of the data. Nevertheless, well-defined input/output relations exist for random data, which are fundamental to a wide range of applications. In such applications, however, an understanding and control of the statistical errors associated with the computed data properties and input/output relationships is essential.

1.4.1 *Basic Descriptive Properties*

Basic statistical properties of importance for describing single stationary random records are

1. Mean and mean square values
2. Probability density functions
3. Autocorrelation functions
4. Autospectral density functions

For the present discussion, it is instructive to define these quantities by words alone, without the use of mathematical equations. After this has been done, they will be illustrated for special cases of interest.

The mean value μ_x and the variance σ_x^2 for a stationary record represent the central tendency and dispersion, respectively, of the data. The mean square value ψ_x^2, which equals the variance plus the square of the mean, constitutes a measure of the combined central tendency and dispersion. The mean value is estimated by simply computing the average of all data values in the record. The mean square value is similarly estimated by computing the average of the squared data values. By first subtracting the mean value estimate from all the data values, the mean square value computation yields a variance estimate.

The probability density function $p(x)$ for a stationary record represents the rate of change of probability with data value. The function $p(x)$ is generally estimated by computing the probability that the instantaneous value of the single record will be in a particular narrow amplitude range centered at various data values, and then dividing by the amplitude range. The total area under the probability density function over all data values will be unity, since this merely indicates the certainty of the fact that the data values must fall between $-\infty$ and $+\infty$. The partial area under the probability density function from $-\infty$ to

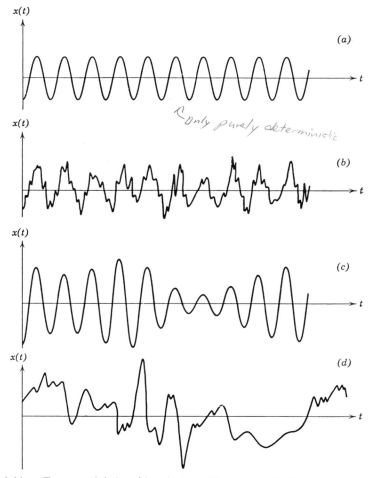

Figure 1.11 Four special time histories. (a) Sine wave. (b) Sine wave plus random noise. (c) Narrow-band random noise. (d) Wide-band random noise.

some given value x represents the probability distribution function, denoted by $P(x)$. The area under the probability density function between any two values x_1 and x_2, given by $P(x_2) - P(x_1)$, defines the probability that any future data values at a randomly selected time will fall within this amplitude interval. Probability density and distribution functions are fully discussed in Chapters 3 and 4.

The autocorrelation function $R_{xx}(\tau)$ for a stationary record is a measure of time-related properties in the data that are separated by fixed time delays. It

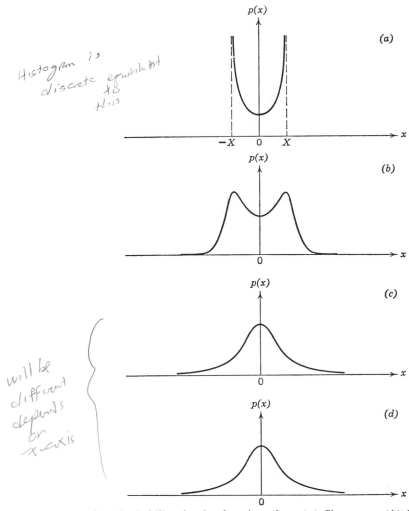

Histogram is discrete equivalent to this

will be different depends on x-axis

Figure 1.12 Probability density function plots. (*a*) Sine wave. (*b*) Sine wave plus random noise. (*c*) Narrow-band random noise. (*d*) Wide-band random noise.

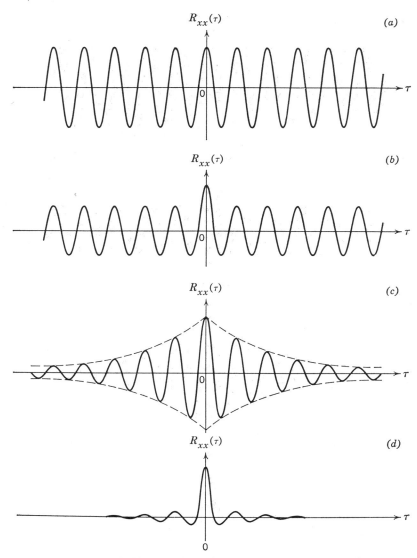

Figure 1.13 Autocorrelation function plots. (*a*) Sine wave. (*b*) Sine wave plus random noise. (*c*) Narrow-band random noise. (*d*) Wide-band random noise.

can be estimated by delaying the record relative to itself by some fixed time delay τ, then multiplying the original record with the delayed record, and averaging the resulting product values over the available record length or over some desired portion of this record length. The procedure is repeated for all time delays of interest.

The autospectral (also called *power* spectral) density function $G_{xx}(f)$ for a stationary record represents the rate of change of mean square value with

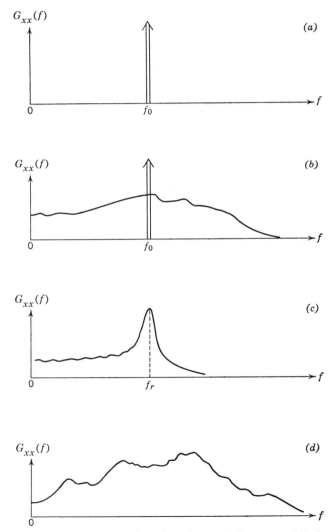

Figure 1.14 Autospectral density function plots. (*a*) Sine wave. (*b*) Sine wave plus random noise. (*c*) Narrow-band random noise. (*d*) Wide-band random noise.

frequency. It is estimated by computing the mean square value in a narrow frequency band at various center frequencies, and then dividing by the frequency band. The total area under the autospectral density function over all frequencies will be the total mean square value of the record. The partial area under the autospectral density function from f_1 to f_2 represents the mean square value of the record associated with that frequency range. Autocorrelation and autospectral density functions are developed in Chapter 5.

Four typical time histories of a sine wave, sine wave plus noise, narrow band noise, and wide band noise are shown in Figure 1.11. Theoretical plots of their probability density functions, autocorrelation functions, and autospectral density functions are shown in Figures 1.12, 1.13, and 1.14, respectively. Equations for all of these plots are given in Chapter 5, together with other theoretical formulas.

For pairs of random records from two different stationary random processes, joint statistical properties of importance are

1. Joint probability density functions
2. Cross-correlation functions
3. Cross-spectral density functions
4. Frequency response functions
5. Coherence functions

The first three functions measure fundamental properties shared by the pair of records in the amplitude, time, or frequency domains. From knowledge of the cross-spectral density function between the pair of records, as well as their individual autospectral density functions, one can compute theoretical linear frequency response functions (gain factors and phase factors) between the two records. Here, the two records are treated as a single-input/single-output problem. The coherence function is a measure of the accuracy of the assumed linear input/output model, and can also be computed from the measured autospectral and cross-spectral density functions. Detailed discussions of these topics appear in Chapters 5, 6 and 7.

Common applications of probability density and distribution functions, beyond a basic probabilistic description of data values, include

1. Evaluation of normality
2. Indication of nonlinear effects
3. Analysis of extreme values

The primary applications of correlation measurements include

1. Detection of periodicities
2. Prediction of signals in noise
3. Measurement of time delays
4. Location of disturbing sources
5. Identification of propagation paths and velocities

Typical applications of spectral density functions include

1. Determination of system properties from input data and output data
2. Prediction of output data from input data and system properties
3. Identification of input data from output data and system properties
4. Specifications of dynamic data for test programs
5. Identification of energy and noise sources
6. Optimum linear prediction and filtering

1.4.2   *Input/Output Relations*

Input/output cases of common interest can usually be considered as combinations of one or more of the following models:

1. Single-input/single-output model
2. Single-input/multiple-output model
3. Multiple-input/single-output model
4. Multiple-input/multiple-output model

In all cases, there may be one or more parallel transmission paths with different time delays between each input point and output point. For multiple input cases, the various inputs may or may not be correlated with each other. Special analysis techniques are required when nonstationary data are involved, as treated in Chapter 12.

A simple single-input/single-output model is shown in Figure 1.15. Here, $x(t)$ and $y(t)$ are the measured input and output stationary random records, and $n(t)$ is unmeasured extraneous output noise. The quantity $H_{xy}(f)$ is the frequency response function of a constant-parameter linear system between $x(t)$ and $y(t)$. Figure 1.16 shows a single-input/multiple-output model that is a simple extension of Figure 1.15 where an input $x(t)$ produces many outputs $y_i(t)$, $i = 1, 2, 3, \ldots$. Any output $y_i(t)$ is the result of $x(t)$ passing through a constant-parameter linear system described by the frequency response function $H_{xi}(f)$. The noise terms $n_i(t)$ represent unmeasured extraneous output noise at the different outputs. It is clear that Figure 1.16 can be considered as a combination of separate single-input/single-output models.

Appropriate procedures for solving single-input models are developed in Chapter 6 using measured autospectral and cross-spectral density functions.

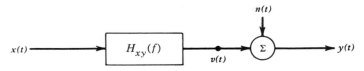

Figure 1.15   Single-input/single-output system with output noise.

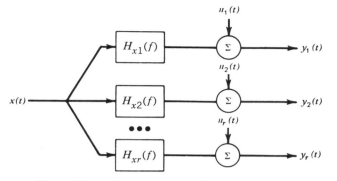

Figure 1.16   Single-input/multiple-output system.

Ordinary coherence functions are defined, which play a key role in both system-identification and source-identification problems. To determine both the gain factor and the phase factor of a desired frequency response function, it is always necessary to measure the cross-spectral density function between the input and output points. A good estimate of the gain factor alone can be obtained from measurements of the input and output autospectral density functions only if there is negligible input and output extraneous noise.

For a well-defined single-input/single-output model where the data are stationary, the system is linear and has constant parameters, and there is no extraneous noise at either the input or output point, the ordinary coherence function will be identically unity for all frequencies. Any deviation from these ideal conditions will cause the coherence function to be less than unity. In practice, measured coherence functions will often be less than unity, and are important in determining the statistical confidence in frequency response function measurements.

Extensions of these ideas can be carried out for general multiple-input/multiple-output problems, which require the definition and proper interpretation of multiple coherence functions and partial coherence functions. These general situations can be considered as combinations of a set of multiple-input/single-output models for a given set of stationary inputs and for different constant-parameter linear systems, as shown in Figure 1.17. Modern procedures for solving multiple-input/output problems are developed in Chapter 7 using conditioned (residual) spectral density functions. These procedures are extensions of classical regression techniques discussed in Chapter 4. In particular, the output autospectral density function in Figure 1.17 is decomposed to show how much of this output spectrum at any frequency is due to any input conditioned on other inputs in a prescribed order.

Basic statistical principles to evaluate random data properties are covered in Chapter 4. Error analysis formulas for bias errors and random errors are developed in Chapters 8 and 9 for various estimates made in analyzing single

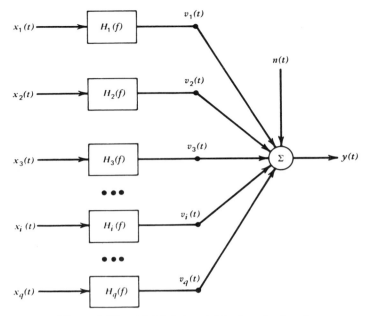

Figure 1.17   Multiple-input/single-output system.

random records and multiple random records. Included are random error formulas for estimates of frequency response functions (both gain factors and phase factors) and estimates of coherence functions (ordinary, multiple, or partial). These computations are easy to apply and should be performed to obtain proper interpretations of measured results.

1.4.3   *Error Analysis Criteria*

Some error analysis criteria for measured quantities will now be defined as background for the material in Chapters 8 and 9. Let a hat ($\hat{}$) symbol over a quantity ϕ, namely, $\hat{\phi}$, denote an estimate of this quantity. The quantity $\hat{\phi}$ will be an estimate of ϕ based on finite time interval or finite number of sample points.

Conceptually, suppose $\hat{\phi}$ can be estimated many times by repeating an experiment or some measurement program. Then, the expected value of $\hat{\phi}$, denoted by $E[\hat{\phi}]$, is something one can estimate. For example, if an experiment is repeated many times to yield results $\hat{\phi}_i$, $i = 1, 2, \ldots, N$, then

$$E[\hat{\phi}] = \frac{1}{N} \sum_{i=1}^{N} \hat{\phi}_i \qquad (1.13)$$

This expected value may or may not equal the true value ϕ. If it does, the estimate $\hat{\phi}$ is said to be unbiased. Otherwise, it is said to be biased. The *bias* of the estimate, denoted $b[\hat{\phi}]$, is equal to the expected value of the estimate minus the true value, that is

$$b[\hat{\phi}] = E[\hat{\phi}] - \phi \tag{1.14}$$

It follows that the bias error is a systematic error that always occurs with the same magnitude in the same direction when measurements are repeated under identical circumstances.

The *variance* of the estimate, denoted $\text{Var}[\hat{\phi}]$, is defined as the expected value of the squared differences from the mean value. In equation form,

$$\text{Var}[\hat{\phi}] = E\left[(\hat{\phi} - E[\hat{\phi}])^2\right] \tag{1.15}$$

The variance describes the *random error* of the estimate, that is, that portion of the error that is not systematic and can occur in either direction with different magnitudes from one measurement to another.

An assessment of the total estimation error is given by the *mean square error*, which is defined as the expected value of the squared differences from the *true* value. The mean square error of $\hat{\phi}$ is indicated by

$$\text{mean square error}[\hat{\phi}] = E\left[(\hat{\phi} - \phi)^2\right] \tag{1.16}$$

It is easy to verify that

$$E\left[(\hat{\phi} - \phi)^2\right] = \text{Var}[\hat{\phi}] + (b[\hat{\phi}])^2 \tag{1.17}$$

In words, the mean square error is equal to the variance plus the square of the bias. If the bias is zero or negligible, then the mean square error and variance are equivalent.

Figure 1.18 illustrates the meaning of the bias (systematic) error and the variance (random) error for the case of testing two guns for possible purchase by shooting each gun at a target. In Figure 1.18(*a*), gun A has a large bias error and small variance error. In Figure 1.18(*b*), gun B has a small bias error but large variance error. As shown, gun A will never hit the target, whereas gun

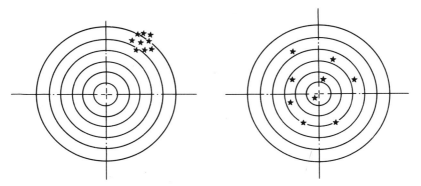

Figure 1.18   Random and bias errors in gun shoots at a target. (*a*) Gun A: Large bias error and small random error. (*b*) Gun B: Small bias error and large random error.

B will occasionally hit the target. Nevertheless, most people would prefer to buy gun A because the bias error can be removed (assuming one knows it is present) by adjusting the sights of the gun, but the random error cannot be removed. Hence, gun A provides the potential for a smaller mean square error.

A final important quantity is the *normalized rms error* of the estimate, denoted by $\varepsilon[\hat{\phi}]$. This error is a dimensionless quantity that is equal to the square root of the mean square error divided by the true value (assumed, of course, to be different from zero). Symbolically,

$$\varepsilon[\hat{\phi}] = \frac{\sqrt{E\left[(\hat{\phi} - \phi)^2\right]}}{\phi} \qquad (1.18)$$

In practice, one should try to make the normalized rms error as small as possible. This will help to guarantee that an arbitrary estimate $\hat{\phi}$ will lie close to the true value ϕ.

1.4.4 *Data Analysis Procedures*

Recommended data analysis procedures are discussed in great detail in Chapters 10–13. Chapter 10 is concerned with data acquisition problems, including data collection, recording, transmission, preparation, and qualification. General steps are outlined for proper data analysis of individual records and collection of records, as would be needed for different applications. The emphasis here is on appropriate methods for analyzing the properties of stationary random data. Digital data analysis techniques discussed in Chapter 11 involve computational procedures to perform trend removal, digital filtering, Fourier series, and fast Fourier transforms on discrete time series data representing sample records from stationary (ergodic) random data. Digital formulas are developed to compute estimates of probability density functions, correlation functions, and spectral density functions for individual records and for associated joint records. Further detailed digital procedures are stated to obtain estimates of all of the quantities described in Chapters 6 and 7 to solve various types of single-input/output problems and multiple-input/output problems. Chapter 12 is devoted to separate methods for nonstationary data analysis, and Chapter 13 concerns Hilbert transform techniques.

PROBLEMS

1.1   For the following functions, which are periodic and which are nonperiodic?
   (a) $x(t) = 3 \sin t + 2 \sin 2t + \sin 3t$.
   (b) $x(t) = 3 \sin t + 2 \sin 2t + \sin \pi t$.
   (c) $x(t) = 3 \sin 4t + 2 \sin 5t + \sin 6t$.
   (d) $x(t) = e^{-t} \sin t$.

1.2   Determine the period of the function defined by

$$x(t) = \sin 11t + \sin 12t$$

In Problems 1.3–1.6, state which properties are always true.

1.3   A stationary random process must
   (a)   be discrete.
   (b)   be continuous.
   (c)   be ergodic.
   (d)   have ensemble averaged properties that are independent of time.
   (e)   have time averaged properties that are equal to the ensemble averaged properties.

1.4   An ergodic random process must
   (a)   be discrete.
   (b)   be continuous.
   (c)   be stationary.
   (d)   have ensemble averaged properties that are independent of time.
   (e)   have time averaged properties that are equal to the ensemble averaged properties.

1.5   A single sample function can be used to find all statistical properties of a random process if the process is
   (a)   deterministic.
   (b)   ergodic.
   (c)   stationary.
   (d)   all of the above.

1.6   The autocorrelation function of a stationary random process
   (a)   must decrease as $|\tau|$ increases.
   (b)   is a function of the time difference only.
   (c)   must approach a constant as $|\tau|$ increases.
   (d)   must always be nonnegative.

1.7   How do the answers to Problem 1.6 change if the stationary random process is restricted not to include any periodic components?

1.8   Derive Equation (1.17).

1.9   An estimate is known to have a mean square error of 0.25 and a bias error of 0.40. Determine the variance of the estimate.

1.10 In Problem 1.9, if the quantity being estimated has a true value of $\phi = 5$, what is the normalized rms error of the estimate?

CHAPTER 2

LINEAR PHYSICAL SYSTEMS

Before the measurement and analysis of random physical data is discussed in greater detail, it is desirable to clarify some pertinent concepts and fundamental definitions related to the dynamic behavior of physical systems. This chapter reviews the theoretical formulas for describing the response characteristics of ideal linear systems, and illustrates the basic ideas for simple physical examples.

2.1 CONSTANT-PARAMETER LINEAR SYSTEMS

An ideal system is one that has *constant parameters* and is *linear* between two clearly defined points of interest called the input or excitation point and the output or response point. A system has constant parameters if all fundamental properties of the system are invariant with respect to time. For example, a simple passive electrical circuit would be a constant-parameter system if the values for the resistance, capacitance, and inductance of all elements did not change from one time to another. A system is linear if the response characteristics are additive and homogeneous. The term *additive* means that the output to a sum of inputs is equal to the sum of the outputs produced by each input individually. The term *homogeneous* means that the output produced by a constant times the input is equal to the constant times the output produced by the input alone. In equation form, if $f(x)$ represents the output to an input x, then the system is linear if for any two inputs x_1, x_2, and constant c,

$$f(x_1 + x_2) = f(x_1) + f(x_2) \qquad \text{additive property} \qquad (2.1a)$$

$$f(cx) = cf(x) \qquad \text{homogeneous property} \qquad (2.1b)$$

26

The constant-parameter assumption is reasonably valid for many physical systems in practice. For example, the fundamental properties of an electrical circuit or a mechanical structure will usually not display significant changes over any time interval of practical interest. There are, of course, exceptions. The value of an electrical resistor may change owing to a high temperature exposure, or the stiffness of a structure may change because of fatigue damage caused by continual vibration. Furthermore, some physical systems are designed to have time-varying parameters that are fundamental to the desired purpose of the system. Electronic communication systems are an obvious example. However, such conditions are generally special cases that can be clearly identified in practice.

A linearity assumption for real systems is somewhat more critical. All physical systems will display nonlinear response characteristics under sufficiently extreme input conditions. For example, an electrical capacitor will ultimately arc as the applied voltage is increased and, hence, will no longer pass a current that is directly proportional to the applied voltage; or a metal cable will ultimately break as the applied load is increased and, hence, will no longer display a strain that is proportional to the applied load. To make the problem more difficult, common nonlinearities usually occur gradually rather than abruptly at one point. For example, the load-strain relationship for the metal cable would actually start deviating from a linear relationship long before the final abrupt break occurs. Nevertheless the response characteristics for many physical systems may be assumed to be linear, at least over some limited range of inputs, without involving unreasonable errors.

***Example 2.1*. Illustration of Nonlinear System.**   Consider a simple square law system where the output is given by

$$y = f(x) = ax^2$$

For any two inputs x_1 and x_2,

$$f(x_1 + x_2) = a(x_1 + x_2)^2 = ax_1^2 + 2ax_1x_2 + ax_2^2$$

but the additive property in Equation (2.1a) requires that

$$f(x_1 + x_2) = ax_1^2 + ax_2^2$$

Furthermore, for an arbitrary constant c,

$$f(cx) = a(cx)^2 = c^2ax^2$$

but the homogeneous property in Equation (2.1b) demands that

$$f(cx) = cax^2$$

Hence the system is not linear, in that it fails to comply with both the additive and homogeneous properties of linear systems.

2.2   BASIC DYNAMIC CHARACTERISTICS

The dynamic characteristics of a constant-parameter linear system can be described by a *weighting function* $h(\tau)$, sometimes called the *unit impulse response function*, which is defined as the output of the system at any time to a unit impulse input applied a time τ before. The usefulness of the weighting function as a description of the system is due to the following fact. For any arbitrary input $x(t)$, the system output $y(t)$ is given by the *convolution integral*

$$y(t) = \int_{-\infty}^{\infty} h(\tau)x(t-\tau)\,d\tau \tag{2.2}$$

That is, the value of the output $y(t)$ is given as a weighted linear (infinite) sum over the entire history of the input $x(t)$.

In order for a constant-parameter linear system to be *physically realizable* (*causal*), it is necessary that the system respond only to past inputs. This implies that

$$h(\tau) = 0 \qquad \text{for } \tau < 0 \tag{2.3}$$

Hence, for physical systems, the effective lower limit of integration in Equation (2.2) is zero rather than $-\infty$.

A constant-parameter linear system is said to be *stable* if every possible bounded input function produces a bounded output function. From Equation (2.2),

$$|y(t)| = \left| \int_{-\infty}^{\infty} h(\tau)x(t-\tau)\,d\tau \right| \le \int_{-\infty}^{\infty} |h(\tau)|\,|x(t-\tau)|\,d\tau \tag{2.4}$$

When the input $x(t)$ is bounded, there exists some finite constant A such that

$$|x(t)| \le A \qquad \text{for all } t \tag{2.5}$$

It follows from Equation (2.4) that

$$|y(t)| \le A \int_{-\infty}^{\infty} |h(\tau)|\,d\tau \tag{2.6}$$

Hence if the constant-parameter linear weighting function $h(\tau)$ is absolutely integrable, that is,

$$\int_{-\infty}^{\infty} |h(\tau)|\,d\tau < \infty \tag{2.7}$$

then the output will be bounded and the system is stable.

Example **2.2. Illustration of Unstable System.**   Consider a simple system with a unit impulse response function of the form

$$h(\tau) = \begin{cases} Ae^{a\tau} & \tau \geq 0 \\ 0 & \tau < 0 \end{cases}$$

Since $h(\tau) = 0$ for $\tau < 0$, the system is physically realizable by definition. However,

$$\int_{-\infty}^{\infty} |h(\tau)| \, d\tau = \int_{0}^{\infty} |Ae^{a\tau}| \, d\tau = \frac{A}{a}(e^{a\infty} - 1)$$

It follows that the system is unstable if $a \geq 0$, but stable if $a < 0$. Specifically, for $a = -b < 0$,

$$\int_{0}^{\infty} |Ae^{-b\tau}| \, d\tau = \frac{A}{-b}(e^{-\infty} - 1) = \frac{A}{b}$$

This completes Example 2.2.

A constant-parameter linear system can also be characterized by a *transfer function H(p)*, which is defined as the Laplace transform of $h(\tau)$. That is,

$$H(p) = \int_{0}^{\infty} h(\tau) e^{-p\tau} \, d\tau \qquad p = a + jb \tag{2.8}$$

The criterion for stability of a constant-parameter linear system (assumed to be physically realizable) takes on an interesting form when considered in terms of the transfer function $H(p)$. Specifically, if $H(p)$ has no poles in the right half of the complex p plane or on the imaginary axis (no poles where $a \geq 0$), then the system is stable. Conversely, if $H(p)$ has at least one pole in the right half of the complex p plane or on the imaginary axis, then the system is unstable.

An important property of constant-parameter linear systems is frequency preservation. Specifically, consider a constant-parameter linear system with a weighting function $h(\tau)$. From Equation (2.2), for any arbitrary input $x(t)$, the nth derivative of the output $y(t)$ with respect to time is given by

$$\frac{d^n y(t)}{dt^n} = \int_{-\infty}^{\infty} h(\tau) \frac{d^n x(t-\tau)}{dt^n} \, d\tau \tag{2.9}$$

Now, assume the input $x(t)$ is sinusoidal, that is,

$$x(t) = X \sin(2\pi ft + \theta) \tag{2.10}$$

The second derivative of $x(t)$ is

$$\frac{d^2 x(t)}{dt^2} = -4\pi^2 f^2 x(t) \tag{2.11}$$

It follows from Equation (2.9) that the second derivative for the output $y(t)$ must be

$$\frac{d^2 y(t)}{dt^2} = -4\pi^2 f^2 y(t) \tag{2.12}$$

Thus $y(t)$ must also be sinusoidal with the same frequency as $x(t)$. This result shows that a constant-parameter linear system cannot cause any frequency translation, but can only modify the amplitude and phase of an applied input.

2.3 FREQUENCY RESPONSE FUNCTIONS

If a constant-parameter linear system is physically realizable and stable, then the dynamic characteristics of the system can be described by a *frequency response function* $H(f)$, which is defined as the Fourier transform of $h(\tau)$. That is,

$$H(f) = \int_0^\infty h(\tau) e^{-j2\pi f\tau} \, d\tau \tag{2.13}$$

Note that the lower limit of integration is zero rather than $-\infty$ since $h(\tau) = 0$ for $\tau < 0$. The frequency response function is simply a special case of the transfer function where, in the exponent $p = a + jb$, $a = 0$, and $b = 2\pi f$. For physically realizable and stable systems, the frequency response function may replace the transfer function with no loss of useful information.

An important relationship for the frequency response function of constant-parameter linear systems is obtained by taking the Fourier transform of both sides of Equation (2.2). Letting $X(f)$ be the Fourier transform of an input $x(t)$ and $Y(f)$ be the Fourier transform of the resulting output $y(t)$, assuming these transforms exist, it follows from Equation (2.2) that

$$Y(f) = H(f) X(f) \tag{2.14}$$

Hence, in terms of the frequency response function of a system and the Fourier transforms of the input and output, the convolution integral in Equation (2.2) reduces to the simple algebraic expression in Equation (2.14).

The frequency response function is generally a complex-valued quantity that may be conveniently thought of in terms of a magnitude and an associated phase angle. This can be done by writing $H(f)$ in complex polar notation as

$$H(f) = |H(f)| e^{-j\phi(f)} \tag{2.15}$$

The absolute value $|H(f)|$ is called the system *gain factor*, and the associated phase angle $\phi(f)$ is called the system *phase factor*. In these terms, the frequency response function takes on a direct physical interpretation as follows. Assume a system is subjected to a sinusoidal input (hypothetically existing over all time) with a frequency f producing an output that, as illustrated in Section 2.2, will also be sinusoidal with the same frequency. The ratio of the output amplitude to the input amplitude is equal to the gain factor $|H(f)|$ of the system, and the phase shift between the output and input is equal to the phase factor $\phi(f)$ of the system.

From physical realizability requirements, the frequency response function, the gain factor, and the phase factor of a constant-parameter linear system satisfy the following symmetry properties:

$$H(-f) = H^*(f)$$

$$|H(-f)| = |H(f)| \tag{2.16}$$

$$\phi(-f) = -\phi(f)$$

Furthermore, if one system described by $H_1(f)$ is followed by a second system described by $H_2(f)$, and there is no loading or feedback between the two systems, then the overall system may be described by $H(f)$, where

$$H(f) = H_1(f)H_2(f)$$

$$|H(f)| = |H_1(f)||H_2(f)| \tag{2.17}$$

$$\phi(f) = \phi_1(f) + \phi_2(f)$$

Thus, on cascading two systems where there is no loading or feedback, the gain factors multiply and the phase factors add.

It is important to note that the frequency response function $H(f)$ of a constant-parameter linear system is a function of only frequency, and is not a function of either time or the system excitation. If the system were nonlinear, $H(f)$ would also be a function of the applied input. If the parameters of the system were not constant, $H(f)$ would also be a function of time.

2.4   ILLUSTRATIONS OF FREQUENCY RESPONSE FUNCTIONS

A clearer understanding of the frequency response function of common physical systems will be afforded by considering some examples. The examples chosen involve simple mechanical and electrical systems because these particular physical systems are generally easier to visualize. The analogous characteristics relating mechanical and electrical systems to other physical systems are noted.

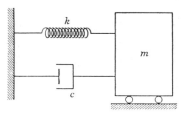

Figure 2.1   Simple mechanical system.

2.4.1   *Mechanical Systems*

Assume a simple mechanical structure can be represented by a lumped parameter system consisting of a mass, a spring, and a dashpot, where the motion of the mass is restricted to translation in only one direction, as shown in Figure 2.1. In this figure, k is a spring constant in pounds/inch, c is a viscous damping coefficient in pound-seconds/inch, and m is a mass in pound-seconds^2/inch.

Before a frequency response function can be determined, it is necessary to define the input and output parameters of interest. There are a number of possibilities for the system in Figure 2.1, as will be illustrated now.

FORCE INPUT AND DISPLACEMENT OUTPUT.   Assume the input of interest is a force applied to the mass, and the output of interest is the resulting displacement of the mass, as illustrated in Figure 2.2. Here, $F(t)$ is an applied force in pounds and $y(t)$ is the resulting output displacement of the mass in inches.

The first step toward establishing an appropriate frequency response function for this system is to determine the equation of motion. This is accomplished by using the relationship from basic mechanics that the sum of all forces acting on the mass must equal zero, as follows:

$$F(t) + F_k(t) + F_c(t) + F_m(t) = 0 \qquad (2.18)$$

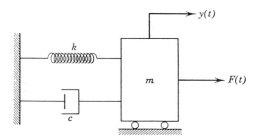

Figure 2.2   Mechanical system with force input.

where

$$F_k(t) = -ky(t) = \text{spring force} \qquad (2.18a)$$

$$F_c(t) = -c\dot{y}(t) = \text{damping force} \qquad (2.18b)$$

$$F_m(t) = -m\ddot{y}(t) = \text{inertial force} \qquad (2.18c)$$

$$\dot{y}(t) = \frac{dy(t)}{dt} = \text{velocity}$$

$$\ddot{y}(t) = \frac{d^2y(t)}{dt^2} = \text{acceleration}$$

Hence the equation of motion for this system is

$$m\ddot{y}(t) + c\dot{y}(t) + ky(t) = F(t) \qquad (2.19)$$

In Section 2.3, the frequency response function is defined as the Fourier transform of the output of the system to a unit impulse. For this case, the output of the system is the displacement $y(t)$ whose Fourier transform is given by

$$Y(f) = \int_0^\infty y(t)e^{-j2\pi ft}\, dt = H(f) \qquad (2.20)$$

It follows that

$$\text{Fourier Transform } [\dot{y}(t)] = j2\pi f H(f) \qquad (2.20a)$$

$$\text{Fourier Transform } [\ddot{y}(t)] = -(2\pi f)^2 H(f) \qquad (2.20b)$$

Now, by taking the Fourier transform of both sides of Equation (2.19) and noting that the Fourier transform for a unit impulse force $F(t) = \delta(t)$ is unity, one obtains the following result:

$$\left[-(2\pi f)^2 m + j2\pi fc + k \right] H(f) = 1 \qquad (2.21a)$$

Thus

$$H(f)_{f-d} = \left[k - (2\pi f)^2 m + j2\pi fc \right]^{-1} \qquad (2.21b)$$

where the subscript $f - d$ is added to indicate that this particular $H(f)$ relates a force input to a displacement output.

It is desirable to write Equation (2.21) in a different form by introducing two definitions.

$$\zeta = \frac{c}{2\sqrt{km}} \tag{2.22a}$$

$$f_n = \frac{1}{2\pi}\sqrt{\frac{k}{m}} \tag{2.22b}$$

The term ζ in Equation (2.22a) is a dimensionless quantity called the *damping ratio*. The term f_n in Equation (2.22b) is called the *undamped natural frequency* and has units of cycles per unit time. When these definitions are substituted into Equation (2.21), the following result is obtained:

$$H(f)_{f-d} = \frac{1/k}{1 - (f/f_n)^2 + j2\zeta f/f_n} \tag{2.23}$$

Writing Equation (2.23) in complex polar notation gives the frequency response function in terms of a gain factor $|H(f)|$ and a phase factor $\phi(f)$ as follows:

$$H(f) = |H(f)|e^{-j\phi(f)} \tag{2.24}$$

where

$$|H(f)|_{f-d} = \frac{1/k}{\sqrt{[1 - (f/f_n)^2]^2 + [2\zeta f/f_n]^2}} \tag{2.24a}$$

$$\phi(f)_{f-d} = \tan^{-1}\left[\frac{2\zeta f/f_n}{1 - (f/f_n)^2}\right] \tag{2.24b}$$

Note that $|H(f)|_{f-d}$ has units of $1/k$ or inches/pound. This particular function is sometimes called a *magnification function*.

Plots of $|H(f)|_{f-d}$ and $\phi(f)_{f-d}$ as defined in Equation (2.24) are presented in Figure 2.3. Three characteristics of these plots are of particular interest. First, the gain factor has a peak at some frequency less than f_n for all cases where $\zeta \le 1/\sqrt{2}$. The frequency at which this peak gain factor occurs is called the *resonance frequency* of the system. Specifically, it can be shown by minimizing the denominator of $|H(f)|_{f-d}$ in Equation (2.24a) that the resonance frequency, denoted by f_r, is given by

$$f_r = f_n\sqrt{1 - 2\zeta^2} \qquad \zeta^2 \le 0.5 \tag{2.25}$$

and that the peak value of the gain factor that occurs at the resonance

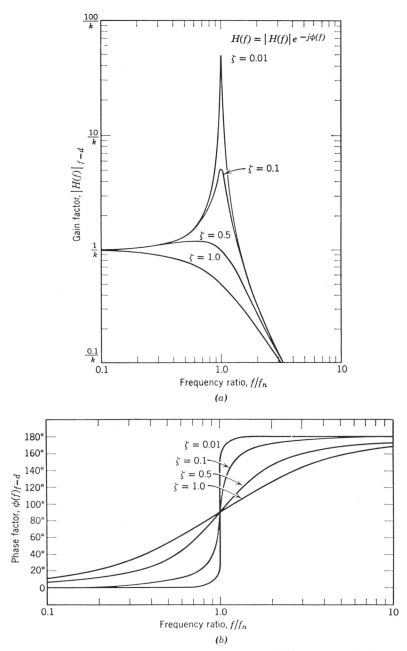

Figure 2.3 Frequency response function of mechanical system with force input. (*a*) Gain factor. (*b*) Phase factor.

frequency is given by

$$|H(f_r)|_{f-d} = \frac{1/k}{2\zeta\sqrt{1-\zeta^2}} \qquad \zeta^2 < 0.5 \qquad (2.26)$$

Second, defining the *half-power point bandwidth* of a system gain factor as $B_r = f_2 - f_1$ where

$$|H(f_1)|^2 = |H(f_2)|^2 = \tfrac{1}{2}|H(f_r)|^2$$

the half-power point bandwidth of the gain factor peak at the resonance frequency may be approximated for light damping by

$$B_r \approx 2\zeta f_r \qquad \zeta \leq 0.1 \qquad (2.27)$$

Third, the phase factor varies from 0 degrees for frequencies much less than f_n to 180 degrees for frequencies much greater than f_n. The exact manner in which $\phi(f)$ varies between these phase angle limits depends on the damping ratio ζ. However, for all values of ζ, the phase $\phi(f)_{f-d} = 90$ degrees for $f = f_n$.

Example 2.3. **Illustration of Resonant System.**   A simple mechanical system like that shown in Figure 2.1 has the following properties:

$$m = 0.1 \text{ lb-sec}^2/\text{in.} \qquad c = 2 \text{ lb-sec}/\text{in.} \qquad k = 1000 \text{ lb}/\text{in.}$$

Determine the undamped natural frequency, the damping ratio, the resonance frequency, and the peak gain factor of the system.

From Equation (2.22) the undamped natural frequency and damping ratio are given by

$$f_n = \frac{1}{2\pi}\sqrt{\frac{k}{m}} = 15.9 \text{ Hz} \qquad \zeta = \frac{c}{2\sqrt{km}} = 0.10$$

The resonance frequency is then given by Equation (2.25) as

$$f_r = f_n\sqrt{1 - 2\zeta^2} = 15.7 \text{ Hz}$$

and the peak gain factor is given by Equation (2.26) as

$$|H(f_r)| = \frac{1/k}{2\zeta\sqrt{1-\zeta^2}} = 0.0050 \text{ in.}/\text{lb}$$

It is common in practice to present gain factors for physical systems in dimensionless terms by multiplying out the stiffness term, that is,

$$k|H(f_r)| = \frac{1}{2\zeta\sqrt{1-\zeta^2}} = 5.0$$

This is often called the *quality factor* of the system, denoted by Q. The reciprocal of Q is usually referred to as the *loss factor* of the system denoted by η. For the system in question, $Q = 5.0$ and $\eta = 0.2$. This completes Example 2.3.

Referring to Section 2.3, the frequency response function $H(f)_{f-d}$ may be interpreted as follows. Assume the applied force in Figure 2.2 is sinusoidal such that $F(t) = F_0\sin 2\pi ft$. Then the output displacement would be given by

$$y(t) = F_0|H(f)|_{f-d}\sin[2\pi ft - \phi(f)_{f-d}] \qquad (2.28)$$

This particular interpretation actually provides another technique for determining the frequency response function. Specifically, one can solve for the output of a system to a sinusoidal input, and determine the frequency response function from the amplitude change and phase shift between the output and input. This will now be illustrated for the present problem.

The output of the system in Figure 2.2 to a sinusoidal input will be given by the particular solution of Equation (2.19) where $F(t)$ is sinusoidal, that is,

$$m\ddot{y}(t) + c\dot{y}(t) + ky(t) = F_0\sin 2\pi ft = \text{Im}\left[F_0 e^{j2\pi ft}\right] \qquad (2.29)$$

where Im[ ] means the imaginary part of [ ]. Now, assume a solution to Equation (2.29) in the general form of a sinusoidal output as follows.

$$y(t) = Y\sin(2\pi ft - \phi) = \text{Im}\left[Ye^{j(2\pi ft - \phi)}\right] \qquad (2.30)$$

When Equation (2.30) is substituted into Equation (2.29), the following relationship is obtained:

$$\text{Im}\left[\left(-(2\pi f)^2 m + j2\pi fc + k\right)Ye^{j(2\pi ft - \phi)}\right] = \text{Im}\left[F_0 e^{j2\pi ft}\right] \qquad (2.31)$$

The particular solution to Equation (2.29) is given from Equations (2.30) and (2.31) as follows:

$$y(t) = \text{Im}\left[\frac{F_0 e^{j2\pi ft}}{k - (2\pi f)^2 m + j2\pi fc}\right] \qquad (2.32)$$

If one uses the definitions from Equation (2.22) and converts to a trigonometric form, the output $y(t)$ becomes

$$y(t) = \frac{F_0 \sin[2\pi ft - \phi(f)]}{k\sqrt{[1 - (f/f_n)^2]^2 + [2\zeta f/f_n]^2}} \qquad (2.33)$$

where

$$\phi(f) = \tan^{-1}\left[\frac{2\zeta f/f_n}{1 - (f/f_n)^2}\right]$$

Hence the output is changed in amplitude by a factor equal to the gain factor defined in Equation (2.24a), and shifted in phase by a factor equal to the phase factor defined in Equation (2.24b).

FOUNDATION DISPLACEMENT INPUT AND DISPLACEMENT OUTPUT. Now consider a different case, where the input of interest is a motion of the foundation and the output of interest is the displacement of the mass, as illustrated in Figure 2.4. Here, $x(t)$ is an applied foundation displacement in inches measured from a mean foundation position, and $y(t)$ is the resulting output displacement of the mass in inches measured from the position of equilibrium.

As before, the equation of motion for the system can be determined from basic principles as follows:

$$F_k(t) + F_c(t) + F_m(t) = 0 \qquad (2.34)$$

where

$$F_k(t) = -k[y(t) - x(t)] = \text{spring force} \qquad (2.34a)$$

$$F_c(t) = -c[\dot{y}(t) - \dot{x}(t)] = \text{damping force} \qquad (2.34b)$$

$$F_m(t) = -m\ddot{y}(t) \qquad\qquad = \text{inertial force} \qquad (2.34c)$$

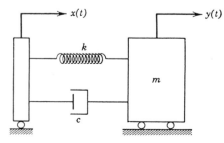

Figure 2.4 Mechanical system with foundation motion input.

Hence the equation of motion for this system is

$$m\ddot{y}(t) + c\dot{y}(t) + ky(t) = kx(t) + c\dot{x}(t) \tag{2.35}$$

Once again the frequency response function of the system will be given by the Fourier transform of the response displacement $y(t)$ for a unit impulse foundation displacement $x(t) = \delta(t)$. By taking the Fourier transform of both sides of Equation (2.35), and noting that Fourier transform $[\dot{\delta}(t)] = j2\pi f$, one obtains the following result:

$$\left[-(2\pi f)^2 m + j2\pi fc + k\right]Y(f) = [k + j2\pi fc] \tag{2.36a}$$

Thus

$$Y(f) = H(f)_{d-d} = \frac{k + j2\pi fc}{k - (2\pi f)^2 m + j2\pi fc} \tag{2.36b}$$

where the subscript $d - d$ means that this particular $H(f)$ relates a displacement input to a displacement output.

Using the definitions from Equation (2.22), the result in Equation (2.36) may be written as

$$H(f)_{d-d} = \frac{1 + j2\zeta f/f_n}{1 - (f/f_n)^2 + j2\zeta f/f_n} \tag{2.37}$$

In complex polar notation, Equation (2.37) reduces to the following gain factor and phase factor:

$$H(f) = |H(f)|e^{-j\phi(f)} \tag{2.38}$$

where

$$|H(f)|_{d-d} = \left(\frac{1 + [2\zeta f/f_n]^2}{[1 - (f/f_n)^2]^2 + [2\zeta f/f_n]^2}\right)^{1/2} \tag{2.38a}$$

$$\phi(f)_{d-d} = \tan^{-1}\left[\frac{2\zeta(f/f_n)^3}{1 - (f/f_n)^2 + 4\zeta^2(f/f_n)^2}\right] \tag{2.38b}$$

Note that $|H(f)|_{d-d}$ is dimensionless. This particular function is often called a *transmissibility function*. Plots of $|H(f)|_{d-d}$ and $\phi(f)_{d-d}$ are presented in Figure 2.5. Note that the gain factor displays a single peak similar to the example for a force input illustrated in Figure 2.3. However, the details of the gain factor as well as the phase factor in Figure 2.5 are quite different from the factors in Figure 2.3.

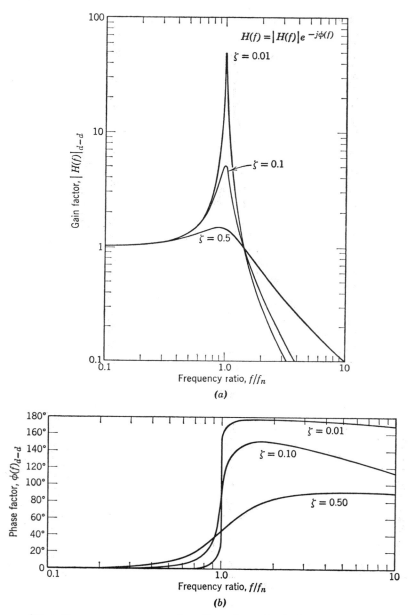

Figure 2.5 Frequency response function of mechanical system with foundation motion input. (*a*) Gain factor. (*b*) Phase factor.

Table 2.1

Summary of Gain Factors for Simple Mechanical System

Values for the Gain Factor $|H(f)|$ of a Simple Mechanical System as a Function of the Input and Output Parameters		Foundation Motion Input (see Figure 2.4 for model)			Force Input (see Figure 2.2 for model)
					Force (in displacement units)
		Displacement $x(t)$ in.	Velocity $\dot{x}(t)$ in./sec	Acceleration $\ddot{x}(t)$ in./sec^2	$x(t) = F(t)/k$ in.
In terms of displacement output, in.	Absolute displacement $y(t)$	$\dfrac{D_1}{D_2}$	$\dfrac{D_1}{2\pi f D_2}$	$\dfrac{D_1}{4\pi^2 f^2 D_2}$	$\dfrac{1}{D_2}$
	Relative displacement $z(t) = y(t) - x(t)$	$\dfrac{f^2}{f_n^2 D_2}$	$\dfrac{f}{2\pi f_n^2 D_2}$	$\dfrac{1}{4\pi^2 f_n^2 D_2}$	
In terms of velocity output, in./sec	Absolute velocity $\dot{y}(t)$	$\dfrac{2\pi f D_1}{D_2}$	$\dfrac{D_1}{D_2}$	$\dfrac{D_1}{2\pi f D_2}$	$\dfrac{2\pi f}{D_2}$
	Relative velocity $\dot{z}(t) = \dot{y}(t) - \dot{x}(t)$	$\dfrac{2\pi f^3}{f_n^2 D_2}$	$\dfrac{f^2}{f_n^2 D_2}$	$\dfrac{f}{2\pi f_n^2 D_2}$	
In terms of acceleration output, in./sec^2	Absolute acceleration $\ddot{y}(t)$	$\dfrac{4\pi^2 f^2 D_1}{D_2}$	$\dfrac{2\pi f D_1}{D_2}$	$\dfrac{D_1}{D_2}$	$\dfrac{4\pi^2 f^2}{D_2}$
	Relative acceleration $\ddot{z}(t) = \ddot{y}(t) - \ddot{x}(t)$	$\dfrac{4\pi^2 f^4}{f_n^2 D_2}$	$\dfrac{2\pi f^3}{f_n^2 D_2}$	$\dfrac{f^2}{f_n^2 D_2}$	

$$D_1 = \sqrt{1 + [2\zeta(f/f_n)]^2} \qquad D_2 = \sqrt{\left[1 - (f/f_n)^2\right]^2 + [2\zeta(f/f_n)]^2}$$

$$f_n = \frac{1}{2\pi}\sqrt{\frac{k}{m}} \qquad \zeta = \frac{c}{2\sqrt{km}}$$

OTHER INPUT AND OUTPUT COMBINATIONS.   The previous two examples indicate how two different frequency response functions are applicable to the same simple mechanical system, depending on the type of input to be considered. Actually, a different frequency response function is generally required for every different combination of input and output parameters that might be desired. For example, the relative displacement output $z(t) = y(t) - x(t)$ to a foundation displacement input $x(t)$ might be of interest for some applications, whereas the absolute acceleration output $\ddot{y}(t)$ to a foundation velocity input $\dot{x}(t)$ would be appropriate for others. A slightly different frequency response function would be required for each case. To illustrate this point, the various possible gain factors of the simple mechanical system in Figure 2.1 for 21 different combinations of input and output parameters are presented in Table 2.1.

Example 2.4. **System Response to Foundation Acceleration.**   A simple mechanical system like that shown in Figure 2.4 is subjected to a foundation motion measured in terms of absolute acceleration. Assuming a system damping ratio of $\zeta = 0.7$, determine the displacement of the mass relative to the foundation at frequencies below the undamped natural frequency of the system.

The appropriate gain factor of this system is given from Table 2.1 as

$$|H(f)| = \frac{1}{4\pi^2 f_n^2 \sqrt{\left[1 - (f/f_n)^2\right]^2 + [1.4 f/f_n]^2}}$$

The values of the appropriately normalized gain factor for various frequency ratios, $(f/f_n) \le 1$, are as follows.

f/f_n	0	0.2	0.4	0.6	0.8	0.9	1.0
$\dfrac{f_n^2 |H(f)|}{0.0253}$	1.00	1.00	0.99	0.95	0.87	0.79	0.72

Note that the gain factor is approximately constant for frequencies well below the undamped natural frequency of the system. It is for this reason that the above system serves as the basis for a type of acceleration transducer, called a *seismic accelerometer*, which is discussed later in Chapter 10.

2.4.2   *Electrical Systems*

Assume a simple electrical circuit can be represented by a lumped parameter system consisting of an inductor, a resistor, and a capacitor. Further assume that the input to the system is a potential difference as shown in Figure 2.6. In this figure, C is a capacitance in farads, R is a resistance in ohms, L is an

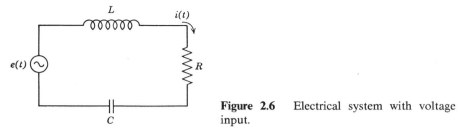

Figure 2.6   Electrical system with voltage input.

inductance in henries, $e(t)$ is an applied potential in volts, and $i(t)$ is the resulting current in amperes. Note that $i(t) = dq(t)/dt$, where $q(t)$ is charge in coulombs.

Assume the input of interest is an applied voltage and the output of interest is the resulting charge. As for the case of mechanical systems in Section 2.4.1, the first step toward establishing a proper frequency response function is to determine the differential equation describing the system. From basic circuit theory, the sum of all potential differences across the circuit elements must equal zero. That is,

$$e(t) + e_C(t) + e_R(t) + e_L(t) = 0 \tag{2.39}$$

where

$$e_C(t) = -\frac{1}{C}q(t) = \text{potential difference across capacitor} \tag{2.39a}$$

$$e_R(t) = -R\dot{q}(t) = \text{potential difference across resistor} \tag{2.39b}$$

$$e_L(t) = -L\ddot{q}(t) = \text{potential difference across inductor} \tag{2.39c}$$

Hence the differential equation for this system is

$$L\ddot{q}(t) + R\dot{q}(t) + \frac{1}{C}q(t) = e(t) \tag{2.40}$$

Note the similarity between Equation (2.40) and the equation of motion for a force excited mechanical system given by Equation (2.19). Using the same analysis procedures outlined in Section 2.4.1, it follows directly that the frequency response function of this simple electrical system is

$$H(f)_{e-q} = \left[\frac{1}{C} - (2\pi f)^2 L + j2\pi f R\right]^{-1} \tag{2.41}$$

where the subscript $e - q$ means that this particular $H(f)$ relates a voltage input to a charge output. Note that $H(f)_{e-q}$ has the units of coulombs/volt.

The plot for $H(f)_{e-q}$ would be identical to the plot for the mechanical frequency response function $H(f)_{f-d}$ presented in Figure 2.3, where the damping ratio ζ and the undamped natural frequency f_n of the electrical circuit are given as follows:

$$\zeta = \frac{R}{2} \sqrt{\frac{C}{L}} \tag{2.42a}$$

$$f_n = \frac{1}{2\pi} \sqrt{\frac{1}{LC}} \tag{2.42b}$$

It should now be clear that a direct analogy may be made between mechanical and electrical systems as presented in Table 2.2.

A more common frequency response function for electrical systems is one that relates a voltage input to a current output. This particular frequency response function is given by

$$H(f)_{e-i} = \left[R + j\left(2\pi fL - \frac{1}{2\pi fC} \right) \right]^{-1} \tag{2.43}$$

where $H(f)_{e-i}$ has the units of amperes/volt. The reciprocal of Equation (2.43), which may be denoted by $H(f)_{i-e}$, is called an *impedance function*:

$$H(f)_{i-e} = R + j\left(2\pi fL - \frac{1}{2\pi fC} \right) \tag{2.44}$$

Table 2.2

Analogous Terms for Mechanical and Electrical Systems

	Electrical System with a Voltage Input	Mechanical System with a Force Input
Input	Voltage, $e(t)$	Force, $F(t)$
Output	Charge, $q(t)$ Current, $i(t) = dq/dt$	Displacement, $y(t)$ Velocity, $v(t) = dy/dt$
Constant parameters	Inductance, L Resistance, R Capacitance, C	Mass, m Damping, c Compliance, $1/k$

Note that the mechanical analogy to Equation (2.44) is given from Table 2.2 by $H(f)_{v-f}$ as follows:

$$H(f)_{v-f} = c + j\left(2\pi fm - \frac{k}{2\pi f}\right) \qquad (2.45)$$

The function in Equation (2.45) is often called a *mechanical impedance function* because of its analogy to the common electrical impedance function.

2.4.3 *Other Systems*

By the same analytical procedures outlined in Section 2.4.1, an appropriate frequency response function can be developed, at least in theory, for any clearly defined constant-parameter linear system that is physically realizable and stable. Moreover, the frequency response functions of different physical systems will often display analogous parameters, just as illustrated for mechanical and electrical systems in Section 2.4.2 and Table 2.2. A summary of analogous characteristics for several common physical systems is presented in Table 2.3.

Table 2.3

Analogous Characteristics for Several Physical Systems

System	Input	Output	Constant Parameters		
Electrical	Voltage	Current	Inductance	Resistance	Capacitance
Mechanical (translational)	Force	Velocity	Mass	Damping	Compliance
Mechanical (rotational)	Torque	Angular velocity	Moment of inertia	Angular damping	Angular compliance
Acoustical	Pressure	Particle velocity	Inertance (acoustical mass)	Acoustical damping	Acoustical capacitance
Thermo	Temperature	Heat flow	—	Thermo resistance	Thermo capacitance
Magnetic	Magneto-motive force	Flux	—	Reluctance	—

2.5   PRACTICAL CONSIDERATIONS

The analytical determination of frequency response functions of physical systems has been illustrated in Section 2.4. To facilitate the development and clarification of basic ideas, examples were limited to simple mechanical and electrical systems. It should not be implied from these examples that the analytical determination of frequency response functions of physical systems is always so easy.

Consider, for example, a mechanical system in the form of a continuous elastic structure where the various parameters (mass, damping, and stiffness) are distributed rather than lumped as hypothetically assumed for the examples in Section 2.4.1. Such a mechanical system would have many different possible input and output points which might be of interest. Furthermore, the frequency response function of each input/output combination would generally display many peaks representing many resonant frequencies, as opposed to a single resonance as illustrated for the examples in Section 2.4.1. For relatively uncomplicated continuous structures such as beams, plates, and shells, appropriate frequency response functions may still be established with reasonable accuracy by direct analytical procedures [References 2.1 and 2.2]. For more complicated structures, as well as fluids and other physical systems, computer modeling procedures such as finite element methods [Reference 2.3] might be used to estimate frequency response functions and other response properties. If the physical system of interest has been constructed and is available for experimental studies, frequency response functions can be estimated by empirical procedures. The most straightforward empirical approach is to subject the system to a sinusoidal input and measure the output amplitude and phase as the input frequency is varied. From Section 2.3, the ratio of the output to input amplitudes at any given frequency equals the gain factor, and the phase of the output relative to the input at any given frequency equals the phase factor. However, the same results can be obtained with substantially less experimental time by applying either random or transient inputs to the system, either natural or artificial, and measuring the system response. The estimation of frequency response functions from random and transient input/output data will be developed in detail in Chapters 6 and 7.

PROBLEMS

2.1   Write a single equation that defines the required conditions for linearity of physical systems.

2.2   To define the dynamic properties of a system by a single-valued weighting function $h(\tau)$, which of the following requirements apply? The system must
   (a)   have constant parameters.
   (b)   be linear.

(c) be physically realizable.

(d) be stable.

2.3 If an input $x(t)$ produces an output $y(t) = x(t)|x(t)|$, prove that the input/output relationship is nonlinear.

2.4 Determine the weighting function of the force excited mechanical system shown in Figure 2.2.

2.5 Determine the frequency response function of a physical system with a weighting function $h(\tau) = Ae^{-a\tau}$, where $a > 0$.

2.6 Assume the mechanical system shown in Figure 2.2 has a spring constant of $k = 10$ lb/in., a viscous damping coefficient of $c = 0.2$ lb-sec/in., and a mass of $m = 0.1$ lb-sec^2/in. Determine the
(a) undamped natural frequency f_n.
(b) damping ratio ζ.
(c) force excited resonant frequency f_r.
(d) peak value of the gain factor $|H(f)|$.

2.7 Assume the mass of the mechanical system shown in Figure 2.1 is displaced from its position of equilibrium and then released. Prove that the time between crossings of the position of equilibrium in the resulting oscillation is $T = \frac{1}{2}[f_n\sqrt{1 - \zeta^2}]^{-1}$ for $\zeta^2 \le 1.0$.

2.8 Prove that the resonance frequency of the force excited mechanical system shown in Figure 2.2 is $f_r = f_n\sqrt{1 - 2\zeta^2}$, $\zeta^2 \le 0.5$, as stated in Equation (2.25).

2.9 The half-power point bandwidth of a resonant physical system is defined as $B_r = f_2 - f_1$, where $|H(f_1)|^2 = |H(f_2)|^2 = \frac{1}{2}|H(f_r)|^2$. Given the force excited mechanical system shown in Figure 2.2, prove that $B_r \approx 2\zeta f_r$ for small ζ, as stated in Equation (2.27).

2.10 Draw the electrical analog circuit for the mechanical system with foundation motion input shown in Figure 2.4. Determine the values of the analogous circuit parameters in terms of the mechanical parameters k, c, and m.

REFERENCES

2.1 Stokey, W. F., "Vibration of Systems Having Distributed Mass and Elasticity," Chapter 7 in *Shock and Vibration Handbook*, 2nd ed. (C. M. Harris and C. E. Crede, Eds.), McGraw-Hill, New York, 1976.

2.2 Hurty, W. G., and Rubinstein, M. F., *Dynamics of Structures*, Prentice-Hall, Englewood Cliffs, New Jersey, 1964.

2.3 Oden, J. I., and Reddy, J. N., *An Introduction to the Mathematical Theory of Finite Elements*, Wiley-Interscience, New York, 1976.

CHAPTER 3

PROBABILITY FUNDAMENTALS

This chapter reviews the fundamental principles of probability theory that are needed as background for the concepts of random process theory developed in later chapters. The material covers random variables, probability distributions, expected values, change of variables, moment generating functions, and characteristic functions for both single and multiple random variables. More detailed developments of probability theory from an engineering viewpoint are presented in References 3.1–3.3.

3.1   ONE RANDOM VARIABLE

The underlying concept in probability theory is that of a *set*, defined as a collection of objects (also called points or elements) about which it is possible to determine whether any particular object is a member of the set. In particular, the possible outcomes of an experiment (or a measurement) represent a set of *points* called the *sample space*. These points may be grouped together in various ways, called *events*, and under suitable conditions *probability functions* may be assigned to each. These probabilities always lie between zero and one, the probability of an impossible event being zero and the probability of the certain event being one. Sample spaces are either finite or infinite.

Consider a sample space of points representing the possible outcomes of a particular experiment (or measurement). A *random variable* $x(k)$ is a set function defined for points k from the sample space; that is, a random variable $x(k)$ is a real number between $-\infty$ and $+\infty$ that is associated to each sample point k that might occur. Stated another way, the random outcome of an experiment, indexed by k, can be represented by a real number $x(k)$, called the random variable. All possible experimental events that might occur constitute a completely additive class of sets, and a probability measure may be assigned to each event.

3.1.1 *Probability Distribution Functions*

Let $x(k)$ denote a random variable of interest. Then for any fixed value of x, the random event $x(k) \leq x$ is defined as the set of possible outcomes k such that $x(k) \leq x$. In terms of the underlying probability measure in the sample space, one may define a *probability distribution function $P(x)$* as the probability which is assigned to the set of points k satisfying the desired inequality $x(k) \leq x$. Observe that the set of points k satisfying $x(k) \leq x$ is a subset of the totality of all points k which satisfy $x(k) \leq \infty$. In notation form,

$$P(x) = \text{Prob}[x(k) \leq x] \qquad (3.1)$$

Clearly,

$$P(a) \leq P(b) \qquad \text{if } a \leq b \qquad (3.2)$$

$$P(-\infty) = 0 \qquad P(\infty) = 1 \qquad (3.3)$$

If the random variable assumes a continuous range of values (which will be assumed hereafter), then a (first-order) *probability density function $p(x)$* may be defined by the differential relation

$$p(x) = \lim_{\Delta x \to 0} \left[\frac{\text{Prob}[x < x(k) \leq x + \Delta x]}{\Delta x} \right] \qquad (3.4)$$

It follows that

$$p(x) \geq 0 \qquad (3.5)$$

$$\int_{-\infty}^{\infty} p(x) \, dx = 1 \qquad (3.6)$$

$$P(x) = \int_{-\infty}^{x} p(\xi) \, d\xi \qquad \frac{dP(x)}{dx} = p(x) \qquad (3.7)$$

To handle discrete cases like Example 3.1, the probability density function $p(x)$ is permitted to include delta functions.

Example 3.1. **Discrete Distribution.** Suppose an experiment consists of tossing a single coin where the two possible outcomes, called heads and tails, are assumed to occur with equal probability ($\frac{1}{2}$). The random variable $x(k)$ for this example takes on only two discrete values, $x(\text{heads})$ and $x(\text{tails})$, to which arbitrary real numbers may be assigned. Specifically, let $x(\text{heads}) = a$ and

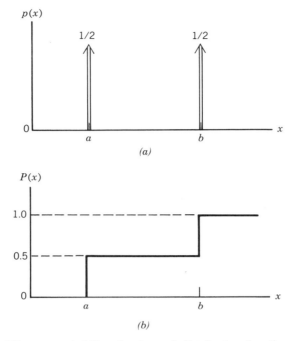

Figure 3.1 Discrete probability density and distribution functions. (a) Probability density function. (b) Probability distribution function.

x(tails) $= b$, where a and b are real numbers with, say $b > a$. With these choices for $x(k)$, it follows that the probability distribution function is

$$P(x) = \begin{cases} 0 & x < a \\ \frac{1}{2} & a \le x < b \\ 1 & x \ge b \end{cases}$$

and the probability density function is given by

$$p(x) = \tfrac{1}{2}\delta(x - a) + \tfrac{1}{2}\delta(x - b)$$

where $\delta(x - a)$ and $\delta(x - b)$ are delta functions, as shown in Figure 3.1.

3.1.2 *Expected Values*

Assume the random variable $x(k)$ may take on values in the range $-\infty$ to $+\infty$. The *mean value* (also called *expected value* or *average value*) of $x(k)$ is obtained by an appropriate limiting operation when each value assumed by $x(k)$ is multiplied by its probability of occurrence. This gives

$$E[x(k)] = \int_{-\infty}^{\infty} xp(x)\, dx = \mu_x \tag{3.8}$$

where $E[\ ]$ represents the expected value over the index k of the term inside the brackets. Similarly, the expected value of any real single-valued continuous function $g(x)$ of the random variable $x(k)$ is given by

$$E[g(x(k))] = \int_{-\infty}^{\infty} g(x)p(x)\, dx \qquad (3.9)$$

where $p(x)$ is the probability density function associated with $x(k)$. In particular, for $g(x) = x^2$, the *mean square value* of $x(k)$ is given by

$$E[x^2(k)] = \int_{-\infty}^{\infty} x^2 p(x)\, dx = \psi_x^2 \qquad (3.10)$$

The *variance* of $x(k)$ is defined by the mean square value of $x(k)$ about its mean value. Here, $g(x) = (x - \mu_x)^2$, and

$$E[(x(k) - \mu_x)^2] = \int_{-\infty}^{\infty} (x - \mu_x)^2 p(x)\, dx = \psi_x^2 - \mu_x^2 = \sigma_x^2 \qquad (3.11)$$

By definition, the *standard deviation* of $x(k)$, denoted by σ_x, is the positive square root of the variance. The standard deviation is measured in the same units as the mean value.

***Example 3.2.* Uniform (Rectangular) Distribution.** Suppose an experiment consists of choosing a point at random in the interval $[a, b]$, including the end points. A continuous random variable $x(k)$ for this example may be defined by the numerical value of the chosen point. The corresponding probability distribution function becomes

$$P(x) = \begin{cases} 0 & x < a \\ \dfrac{x - a}{b - a} & a \le x \le b \\ 1 & x > b \end{cases}$$

Hence the probability density function is given by

$$p(x) = \begin{cases} (b - a)^{-1} & a \le x \le b \\ 0 & \text{otherwise} \end{cases}$$

For this example, from Equations (3.9) and (3.11), the mean value and variance are given by

$$\mu_x = \frac{a + b}{2} \qquad \sigma_x^2 = \frac{(b - a)^2}{12}$$

Plots of $P(x)$ and $p(x)$ are shown in Figure 3.2.

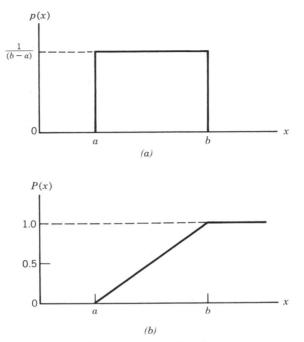

Figure 3.2   Uniform probability density and distribution functions. (a) Probability density function. (b) Probability distribution function.

3.1.3   *Change of Variables*

Suppose that $x(k)$ is a random variable with probability density function $p(x)$, and that $g(x)$ is any real single-valued continuous function of x. Consider first the case where the inverse function $x(g)$ is also a real single-valued continuous function of g. The probability density function $p(g)$ associated with the random variable $g[x(k)] = g(k)$ may be determined from the probability density function $p(x)$ of $x(k)$ and from the derivative (dg/dx), assuming the derivative exists and is not equal to zero, as follows:

$$\frac{\text{Prob}[\,g < g(x(k)) \le g + \Delta g\,]}{\Delta g} = \frac{\text{Prob}[\,x < x(k) \le x + \Delta x\,]}{\Delta g}$$

$$= \frac{\text{Prob}[\,x < x(k) \le x + \Delta x\,]}{\Delta x} \cdot \frac{\Delta x}{\Delta g}$$

$$(3.12)$$

Hence in the limit when $dg/dx \neq 0$,

$$p(g) = p(x)\left|\frac{dx}{dg}\right| = \frac{p(x)}{|dg/dx|} \tag{3.13}$$

When using this formula, it is necessary to replace the variable x on the right-hand side by its equivalent g.

Now consider the case where the inverse function $x(g)$ is a real n-valued function of g, where n is an integer and all of the n values have equal probability. Here

$$p(g) = \frac{np(x)}{|dg/dx|} \tag{3.14}$$

Example 3.3. Sine Wave Distribution.   A sine wave of fixed amplitude X and fixed frequency f_0 may be considered to be a random variable if its initial phase angle $\theta = \theta(k)$ is a random variable. In particular, consider t to be fixed at some value t_0, and let the sine wave random variable be represented by

$$x(k) = x(\theta) = X\sin[2\pi f_0 t_0 + \theta(k)]$$

Suppose that $\theta(k)$ has a uniform probability density function $p(\theta)$ given by

$$p(\theta) = \begin{cases} (2\pi)^{-1} & 0 \leq \theta \leq 2\pi \\ 0 & \text{otherwise} \end{cases}$$

What is the sine wave probability density function $p(x)$ of $x(k)$?

For this example, the direct function $x(\theta)$ is single-valued, but the inverse function $\theta(x)$ is double-valued. From Equation (3.14), with θ replacing x, and x replacing g,

$$p(x) = \frac{2p(\theta)}{dx/d\theta} \qquad \text{for } \frac{dx}{d\theta} \neq 0$$

where

$$\frac{dx}{d\theta} = X\cos(2\pi f_0 t_0 + \theta) = X\sqrt{1 - \sin^2(2\pi f_0 t_0 + \theta)} = \sqrt{X^2 - x^2}$$

Thus

$$p(x) = \begin{cases} \left(\pi\sqrt{X^2 - x^2}\right)^{-1} & |x| < X \\ 0 & |x| \geq X \end{cases}$$

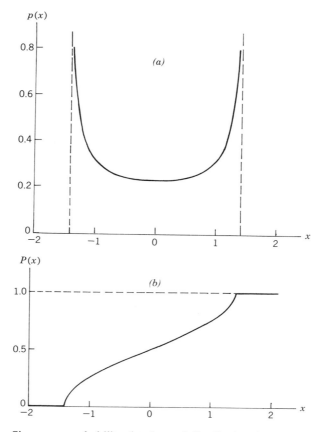

Figure 3.3 Sine wave probability density and distribution functions. (*a*) Probability density function. (*b*) Probability distribution function.

The corresponding sine wave probability distribution function is given by

$$P(x) = \begin{cases} 0 & x < -X \\ \int_{-X}^{x} p(\xi)\, d\xi = \dfrac{1}{\pi}\left(\dfrac{\pi}{2} + \sin^{-1}\dfrac{x}{X}\right) & -X \le x \le X \\ 1 & x > X \end{cases}$$

Plots of $P(x)$ and $p(x)$ are shown in Figure 3.3.

3.1.4 *Moment Generating and Characteristic Functions*

The *moment generating function* $m(s)$ of $x(k)$ is defined by letting $g(x) = \exp(sx)$ in Equation (3.9), namely,

$$m(s) = E[e^{sx}] = \int_{-\infty}^{\infty} e^{sx} p(x)\, dx \tag{3.15}$$

Now

$$m(0) = \int_{-\infty}^{\infty} p(x)\, dx = 1$$

Assuming all derivatives exist, then

$$m'(s) = \frac{dm(s)}{ds} = \int x e^{sx} p(x)\, dx$$

$$m''(s) = \frac{d^2 m(s)}{ds^2} = \int x^2 e^{sx} p(x)\, dx$$

and so on. Hence at $s = 0$,

$$E[x] = \int x p(x)\, dx = m'(0)$$

$$E[x^2] = \int x^2 p(x)\, dx = m''(0)$$

(3.16)

and so on. For any integer n, the *moments*

$$E[x^n] = \int_{-\infty}^{\infty} x^n p(x)\, dx = m^{(n)}(0) \tag{3.17}$$

where $m^{(n)}$ denotes the nth derivative of $m(s)$.

The *characteristic function* $C(f)$ of $x(k)$ is defined by letting $g(x) = \exp(j2\pi fx)$ in Equation (3.9), namely,

$$C(f) = E[e^{j2\pi fx}] = \int_{-\infty}^{\infty} p(x) e^{j2\pi fx}\, dx \tag{3.18}$$

Thus $C(f)$ has the form of an inverse Fourier transform, where $p(x)$ is the Fourier transform of $C(f)$. Assuming all integrals exist,

$$p(x) = \int_{-\infty}^{\infty} C(f) e^{-j2\pi fx}\, df \tag{3.19}$$

If $p(x) = \delta(x)$, a delta function, then

$$C(f) = \int_{-\infty}^{\infty} \delta(x) e^{j2\pi fx}\, dx = 1$$

$$\delta(x) = \int_{-\infty}^{\infty} e^{-j2\pi fx}\, df \tag{3.20}$$

Note that $C(f)$ is the same as $m(s)$ when $s = j2\pi f$, namely,

$$C(f) = m(j2\pi f) \tag{3.21}$$

3.1.5   Chebyshev's Inequality

Suppose that $x(k)$ is an arbitrary random variable with mean value μ_x, mean square value ψ_x^2, and variance σ_x^2. Suppose that its probability density function, which may be unknown, is $p(x)$. Then

$$\psi_x^2 = \int_{-\infty}^{\infty} x^2 p(x)\, dx \geq \int_{|x| \geq \varepsilon} x^2 p(x)\, dx \geq \varepsilon^2 \int_{|x| \geq \varepsilon} p(x)\, dx$$

since the integrand is nonnegative and since $x^2 \geq \varepsilon^2$ at every point in the right-hand region of integration. This proves that

$$\text{Prob}\big[|x(k)| \geq \varepsilon\big] = \int_{|x| \geq \varepsilon} p(x)\, dx \leq \frac{\psi_x^2}{\varepsilon^2} \tag{3.22}$$

Now replace $x(k)$ by $x(k) - \mu_x$. Then ψ_x^2 is replaced by σ_x^2 and Equation (3.22) becomes

$$\text{Prob}\big[|x(k) - \mu_x| \geq \varepsilon\big] \leq \frac{\sigma_x^2}{\varepsilon^2} \tag{3.22a}$$

In particular, if $\varepsilon = c\sigma_x$, then

$$\text{Prob}\big[|x(k) - \mu_x| \geq c\sigma_x\big] \leq \frac{1}{c^2} \tag{3.22b}$$

which is equivalent to

$$\text{Prob}\big[|x(k) - \mu_x| \leq c\sigma_x\big] \geq 1 - \frac{1}{c^2} \tag{3.22c}$$

Any of the forms of Equation (3.22) is known as *Chebyshev's inequality*.

Example 3.4. **Illustration of Probability Intervals.**   Consider a random variable $x(k)$ with an unknown probability density function. Using the Chebyshev inequality in Equation (3.22b) with $c = 2$ and $c = 3$, the following probability statements apply:

$$\text{Prob}\big[|x(k) - \mu_x| \geq 2\sigma_x\big] \leq 0.250$$

$$\text{Prob}\big[|x(k) - \mu_x| \geq 3\sigma_x\big] \leq 0.111$$

These relatively weak results should be compared to situations where $x(k)$ follows a Gaussian distribution. From Table A.2 in Appendix A, for Gaussian data, one obtains the stronger results

$$\text{Prob}\big[|x(k) - \mu_x| \geq 2\sigma_x\big] < 0.050$$

$$\text{Prob}\big[|x(k) - \mu_x| \geq 3\sigma_x\big] < 0.003$$

Here, one can state that 95% of the values will lie within $\pm 2\sigma$ of the mean value, whereas for an arbitrary distribution one can state that this will occur for only 75% of the values.

3.2 TWO RANDOM VARIABLES

Consider next two random variables $x(k)$ and $y(k)$, where k represents points in a suitable sample space. Let $P(x)$ and $P(y)$ be two distinct distribution functions associated with $x(k)$ and $y(k)$, respectively. The *joint probability distribution function* $P(x, y)$ is defined to be the probability that is associated with the subset of points k in the sample space satisfying simultaneously both of the inequalities $x(k) \leq x$ and $y(k) \leq y$. The totality of all points k satisfies the inequalities $x(k) \leq \infty$ and $y(k) \leq \infty$. In notation form,

$$P(x, y) = \text{Prob}\big[x(k) \leq x \text{ and } y(k) \leq y\big] \tag{3.23}$$

Clearly,

$$P(-\infty, y) = 0 = P(x, -\infty) \qquad P(\infty, \infty) = 1 \tag{3.24}$$

As before, assuming the random variables to be continuous, the *joint probability density function* $p(x, y)$ is defined by

$$p(x, y) = \lim_{\substack{\Delta x \to 0 \\ \Delta y \to 0}} \left[\frac{\text{Prob}\big[x < x(k) \leq x + \Delta x \text{ and } y < y(k) \leq y + \Delta y\big]}{\Delta x \, \Delta y} \right]$$

$$\tag{3.25}$$

It follows that

$$p(x, y) \geq 0 \tag{3.26}$$

$$\iint\limits_{-\infty}^{\infty} p(x, y) \, dx \, dy = 1 \tag{3.27}$$

$$P(x, y) = \int_{-\infty}^{y} \int_{-\infty}^{x} p(\xi, \eta) \, d\xi \, d\eta \qquad \frac{\partial}{\partial y} \left[\frac{\partial P(x, y)}{\partial x} \right] = p(x, y)$$

$$\tag{3.28}$$

The probability density functions of $x(k)$ and $y(k)$ individually are obtained from the joint probability density function by

$$p(x) = \int_{-\infty}^{\infty} p(x, y) \, dy$$

$$p(y) = \int_{-\infty}^{\infty} p(x, y) \, dx \tag{3.29}$$

Now if

$$p(x, y) = p(x)p(y) \tag{3.30}$$

then the two random variables $x(k)$ and $y(k)$ are said to be *statistically independent*. It follows for statistically independent variables that

$$P(x, y) = P(x)P(y) \tag{3.31}$$

3.2.1   *Expected Values and Correlation Coefficient*

The *expected value* of any real single-valued continuous function $g(x, y)$ of the two random variables $x(k)$ and $y(k)$ is given by

$$E[g(x, y)] = \iint_{-\infty}^{\infty} g(x, y)p(x, y) \, dx \, dy \tag{3.32}$$

For example, if $g(x, y) = (x(k) - \mu_x)(y(k) - \mu_y)$, where μ_x and μ_y are the mean values of $x(k)$ and $y(k)$, respectively, this defines the *covariance* C_{xy} between $x(k)$ and $y(k)$. That is,

$$C_{xy} = E\left[(x(k) - \mu_x)(y(k) - \mu_y)\right] = E[x(k)y(k)] - E[x(k)]E[y(k)]$$

$$= \iint_{-\infty}^{\infty} (x - \mu_x)(y - \mu_y)p(x, y) \, dx \, dy \tag{3.33}$$

Note that $C_{xx} = \sigma_x^2$, the variance of $x(k)$, as defined in Equation (3.11).

A simple relation exists between the covariance of $x(k)$ and $y(k)$ and the standard deviations of $x(k)$ and $y(k)$ as expressed by the inequality

$$|C_{xy}| \leq \sigma_x \sigma_y \tag{3.34}$$

Thus the magnitude of the covariance between $x(k)$ and $y(k)$ is less than or equal to the product of the standard deviation of $x(k)$ multiplied by the standard deviation of $y(k)$. This is proved later in Section 5.1.3.

It follows from the above result that the normalized quantity

$$\rho_{xy} = \frac{C_{xy}}{\sigma_x \sigma_y} \qquad (3.35)$$

known as the *correlation coefficient*, will lie between -1 and $+1$. Random variables $x(k)$ and $y(k)$ whose correlation coefficient is zero are said to be *uncorrelated*. This concept is quite distinct from the previous definition of *independent* random variables. Note that if $x(k)$ and $y(k)$ are independent random variables, then, from Equations (3.30) and (3.32),

$$E[x(k)y(k)] = \int\!\!\!\int_{-\infty}^{\infty} xy p(x, y)\, dx\, dy$$

$$= \int_{-\infty}^{\infty} xp(x)\, dx \int_{-\infty}^{\infty} yp(y)\, dy = E[x(k)]E[y(k)]$$

$$(3.36)$$

Hence C_{xy} and, in turn, ρ_{xy} equal zero so that *independent random variables are also uncorrelated*. The converse statement is not true in general; that is, *uncorrelated random variables are not necessarily independent*. For physically important situations involving two or more normally (Gaussian) distributed random variables, however, being mutually uncorrelated does imply independence. This is proved later in Section 3.3.4.

3.2.2   *Distribution for Sum of Two Random Variables*

Suppose $x(k)$ and $y(k)$ are two random variables with a joint probability density function $p(x, y)$. Determine the probability density function $p(z)$ of the sum random variable

$$z(k) = x(k) + y(k)$$

For each fixed value of x, the corresponding $y = z - x$. This gives

$$p(x, y) = p(x, z - x)$$

For each fixed value of z, the values of x may range from $-\infty$ to ∞. Hence,

$$p(z) = \int_{-\infty}^{\infty} p(x, z - x)\, dx \qquad (3.37)$$

which shows that the desired sum probability density function requires knowledge of the input joint probability density function. If $x(k)$ and $y(k)$ are independent random variables with probability density functions $p_1(x)$ and $p_2(y)$, respectively, then $p(x, y) = p_1(x)p_2(y) = p_1(x)p_2(z - x)$, and

$$p(z) = \int_{-\infty}^{\infty} p_1(x)p_2(z - x)\, dx \qquad (3.38)$$

***Example 3.5.* Sum of Two Independent Uniformly Distributed Variables.** Suppose two independent random variables x and y satisfy

$$p_1(x) = \frac{1}{a} \qquad 0 \le x \le a \qquad \text{otherwise zero}$$

$$p_2(y) = \frac{1}{a} \qquad 0 \le y \le a \qquad \text{otherwise zero}$$

Find the probability density function $p(z)$ for their sum $z = x + y$.

The probability density function $p_2(y) = p_2(z - x)$ for $0 \le z - x \le a$, which may be written as $z - a \le x \le z$. Hence x cannot exceed z. Also $p_1(x)$

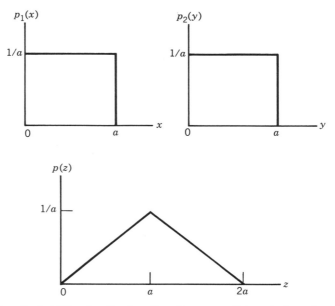

Figure 3.4   Probability density function for sum of two independent uniformly distributed variables.

requires $0 \leq x \leq a$. It follows from Equation (3.38) that

$$
p(z) = \begin{cases}
\int_0^z \left(\frac{1}{a}\right)^2 dx = \dfrac{z}{a^2} & 0 \leq z \leq a \\[2ex]
\int_{z-a}^a \left(\frac{1}{a}\right) dx = \dfrac{2a - z}{a^2} & a \leq z \leq 2a \\[2ex]
0 & \text{otherwise}
\end{cases}
$$

Plots of $p_1(x)$, $p_2(y)$, and $p(z)$ are shown in Figure 3.4. Observe that the sum of two independent random variables with uniform probability density functions will have a triangular probability density function. It is straightforward to verify that the probability density function for the sum of four independent random variables, each with uniform probability density function, will begin to resemble a Gaussian form, as predicted by the central limit theorem in the next section.

3.2.3 *Moment Generating and Characteristic Functions*

The *joint moment generating function* $m(s, t)$ of $x(k)$ and $y(k)$ is defined by letting $g(x, y) = \exp(sx + ty)$ in Equation (3.32), namely,

$$
m(s, t) = E[e^{sx+ty}] = \iint\limits_{-\infty}^{\infty} e^{sx+ty} p(x, y) \, dx \, dy \tag{3.39}
$$

Now

$$
m(0,0) = \iint\limits_{-\infty}^{\infty} p(x, y) \, dx \, dy = 1
$$

Assuming all partial derivatives exist, it follows that

$$
\frac{\partial m(s, t)}{\partial s} = \int\int x e^{sx+ty} p(x, y) \, dx \, dy
$$

$$
\frac{\partial m(s, t)}{\partial t} = \int\int y e^{sx+ty} p(x, y) \, dx \, dy
$$

$$
\frac{\partial^2 m(s, t)}{\partial s^2} = \int\int x^2 e^{sx+ty} p(x, y) \, dx \, dy
$$

$$
\frac{\partial^2 m(s, t)}{\partial t^2} = \int\int y^2 e^{sx+ty} p(x, y) \, dx \, dy
$$

$$
\frac{\partial^2 m(s, t)}{\partial s \, \partial t} = \int\int xy e^{sx+ty} p(x, y) \, dx \, dy
$$

and so on. Hence at $s = 0$ and $t = 0$,

$$E[x] = \int\int xp(x, y)\, dx\, dy = \frac{\partial m(s, t)}{\partial s}$$

$$E[y] = \int\int yp(x, y)\, dx\, dy = \frac{\partial m(s, t)}{\partial t}$$

(3.40)

$$E[x^2] = \int\int x^2 p(x, y)\, dx\, dy = \frac{\partial^2 m(s, t)}{\partial s^2}$$

(3.41)

$$E[y^2] = \int\int y^2 p(x, y)\, dx\, dy = \frac{\partial^2 m(s, t)}{\partial t^2}$$

$$E[xy] = \int\int xyp(x, y)\, dx\, dy = \frac{\partial^2 m(s, t)}{\partial s\, \partial t}$$

(3.42)

and so on. In general, at $s, t = 0$, the *mixed moments*

$$E[x^r y^n] = \int\int_{-\infty}^{\infty} x^r y^n p(x, y)\, dx\, dy = \frac{\partial^{r+n} m(s, t)}{\partial s^r\, \partial t^n}$$

(3.43)

The *joint characteristic function* $C(f, g)$ of $x(k)$ and $y(k)$ is defined by letting $g(x, y) = \exp[\,j2\pi(fx + gy)]$ in Equation (3.32), namely,

$$C(f, g) = E[e^{j2\pi(fx+gy)}] = \int\int_{-\infty}^{\infty} p(x, y) e^{j2\pi(fx+gy)}\, dx\, dy$$

(3.44)

Thus $C(f, g)$ has the form of an inverse double Fourier transform, where $p(x, y)$ is the double Fourier transform of $C(f, g)$. Assuming all integrals exist,

$$p(x, y) = \int\int_{-\infty}^{\infty} C(f, g) e^{-j2\pi(fx+gy)}\, df\, dg$$

(3.45)

Note that $C(f, g)$ is the same as $m(s, t)$ when $s = j2\pi f$ and $t = j2\pi g$, namely,

$$C(f, g) = m(j2\pi f, j2\pi g)$$

(3.46)

3.3 GAUSSIAN (NORMAL) DISTRIBUTION

A random variable $x(k)$ is said to follow a *Gaussian* (or *normal*) distribution if its probability density function is given by

$$p(x) = (b\sqrt{2\pi})^{-1}\exp\left[-\frac{(x-a)^2}{2b^2}\right] \tag{3.47}$$

where a is any real constant and b is any positive constant. It may be verified that a and b constitute the mean value and standard deviation of the random variable $x(k)$ since

$$E[x(k)] = \int_{-\infty}^{\infty} xp(x)\,dx = a = \mu_x$$

$$E\left[(x(k) - a)^2\right] = \int_{-\infty}^{\infty} (x-a)^2 p(x)\,dx = b^2 = \sigma_x^2$$

Thus the normal probability density function should be expressed by

$$p(x) = \left(\sigma_x\sqrt{2\pi}\right)^{-1}\exp\left[-\frac{(x-\mu_x)^2}{2\sigma_x^2}\right] \tag{3.48a}$$

The normal probability distribution function is, by definition,

$$P(x) = \left(\sigma_x\sqrt{2\pi}\right)^{-1}\int_{-\infty}^{x}\exp\left[-\frac{(\xi-\mu_x)^2}{2\sigma_x^2}\right]d\xi \tag{3.48b}$$

Without loss of generality, it will now be assumed that all mean values are zero. For one variable $x(k)$, the normal probability density function becomes

$$p(x) = \left(\sigma_x\sqrt{2\pi}\right)^{-1}\exp\left(-\frac{x^2}{2\sigma_x^2}\right) \tag{3.49a}$$

The associated normal probability distribution function becomes

$$P(x) = \left(\sigma_x\sqrt{2\pi}\right)^{-1}\int_{-\infty}^{x}\exp\left(-\frac{\xi^2}{2\sigma_x^2}\right)d\xi \tag{3.49b}$$

Further discussions of the normal distribution and its history are available from Reference 3.4. Applications of the normal distribution to statistical data analysis problems are outlined in Chapter 4. Standardized plots are shown in Figure 4.1. A summary of other probability density functions that are often used in theoretical studies is presented in Table 3.1.

Table 3.1

Type	Probability Density Function
Discrete	$p(x) = A\,\delta(x - a) + B\,\delta(x - b) + \cdots + N\,\delta(x - n)$ where $A + B + \cdots + N = 1$
Uniform (rectangular)	$p(x) = (b - a)^{-1}, \quad a \leq x \leq b; \quad$ otherwise zero
Sine wave	$p(x) = \left(\pi\sqrt{X^2 - x^2}\,\right)^{-1}, \quad |x| < X; \quad$ otherwise zero
Gaussian (normal)	$p(x) = \left(\sigma_x\sqrt{2\pi}\,\right)^{-1} e^{-(x - \mu_x)^2/2\sigma_x^2}$
Rayleigh	$p(x) = \dfrac{x}{c^2}\, e^{-x^2/2c^2}, \quad x \geq 0; \quad$ otherwise zero
Maxwell	$p(x) = \dfrac{x^2}{c^3}\sqrt{\dfrac{2}{\pi}}\, e^{-x^2/2c^2}, \quad x \geq 0; \quad$ otherwise zero
Truncated	Assume original $p_1(x)$ defined over $(-\infty, \infty)$. Truncated $p(x) = Cp_1(x), \quad a \leq x \leq b; \quad$ otherwise zero where $\displaystyle\int_{-\infty}^{\infty} p(x)\, dx = C\int_{a}^{b} p_1(x)\, dx = 1$
Clipped	Assume original $p_1(x)$ defined over $(-\infty, \infty)$. Clipped $\begin{aligned} p(x) &= p_1(x) & a < x < b \\ &= A\,\delta(x - a) & x = a \\ &= B\,\delta(x - b) & x = b \\ &= 0 & x < a \text{ or } x > b \end{aligned}$ where $\displaystyle\int_{-\infty}^{\infty} p(x)\, dx = \int_{a}^{b} p_1(x)\, dx + A + B = 1$

3.3.1 *Central Limit Theorem*

The importance of the normal distribution in physical problems may be attributed in part to the *central limit theorem* [References 3.2 and 3.5], which asserts that this distribution will result quite generally from the sum of a large number of independent random variables acting together. To be a bit more specific, let $x_1(k), x_2(k), \ldots, x_N(k)$ be N mutually independent random variables whose individual distributions are not specified and may be different. Let μ_i and σ_i^2 be the mean value and variance of each random variable $x_i(k)$, $i = 1, 2, \ldots, N$. Consider the sum random variable

$$x(k) = \sum_{i=1}^{N} a_i x_i(k) \tag{3.50}$$

where a_i are arbitrary fixed constants. Now, the mean value μ_x and variance σ_x^2 become

$$\mu_x = E[x(k)] = E\left[\sum_{i=1}^{N} a_i x_i(k)\right] = \sum_{i=1}^{N} a_i E[x_i(k)] = \sum_{i=1}^{N} a_i \mu_i \tag{3.51}$$

$$\sigma_x^2 = E\left[(x(k) - \mu_x)^2\right] = E\left[\sum_{i=1}^{N} a_i(x_i(k) - \mu_i)\right]^2 = \sum_{i=1}^{N} a_i^2 \sigma_i^2$$

The last expression is a result of the mutual independence of $x_i(k)$ with $x_j(k)$ for $i \neq j$. The central limit theorem states that under fairly common conditions, the sum random variable $x(k)$ will be normally distributed as $N \to \infty$ with the above mean value μ_x and variance σ_x^2.

3.3.2 *Joint Gaussian (Normal) Distribution*

For two variables $x(k)$ and $y(k)$ with zero mean values and equal variances $\sigma_1^2 = \sigma_2^2 = \sigma^2$, their joint Gaussian (normal) probability density function is defined by

$$p(x, y) = \frac{1}{2\pi\sigma^2\sqrt{1 - \rho^2}} \exp\left(\frac{-[x^2 - 2\rho xy + y^2]}{2\sigma^2(1 - \rho^2)}\right) \tag{3.52}$$

where

$$\rho = \frac{E[xy]}{\sigma_1 \sigma_2} = \frac{E[xy]}{\sigma^2} \tag{3.53}$$

It is instructive to verify Equation (3.49) using Equations (3.29) and (3.52). Specifically, consider

$$p(x) = \int_{-\infty}^{\infty} p(x, y)\, dy = \frac{1}{2\pi\sigma^2\sqrt{1 - \rho^2}} \int_{-\infty}^{\infty} \exp\left(\frac{-[x^2 - 2\rho xy + y^2]}{2\sigma^2(1 - \rho^2)}\right) dy$$

Now

$$x^2 - 2\rho xy + y^2 = (1 - \rho^2)x^2 + (y - \rho x)^2$$

so the exponential term can be written

$$\exp\left(\frac{-[x^2 - 2\rho xy + y^2]}{2\sigma^2(1 - \rho^2)}\right) = \exp\left(-\frac{x^2}{2\sigma^2}\right)\exp\left[\frac{-(y - \rho x)^2}{2\sigma^2(1 - \rho^2)}\right]$$

Hence

$$p(x) = \frac{\exp(-x^2/2\sigma^2)}{2\pi\sigma^2\sqrt{1 - \rho^2}} \int_{-\infty}^{\infty} \exp\left[\frac{-(y - \rho x)^2}{2\sigma^2(1 - \rho^2)}\right] dy$$

Now let

$$u = \frac{y - \rho x}{\sigma\sqrt{2(1 - \rho^2)}} \qquad du = \frac{dy}{\sigma\sqrt{2(1 - \rho^2)}}$$

Then

$$p(x) = \frac{\exp(-x^2/2\sigma^2)}{\sigma\pi\sqrt{2}} \int_{-\infty}^{\infty} \exp(-u^2)\, du = \frac{\exp(-x^2/2\sigma^2)}{\sigma\sqrt{2\pi}}$$

in agreement with Equation (3.49).

3.3.3   *Moment Generating and Characteristic Functions*

From Equations (3.15) and (3.49), the moment generating function for the zero mean value Gaussian variable is

$$m(s) = \exp(\sigma^2 s^2/2) \tag{3.54}$$

One can now verify from Equation (3.17) that

$$E[x^{2n-1}] = 0$$
$$E[x^{2n}] = 1 \cdot 3 \cdot 5 \cdots (2n - 1)\sigma^{2n} \qquad n = 1, 2, 3, \ldots \tag{3.55}$$

Thus, all odd-order moments are zero while even-order moments become

$$E[x^2] = \sigma^2 \qquad E[x^4] = 3\sigma^4$$
$$E[x^6] = 15\sigma^6 \qquad E[x^8] = 105\sigma^8 \tag{3.56}$$

and so on.

From Equation (3.21), the characteristic function for the zero mean value Gaussian variable is

$$C(f) = \exp(-2\pi^2\sigma^2 f^2) \tag{3.57}$$

The joint moment generating function from Equation (3.39) for the two zero mean value Gaussian variables satisfying Equation (3.52) is

$$m(s, t) = \exp\left[\frac{\sigma^2}{2}(s^2 + 2\rho st + t^2)\right] \tag{3.58}$$

It follows directly from Equation (3.43) that

$$E[x^m y^n] = 0 \qquad\qquad \text{if } r + n \text{ is odd}$$
$$E[x^2] = E[y^2] = \sigma^2 \qquad E[xy] = \rho\sigma^2 \tag{3.59}$$
$$E[x^3 y] = E[xy^3] = 3\rho\sigma^4$$
$$E[x^2 y^2] = \sigma^4(1 + 2\rho^2)$$

and so on.

The joint characteristic function from Equation (3.46) for the two zero mean value Gaussian variables satisfying Equation (3.52) is

$$C(f, g) = \exp\{-2\pi^2\sigma^2[f^2 + 2\rho fg + g^2]\} \tag{3.60}$$

3.3.4 N-Dimensional Gaussian (Normal) Distribution

Consider now N random variables $x_1(k), x_2(k), \ldots, x_N(k)$, which may be correlated. Denote their respective mean values, variances, and covariances by

$$\mu_i = E[x_i(k)]$$
$$\sigma_i^2 = E[(x_i(k) - \mu_i)^2] \tag{3.61}$$
$$C_{ij} = E[(x_i(k) - \mu_i)(x_j(k) - \mu_j)] \qquad C_{ii} = \sigma_i^2$$

Their joint distribution is said to follow an *N-dimensional Gaussian (normal) distribution* if the associated *N*-fold probability density function is given by

$$p(x_1, x_2, \ldots, x_N) = \frac{\exp\left[(-1/2|\mathbf{C}|)\Sigma^N_{i,j=1}|\mathbf{C}_{ij}|(x_i - \mu_i)(x_j - \mu_j)\right]}{(2\pi)^{N/2}|\mathbf{C}|^{1/2}} \tag{3.62}$$

where **C** is the covariance matrix of the C_{ij} defined below, $|\mathbf{C}|$ is the determinant of **C**, and $|\mathbf{C}_{ij}|$ is the cofactor of C_{ij} in determinant $|\mathbf{C}|$. To be explicit,

$$\mathbf{C} = \begin{bmatrix} C_{11} & C_{12} & \cdots & C_{1N} \\ C_{21} & C_{22} & \cdots & C_{2N} \\ \vdots & \vdots & & \vdots \\ C_{N1} & C_{N2} & \cdots & C_{NN} \end{bmatrix} \tag{3.63}$$

and the cofactor $|\mathbf{C}_{ij}|$ of any element C_{ij} is defined to be the determinant of order $N - 1$ formed by omitting the ith row and jth column of **C**, multiplied by $(-1)^{i+j}$.

The outstanding feature of the *N*-dimensional normal distribution is that all of its properties are determined solely from knowledge of the various mean values μ_i and covariances C_{ij}. For $N = 1$, this function reduces to

$$p(x_1) = (\sigma_1\sqrt{2\pi})^{-1}\exp\left[-\frac{(x_1 - \mu_1)^2}{2\sigma_1^2}\right] \tag{3.64}$$

which is the normal probability density function defined previously in Equation (3.48).

For $N = 2$, there results the joint normal probability density function

$$p(x_1, x_2) = \frac{\exp\left\{\left[\frac{-1}{2(1 - \rho_{12}^2)}\right]\left[\left(\frac{x_1 - \mu_1}{\sigma_1}\right)^2 - 2\rho_{12}\left(\frac{x_1 - \mu_1}{\sigma_1}\right)\left(\frac{x_2 - \mu_2}{\sigma_2}\right) + \left(\frac{x_2 - \mu_2}{\sigma_2}\right)^2\right]\right\}}{2\pi\sigma_1\sigma_2\sqrt{1 - \rho_{12}^2}} \tag{3.65}$$

where $\rho_{12} = C_{12}/\sigma_1\sigma_2$ is the correlation coefficient between $x_1(k)$ and $x_2(k)$. Observe that when $x_1(k)$ and $x_2(k)$ are uncorrelated so that $\rho_{12} = 0$, one obtains

$$p(x_1, x_2) = p(x_1)p(x_2) \qquad (3.66)$$

which shows that $x_1(k)$ and $x_2(k)$ are also independent. This result is not true for arbitrary distributions.

Similar formulas may be written down for higher-order cases where $N = 3, 4, 5, \ldots$. For arbitrary N, it follows quite easily that if all different pairs of normally distributed random variables are mutually uncorrelated (that is, $\rho_{ij} = 0$ whenever $i \neq j$), then these random variables are mutually independent in the probability sense. That is,

$$p(x_1, x_2, \ldots, x_N) = p(x_1)p(x_2) \cdots p(x_N) \qquad (3.67)$$

The importance of the N-dimensional normal distribution in physical problems, analogous to the common one-dimensional normal distribution, is due in part to the *multidimensional central limit theorem* [References 3.2 and 3.5]. This theorem yields the result that the vector sum of a large number of mutually independent N-dimensional random variables approaches an N-dimensional normal distribution under fairly general conditions.

For any N-dimensional Gaussian probability density function represented by Equation (3.62), it follows that

$$\int\!\!\!\int\limits_{-\infty}^{\infty} \cdots \int p(x_1, x_2, \ldots, x_N) \, dx_1 \, dx_2 \, \cdots \, dx_N = 1 \qquad (3.68)$$

Also, the expected value of any real single-valued continuous function $g(x_1, x_2, \ldots, x_N)$ of the N random variables involved is given by

$$E[g(x_1, x_2, \ldots, x_N)] = \int\!\!\!\int\limits_{-\infty}^{\infty} \cdots \int g(x_1, x_2, \ldots, x_N) p(x_1, x_2, \ldots, x_N)$$

$$\times dx_1 \, dx_2 \, \cdots \, dx_N \qquad (3.69)$$

When $g(x_1, x_2, \ldots, x_N) = \exp(s_1 x_1 + s_2 x_2 + \cdots + s_N x_N)$, this defines from Equation (3.62) the Nth-*order moment generating function*, namely,

$$m(s_1, s_2, \ldots, s_N) = E[\exp(s_1 x_1 + s_2 x_2 + \cdots + s_N x_N)] \qquad (3.70)$$

Consider the case of four Gaussian random variables, x_1, x_2, x_3, and x_4, with *zero mean values* and equal variances $\sigma_i^2 = \sigma^2$. Their respective covariances C_{ij} now satisfy $C_{ij} = \rho_{ij}\sigma^2$, where

$$C_{12} = E[x_1 x_2] \qquad C_{23} = E[x_2 x_3]$$

$$C_{13} = E[x_1 x_3] \qquad C_{24} = E[x_2 x_4] \qquad (3.71)$$

$$C_{14} = E[x_1 x_4] \qquad C_{34} = E[x_3 x_4]$$

Let $p(x_1, x_2, x_3, x_4)$ be given by Equation (3.62). Then the *fourth-order moment generating function* becomes

$$m(s_1, s_2, s_3, s_4) = E\left[\exp(s_1 x_1 + s_2 x_2 + s_3 x_3 + s_4 x_4)\right]$$

$$= \exp\left[\frac{\sigma^2}{2}(s_1^2 + s_2^2 + s_3^2 + s_4^2 + 2\rho_{12} s_1 s_2 + 2\rho_{13} s_1 s_3\right.$$

$$\left. + 2\rho_{14} s_1 s_4 + 2\rho_{23} s_2 s_3 + 2\rho_{24} s_2 s_4 + 2\rho_{34} s_3 s_4)\right]$$

One can directly verify that the fourth-order moment is given by the fourth partial derivative

$$E[x_1 x_2 x_3 x_4] = \frac{\partial^4 m(s_1, s_2, s_3, s_4)}{\partial s_1 \, \partial s_2 \, \partial s_3 \, \partial s_4}\Bigg|_{s_1 = s_2 = s_3 = s_4 = 0}$$

It is now straightforward but tedious to perform the partial derivatives and to set $s_1 = s_2 = s_3 = s_4 = 0$ so as to derive

$$E[x_1 x_2 x_3 x_4] = C_{12} C_{34} + C_{13} C_{24} + C_{14} C_{23} \qquad (3.72)$$

This shows that the fourth-order moment is the sum of $3 = (3 \cdot 1)$ different pairs of second-order moments (covariances) contained therein.

A similar derivation for the sixth-order moment of zero mean value Gaussian data gives the result that the sixth-order moment is the product of $15 = (5 \cdot 3 \cdot 1)$ different triplets of second-order moments contained therein, namely,

$$E[x_1 x_2 x_3 x_4 x_5 x_6] = C_{12}[C_{34} C_{56} + C_{35} C_{46} + C_{36} C_{45}]$$

$$+ C_{13}[C_{24} C_{56} + C_{25} C_{46} + C_{26} C_{45}]$$

$$+ C_{14}[C_{23} C_{56} + C_{25} C_{36} + C_{26} C_{35}]$$

$$+ C_{15}[C_{23} C_{46} + C_{24} C_{36} + C_{26} C_{34}]$$

$$+ C_{16}[C_{23} C_{45} + C_{24} C_{35} + C_{25} C_{34}] \qquad (3.73)$$

In general, if n is an even integer, then $E[x_1 x_2 \cdots x_N]$ consists of $(N - 1)$ $(N - 3) \cdots (3)(1)$ different products of all possible C_{ij} terms.

On the other hand, all odd-order moments of Gaussian random variables with zero mean values will be zero, that is,

$$E[x_1 x_2 \cdots x_N] = 0 \qquad \text{if } N \text{ is odd} \tag{3.74}$$

All of these relations apply not only to the original random variables x_i, but also to any linear transformations, such as their Fourier transforms.

More complicated expressions occur when mean values are not zero. For example, consider four Gaussian random variables x_1, x_2, x_3, x_4 with equal *nonzero mean values* $\mu \neq 0$ and equal variances σ^2. Now, in place of Equation (3.71), the covariances C_{ij} for $i \neq j$ are given by

$$C_{ij} = E\big[(x_i - \mu)(x_j - \mu)\big] \tag{3.75}$$

When $i = j$,

$$C_{ii} = E\big[(x_i - \mu)^2\big] = \sigma^2 = \psi^2 - \mu^2 \tag{3.76}$$

The result in Equation (3.72) applies to the four variables $(x_1 - \mu), (x_2 - \mu),$ $(x_3 - \mu), (x_4 - \mu)$. Hence, from Equations (3.72) and (3.74), it follows that

$$E[x_1 x_2 x_3 x_4] = C_{12}C_{34} + C_{13}C_{24} + C_{14}C_{23} + \mu^4$$

$$+ \mu^2[C_{12} + C_{13} + C_{14} + C_{23} + C_{24} + C_{34}] \tag{3.77}$$

As a special case,

$$E\big[x_1^2 x_2^2\big] = \psi^4 + 2C_{12}^2 + 4\mu^2 C_{12} \tag{3.78}$$

The covariance term

$$C_{12} = E\big[(x_1 - \mu)(x_2 - \mu)\big] = E[x_1 x_2] - \mu^2$$

$$= R_{12} - \mu^2 \tag{3.79}$$

where

$$R_{12} = E[x_1 x_2] \tag{3.80}$$

Thus, Equation (3.78) can also be written

$$E\big[x_1^2 x_2^2\big] = \psi^4 + 2\big[R_{12}^2 - \mu^4\big] \tag{3.81}$$

and Equation (3.77) is the same as

$$E[x_1x_2x_3x_4] = R_{12}R_{34} + R_{13}R_{24} + R_{14}R_{23} - 2\mu^4 \tag{3.82}$$

Similar extensions occur for sixth-order moments when $\mu \neq 0$.

PROBLEMS

3.1 Consider a random variable x with a probability distribution function given by

$$P(x) = \begin{cases} 0 & x \leq 0 \\ x^n & 0 < x \leq 1 \\ 1 & x > 1 \end{cases}$$

    (a) Determine the probability density function $p(x)$.
    (b) Determine the mean value and variance of x.

3.2 Consider a random variable x with a probability density function given by

$$p(x) = \begin{cases} 4x^3 & 0 < x < 1 \\ 0 & \text{otherwise} \end{cases}$$

    (a) Determine the probability distribution function $P(x)$.
    (b) Determine the mean value and variance of x.

3.3 A random variable x is uniformly distributed such that the probability density function is given by

$$p(x) = \begin{cases} 1 & 0 < x < 1 \\ 0 & \text{elsewhere} \end{cases}$$

Find the probability density function of the random variable $y = 2x + 1$.

3.4 Assume a random variable x is normally distributed with a mean value of zero and a variance of unity. Determine the probability density function of the random variable $y = x^2$.

3.5 Assume a computer generates random digits $(0, 1, 2, 3, 4, 5, 6, 7, 8, 9)$ with equal probabilities. Let T be a random variable representing the sum of N digits. Determine the mean value and variance of T.

3.6 A manufacturer produces shafts and bearings that are ultimately assembled by placing a shaft into a bearing. The shafts are produced with an outside diameter s that is normally distributed with a mean value of $\mu_s = 1.0$ in. and a standard deviation of $\sigma_s = 0.003$ in. The inside diameter b of the manufactured bearings is also normally distributed with a mean value of $\mu_b = 1.01$ in. and a standard deviation of $\sigma_b = 0.004$ in. If the assembler selects a shaft and a bearing at random, what is the probability that a selected shaft will not fit inside the bearing? (Assume the shafts and bearings are perfectly circular and a fit occurs if $s < b$.)

3.7   In Problem 3.6, the manufacturer would like to see the randomly selected shafts and bearings fit on each assembly attempt with a clearance of no more than 0.02 inches. The standard deviation associated with the shaft diameter s cannot be altered, but the standard deviation of the inside diameter b of the bearings can be changed. Assuming a fit with the stated clearance is desired on at least 99% of the assembly attempts, what is the desired standard deviation of the diameter b?

3.8   Consider a random variable x with a Poisson distribution defined by $p(x) = \mu^x e^{-\mu}/x!$. Determine the mean value and variance of x using the moment generating function.

3.9   A manufacturer produces washers with a thickness d that has a mean value of $\mu_d = 0.05$ in. and a standard deviation of $\sigma_d = 0.005$. If $N = 25$ washers are selected at random and stacked on top of one another, determine the probability that the height of the stack will be between 1.20 and 1.30 in., assuming
(a)   the thickness d is normally distributed.
(b)   the thickness d has an unknown distribution function.

3.10   Two independent random variables x and y have probability density functions given by

$$p(x) = \frac{1}{2\sqrt{\pi}} e^{-(x-1)^2/4} \qquad p(y) = \frac{1}{2\sqrt{\pi}} e^{-(y+1)^2/4}$$

Determine the probability density functions of the random variables
(a) $u = x + y$.
(b) $v = x - y$.

REFERENCES

3.1   Papoulis, A., *Probability, Random Variables, and Stochastic Processes*, McGraw-Hill, New York, 1965.

3.2   Loeve, M. M., *Probability Theory*, 4th ed., Springer-Verlag, New York, 1977.

3.3   Laha, R. G., and Rohatgi, V. K., *Probability Theory*, Wiley, New York, 1979.

3.4   Patel, J. K., and Read, C. B., *Handbook of the Normal Distribution*, Dekker, New York, 1982.

3.5   Laning, J. H., Jr., and Battin, R. H., *Random Processes in Automatic Control*, McGraw-Hill, New York, 1956.

CHAPTER 4

STATISTICAL PRINCIPLES

Beyond the basic ideas of probability theory discussed in Chapter 3, the measurement and analysis of random data involve uncertainties and estimation errors that must be evaluated by statistical techniques. This chapter reviews and illustrates various statistical ideas that have wide applications to commonly occurring data evaluation problems. The intent is to provide the reader with a minimum background in terminology and certain techniques of engineering statistics that are relevant to discussions in later chapters. More detailed treatments of applied statistics with engineering applications are available from References 4.1–4.4.

4.1 SAMPLE VALUES AND PARAMETER ESTIMATION

Consider a random variable x, as defined in Section 3.1, where the index k of the sample space is omitted for simplicity in notation. Further consider the two basic parameters of x which specify its central tendency and dispersion; namely, the mean value and variance, respectively. From Equations (3.8) and (3.11), the mean value and variance are given by

$$\mu_x = E[x] = \int_{-\infty}^{\infty} xp(x)\, dx \tag{4.1}$$

$$\sigma_x^2 = E\left[(x - \mu_x)^2\right] = \int_{-\infty}^{\infty} (x - \mu_x)^2 p(x)\, dx \tag{4.2}$$

where $p(x)$ is the probability density function of the variable x. These two parameters of x cannot, of course, be precisely determined in practice since an exact knowledge of the probability density function will not generally be available. Hence one must be content with estimates of the mean value and variance based on a finite number of observed values.

One possible method (there are others) for estimating the mean value and variance of x based on N independent observations would be as follows:

$$\bar{x} = \hat{\mu}_x = \frac{1}{N} \sum_{i=1}^{N} x_i \tag{4.3}$$

$$s_b^2 = \hat{\sigma}_x^2 = \frac{1}{N} \sum_{i=1}^{N} (x - \bar{x})^2 \tag{4.4}$$

Here, $\bar{x}$ and s_b^2 are the *sample mean* and *sample variance*, respectively. The hats (^) over $\hat{\mu}_x$ and $\hat{\sigma}_x^2$ indicate that these sample values are being used as *estimators* for the mean value and variance of x. The subscript on s_b^2 means that this is a biased variance estimate (to be discussed later). The number of observations used to compute the estimates (sample values) is called the *sample size*.

The specific sample values in Equations (4.3) and (4.4) are not the only quantities that might be used to estimate the mean value and variance of a random variable x. For example, reasonable estimates of the mean value and variance would also be obtained by dividing the summations in Equations (4.3) and (4.4) by $N - 1$ instead of N. Estimators are never clearly right or wrong since they are defined somewhat arbitrarily. Nevertheless, certain estimators can be judged as being "good" estimators or "better" estimators than others.

Three principal factors can be used to establish the quality or "goodness" of an estimator. First, it is desirable that the expected value of the estimator be equal to the parameter being established. That is,

$$E[\hat{\phi}] = \phi \tag{4.5}$$

where $\hat{\phi}$ is an estimator for the parameter ϕ. If this is true, the estimator is said to be *unbiased*. Second, it is desirable that the mean square error of the estimator be smaller than for other possible estimators. That is,

$$E\left[(\hat{\phi}_1 - \phi)^2\right] \le E\left[(\hat{\phi}_i - \phi)^2\right] \tag{4.6}$$

where $\hat{\phi}_1$ is the estimator of interest and $\hat{\phi}_i$ is any other possible estimator. If this is true, the estimator is said to be more *efficient* than other possible estimators. Third, it is desirable that the estimator approach the parameter being estimated with a probability approaching unity as the sample size becomes large. That is, for any $\varepsilon > 0$,

$$\lim_{N \to \infty} \text{Prob}[|\hat{\phi} - \phi| \ge \varepsilon] = 0 \tag{4.7a}$$

If this is true, the estimator is said to be *consistent*. It follows from the Chebyshev inequality of Equation (3.22) that a sufficient (but not necessary) condition to meet the requirements of Equation (4.7a) is given by

$$\lim_{N \to \infty} E\left[(\hat{\phi} - \phi)^2\right] = 0 \tag{4.7b}$$

Note that the requirements stated in Equation (4.7) are simply convergence requirements in (a) probability and (b) the mean square sense, as defined later in Section 5.3.4.

Consider the example of the mean value estimator given by Equation (4.3). The expected value of the sample mean $\bar{x}$ is

$$E[\bar{x}] = E\left[\frac{1}{N}\sum_{i=1}^{N} x_i\right] = \frac{1}{N}E\left[\sum_{i=1}^{N} x_i\right] = \frac{1}{N}(N\mu_x) = \mu_x \tag{4.8}$$

Hence from Equation (4.5), the estimator $\hat{\mu}_x = \bar{x}$ is unbiased. The mean square error of the sample mean $\bar{x}$ is given by

$$E\left[(\bar{x} - \mu_x)^2\right] = E\left[\left(\frac{1}{N}\sum_{i=1}^{N} x_i - \mu_x\right)^2\right] = \frac{1}{N^2}E\left[\left(\sum_{i=1}^{N}(x_i - \mu_x)\right)^2\right]$$

From Section 3.2.1, since the observations x_i are independent, the cross product terms in the last expression will have an expected value of zero. It then follows that

$$E\left[(\bar{x} - \mu_x)^2\right] = \frac{1}{N^2}E\left[\sum_{i=1}^{N}(x_i - \mu_x)^2\right] = \frac{1}{N^2}(N\sigma_x^2) = \frac{\sigma_x^2}{N} \tag{4.9}$$

Hence from Equation (4.7b), the estimator $\hat{\mu}_x = \bar{x}$ is consistent. It can be shown that the estimator is also efficient.

Now consider the example of the variance estimator given by Equation (4.4). The expected value of the sample variance s_b^2 is

$$E\left[s_b^2\right] = E\left[\frac{1}{N}\sum_{i=1}^{N}(x_i - \bar{x})^2\right] = \frac{1}{N}E\left[\sum_{i=1}^{N}(x_i - \bar{x})^2\right]$$

However,

$$\sum_{i=1}^{N}(x_i - \bar{x})^2 = \sum_{i=1}^{N}(x_i - \mu_x + \mu_x - \bar{x})^2$$

$$= \sum_{i=1}^{N}(x_i - \mu_x)^2 - 2(\bar{x} - \mu_x)\sum_{i=1}^{N}(x_i - \mu_x) + \sum_{i=1}^{N}(\bar{x} - \mu_x)^2$$

$$= \sum_{i=1}^{N}(x_i - \mu_x)^2 - 2(\bar{x} - \mu_x)N(\bar{x} - \mu_x) + N(\bar{x} - \mu_x)^2$$

$$= \sum_{i=1}^{N}(x_i - \mu_x)^2 - N(\bar{x} - \mu_x)^2 \tag{4.10}$$

Since $E[(x_i - \mu_x)^2] = \sigma_x^2$ and $E[(\bar{x} - \mu_x)^2] = \sigma_x^2/N$, it follows that

$$E\left[s_b^2\right] = \frac{1}{N}\left(N\sigma_x^2 - \sigma_x^2\right) = \frac{(N-1)}{N}\sigma_x^2 \qquad (4.11)$$

Hence the estimator $\hat{\sigma}_x^2 = s_b^2$ is *biased*. Although the sample variance s_b^2 is a biased estimator for σ_x^2, it is a consistent and efficient estimator.

From the results in Equation (4.11), it is clear that an unbiased estimator for σ_x^2 may be obtained by computing a slightly different sample variance as follows

$$s^2 = \hat{\sigma}_x^2 = \frac{1}{N-1}\sum_{i=1}^{N}(x_i - \bar{x})^2 \qquad (4.12)$$

The quantity defined in Equation (4.12) is an *unbiased* estimator for σ_x^2. For this reason, the sample variance defined in Equation (4.12) is often considered a "better" estimator than the sample variance given by Equation (4.4). The sample variance defined in Equation (4.12) will be used henceforth as an estimator for the variance of a random variable.

4.2 IMPORTANT PROBABILITY DISTRIBUTION FUNCTIONS

Examples of several theoretical probability distribution functions are given in Chapter 3. The most important of these distribution functions from the viewpoint of applied statistics is the Gaussian (*normal*) distribution. There are three other distribution functions associated with normally distributed random variables that have wide applications as statistical tools. These are the χ^2 distribution, the t distribution, and the F distribution. Each of these three, along with the normal distribution, will now be defined and discussed. Applications for each as an analysis tool will be covered in later sections.

4.2.1 *Normal Distribution*

The probability density and distribution functions of a normally distributed random variable x are defined by Equation (3.48) in Section 3.3. A more convenient form of the normal distribution is obtained by using the standardized variable z given by

$$z = \frac{x - \mu_x}{\sigma_x} \qquad (4.13)$$

When Equation (4.13) is substituted into Equation (3.48), standardized normal density and distribution functions with zero mean and unit variance ($\mu_z = 0$;

$\sigma_z^2 = 1$) are obtained as follows:

$$p(z) = (\sqrt{2\pi})^{-1} e^{-z^2/2} \tag{4.14a}$$

$$P(z) = (\sqrt{2\pi})^{-1} \int_{-\infty}^{z} e^{-\xi^2/2}\, d\xi \tag{4.14b}$$

It is desirable for later applications to denote the value of z that corresponds with a specific probability $P(z) = 1 - \alpha$ by z_α. That is,

$$P(z_\alpha) = \int_{-\infty}^{z_\alpha} p(z)\, dz = \text{Prob}[z \le z_\alpha] = 1 - \alpha \tag{4.15a}$$

or

$$1 - P(z_\alpha) = \int_{z_\alpha}^{\infty} p(z)\, dz = \text{Prob}[z > z_\alpha] = \alpha \tag{4.15b}$$

The value of z_α that satisfies Equation (4.15) is called the 100α *percentage point* of the normal distribution.

The standardized normal probability density function $p(z)$ is unimodal, monotonic about the mode, and symmetric with inflection points at ± 1, as illustrated in Figure 4.1(a). The corresponding distribution function $P(z)$ is illustrated in Figure 4.1(b). A limited tabulation is presented for the ordinates of the standardized normal density function in Table A.1, and for the areas under the standardized normal density function in Table A.2 in Appendix A.

4.2.2 Chi-Square Distribution

Let $z_1, z_2, z_3, \ldots, z_n$ be n independent random variables, each of which has a normal distribution with zero mean and unit variance. Let a new random variable be defined as

$$\chi_n^2 = z_1^2 + z_2^2 + z_3^2 + \cdots + z_n^2 \tag{4.16}$$

The random variable χ_n^2 is the chi-square variable with n degrees of freedom. The number of *degrees of freedom* n represents the number of independent or "free" squares entering into the expression. From Reference 4.2, the probability density function of χ_n^2 is given by

$$p(\chi^2) = [2^{n/2}\Gamma(n/2)]^{-1}(\chi^2)^{((n/2)-1)} e^{-\chi^2/2} \qquad \chi^2 \ge 0 \tag{4.17}$$

where $\Gamma(n/2)$ is the gamma function. The corresponding distribution function of χ_n^2, given by the integral of Equation (4.17) from $-\infty$ to a specific value of χ_n^2, is called the *chi-square distribution with n degrees of freedom*. The 100α

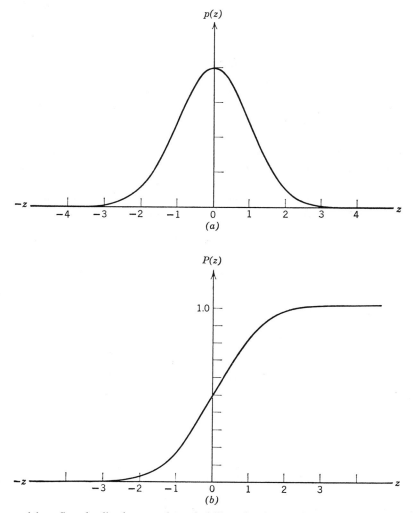

Figure 4.1 Standardized normal probability density and distribution functions. (*a*) Probability density function. (*b*) Probability distribution function.

percentage point of the χ^2 distribution will be denoted by $\chi^2_{n;\,\alpha}$. That is,

$$\int_{\chi^2_{n;\,\alpha}}^{\infty} p(\chi^2)\, d\chi^2 = \mathrm{Prob}\big[\chi^2_n > \chi^2_{n;\,\alpha}\big] = \alpha \qquad (4.18)$$

The mean value and variance of the variable χ^2_n are

$$E\big[\chi^2_n\big] = \mu_{\chi^2} = n \qquad (4.19)$$

$$E\Big[\big(\chi^2_n - \mu_{\chi^2}\big)^2\Big] = \sigma^2_{\chi^2} = 2n \qquad (4.20)$$

A limited tabulation of percentage points for the χ^2 distribution function is presented in Table A.3 in Appendix A.

Several features of the chi-square distribution should be noted. First, the chi-square distribution is actually a special case of the more general gamma distribution function. Second, the square root of chi-square with two degrees of freedom (χ_2) constitutes an important special case called the *Rayleigh distribution function*. The Rayleigh distribution has wide applications to two-dimensional target problems and is also the limiting distribution function of *peak values* for a narrow-band Gaussian random signal as the bandwidth approaches zero. Third, the square root of chi-square with three degrees of freedom (χ_3) constitutes another important special case, which is called the *Maxwell distribution function*. The Maxwell distribution has applications to three-dimensional target problems. Fourth, a chi-square distribution approaches a normal distribution as the number of degrees of freedom becomes large. Specifically, for $n > 30$, the quantity $\sqrt{2\chi_n^2}$ is distributed approximately as a normal variable with a mean of $\mu = \sqrt{2n-1}$ and a variance of $\sigma^2 = 1$.

4.2.3   Student's *t* Distribution   *(Pseudo Normal Dist.)*

Let y and z be independent random variables such that y has a χ_n^2 distribution function and z has a normal distribution function with zero mean and unit variance. Let a new random variable be defined as

$$t_n = \frac{z}{\sqrt{y/n}} \tag{4.21}$$

The random variable t_n is Student's t variable with n degrees of freedom. From Reference 4.2, the probability density function of t_n is given by

$$p(t) = \frac{\Gamma[(n+1)/2]}{\sqrt{\pi n}\,\Gamma(n/2)}\left[1 + \frac{t^2}{n}\right]^{-(n+1)/2} \tag{4.22}$$

The corresponding distribution function of t_n, given by the integral of Equation (4.22) from $-\infty$ to a specific value of t_n, is called the *t distribution with n degrees of freedom*. The 100α percentage point of the t distribution will be denoted by $t_{n;\,\alpha}$. That is,

$$\int_{t_{n;\,\alpha}}^{\infty} p(t)\, dt = \text{Prob}[t_n > t_{n;\,\alpha}] = \alpha \tag{4.23}$$

The mean value and variance of the variable t_n are

$$E[t_n] = \mu_t = 0 \qquad \text{for } n > 1 \tag{4.24}$$

$$E\left[(t_n - \mu_t)^2\right] = \sigma_t^2 = \frac{n}{n-2} \qquad \text{for } n > 2 \tag{4.25}$$

A limited tabulation of percentage points for the t distribution function is presented in Table A.4 in Appendix A. It should be noted that the t distribution approaches a standardized normal distribution as the number of degrees of freedom n becomes large.

4.2.4 The F Distribution *(Distribution of variances) of*

Let y_1 and y_2 be independent random variables such that y_1 has a χ^2 distribution function with n_1 degrees of freedom and y_2 has a χ^2 distribution function with n_2 degrees of freedom. Let a new random variable be defined as

$$F_{n_1, n_2} = \frac{y_1/n_1}{y_2/n_2} = \frac{y_1 n_2}{y_2 n_1} \tag{4.26}$$

The random variable F_{n_1, n_2} is the F variable with n_1 and n_2 degrees of freedom. From Reference 4.2, the probability density function of F_{n_1, n_2} is given by

$$p(F) = \frac{\Gamma[(n_1 + n_2)/2](n_1/n_2)^{n_1/2} F^{(n_1/2) - 1}}{\Gamma(n_1/2)\Gamma(n_2/2)[1 + (n_1 F/n_2)]^{(n_1 + n_2)/2}} \qquad F \geq 0 \quad (4.27)$$

The corresponding distribution function of F_{n_1, n_2}, given by the integral of Equation (4.27) from $-\infty$ to a specific value of F_{n_1, n_2}, is called the F *distribution with n_1 and n_2 degrees of freedom*. The 100α percentage point of the F distribution will be denoted by $F_{n_1, n_2; \alpha}$. That is,

$$\int_{F_{n_1, n_2; \alpha}}^{\infty} p(F)\, dF = \text{Prob}\left[F_{n_1, n_2} > F_{n_1, n_2; \alpha} \right] = \alpha \tag{4.28}$$

The mean value and variance of F_{n_1, n_2} are

$$E\left[F_{n_1, n_2} \right] = \mu_F = \frac{n_2}{n_2 - 2} \qquad \text{for } n_2 > 2 \quad (4.29)$$

$$E\left[(F_{n_1, n_2} - \mu_F)^2 \right] = \sigma_F^2 = \frac{2n_2^2(n_1 + n_2 - 2)}{n_1(n_2 - 2)^2(n_2 - 4)} \qquad \text{for } n_2 > 4 \quad (4.30)$$

A limited tabulation of percentage points for the F distribution function is presented in Tables A.5(a), (b), and (c) in Appendix A. It should be noted that the statistic t_n^2, the square of the variable defined in Equation (4.21), has an F distribution with $n_1 = 1$ and $n_2 = n$ degrees of freedom.

4.3 SAMPLING DISTRIBUTIONS AND ILLUSTRATIONS

Consider a random variable x with a probability distribution function $P(x)$. Let $x_1, x_2, \ldots, x_N$ be a sample of N observed values of x. Any quantity computed from these sample values will also be a random variable. For example, consider the mean value $\bar{x}$ of the sample. If a series of different samples of size N were selected from the same random variable x, the value of $\bar{x}$ computed from each sample would generally be different. Hence, $\bar{x}$ is also a random variable with a probability distribution function $P(\bar{x})$. This probability distribution function is called the *sampling distribution* of $\bar{x}$.

Some of the more common sampling distributions which often arise in practice will now be considered. These involve the probability distribution functions defined and discussed in Section 4.2. The use of these sampling distributions to establish confidence intervals and perform hypothesis tests is illustrated in Sections 4.4–4.8.

4.3.1 *Distribution of Sample Mean with Known Variance*

Consider the mean value of a sample of N independent observations from a random variable x as follows:

$$\bar{x} = \frac{1}{N} \sum_{i=1}^{N} x_i \qquad (4.31)$$

First, consider the case where the random variable x is normally distributed with a mean value of μ_x and a known variance of σ_x^2. From Section 3.3.1, the sampling distribution of the sample mean $\bar{x}$ will also be normally distributed. From Equation (4.8), the mean value of the sampling distribution of $\bar{x}$ is

$$\mu_{\bar{x}} = \mu_x \qquad (4.32)$$

and from Equation (4.9), the variance of the sampling distribution of $\bar{x}$ is

$$\sigma_{\bar{x}}^2 = \frac{\sigma_x^2}{N} \qquad (4.33)$$

Hence, from Equation (4.13), the following sampling distribution applies for the sample mean $\bar{x}$:

$$\frac{(\bar{x} - \mu_x)\sqrt{N}}{\sigma_x} = z \qquad (4.34)$$

where z has a standardized normal distribution, as defined in Section 4.2.1. It follows that a probability statement concerning future values of the sample mean may be made as follows.

$$\text{Prob}\left[\bar{x} > \left(\frac{\sigma_x z_\alpha}{\sqrt{N}} + \mu_x\right)\right] = \alpha. \tag{4.35}$$

Now, consider the case where the random variable x is not normally distributed. From the practical implications of the central limit theorem, the following result occurs. As the sample size N becomes large, the *sampling distribution of the sample mean $\bar{x}$ approaches a normal distribution regardless of the distribution of the original variable x*. In practical terms, a normality assumption for the sampling distribution of $\bar{x}$ becomes reasonable in many cases for $N > 4$ and quite accurate in most cases for $N > 10$. Hence, for reasonably large sample sizes, Equation (4.34) applies to the sampling distribution of $\bar{x}$ computed for any random variable x, regardless of its probability distribution function.

4.3.2 *Distribution of Sample Variance*

Consider the variance of a sample of N independent observations from a random variable x as follows.

$$s^2 = \frac{1}{N-1} \sum_{i=1}^{N} (x_i - \bar{x})^2 \tag{4.36}$$

If the variable x is normally distributed with a mean of μ_x and a variance of σ_x^2, it is shown in Reference 4.1 that

$$\sum_{i=1}^{N} (x_i - \bar{x})^2 = \sigma_x^2 \chi_n^2 \qquad n = N - 1$$

where χ_n^2 has a chi-square distribution with $n = N - 1$ degrees of freedom, as defined in Section 4.2.2. Hence, the sampling distribution of the sample variance s^2 is given by

$$\frac{ns^2}{\sigma_x^2} = \chi_n^2 \qquad n = N - 1 \tag{4.37}$$

It follows that a probability statement concerning future values of the sample variance s^2 may be made as follows:

$$\text{Prob}\left[s^2 > \frac{\sigma_x^2 \chi_{n;\,\alpha}^2}{n}\right] = \alpha \tag{4.38}$$

4.3.3 *Distribution of Sample Mean with Unknown Variance*

Consider the mean value of a sample of N independent observations from a random variable x, as given by Equation (4.31). If the variable x is normally distributed with a mean value of μ_x and an unknown variance, it is seen from Equations (4.21) and (4.37) that

$$\frac{(\bar{x} - \mu_x)}{s/\sqrt{N}} = \frac{\sigma_x z/\sqrt{N}}{\sqrt{\sigma_x^2 \chi_n^2/n}/\sqrt{N}} = \frac{z}{\sqrt{\chi_n^2/n}} = t_n$$

where t_n has a t distribution with $n = N - 1$ degrees of freedom, as defined in Section 4.2.3. Hence the sampling distribution of the sample mean $\bar{x}$ when σ_x^2 is unknown is given by

$$\frac{(\bar{x} - \mu_x)\sqrt{N}}{s} = t_n \qquad n = N - 1 \tag{4.39}$$

It follows that a probability statement concerning future values of the sample mean $\bar{x}$ may be made as follows.

$$\text{Prob}\left[\bar{x} > \left(\frac{s t_{n;\,x}}{\sqrt{N}} + \mu_x\right)\right] = \alpha \tag{4.40}$$

4.3.4 *Distribution of Ratio of Two Sample Variances*

Consider the variances of two samples, one consisting of N_x independent observations of a random variable x, and the other consisting of N_y independent observations of a random variable y, as given by Equation (4.36). If the variable x is normally distributed with a mean value of μ_x and a variance of σ_x^2, and the variable y is normally distributed with a mean value of μ_y and a variance σ_y^2, it is seen from Equations (4.26) and (4.37) that

$$\frac{s_x^2/\sigma_x^2}{s_y^2/\sigma_y^2} = \frac{\sigma_x^2 \chi_{n_x}^2/n_x \sigma_x^2}{\sigma_y^2 \chi_{n_y}^2/n_y \sigma_y^2} = F_{n_x,\,n_y}$$

where $F_{n_x,\,n_y}$ has an F distribution with $n_x = N_x - 1$ and $n_y = N_y - 1$ degrees of freedom, as defined in Section 4.2.4. Hence the sampling distribution of the ratio of the sample variances s_x^2 and s_y^2 is given by

$$\frac{s_x^2/\sigma_x^2}{s_y^2/\sigma_y^2} = F_{n_x,\,n_y} \qquad \begin{array}{l} n_x = N_x - 1 \\ n_y = N_y - 1 \end{array} \tag{4.41}$$

It follows that a probability statement concerning future values of the ratio of the sample variances s_x^2 and s_y^2 may be made as follows:

$$\text{Prob}\left[\frac{s_x^2}{s_y^2} > \frac{\sigma_x^2}{\sigma_y^2}F_{n_x, n_y; \alpha}\right] = \alpha \qquad (4.42)$$

Note that if the two samples are obtained from the same random variable, $x = y$, then Equation (4.41) reduces to

$$\frac{s_1^2}{s_2^2} = F_{n_1, n_2} \qquad \begin{array}{l} n_1 = N_1 - 1 \\ n_2 = N_2 - 1 \end{array} \qquad (4.43)$$

4.4 CONFIDENCE INTERVALS

The use of sample values as estimators for parameters of random variables is discussed in Section 4.1. However, those procedures result only in point estimates for a parameter of interest; no indication is provided as to how closely a sample value estimates the parameter. A more meaningful procedure for estimating parameters of random variables involves the estimation of an interval, as opposed to a single point value, which will include the parameter being estimated with a known degree of uncertainty. For example, consider the case where the sample mean $\bar{x}$ computed from N independent observations of a random variable x is being used as an estimator for the mean value μ_x. It is usually more desirable to estimate μ_x in terms of some interval, such as $\bar{x} \pm d$, where there is a specified uncertainty that μ_x falls within that interval. Such intervals can be established if the sampling distributions of the estimator in question is known.

Continuing with the example of a mean value estimate, it is shown in Section 4.3 that probability statements can be made concerning the value of a sample mean $\bar{x}$ as follows:

$$\text{Prob}\left[z_{1-\alpha/2} < \frac{(\bar{x} - \mu_x)\sqrt{N}}{\sigma_x} \le z_{\alpha/2}\right] = 1 - \alpha \qquad (4.44)$$

The above probability statement is technically correct *before* the sample has been collected and $\bar{x}$ has been computed. After the sample has been collected, however, the value of $\bar{x}$ is a fixed number rather than a random variable. Hence it can be argued that the probability statement in Equation (4.44) no longer applies since the quantity $(\bar{x} - \mu_x)\sqrt{N}/\sigma_x$ either *does* or *does not* fall within the indicated limits. In other words, after a sample has been collected, a technically correct probability statement would be as follows:

$$\text{Prob}\left[z_{1-\alpha/2} < \frac{(\bar{x} - \mu_x)\sqrt{N}}{\sigma_x} \le z_{\alpha/2}\right] = \begin{cases} 0 \\ 1 \end{cases} \qquad (4.45)$$

Whether the correct probability is zero or unity is usually not known. As the value of α becomes small (as the interval between $z_{1-\alpha/2}$ and $z_{\alpha/2}$ becomes wide), however, one would tend to guess that the probability is more likely to be unity than zero. In slightly different terms, if many different samples were repeatedly collected and a value of $\bar{x}$ were computed for each sample, one would tend to expect the quantity in Equation (4.45) to fall within the noted interval for about $1 - \alpha$ of the samples. In this context, a statement can be made about an interval within which one would expect to find the quantity $(\bar{x} - \mu_x)\sqrt{N}/\sigma_x$ with a small degree of uncertainty. Such statements are called *confidence statements*. The interval associated with a confidence statement is called a *confidence interval*. The degree of trust associated with the confidence statement is called the *confidence coefficient*.

For the case of the mean value estimate, a confidence interval can be established for the mean value μ_x based on the sample value $\bar{x}$ by rearranging terms in Equation (4.45) as follows:

$$\left[\bar{x} - \frac{\sigma_x z_{\alpha/2}}{\sqrt{N}} \le \mu_x < \bar{x} + \frac{\sigma_x z_{\alpha/2}}{\sqrt{N}} \right] \qquad (4.46a)$$

Furthermore, if σ_x is unknown, a confidence interval can still be established for the mean value μ_x based on the sample values $\bar{x}$ and s by rearranging terms in Equation (4.39) as follows:

$$\left[\bar{x} - \frac{s t_{n;\,\alpha/2}}{\sqrt{N}} \le \mu_x < \bar{x} + \frac{s t_{n;\,\alpha/2}}{\sqrt{N}} \right] \qquad n = N - 1 \qquad (4.46b)$$

Equation (4.46) uses the fact that $z_{1-\alpha/2} = -z_{\alpha/2}$ and $t_{n;1-\alpha/2} = -t_{n;\,\alpha/2}$. The confidence coefficient associated with the intervals is $1 - \alpha$. Hence the confidence statement would be as follows: The true mean value μ_x falls within the noted interval with a confidence coefficient of $1 - \alpha$, or, in more common terminology, with a confidence of $100(1 - \alpha)\%$. Similar confidence statements can be established for any parameter estimates where proper sampling distributions are known. For example, from Equation (4.37), a $1 - \alpha$ confidence interval for the variance σ_x^2 based on a sample variance s^2 from a sample of size N is

$$\left[\frac{ns^2}{\chi_{n;\,\alpha/2}^2} \le \sigma_x^2 < \frac{ns^2}{\chi_{n;\,1-\alpha/2}^2} \right] \qquad n = N - 1 \qquad (4.47)$$

Example 4.1. Illustration of Confidence Intervals. Assume a sample of $N = 31$ independent observations are collected from a normally distributed random variable x with the following results:

60	61	47	56	61	63
65	69	54	59	43	61
55	61	56	48	67	65
60	58	57	62	57	58
53	59	58	61	67	62
54					

Determine a 90% confidence interval for the mean value and variance of the random variable x.

From Equation (4.46b), a $1 - \alpha$ confidence interval for the mean value μ_x based on the sample mean x and the sample variance s^2 for a sample size of $N = 31$ is given by

$$\left[\left(\bar{x} - \frac{st_{30;\,\alpha/2}}{\sqrt{31}}\right) \le \mu_x < \left(\bar{x} + \frac{st_{30;\,\alpha/2}}{\sqrt{31}}\right)\right]$$

From Table A.4, for $\alpha = 0.10$, $t_{30;\,\alpha/2} = t_{30;\,0.05} = 1.697$, so the interval reduces to

$$\left[(\bar{x} - 0.3048s) \le \mu_x < (\bar{x} + 0.3048s)\right]$$

From Equation (4.47), a $1 - \alpha$ confidence interval for the variance σ_x^2 based on the sample variance s^2 for a sample size of $N = 31$ is given by

$$\left[\frac{30s^2}{\chi_{30;\,\alpha/2}^2} \le \sigma_x^2 < \frac{30s^2}{\chi_{30;\,1-\alpha/2}^2}\right]$$

From Table A.3, for $\alpha = 0.10$, $\chi_{30;\,\alpha/2}^2 = \chi_{30;\,0.05}^2 = 43.77$ and $\chi_{30;\,1-\alpha/2}^2 = \chi_{30;\,0.95}^2 = 18.49$, so the interval reduces to

$$\left[0.6854s^2 \le \sigma_x^2 < 1.622s^2\right]$$

It now remains to calculate the sample mean and variance, and substitute these values into the interval statements. From Equation (4.3), the sample mean is

$$\bar{x} = \frac{1}{N}\sum_{i=1}^{N} x_i = 58.61$$

From Equation (4.12), the sample variance is

$$s^2 = \frac{1}{N-1}\sum_{i=1}^{N}(x_i - \bar{x})^2 = \frac{1}{N-1}\left\{\sum_{i=1}^{N}x_i^2 - N(\bar{x})^2\right\} = 33.43$$

Hence the 90% confidence intervals for the mean value and variance of the random variable x are as follows:

$$[56.85 \leq \mu_x < 60.37]$$

$$[22.91 \leq \sigma_x^2 < 54.22]$$

4.5 HYPOTHESIS TESTS

Consider the case where a given estimator $\hat{\phi}$ is computed from a sample of N independent observations of a random variable x. Assume that there is reason to believe that the true parameter ϕ being estimated has a specific value ϕ_0. Now, even if $\phi = \phi_0$, the sample value $\hat{\phi}$ will probably not come out exactly equal to ϕ_0 because of the sampling variability associated with $\hat{\phi}$. Hence the following question arises. If it is hypothesized that $\phi = \phi_0$, how much difference between $\hat{\phi}$ and ϕ_0 must occur before the hypothesis should be rejected as being invalid? This question can be answered in statistical terms by considering the probability of any noted difference between $\hat{\phi}$ and ϕ_0 based upon the sampling distribution of $\hat{\phi}$. If the probability of a given difference is small, the difference would be considered significant and the hypothesis that $\phi = \phi_0$ would be rejected. If the probability of a given difference is not small, the difference would be accepted as normal statistical variability and the hypothesis that $\phi = \phi_0$ would be accepted.

The preceding discussion outlines the simplest form of a statistical procedure called hypothesis testing. To clarify the general technique, assume that a sample value $\hat{\phi}$, which is an estimate of a parameter ϕ, has a probability density function of $p(\hat{\phi})$. Now, if a hypothesis that $\phi = \phi_0$ is true, then $p(\hat{\phi})$ would have a mean value of ϕ_0 as illustrated in Figure 4.2. The probability that $\hat{\phi}$ would fall below the lower level $\phi_{1-\alpha/2}$ is

$$\text{Prob}\left[\hat{\phi} \leq \phi_{1-\alpha/2}\right] = \int_{-\infty}^{\phi_{1-\alpha/2}} p(\hat{\phi}) \, d\hat{\phi} = \frac{\alpha}{2} \qquad (4.48a)$$

The probability that $\hat{\phi}$ would fall above the upper value $\phi_{\alpha/2}$ is

$$\text{Prob}\left[\hat{\phi} > \phi_{\alpha/2}\right] = \int_{\phi_{\alpha/2}}^{\infty} p(\hat{\phi}) \, d\hat{\phi} = \frac{\alpha}{2} \qquad (4.48b)$$

Hence the probability that $\hat{\phi}$ would be outside the range between $\phi_{1-\alpha/2}$ and $\phi_{\alpha/2}$ is α. Now let α be small so that it is very unlikely that $\hat{\phi}$ would fall outside the range between $\phi_{1-\alpha/2}$ and $\phi_{\alpha/2}$. If a sample were collected and a value of $\hat{\phi}$ were computed that in fact fell outside the range between $\phi_{1-\alpha/2}$ and $\phi_{\alpha/2}$, there would be strong reason to question the original hypothesis that

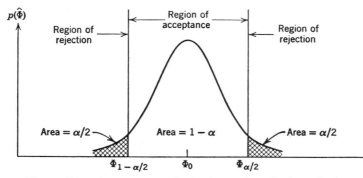

Figure 4.2   Acceptance and rejection regions for hypothesis tests.

$\phi = \phi_0$ since such a value for $\hat{\phi}$ would be very unlikely if the hypothesis were true. Hence the hypothesis that $\phi = \phi_0$ would be rejected. On the other hand, if the value for $\hat{\phi}$ fell within the range between $\phi_{1-\alpha/2}$ and $\phi_{\alpha/2}$, there would be no strong reason to question the original hypothesis. Hence the hypothesis that $\phi = \phi_0$ would be accepted.

The small probability α used for the hypothesis test is called the *level of significance* of the test. The range of values of $\hat{\phi}$ for which the hypothesis will be rejected is called the *region of rejection* or *critical region*. The range of values of $\hat{\phi}$ for which the hypothesis will be accepted is called the *region of acceptance*. The simple hypothesis test outlined above is called a *two-sided test* because, if the hypothesis is not true, the value of ϕ could be either greater or less than ϕ_0. Hence it is necessary to test for significant differences between ϕ and ϕ_0 in both directions. In other cases a *one-sided test* might be sufficient. For example, let it be hypothesized that $\phi \geq \phi_0$. For this case, the hypothesis would be false only if ϕ were less than ϕ_0. Thus the test would be performed using the lower side of the probability density function $p(\hat{\phi})$.

Two possible errors can occur when a hypothesis test is performed. First, the hypothesis might be rejected when in fact it is true. This possible error is called a *Type I Error*. Second, the hypothesis might be accepted when in fact it is false. This possible error is called a *Type II Error*. From Figure 4.2, a Type I Error would occur if the hypothesis were true and $\hat{\phi}$ fell in the region of rejection. It follows that the probability of a Type I Error is equal to α, the level of significance of the test.

In order to establish the probability of a Type II Error, it is necessary to specify some deviation of the true parameter ϕ from the hypothesized parameter ϕ_0 that one desires to detect. For example, assume that the true parameter actually has a value of either $\phi = \phi_0 + d$ or $\phi = \phi_0 - d$, as illustrated in Figure 4.3. If it is hypothesized that $\phi = \phi_0$ when in fact $\phi = \phi_0 \pm d$, the probability that $\hat{\phi}$ would fall inside the acceptance region between $\phi_{1-\alpha/2}$ and $\phi_{\alpha/2}$ is β. Hence the probability of a Type II Error is β for detecting a difference of $\pm d$ from the hypothesized value ϕ_0.

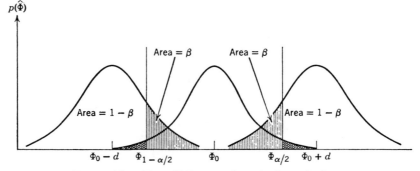

Figure 4.3   Type II Error regions for hypothesis tests.

The probability $1 - \beta$ is called the *power of the test*. Clearly, for any given sample size N, the probability of a Type I Error can be reduced by reducing the level of significance α. However, this will increase the probability β of a Type II Error (reduce the power of the test). The only way to reduce both α and β is to increase the sample size N for the estimate $\hat{\phi}$. These ideas form the basis for selecting the necessary sample sizes for statistical experiments.

Example 4.2. Illustration of Hypothesis Test Design.   Assume there is reason to believe that the mean value of a random variable x is $\mu_x = 10$. Further assume that the variance of x is known to be $\sigma_x^2 = 4$. Determine the proper sample size to test the hypothesis that $\mu_x = 10$ at the 5% level of significance, where the probability of a Type II Error is to be 5% for detecting a difference of 10% from the hypothesized value. Determine the region of acceptance to be used for the test.

An unbiased estimate for μ_x is given by the sample mean value $\bar{x}$ as defined in Equation (4.3). The appropriate sampling distribution of $\bar{x}$ is given by Equation (4.34) as

$$\bar{x} = \frac{\sigma_x}{\sqrt{N}} z + \mu_x$$

where z is normally distributed with zero mean and unit variance. Note that this sampling distribution of $\bar{x}$ is precise if x is normally distributed and is still a good approximation if x is not normally distributed.

The upper and lower limits of the acceptance region for the hypothesis test are as follows:

$$\text{upper limit} = \frac{\sigma_x}{\sqrt{N}} z_{\alpha/2} + \mu_x$$

$$\text{lower limit} = \frac{\sigma_x}{\sqrt{N}} z_{1-\alpha/2} + \mu_x$$

Now if the true mean value were in fact $\mu'_x = \mu_x \pm d$, a Type II Error would occur with probability β if the sample value $\bar{x}$ fell below the upper limit or above the lower limit. In terms of the sampling distributions of $\bar{x}$ with a mean value $\mu'_x = \mu_x + d$ or $\mu'_x = \mu_x - d$,

$$\text{upper limit} = \frac{\sigma_x}{\sqrt{N}} z_{1-\beta} + \mu_x + d$$

$$\text{lower limit} = \frac{\sigma_x}{\sqrt{N}} z_\beta + \mu_x - d$$

Hence the following equalities apply:

$$\frac{\sigma_x}{\sqrt{N}} z_{\alpha/2} + \mu_x = \frac{\sigma_x}{\sqrt{N}} z_{1-\beta} + \mu_x + d$$

$$\frac{\sigma_x}{\sqrt{N}} z_{1-\alpha/2} + \mu_x = \frac{\sigma_x}{\sqrt{N}} z_\beta + \mu_x - d$$

These relationships both reduce to

$$z_{\alpha/2} = z_{1-\beta} + \frac{\sqrt{N}}{\sigma_x} d = -z_\beta + \frac{\sqrt{N}}{\sigma_x} d$$

It follows that the required sample size is given by

$$N = \left[\frac{\sigma_x (z_{\alpha/2} + z_\beta)}{d} \right]^2$$

For the specific values in this example ($\sigma_x = 2$, $z_{\alpha/2} = 1.96$, $z_\beta = 1.645$, $d = 0.1(10) = 1$), the required sample size is

$$N = 52$$

The region of acceptance for the hypothesis test will be

$$\text{upper limit} = \frac{\sigma_x}{\sqrt{N}} z_{\alpha/2} + \mu_x = 10.54$$

$$\text{lower limit} = \frac{\sigma_x}{\sqrt{N}} z_{1-\alpha/2} + \mu_x = 9.46$$

4.6   CHI-SQUARE GOODNESS-OF-FIT TEST

A special type of hypothesis test that is often used to test the equivalence of a probability density function of sampled data to some theoretical density function is called the chi-square goodness-of-fit test. The general procedure involves the use of a statistic with an approximate chi-square distribution as a measure of the discrepancy between an observed probability density function and the theoretical density function. A hypothesis of equivalence is then tested by studying the sampling distribution of this statistic.

To be more specific, consider a sample of N independent observations from a random variable x with a probability density function of $p(x)$. Let the N observations be grouped into K intervals, called *class intervals*, which together form a *frequency histogram*. The number of observations falling within the ith class interval is called the *observed frequency* in the ith class, and will be denoted by f_i. The number of observations that would be expected to fall within the ith class interval if the true probability density function of x were $p_0(x)$ is called the *expected frequency* in the ith class interval, and will be denoted by F_i. Now, the discrepancy between the observed frequency and the expected frequency within each class interval is given by $f_i - F_i$. To measure the total discrepancy for all class intervals, the squares of the discrepancies in each interval are normalized by the associated expected frequencies and summed to obtain the sample statistic

$$X^2 = \sum_{i=1}^{K} \frac{(f_i - F_i)^2}{F_i} \tag{4.49}$$

It is shown in Reference 4.1 that the distribution of X^2 in Equation (4.49) is approximately the same as for χ_n^2 discussed in Section 4.2.2. The number of degrees of freedom n in this case is equal to K minus the number of different independent linear restrictions imposed on the observations. There is one such restriction due to the fact that the frequency in the last class interval is determined once the frequencies in the first $K - 1$ class intervals are known. If the comparison is made by fitting the expected theoretical density function to the frequency histogram for the observed data, then one additional constraint results from each independent parameter of the theoretical density function which must be computed to make the fit. For example, if the expected theoretical density function is a normal density function with unknown mean and variance, then two additional constraints are involved, since two parameters (a mean and variance) must be computed to fit a normal density function. Hence for the common case where the chi-square goodness-of-fit test is used as a test for normality, the number of degrees of freedom for X^2 in Equation (4.49) is $n = K - 3$.

Having established the proper degrees of freedom for X^2, a hypothesis test may be performed as follows. Let it be hypothesized that the variable x has a probability density function of $p(x) = p_0(x)$. After grouping the sampled observations into K class intervals and computing the expected frequency for each interval assuming $p(x) = p_0(x)$, compute X^2 as indicated in Equation (4.49). Since any deviation of $p(x)$ from $p_0(x)$ will cause X^2 to increase, a one-sided (upper tail) test is used. The region of acceptance is

$$X^2 \leq \chi^2_{n;\,\alpha} \qquad (4.50)$$

where the value of $\chi^2_{n;\,\alpha}$ is available from Table A.3. If the sample value X^2 is greater than $\chi^2_{n;\,\alpha}$, the hypothesis that $p(x) = p_0(x)$ is rejected at the α level of significance. If X^2 is less than or equal to $\chi^2_{n;\,\alpha}$, the hypothesis is accepted at the α level of significance.

There are two basic ways to apply the chi-square goodness-of-fit test. The first way is to select class intervals in a manner that will provide equal expected frequencies within each interval. Excluding a uniform distribution hypothesis, this procedure will result in different interval widths from one class interval to another. The second way is to select class intervals of equal width. Again, except for the uniform distribution hypothesis, this procedure will result in different expected frequencies from one class interval to another. Chi-square tests for normality are usually performed using the constant interval width approach. Given sample data with a standard deviation of s, a class interval width of $\Delta x \simeq 0.4s$ is often used. A more fundamental requirement is that the expected frequencies in all class intervals must be sufficiently large to make Equation (4.49) an acceptable approximation to χ^2_n. A common recommendation is that $F_i > 3$ in all intervals. In a normality test where the expected frequencies diminish on the tails of the distribution, this requirement is complied with by letting the first and last intervals extend to $-\infty$ and $+\infty$, respectively, such that $F_1, F_K > 3$.

***Example 4.3*. Illustration of Test for Normality.**   A sample of $N = 200$ independent observations of the digitized output of a thermal noise generator are presented in Table 4.1. The sample values have been rank ordered from the smallest to largest value for convenience. Test the noise generator output for normality by performing a chi-square goodness-of-fit test at the $\alpha = 0.05$ level of significance.

The calculations required to perform the test are summarized in Table 4.2. For an interval width of $\Delta x = 0.4s$, the standardized values of the normal distribution that define the class interval boundaries are as shown under z_α in the table. These interval boundaries are converted to volts in the next column. From Table A.2 in Appendix A, the probability P that a sample value will fall in each class interval is determined using the z_α values. The product of P and the sample size N yields the expected frequency in each interval as listed under F in Table 4.2. Note that the first and last class intervals are selected so that $F > 3$. A total of 12 class intervals result. The observed frequencies are now counted using the interval boundaries in volts as indicated in Table 4.1. The normalized squared discrepancies between the expected and observed frequencies are then calculated and summed to obtain $X^2 = 2.43$. Note that the appropriate degrees-of-freedom is $n = K - 3 = 9$. The acceptance region for

Table 4.1

Sample Observations Arranged in Increasing Order

−7.6	−3.8	−2.5	−1.6	−0.7	0.2	1.1	2.0	3.4	4.6
−6.9	−3.8	−2.5	−1.6	−0.7	0.2	1.1	2.1	3.5	4.8
−6.6	−3.7	−2.4	−1.6	−0.6	0.2	1.2	2.3	3.5	4.8
−6.4	−3.6	−2.3	−1.5	−0.6	0.3	1.2	2.3	3.6	4.9
−6.2	−3.5	−2.3	−1.5	−0.5	0.3	1.3	2.3	3.6	5.0
−6.1	−3.4	−2.3	−1.4	−0.5	0.3	1.3	2.4	3.6	5.2
−6.0	−3.4	−2.2	−1.4	−0.4	0.4	1.3	2.4	3.7	5.3
−5.7	−3.4	−2.2	−1.2	−0.4	0.4	1.4	2.5	3.7	5.4
−5.6	−3.3	−2.1	−1.2	−0.4	0.5	1.5	2.5	3.7	5.6
−5.5	−3.2	−2.1	−1.2	−0.3	0.5	1.5	2.6	3.7	5.9
−5.4	−3.2	−2.0	−1.1	−0.3	0.6	1.6	2.6	3.8	6.1
−5.2	−3.1	−2.0	−1.1	−0.2	0.6	1.6	2.6	3.8	6.3
−4.8	−3.0	−1.9	−1.0	−0.2	0.7	1.6	2.7	3.9	6.3
−4.6	−3.0	−1.9	−1.0	−0.2	0.8	1.7	2.8	4.0	6.5
−4.4	−2.9	−1.8	−1.0	−0.1	0.9	1.8	2.8	4.2	6.9
−4.4	−2.9	−1.8	−0.9	−0.0	0.9	1.8	2.9	4.2	7.1
−4.3	−2.9	−1.8	−0.9	0.0	1.0	1.8	3.1	4.3	7.2
−4.1	−2.7	−1.7	−0.8	0.1	1.0	1.9	3.2	4.3	7.4
−4.0	−2.6	−1.7	−0.8	0.1	1.1	1.9	3.2	4.4	7.9
−3.8	−2.6	−1.6	−0.7	0.2	1.1	2.0	3.3	4.4	9.0

the test is found in Table A.3 to be $X^2 \leq \chi^2_{9;\,0.05} = 16.92$. Hence, the hypothesis of normality is accepted at the $\alpha = 0.05$ level of significance.

4.7 STATISTICAL INDEPENDENCE AND TREND TESTS

Situations often arise in data analysis where it is desired to establish if a sequence of observations or parameter estimates are statistically independent or include an underlying trend. This is particularly true in the analysis of nonstationary data discussed later in Chapter 12. Since the observations or parameter estimates of interest may have a wide range of probability distribution functions, it is convenient to perform such evaluations with *distribution-free* or *nonparametric* procedures, where no assumption is made concerning the probability distribution of the data being evaluated. Two such procedures, useful for evaluations of statistical independence and underlying trends, are the *run test* and the *reverse arrangements test*.

Table 4.2

Calculations for Goodness-of-Fit Test

Interval Number	Upper Limit of Interval		P	$F = NP$	f	$\|F - f\|$	$\dfrac{(F - f)^2}{F}$
	z_α	$x = sz + \bar{x}$					
1	-2.0	-6.36	0.0228	4.5	4	0.5	0.06
2	-1.6	-5.04	0.0320	6.4	8	1.6	0.40
3	-1.2	-3.72	0.0603	12.1	10	2.1	0.36
4	-0.8	-2.40	0.0968	19.4	21	1.6	0.13
5	-0.4	-1.08	0.1327	26.5	29	2.5	0.24
6	0	0.24	0.1554	31.1	31	0.1	0.00
7	0.4	1.56	0.1554	31.1	27	4.1	0.54
8	0.8	2.88	0.1327	26.5	25	1.5	0.08
9	1.2	4.20	0.0968	19.4	20	0.6	0.02
10	1.6	5.52	0.0603	12.1	13	0.9	0.07
11	2.0	6.84	0.0320	6.4	6	0.4	0.03
12	∞	∞	0.0228	4.5	6	1.5	0.50
			1.000	200	200		2.43
$N = 200$		$\bar{x} = 0.24$	$s = 3.30$	$n = K - 3 = 9$			$X^2 = 2.43$

4.7.1 Run Test

Consider a sequence of N observed values of a random variable x where each observation is classified into one of two mutually exclusive categories, which may be identified simply by plus $(+)$ or minus $(-)$. The simplest example would be a sequence of flips of a coin, where each observation is either a head $(+)$ or a tail $(-)$. A second example might be a sequence of measured values x_i, $i = 1, 2, 3, \ldots, N$, with a mean value $\bar{x}$, where each observation is $x_i \geq \bar{x}(+)$ or $x_i < \bar{x}(-)$. A third example might be a simultaneous sequence of two sets of measured values x_i and y_i, $i = 1, 2, 3, \ldots, N$, where each observation is $x_i \geq y_i(+)$ or $x_i < y_i(-)$. In any case, the sequence of plus and minus observations might be as follows.

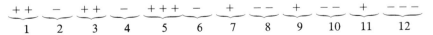

A run is defined as a sequence of identical observations that is followed and preceded by a different observation or no observation at all. In this example there are $r = 12$ runs in the sequence of $N = 20$ observations.

The number of runs that occur in a sequence of observations gives an indication as to whether or not results are independent random observations of the same random variable. Specifically, if a sequence of N observations are independent observations of the same random variable, that is, the probability of a $(+)$ or $(-)$ result does not change from one observation to the next, then the sampling distribution of the number of runs in the sequence is a random variable r with a mean value and variance as follows [Reference 4.1]:

$$\mu_r = \frac{2N_1 N_2}{N} + 1 \tag{4.51}$$

$$\sigma_r^2 = \frac{2N_1 N_2 (2N_1 N_2 - N)}{N^2(N-1)} \tag{4.52}$$

Here, N_1 is the number of $(+)$ observations and N_2 is the number of $(-)$ observations. For the special case where $N_1 = N_2 = N/2$, Equations (4.51) and (4.52) reduce to

$$\mu_r = \frac{N}{2} + 1 \tag{4.53}$$

$$\sigma_r^2 = \frac{N(N-2)}{4(N-1)} \tag{4.54}$$

A limited tabulation of 100α percentage points for the distribution function of runs is presented in Table A.6 in the Appendix.

Perhaps the most direct application of runs to data evaluation problems involves the testing of a single sequence of observations for a trend. Assume there is reason to suspect an underlying trend in a sequence of observations; that is, there is reason to believe that the probability of a $(+)$ or $(-)$ is changing from one observation to the next. The existence of a trend can be tested for as follows. Let it be hypothesized that there is no trend by assuming that the sequence of N observations are independent observations of the same random variable. Then, assuming the number of $(+)$ observations equals the number of $(-)$ observations, the number of runs in the sequence will have a sampling distribution as given in Table A.6. The hypothesis can be tested at any desired level of significance α by comparing the observed runs to the interval between $r_{n; 1-\alpha/2}$ and $r_{n; \alpha/2}$ where $n = N/2$. If the observed number of runs falls outside the interval, the hypothesis would be rejected at the α level of significance. Otherwise, the hypothesis would be accepted.

Example 4.4. **Illustration of Run Test.** Assume a sequence of $N = 20$ observations of a random variable produces results as noted below.

(1) 5.5	(6) 5.7	(11) 6.8	(16) 5.4
(2) 5.1	(7) 5.0	(12) 6.6	(17) 6.8
(3) 5.7	(8) 6.5	(13) 4.9	(18) 5.8
(4) 5.2	(9) 5.4	(14) 5.4	(19) 6.9
(5) 4.8	(10) 5.8	(15) 5.9	(20) 5.5

Determine if the observations are independent by testing the runs which occur in the variation of the observations about their median value. Perform the test at the $\alpha = 0.05$ level of significance.

By visual inspection of the data, it is seen that $x = 5.6$ is the median value of the 20 observations. Let all observations with a value greater than 5.6 be identified by $(+)$ and all with a value less than 5.6 be identified by $(-)$. The result is

$$
\underbrace{--}_{1}\ \underbrace{+}_{2}\ \underbrace{--}_{3}\ \underbrace{+}_{4}\ \underbrace{-}_{5}\ \underbrace{+}_{6}\ \underbrace{-}_{7}\ \underbrace{+++}_{8}\ \underbrace{--}_{9}\ \underbrace{+}_{10}\ \underbrace{-}_{11}\ \underbrace{+++}_{12}\ \underbrace{-}_{13}
$$

Hence there are 13 runs represented by the sequence of 20 observations. Let it be hypothesized that the observations are independent. The acceptance region for this hypothesis is

$$
[r_{10;\,1-\alpha/2} < r \le r_{10;\,\alpha/2}]
$$

From Table A.6, for $\alpha = 0.05$, $r_{10;\,1-\alpha/2} = r_{10;\,0.975} = 6$ and $r_{10;\,\alpha/2} = r_{10;\,0.025} = 15$. The hypothesis is accepted since $r = 13$ falls within the range between 6 and 15. That is, there is no reason to question that the observations are independent, which means there is no evidence of an underlying trend.

4.7.2   *Reverse Arrangements Test*

Consider a sequence of N observations of a random variable x, where the observations are denoted by x_i, $i = 1, 2, 3, \ldots, N$. Now, count the number of times that $x_i > x_j$ for $i < j$. Each such inequality is called a reverse arrangement. The total number of reverse arrangements is denoted by A.

A general definition for A is as follows. From the set of observations $x_1, x_2, \ldots, x_N$, define

$$
h_{ij} = \begin{cases} 1 & \text{if } x_i > x_j \\ 0 & \text{otherwise} \end{cases} \tag{4.55}
$$

Then

$$
A = \sum_{i=1}^{N-1} A_i \tag{4.56}
$$

where

$$
A_i = \sum_{j=i+1}^{N} h_{ij} \tag{4.57}
$$

For example,

$$A_1 = \sum_{j=2}^{N} h_{1j} \qquad A_2 = \sum_{j=3}^{N} h_{2j} \qquad A_3 = \sum_{j=4}^{N} h_{3j} \qquad \text{etc.}$$

To help clarify the meaning of reverse arrangements, consider the following sequence of $N = 8$ observations:

$$x_1 = 5, \quad x_2 = 3, \quad x_3 = 8, \quad x_4 = 9, \quad x_5 = 4, \quad x_6 = 1, \quad x_7 = 7, \quad x_8 = 5$$

In the above sequence $x_1 > x_2$, $x_1 > x_5$, and $x_1 > x_6$, which gives $A_1 = 3$ reverse arrangements for x_1. Now, choosing x_2 and comparing it against subsequent observations (that is, for $i = 2$ and $i < j = 3, 4, \ldots, 8$), one notes $x_2 > x_6$ only, so that the number of reverse arrangements for x_2 is $A_2 = 1$. Continuing on, it is seen that $A_3 = 4$, $A_4 = 4$, $A_5 = 1$, $A_6 = 0$, and $A_7 = 1$. The total number of reverse arrangements is, therefore,

$$A = A_1 + A_2 + \cdots + A_7 = 3 + 1 + 4 + 4 + 1 + 0 + 1 = 14$$

If the sequence of N observations are independent observations of the same random variable, then the number of reverse arrangements is a random variable A, with a mean variable and variance as follows [Reference 4.4]:

$$\mu_A = \frac{N(N-1)}{4} \tag{4.58}$$

$$\sigma_A^2 = \frac{2N^3 + 3N^2 - 5N}{72} = \frac{N(2N+5)(N-1)}{72} \tag{4.59}$$

A limited tabulation of 100α percentage points for the distribution function of A is presented in Table A.7.

The reverse arrangements test may be applied in basically the same way as the run test. Generally speaking, it is more powerful than the run test for detecting monotonic trends in a sequence of observations. This test is not powerful, however, for detecting fluctuating trends.

Example 4.5. Illustration of Reverse Arrangements Test. Test the sequence of $N = 20$ observations in Example 4.4 for a trend at the $\alpha = 0.05$ level of significance. The number of reverse arrangements in the observations is as follows.

$A_1 = 8$	$A_6 = 6$	$A_{11} = 7$	$A_{16} = 0$
$A_2 = 3$	$A_7 = 1$	$A_{12} = 6$	$A_{17} = 2$
$A_3 = 8$	$A_8 = 8$	$A_{13} = 0$	$A_{18} = 1$
$A_4 = 3$	$A_9 = 1$	$A_{14} = 0$	$A_{19} = 1$
$A_5 = 0$	$A_{10} = 4$	$A_{15} = 3$	

The total number of reverse arrangements is $A = 62$.

Let it be hypothesized that the observations are independent observations of a random variable x, where there is no trend. The acceptance region for this hypothesis is

$$\left[A_{20;\,1-\alpha/2} < A \leq A_{20;\,\alpha/2} \right]$$

From Table A.7, for $\alpha = 0.05$, $A_{20;\,1-\alpha/2} = A_{20;\,0.975} = 64$ and $A_{20;\,\alpha/2} = A_{20;\,0.025} = 125$. Hence the hypothesis is rejected at the 5% level of significance since $A = 62$ does not fall within the range between 64 and 125. Note that a hypothesis of independence for this same sequence of observations was accepted by the run test in Example 4.4. This illustrates the difference in sensitivity between the two testing procedures.

4.8 CORRELATION AND REGRESSION PROCEDURES

Techniques of correlation and regression analysis are fundamental to much of the material developed in this book. The concept of correlation between two random variables has already been introduced in Chapter 3 and will be expanded on in Chapter 5. The concept of linear regression is basic to the techniques of frequency response function estimation from input/output data, as formulated in Chapters 6 and 7. The material in these chapters, however, is developed in a frequency domain context that may obscure associations with more familiar classical presentations. Hence a brief review of correlation and regression concepts from the viewpoint of elementary statistics may be helpful as an introduction to this later material.

4.8.1 *Linear Correlation Analysis*

For a wide class of problems, a matter of primary interest is whether or not two or more random variables are interrelated. For example, is there a relationship between cigarette smoking and life expectancy, or between measured aptitude and academic success. In an engineering context, such problems often reduce to detecting a relationship between some assumed excitation and an observed response of a physical system of interest. The existence of such interrelationships and their relative strength can be measured in terms of a correlation coefficient ρ as defined in Section 3.2.1. For the simple case of two random variables x and y, the correlation coefficient is given by Equation (3.35) as

$$\rho_{xy} = \frac{C_{xy}}{\sigma_x \sigma_y} \tag{4.60}$$

where C_{xy} is the covariance of x and y as defined in Equation (3.33).

Now assume the random variables x and y are sampled to obtain N pairs of observed values. The correlation coefficient may be estimated from the sample data by

$$r_{xy} = \hat{\rho}_{xy} = \frac{s_{xy}}{s_x s_y} = \frac{\sum_{i=1}^{N}(x_i - \bar{x})(y_i - \bar{y})}{\left[\sum_{i=1}^{N}(x_i - \bar{x})^2 \sum_{i=1}^{N}(y_i - \bar{y})^2\right]^{1/2}}$$

$$= \frac{\sum_{i=1}^{N} x_i y_i - N\bar{x}\bar{y}}{\left[\left(\sum_{i=1}^{N} x_i^2 - N\bar{x}^2\right)\left(\sum_{i=1}^{N} y_i^2 - N\bar{y}^2\right)\right]^{1/2}} \tag{4.61}$$

Like ρ_{xy}, the sample correlation coefficient r_{xy} will lie between -1 and $+1$, and will have a bounding value only when the observations display a perfect linear relationship. A nonlinear relationship and/or data scatter, whether it be due to measurement errors or imperfect correlation of the variables, will force the value of r_{xy} toward zero, as illustrated in Figure 4.4.

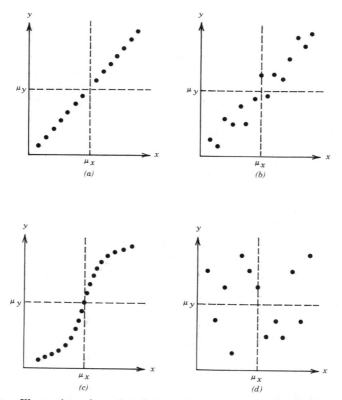

Figure 4.4 Illustration of varying degrees of correlation. (a) Perfect linear correlation. (b) Moderate linear correlation. (c) Nonlinear correlation. (d) No correlation.

To evaluate the accuracy of the estimate r_{xy}, it is convenient to work with a particular function of r_{xy} given by

$$w = \tfrac{1}{2}\ln\left[\frac{1 + r_{xy}}{1 - r_{xy}}\right]$$ (4.62)

From Reference 4.2, the random variable w has an approximately normal distribution with a mean and variance of

$$\mu_w = \tfrac{1}{2}\ln\left[\frac{1 + \rho_{xy}}{1 - \rho_{xy}}\right]$$ (4.63)

$$\sigma_w^2 = \frac{1}{N - 3}$$ (4.64)

Using the above relationships, confidence intervals for ρ_{xy} based on a sample estimate r_{xy} may be readily established as outlined in Section 4.4.

Because of the variability of correlation estimates, it is usually desirable to verify that a nonzero value of the sample correlation coefficient indeed reflects the existence of a statistically significant correlation between the variables of interest. This may be accomplished by testing the hypothesis that $\rho_{xy} = 0$, where a significant correlation is indicated if the hypothesis is rejected. From Equations (4.63) and (4.64), the sampling distribution of w given $\rho_{xy} = 0$ is normal with a mean of $\mu_w = 0$ and a variance of $\sigma_w^2 = 1/(N - 3)$. Hence the acceptance region for the hypothesis of zero correlation is given by

$$\left[-z_{\alpha/2} \leq \frac{\sqrt{N - 3}}{2}\ln\left[\frac{1 + r_{xy}}{1 - r_{xy}}\right] < z_{\alpha/2}\right]$$ (4.65)

where z is the standardized normal variable. Values outside the above interval would constitute evidence of statistical correlation at the α level of significance.

Example 4.6. Illustration of Linear Correlation Analysis.   The heights and weights of $N = 25$ male university students selected at random are presented in Table 4.3. Is there reason to believe that the height and weight of male students are correlated at the $\alpha = 0.05$ level of significance?

Let x be height and y be weight. From the data in Table 4.3, the following values needed in Equation (4.55) are calculated:

$$\sum_{i=1}^{N} x_i y_i = 299{,}056 \qquad \sum_{i=1}^{N} x_i^2 = 124{,}986 \qquad \sum_{i=1}^{N} y_i^2 = 723{,}604$$

$$\bar{x} = \frac{1}{N}\sum_{i=1}^{N} x_i = \frac{1766}{25} = 70.64 \qquad \bar{y} = \frac{1}{N}\sum_{i=1}^{N} y_i = \frac{4224}{25} = 168.96$$

Table 4.3

Height and Weight Data for Male Students

	x = height in inches								y = weight in pounds			
x	70	74	70	65	69	73	72	69	72	76	74	72
y	140	210	148	145	182	165	155	170	174	155	185	185
x	68	70	71	68	73	65	73	74	64	72	72	67	73
y	165	220	185	180	170	135	175	180	150	170	165	145	170

Substituting the above values into Equation (4.61) yields the estimated correlation coefficient as follows:

$$r_{xy} = \frac{299{,}056 - (25)(70.64)(168.96)}{\left[(124{,}986 - 25(70.64)^2)(723{,}604 - 25(168.96)^2)\right]^{1/2}}$$

$$= 0.44$$

From Equation (4.62), the quantity $w = 0.472$; thus $\sqrt{N-3}\,w = 2.21$. Now using Equation (4.65), the hypothesis that $\rho_{xy} = 0$ is rejected at the 5% level of significance since $\sqrt{N-3}\,w = 2.21$ falls outside the acceptance region bounded by $\pm z_{\alpha/2} = \pm 1.96$. Hence, there is reason to believe that significant correlation exists between the height and weight of male students.

4.8.2 Linear Regression Analysis

Correlation analysis can establish the degree to which two or more random variables are interrelated. Beyond this, however, a model for the relationship may be desired so that predictions can be made for one variable based on specific values of other variables. For instance, a significant linear relationship between the height and weight of male university students is indicated by the correlation analysis of data presented in Example 4.6. A logical second step would be to evaluate the relationship further so that the weight of students can be predicted for any given height. Procedures for dealing with problems of this type come under the heading of regression analysis.

Consider the simple case of two correlated random variables x and y. Referring again to Example 4.6, x might be student height and y student weight. A linear relationship between the two variables would suggest that for a given value of x, a value of y would be predicted by

$$\tilde{y} = A + Bx \tag{4.66}$$

where A and B are the intercept and slope, respectively, of a straight line. For the case of data that display perfect linear correlation ($r_{xy} = 1$), the predicted value $\tilde{y}_i$ would always equal the observed value y_i for any given x_i. In practice, however, data usually do not display a perfect linear relationship. There generally is some scatter due to extraneous random effects, and perhaps distortion due to nonlinearities, as illustrated in Figure 4.4. Nevertheless, if a linear relationship is assumed and unlimited data are available, appropriate values of A and B can be determined that will predict the expected value of y_i for any given x_i. That is, $\tilde{y}_i$ will not necessarily equal the observed value y_i associated with the corresponding x_i, but it will be an average for all such observed values.

The accepted procedure for determining the coefficients in Equation (4.66) is to select those values of A and B that minimize the sum of the squared deviations of the observed values from the predicted values of y. This procedure is called a *least squares fit*. Specifically, noting that the deviation of the observed values from the predicted values is

$$y_i - \tilde{y}_i = y_i - (A + Bx_i) \tag{4.67}$$

it follows that the sum of the squared deviations is given by

$$Q = \sum_{i=1}^{N} (y_i - A - Bx_i)^2 \tag{4.68}$$

Hence, a least squares fit is provided by those values of A and B that make

$$\frac{\partial Q}{\partial A} = \frac{\partial Q}{\partial B} = 0 \tag{4.69}$$

In practice, the available data will be limited to a sample of N pairs of observed values for x and y. This means that Equation (4.69) will yield only estimates of A and B, to be denoted by a and b, respectively. Substituting Equation (4.68) into Equation (4.69) and solving for the estimates of A and B yields

$$a = \bar{y} - b\bar{x} \tag{4.70a}$$

$$b = \frac{\sum_{i=1}^{N}(x_i - \bar{x})y_i}{\sum_{i=1}^{N}(x_i - \bar{x})^2} = \frac{\sum_{i=1}^{N}x_i y_i - N\bar{x}\bar{y}}{\sum_{i=1}^{N}x_i^2 - N\bar{x}^2} \tag{4.70b}$$

These estimates can now be used to write a prediction model for y given x as follows:

$$\hat{y} = a + bx = (\bar{y} - b\bar{x}) + bx = \bar{y} + b(x - \bar{x}) \tag{4.71}$$

The straight line defined by Equation (4.71) is called the *linear regression line for y on x*. By switching the dependent and independent variables in Equation (4.70), a regression line for x on y could also be calculated. Specifically,

$$\hat{x} = \bar{x} + b'(y - \bar{y}) \tag{4.72}$$

where

$$b' = \frac{\sum_{i=1}^{N} x_i y_i - N\bar{x}\bar{y}}{\sum_{i=1}^{N} y_i^2 - N\bar{y}^2} \tag{4.73}$$

Comparing the product of Equations (4.70b) and (4.73) to Equation (4.61), it is seen that the slopes of the regression lines for y on x and x on y are related to the sample correlation coefficient of x and y by

$$r_{xy} = [bb']^{\frac{1}{2}} \tag{4.74}$$

Now consider the accuracy of the estimates a and b given by Equation (4.70). Assuming a normal distribution of y given x, it is shown in Reference 4.2 that a and b are unbiased estimates of A and B, respectively, with sampling distributions related to the t distribution as follows.

$$\frac{a - A}{\left(\dfrac{1}{N} + \dfrac{\bar{x}^2}{\sum_{i=1}^{N}(x_i - \bar{x})^2} \right)^{\frac{1}{2}}} = s_{y|x} t_{N-2} \tag{4.75}$$

$$\frac{b - B}{\left(\dfrac{1}{\sum_{i=1}^{N}(x_i - \bar{x})^2} \right)^{\frac{1}{2}}} = s_{y|x} t_{N-2} \tag{4.76}$$

Of particular interest is the sampling distribution of $\hat{y}$ associated with a specific value of $x = x_0$. This is given by

$$\frac{\hat{y} - \tilde{y}}{\left(\dfrac{1}{N} + \dfrac{(x_0 - \bar{x})^2}{\sum_{i=1}^{N}(x_i - \bar{x})^2} \right)^{\frac{1}{2}}} = s_{y|x} t_{N-2} \tag{4.77}$$

In Equations (4.75)–(4.77), the term $s_{y|x}$ is the sample standard deviation of the observed values of y_i about the prediction $\hat{y}_i = a + bx_i$, and is given by

$$s_{y|x} = \left[\frac{\sum_{i=1}^{N}(y_i - \hat{y}_i)^2}{N - 2} \right]^{\frac{1}{2}} = \left[\left(\frac{n - 1}{n - 2} \right) s_y^2 (1 - r_{xy}^2) \right]^{\frac{1}{2}} \qquad (4.78)$$

The above relationships provide a basis for establishing confidence intervals for A, B, and $\bar{y}$ based on the estimates a, b, and $\hat{y}$.

***Example 4.7*. Illustration of Linear Regression Analysis.** Using the data presented in Table 4.3 for Example 4.6, determine a regression line that will provide a linear prediction for the average weight of male university students as a function of their height. Determine a 95% confidence interval for the average weight of male students who are 70 in. tall.

As in Example 4.6, let x be height and y be weight. The values needed to determine the slope and intercept of the regression line for y on x have already been calculated in Example 4.6. Substituting these values into Equation (4.70) yields the estimated slope and intercept as follows:

$$b = \frac{299{,}056 - (25)(70.64)(168.96)}{168.96 - (25)(70.64)^2} = 2.85$$

$$a = 168.96 - (2.85)(70.64) = -32.6$$

Hence the regression line estimating the average weight of male university students given height is

$$\hat{y} = -32.6 + 2.85x$$

which yields an estimated weight of $\hat{y} = 167.1$ lb for a height of $x = 70$ in.

To establish a confidence interval for the average weight $\bar{y}$ based on the estimate $\hat{y} = 167.1$ lb, it is necessary to calculate $s_{y|x}$ given by Equation (4.78). A more convenient equation for $s_{y|x}$ from the computational viewpoint is

$$s_{y|x} = \left[\frac{1}{N - 2} \left(\sum_{i=1}^{N}(y_i - \bar{y})^2 - \frac{[\sum_{i=1}^{N}(x_i - \bar{x})(y_i - \bar{y})]^2}{\sum_{i=1}^{N}(x_i - \bar{x})^2} \right) \right]^{\frac{1}{2}}$$

where the terms in the above expression are further simplified for computational purposes by noting that

$$\sum_{i=1}^{N}(v_i - \bar{v})^2 = \sum_{i=1}^{N} v_i^2 - N\bar{v}^2 \qquad \sum_{i=1}^{N}(x_i - \bar{x})(y_i - \bar{y}) = \sum_{i=1}^{N} x_i y_i - N\bar{x}\bar{y}$$

Substitution of the appropriate values into these expressions yields

$$s_{y|x} = \left[\frac{1}{23} \left(9917 - \frac{(673)^2}{236} \right) \right]^{\frac{1}{2}} = 18.65$$

Then, from Equation (4.77), a 95% confidence interval for the average weight of male university students with a height of 70 in. is

$$\hat{y} \pm s_{y|x} t_{N-2;\, \alpha/2} \left[\frac{1}{N} + \frac{(x_0 - \bar{x})^2}{\sum\limits_{i=1}^{N} (x_i - \bar{x})^2} \right]^{\frac{1}{2}}$$

$$= 167.2 \pm (18.65) t_{23;\, 0.025} \left[\frac{1}{25} + \frac{(70 - 70.64)^2}{236} \right]^{\frac{1}{2}}$$

$$= 167.2 \pm 7.9 = 159.3 \text{ to } 175.1 \text{ lb}$$

This concludes Example 4.7.

The techniques of correlation and regression analysis are readily extended for applications involving more than two random variables. As noted earlier, such extensions are fundamental to the analysis of multiple-input/output problems developed in Chapter 7. Hence, further discussions of this subject are deferred to that chapter.

PROBLEMS

4.1   Given a random variable x with a probability density function of

$$p(x) = \frac{1}{2\sqrt{2\pi}} e^{-(x-1)^2/8}$$

What are the mean value and variance of x?

4.2   Given two independent random variables, x and y, with mean values of μ_x and μ_y, and variances of σ_x^2 and σ_y^2, determine the
(a)   mean value of the product xy.
(b)   variance of the difference $x - y$.

4.3   Given the random variable $y = cx$ where c is a constant and x is a random variable with a mean value and variance of μ_x and σ_x^2, respectively, prove the following relationships are true.
(a) $\mu_y = c\mu_x$.
(b) $\sigma_y^2 = c^2 \sigma_x^2$.

4.4 Given four independent standardized normally distributed random variables, z_1, z_2, z_3, and z_4, define the distribution functions of the following combinations of these variables. For each case, specify the associated degrees of freedom or mean value and variance, as appropriate.

(a) $z_1^2 + z_2^2 + z_3^2 + z_4^2$.

(b) $z_1 + z_2 - z_3 - z_4$.

(c) $\dfrac{z_4}{\left\{ \left[z_1^2 + z_2^2 + z_3^2 \right] / 3 \right\}^{1/2}}$.

(d) $\dfrac{\left[z_1^2 + z_2^2 + z_3^2 \right] / 3}{z_4^2}$.

4.5 What distribution function would be used to establish confidence intervals for the following parameters of two independent normally distributed random variables, x and y?

(a) interval for μ_x based on a sample mean $\bar{x}$ and known variance σ_x^2.

(b) interval for σ_x^2 / σ_y^2 based on a ratio of sample variances s_x^2 / s_y^2.

(c) interval for σ_x^2 based on a sample variance s_x^2.

(d) interval for μ_x based on a sample mean $\bar{x}$ and sample variance s_x^2.

4.6 Assume a time sequence of $N = 100$ measurements is made of a random variable x. Determine the acceptance region at the 2 percent level of significance to test the hypothesis that

(a) there is no underlying time trend in the variable x.

(b) the 100 measurements of x represent statistically independent observations.

4.7 Given a sample of N independent observations of a random variable x with a known mean value of zero, an *efficient* estimator for the variance of x is

$$s^2 = \frac{1}{N} \sum_{i=1}^{N} x_i^2$$

(a) Prove the above estimator is unbiased.

(b) Write an expression relating the above estimator to a chi-square variable with the appropriate degrees of freedom specified.

(c) What is the variance of the above estimator? (*Hint*: the variance of χ_n^2 is $2n$.)

4.8 The normalized random error (coefficient of variation) ε_r of an unbiased parameter estimate $\hat{\phi}$ is defined as the ratio of the standard deviation of the estimate to the expected value of the estimate, that is, $\varepsilon_r = \sigma_{\hat{\phi}} / \mu_{\hat{\phi}}$. Determine the normalized random error of a variance estimate s^2 computed from $N = 200$ sample observations using Equation (4.12).

4.9 A correlation study is performed using a sample of $N = 7$ pairs of observations $(x_1 y_1, x_2 y_2, \ldots, x_7 y_7)$. A sample correlation coefficient of $r_{xy} = 0.77$ is calculated. Test the hypothesis that ρ_{xy} is greater than zero at the $\alpha = 0.01$ level of significance.

4.10 Assume the sample mean values of two correlated random variables are $\bar{x} = 1$ and $\bar{y} = 2$. Further assume that the sample correlation coefficient is $r_{xy} = 0.5$. If the regression line for y on x is given by $\hat{y} = 1 + x$,
   (a) what is the slope b' of the regression line for x on y?
   (b) what is the equation for the regression line for x on y ($\hat{x} = a' + b'y$)?

REFERENCES

4.1 Brownlee, K. A., *Statistical Theory and Methodology in Science and Engineering*, 2nd ed., Wiley, New York, 1965.

4.2 Guttman, I., Wilks, S. S., and Hunter, J. S., *Introductory Engineering Statistics*, 3rd ed., Wiley, New York, 1982.

4.3 Johnson, N. L., and Leone, F. C., *Statistics and Experimental Design in Engineering and the Physical Sciences*, 2nd ed., Wiley, New York, 1977.

4.4 Kendall, M. G., and Stuart, A., *The Advanced Theory of Statistics*, Vol. 2, Hafner, New York, 1961.

CHAPTER 5

STATIONARY RANDOM PROCESSES

This chapter discusses elementary and advanced concepts from stationary random processes theory to form a foundation for applications to analysis and measurement problems as contained in References 5.1 and 5.2. Material includes theoretical definitions for stationary random processes together with basic properties for correlation and spectral density functions. Results are stated for ergodic random processes, Gaussian random processes, and derivative random processes. Nonstationary random processes are covered in Chapter 12.

5.1 BASIC CONCEPTS

A *random process* $\{x_k(t)\}$, $-\infty < t < \infty$ (also called a *time series* or *stochastic process*), denoted by the symbol { }, is an ensemble of real-valued (or complex-valued) functions, which can be characterized through its probability structure. For convenience, the variable t will be interpreted as time in the following discussion. Each particular function $x_k(t)$, where t is variable and k is fixed, is called a *sample function*. In practice, a sample function (or some time history record of finite length from a sample function) may be thought of as the observed result of a single experiment. The possible number of experiments represents a sample space of index k, which may be countable or uncountable. For any number N and any fixed times $t_1, t_2, \ldots, t_N$, the quantities $x_k(t_1), x_k(t_2), \ldots, x_k(t_N)$, represent N random variables over the index k. It is required that there exist a well-defined N-dimensional probability distribution function for every N. An ensemble of sample functions forming a random process is illustrated in Figure 1.10.

A particular sample function $x_k(t)$, in general, would not be suitable for representing the entire random process $\{x_k(t)\}$ to which it belongs. Under

109

certain conditions to be described later, however, it turns out that for the special class of ergodic random processes, it is possible to derive desired statistical information about the entire random process from appropriate analysis of a single arbitrary sample function. For the situation of a pair of random processes $\{x_k(t)\}$ and $\{y_k(t)\}$, the corresponding problem is to estimate joint statistical properties of the two random processes from proper analysis of an arbitrary pair of sample functions $x_k(t)$ and $y_k(t)$.

Consider two arbitrary random processes $\{x_k(t)\}$ and $\{y_k(t)\}$. The first statistical quantities of interest are the ensemble *mean values* at arbitrary fixed values of t, where $x_k(t)$ and $y_k(t)$ are random variables over the index k. These are defined as in Equation (3.8) by

$$\mu_x(t) = E[x_k(t)]$$
$$\mu_y(t) = E[y_k(t)]$$

(5.1)

In general, these mean values are different at different times, and must be calculated separately for every t of interest. That is,

$$\mu_x(t_1) \neq \mu_x(t_2) \qquad \text{if } t_1 \neq t_2$$
$$\mu_y(t_1) \neq \mu_y(t_2) \qquad \text{if } t_1 \neq t_2$$

(5.2)

The next statistical quantities of interest are the *covariance functions* at arbitrary fixed values of $t_1 = t$ and $t_2 = t + \tau$. These are defined by

$$C_{xx}(t, t + \tau) = E\big[(x_k(t) - \mu_x(t))(x_k(t + \tau) - \mu_x(t + \tau))\big]$$
$$C_{yy}(t, t + \tau) = E\big[(y_k(t) - \mu_y(t))(y_k(t + \tau) - \mu_y(t + \tau))\big]$$
$$C_{xy}(t, t + \tau) = E\big[(x_k(t) - \mu_x(t))(y_k(t + \tau) - \mu_y(t + \tau))\big]$$

(5.3)

In general, these quantities are different for different combinations of t_1 and t_2. Observe that at $\tau = 0$ $(t_1 = t_2 = t)$,

$$C_{xx}(t, t) = E\big[(x_k(t) - \mu_x(t))^2\big] = \sigma_x^2(t)$$
$$C_{yy}(t, t) = E\big[(y_k(t) - \mu_y(t))^2\big] = \sigma_y^2(t)$$
$$C_{xy}(t, t) = E\big[(x_k(t) - \mu_x(t))(y_k(t) - \mu_y(t))\big] = C_{xy}(t)$$

(5.4)

Thus the covariance functions $C_{xx}(t, t)$ and $C_{yy}(t, t)$ represent the ordinary variances of $\{x_k(t)\}$ and $\{y_k(t)\}$ at a fixed value of t, whereas $C_{xy}(t, t)$ represents the covariance between $\{x_k(t)\}$ and $\{y_k(t)\}$. As before, different results would generally be obtained for different values of t.

Other statistical quantities can be defined over the ensemble that involve fixing three or more times. The probability structure of the random processes is thus described in finer and finer detail. If $\{x_k(t)\}$ and $\{y_k(t)\}$ form a two-dimensional Gaussian distribution at a fixed value of t, however, then $\{x_k(t)\}$ and $\{y_k(t)\}$ are separately Gaussian. The mean values and covariance functions listed above then provide a complete description of the underlying probability structure. For this reason, the main emphasis in this chapter is on only these two statistical quantities and their relationships to spectral density functions.

If the mean values $\mu_x(t)$ and $\mu_y(t)$, together with the covariance functions $C_{xx}(t, t + \tau)$, $C_{yy}(t, t + \tau)$, and $C_{xy}(t, t + \tau)$, yield the same results for all fixed values of t (that is, are independent of time translations), then the random processes $\{x_k(t)\}$ and $\{y_k(t)\}$ are said to be *weakly stationary*. If all possible probability distributions involving $\{x_k(t)\}$ and $\{y_k(t)\}$ are independent of time translations, then the processes are said to be *strongly stationary*. Since the mean values and covariance functions are consequences only of the first- and second-order probability distributions, it follows that the class of strongly stationary random processes is a subclass of the class of weakly stationary random processes. For Gaussian random processes, however, weak stationarity implies strong stationarity since all possible probability distributions may be derived from the mean values and covariance functions. Thus, for Gaussian random processes, these two stationary concepts coincide.

5.1.1   *Correlation (Covariance) Functions*

For stationary random processes $\{x_k(t)\}$ and $\{y_k(t)\}$, which will be considered henceforth in this chapter, the mean values become constants independent of t. That is, for all t,

$$\mu_x = E[x_k(t)] = \int_{-\infty}^{\infty} xp(x)\, dx$$

$$\mu_y = E[y_k(t)] = \int_{-\infty}^{\infty} yp(y)\, dy$$

(5.5)

where $p(x)$ and $p(y)$ are the probability density functions associated with the random variables $x_k(t)$ and $y_k(t)$, respectively. The covariance functions of stationary random processes are also independent of t.

For arbitrary fixed t and τ, define

$$R_{xx}(\tau) = E[x_k(t)x_k(t + \tau)]$$

$$R_{yy}(\tau) = E[y_k(t)y_k(t + \tau)]$$

$$R_{xy}(\tau) = E[x_k(t)y_k(t + \tau)]$$

(5.6)

where R is introduced instead of C to distinguish these expressions from the covariance functions defined in Equation (5.3). For nonzero mean values, R is different from C. The quantities $R_{xx}(\tau)$ and $R_{yy}(\tau)$ are called the *autocorrelation functions* of $\{x_k(t)\}$ and $\{y_k(t)\}$, respectively, whereas $R_{xy}(\tau)$ is called the *cross-correlation function* between $\{x_k(t)\}$ and $\{y_k(t)\}$.

A necessary and sufficient condition that $R_{xx}(\tau)$ be the autocorrelation function of a weakly stationary random process $\{x_k(t)\}$ is that $R_{xx}(\tau) = R_{xx}(-\tau)$, and that $R_{xx}(\tau)$ be a nonnegative definite function. One can prove also that $R_{xx}(\tau)$ will be a continuous function of τ if it is continuous at the origin. Similarly the cross-correlation function $R_{xy}(\tau)$ will be continuous for all τ if $R_{xx}(\tau)$ or $R_{yy}(\tau)$ is continuous at the origin [Reference 5.3].

For a pair of stationary random processes $\{x_k(t)\}$ and $\{y_k(t)\}$, the joint probability density function $p(x_1, x_2)$ of the pair of random variables $x_1 = x_k(t)$ and $x_2 = x_k(t + \tau)$ is independent of t, and the joint probability density function $p(y_1, y_2)$ associated with the pair of random variables $y_1 = y_k(t)$ and $y_2 = y_k(t + \tau)$ is independent of t. This is also true for the joint probability density function $p(x_1, y_2)$ associated with the pair of random variables $x_1 = x_k(t)$ and $y_2 = y_k(t + \tau)$. In terms of these probability density functions

$$R_{xx}(\tau) = \iint_{-\infty}^{\infty} x_1 x_2 p(x_1, x_2)\, dx_1\, dx_2$$

$$R_{yy}(\tau) = \iint_{-\infty}^{\infty} y_1 y_2 p(y_1, y_2)\, dy_1\, dy_2 \qquad (5.7)$$

$$R_{xy}(\tau) = \iint_{-\infty}^{\infty} x_1 y_2 p(x_1, y_2)\, dx_1\, dy_2$$

For arbitrary values of μ_x and μ_y, the covariance functions are related to the correlation functions by the equations

$$C_{xx}(\tau) = R_{xx}(\tau) - \mu_x^2$$

$$C_{yy}(\tau) = R_{yy}(\tau) - \mu_y^2 \qquad (5.8)$$

$$C_{xy}(\tau) = R_{xy}(\tau) - \mu_x \mu_y$$

Thus correlation functions are identical with covariance functions when the mean values are zero. Note that, by definition, two stationary random processes are uncorrelated if $C_{xy}(\tau) = 0$ for all τ. This occurs, from Equation (5.8), whenever $R_{xy}(\tau) = \mu_x \mu_y$ for all τ. Hence, the two processes will be uncorrelated when $R_{xy}(\tau) = 0$ for all τ only if also either μ_x or μ_y equals zero.

From the stationary hypothesis, it follows that the autocorrelation functions $R_{xx}(\tau)$ and $R_{yy}(\tau)$ are even functions of τ. That is,

$$R_{xx}(-\tau) = R_{xx}(\tau)$$

$$R_{yy}(-\tau) = R_{yy}(\tau)$$

(5.9)

while the cross-correlation function is neither odd nor even, but satisfies the relation

$$R_{xy}(-\tau) = R_{yx}(\tau) \tag{5.10}$$

Equation (5.10) can be proved as follows. By definition

$$R_{xy}(-\tau) = E[x(t)y(t - \tau)]$$

where the dependence on k is omitted to simplify the notation. Since results are invariant with respect to translations in time, one can replace t by $t + \tau$ wherever t appears prior to taking the expected value. Hence

$$R_{xy}(-\tau) = E[x(t + \tau)y(t + \tau - \tau)]$$

$$= E[y(t)x(t + \tau)] = R_{yx}(\tau)$$

which completes the proof. When $x = y$, one obtains

$$R_{xx}(-\tau) = R_{xx}(\tau) \qquad R_{yy}(-\tau) = R_{yy}(\tau)$$

showing that Equation (5.9) is a special case of Equation (5.10).

The correlation properties of stationary random processes $\{x_k(t)\}$ and $\{y_k(t)\}$, which are described by the four functions $R_{xx}(\tau)$, $R_{yy}(\tau)$, $R_{xy}(\tau)$ and $R_{yx}(\tau)$, need be calculated only for values of $\tau \geq 0$, since the relations of Equations (5.9) and (5.10) yield results for $\tau < 0$.

5.1.2 Examples of Autocorrelation Functions

Examples of special autocorrelation functions that are useful in theoretical studies are given in Table 5.1.

Example 5.1. Autocorrelation Function of Sine Wave Process. Suppose $\{x_k(t)\} = \{X \sin[2\pi f_0 t + \theta(k)]\}$ is a sine wave process in which X and f_0 are constants and $\theta(k)$ is a random variable with a uniform probability density function $p(\theta)$ over $(0, 2\pi)$. Determine the autocorrelation function $R_{xx}(\tau)$.

Table 5.1

Special Autocorrelation Functions

Type	Autocorrelation Function
Constant	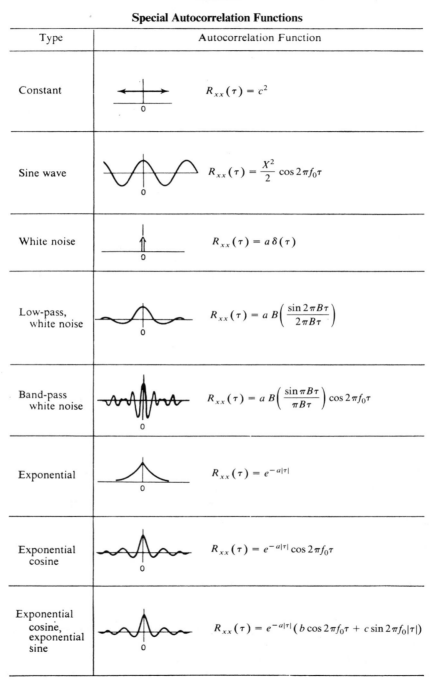 $R_{xx}(\tau) = c^2$
Sine wave	$R_{xx}(\tau) = \dfrac{X^2}{2}\cos 2\pi f_0 \tau$
White noise	$R_{xx}(\tau) = a\,\delta(\tau)$
Low-pass, white noise	$R_{xx}(\tau) = a\,B\left(\dfrac{\sin 2\pi B\tau}{2\pi B\tau}\right)$
Band-pass white noise	$R_{xx}(\tau) = a\,B\left(\dfrac{\sin \pi B\tau}{\pi B\tau}\right)\cos 2\pi f_0 \tau$
Exponential	$R_{xx}(\tau) = e^{-a|\tau|}$
Exponential cosine	$R_{xx}(\tau) = e^{-a|\tau|}\cos 2\pi f_0 \tau$
Exponential cosine, exponential sine	$R_{xx}(\tau) = e^{-a|\tau|}\left(b\cos 2\pi f_0\tau + c\sin 2\pi f_0|\tau|\right)$

Here, for any fixed value of t, the random variable

$$x_k(t) = X \sin[2\pi f_0 t + \theta(k)] = x_1(\theta)$$

$$x_k(t + \tau) = X \sin[2\pi f_0(t + \tau) + \theta(k)] = x_2(\theta)$$

From Equation (5.6),

$$R_{xx}(\tau) = E[x_k(t)x_k(t + \tau)] = E[x_1(\theta)x_2(\theta)]$$

with

$$p(\theta) = (2\pi)^{-1} \qquad 0 \le \theta \le 2\pi \qquad \text{otherwise zero}$$

Hence

$$R_{xx}(\tau) = \frac{X^2}{2\pi} \int_0^{2\pi} \sin(2\pi f_0 t + \theta)\sin[2\pi f_0(t + \tau) + \theta] \, d\theta$$

$$= \frac{X^2}{2} \cos 2\pi f_0 \tau$$

giving the autocorrelation function of a sine wave stationary random process as pictured in Table 5.1.

Example 5.2. Autocorrelation Function of Rectangular Wave Process.
Consider a physical situation where a sample function $x_k(t)$ from a random rectangular wave process $\{x_k(t)\}$ is restricted so as to assume only values of c or $-c$, where the number of changes of sign in an interval $(t, t + \tau)$ occurs at random and independent times with an average density of λ. Assume also that what happens inside an interval $(t, t + \tau)$ is independent of what happens outside the interval. Define

$$A_n = \text{Event}[\text{exactly } n \text{ changes of sign fall inside } (t, t + \tau)]$$

This type of physical situation follows a Poisson distribution [Reference 5.1], where the probability of event A_n is

$$P(A_n) = \frac{(\lambda|\tau|)^n}{n!} e^{-\lambda|\tau|}$$

Determine the autocorrelation function of $\{x_k(t)\}$.

The autocorrelation function may be calculated as follows. An individual product term $x_k(t)x_k(t + \tau)$ equals c^2 if $x_k(t)$ and $x_k(t + \tau)$ are of the same sign, and it equals $-c^2$ if they are of opposite sign. The total probability for c^2

is given by the sum $P(A_0) + P(A_2) + P(A_4) + \cdots$, and the total probability for $-c^2$ is given by the sum $P(A_1) + P(A_3) + P(A_5) + \cdots$. Hence

$$R_{xx}(\tau) = E[x_k(t)x_k(t + \tau)] = c^2 \sum_{n=0}^{\infty} (-1)^n P(A_n)$$

$$= c^2 e^{-\lambda|\tau|} \sum_{n=1}^{\infty} (-1)^n \frac{(\lambda|\tau|)^n}{n!} = c^2 e^{-2\lambda|\tau|}$$

This exponential function is pictured in Table 5.1 with $a = 2\lambda$ and $c^2 = 1$.

Example 5.3. Autocorrelation Function of Sum of Two Stationary Processes. Assume that a random process $\{y_k(t)\}$ is the sum of two stationary processes $\{x_{1,k}(t)\}$ and $\{x_{2,k}(t)\}$ such that each sample function

$$y_k(t) = a_1 x_{1,k}(t) + a_2 x_{2,k}(t)$$

where a_1 and a_2 are constants. Assume also that $\{x_{1,k}(t)\}$ and $\{x_{2,k}(t)\}$ may be correlated. Determine the autocorrelation function $R_{yy}(\tau)$.

From Equation (3.44), one obtains

$$R_{yy}(\tau) = E[y_k(t)y_k(t + \tau)]$$

$$= E[(a_1 x_{1,k}(t) + a_2 x_{2,k}(t))(a_1 x_{1,k}(t + \tau) + a_2 x_{2,k}(t + \tau))]$$

$$= a_1^2 E[x_{1,k}(t)x_{1,k}(t + \tau)] + a_1 a_2 E[x_{1,k}(t)x_{2,k}(t + \tau)]$$

$$+ a_1 a_2 E[x_{2,k}(t)x_{1,k}(t + \tau)] + a_2^2 E[x_{2,k}(t)x_{2,k}(t + \tau)]$$

$$= a_1^2 R_{x_1 x_1}(\tau) + a_1 a_2 [R_{x_1 x_2}(\tau) + R_{x_2 x_1}(\tau)] + a_2^2 R_{x_2 x_2}(\tau)$$

Thus the sum autocorrelation function requires knowledge of the input cross-correlation functions as well as their autocorrelation functions.

Example 5.4. Uncorrelated Dependent Random Variables. Assume that two random variables x and y are such that $x = \cos\phi$ and $y = \sin\phi$, where ϕ is uniformly distributed from 0 to 2π. Here x and y are related since

$$y = \sqrt{1 - x^2}$$

It follows that

$$p(x, y) \neq p(x)p(y)$$

showing that x and y are statistically dependent. However, the covariance between x and y is

$$C_{xy} = E[xy] - E[x]E[y]$$

$$= E[\cos\phi \sin\phi] - E[\cos\phi]E[\sin\phi]$$

$$= \tfrac{1}{2}E[\sin 2\phi] = 0$$

Hence x and y are uncorrelated.

5.1.3   Correlation Coefficient Functions

The cross-correlation function is bounded by the *cross-correlation inequality*

$$|R_{xy}(\tau)|^2 \le R_{xx}(0)R_{yy}(0) \tag{5.11}$$

which may be proved as follows. For any real constants a and b, the expected value

$$E\left[(ax(t) + by(t + \tau))^2\right] \ge 0$$

since only nonnegative quantities are being considered. This is equivalent to

$$a^2 R_{xx}(0) + 2abR_{xy}(\tau) + b^2 R_{yy}(0) \ge 0$$

Hence, assuming $b \ne 0$,

$$\left(\frac{a}{b}\right)^2 R_{xx}(0) + 2\left(\frac{a}{b}\right)R_{xy}(\tau) + R_{yy}(0) \ge 0$$

This is a quadratic equation in a/b without real different roots since one side is nonnegative. Therefore, the discriminant of this quadratic equation in a/b must be nonpositive. That is,

$$\text{Discriminant} = 4R_{xy}^2(\tau) - 4R_{xx}(0)R_{yy}(0) \le 0$$

Thus

$$R_{xy}^2(\tau) = |R_{xy}(\tau)|^2 \le R_{xx}(0)R_{yy}(0)$$

This completes the proof.

By considering $x(t) - \mu_x$ and $y(t + \tau) - \mu_y$ instead of $x(t)$ and $y(t + \tau)$, the same proof gives the cross-covariance inequality

$$|C_{xy}(\tau)|^2 \le C_{xx}(0)C_{yy}(0) \tag{5.12}$$

Noting that

$$|R_{xx}(\tau)| \le R_{xx}(0) \qquad |C_{xx}(\tau)| \le C_{xx}(0) \qquad (5.13)$$

it follows that the maximum possible values of $R_{xx}(\tau)$ and $C_{xx}(\tau)$ occur at $\tau = 0$ and correspond to the mean square value and variance, respectively, of the data. That is,

$$R_{xx}(0) = E\left[x_k^2(t)\right] = \psi_x^2 \qquad C_{xx}(0) = \sigma_x^2$$

$$R_{yy}(0) = E\left[y_k^2(t)\right] = \psi_y^2 \qquad C_{yy}(0) = \sigma_y^2$$

$$(5.14)$$

Hence, Equation (5.12) takes the form

$$|C_{xy}(\tau)|^2 \le \sigma_x^2 \sigma_y^2 \qquad (5.15)$$

The *correlation coefficient function* (*normalized cross-covariance function*) may now be defined by

$$\rho_{xy}(\tau) = \frac{C_{xy}(\tau)}{\sigma_x \sigma_y} \qquad (5.16)$$

and satisfies for all τ

$$-1 \le \rho_{xy}(\tau) \le 1 \qquad (5.17)$$

If either u_x or u_y is equal to zero, then $\rho_{xy}(\tau)$ becomes

$$\rho_{xy}(\tau) = \frac{R_{xy}(\tau)}{\sigma_x \sigma_y} \qquad (5.18)$$

since $C_{xy}(\tau) = R_{xy}(\tau)$ in these cases. The function $\rho_{xy}(\tau)$ measures the degree of linear dependence between $\{x_k(t)\}$ and $\{y_k(t)\}$ for a displacement of τ in $\{y_k(t)\}$ relative to $\{x_k(t)\}$. It is essentially a generalization of the correlation coefficient used in classical statistics as discussed earlier in Sections 3.2.1 and 4.8.1.

5.1.4 *Cross-Correlation Function for Time Delay*

Assume a transmitted signal is represented by a zero mean value stationary random signal $x(t)$. Let the received signal be represented by another zero mean value stationary random signal $y(t)$ such that

$$y(t) = \alpha x(t - \tau_0) + n(t) \qquad (5.19)$$

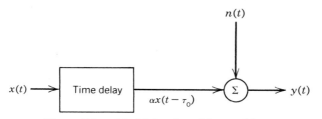

Figure 5.1   Model for time-delay problem.

The quantity α is a constant attenuation factor, the quantity $\tau_0 = (d/c)$ is a constant time delay equal to a distance d divided by a velocity of propagation c, and $n(t)$ represents uncorrelated zero mean value noise at the output, as illustrated in Figure 5.1.

For this problem, the cross-correlation function between $x(t)$ and $y(t)$ is given by

$$R_{xy}(\tau) = E[x(t)y(t+\tau)] = E[x(t)\{\alpha x(t+\tau-\tau_0) + n(t+\tau)\}]$$

$$= \alpha E[x(t)x(t+\tau-\tau_0)] = \alpha R_{xx}(\tau-\tau_0) \tag{5.20}$$

Thus $R_{xy}(\tau)$ is merely the autocorrelation function $R_{xx}(\tau)$ displaced by the time delay τ_0 and multiplied by the attenuation factor α. The peak value of $R_{xy}(\tau)$ occurs at $\tau = \tau_0$, namely,

$$R_{xy}(\tau)_{\text{peak}} = R_{xy}(\tau_0) = \alpha R_{xx}(0) = \alpha\sigma_x^2 \tag{5.21}$$

This result is pictured in Figure 5.2. Note that measurement of the value τ_0 where the peak occurs plus knowledge of either the distance d or the velocity

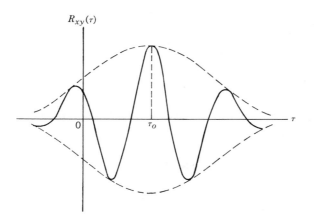

Figure 5.2   Typical cross-correlation function for time-delay problem.

of propagation c will yield the other quantity since $d = c\tau_0$. See Reference 5.2 for physical illustrations of time-delay measurement problems.

Again assuming $x(t)$ and $n(t)$ have zero mean values, the correlation coefficient function at $\tau = \tau_0$ from Equation (5.18) is given by

$$\rho_{xy}(\tau_0) = \frac{R_{xy}(\tau_0)}{\sigma_x \sigma_y} = \alpha \frac{\sigma_x}{\sigma_y} \tag{5.22}$$

Thus, measurement of $\rho_{xy}(\tau_0)$ yields the attenuation factor α by

$$\alpha = \rho_{xy}(\tau_0)\left[\sigma_y / \sigma_x\right] \tag{5.23}$$

Also, the variance in $y(t)$, for uncorrelated $x(t)$ and $n(t)$, is

$$\sigma_y^2 = E\left[y^2(t)\right] = \alpha^2 \sigma_x^2 + \sigma_n^2 \tag{5.24}$$

where the two components are

$$\alpha^2 \sigma_x^2 = \rho_{xy}^2 \sigma_y^2 = \text{variance in } y(t) \text{ due to } x(t)$$

$$\sigma_n^2 = \left(1 - \rho_{xy}^2\right)\sigma^2 = \text{variance in } y(t) \text{ due to } n(t) \tag{5.25}$$

Equation (5.24) is a special case at $\tau = 0$ of the result

$$R_{yy}(\tau) = E\left[y(t)y(t + \tau)\right] = \alpha^2 R_{xx}(\tau) + R_{nn}(\tau) \tag{5.26}$$

5.2 SPECTRAL DENSITY FUNCTIONS

Spectral density functions can be defined in three different equivalent ways as will be proved in later sections:

(a) via correlation functions
(b) via finite Fourier transforms
(c) via filtering-squaring-averaging operations

Important relations will also be developed for these functions that are needed for many applications.

5.2.1 Spectra via Correlation Functions

The first way to define spectral density functions is a (historical) mathematical method where a single Fourier transform is taken of a previously calculated correlation function. When mean values are removed, this (infinite) Fourier transform will usually exist even though the (infinite) Fourier transform of the original stationary random data does not exist. This approach yields *two-sided* spectral density functions, denoted $S(f)$, which are defined for f over $(-\infty, \infty)$.

Specifically, assume that the autocorrelation and cross-correlation functions $R_{xx}(\tau)$, $R_{yy}(\tau)$, and $R_{xy}(\tau)$ exist, as defined in Equation (5.6). Further assume that the integrals of their absolute values are finite, namely,

$$\int_{-\infty}^{\infty} |R(\tau)| \, d\tau < \infty$$

This will always be true in practice for finite record lengths. Then Fourier transforms of $R(\tau)$ will exist as defined by

$$S_{xx}(f) = \int_{-\infty}^{\infty} R_{xx}(\tau) e^{-j2\pi f\tau} \, d\tau$$

$$S_{yy}(f) = \int_{-\infty}^{\infty} R_{yy}(\tau) e^{-j2\pi f\tau} \, d\tau \qquad (5.27)$$

$$S_{xy}(f) = \int_{-\infty}^{\infty} R_{xy}(\tau) e^{-j2\pi f\tau} \, d\tau$$

Such integrals over finite record lengths always exist. The quantities $S_{xx}(f)$ and $S_{yy}(f)$ are called the *autospectral density functions* of $\{x_k(t)\}$ and $\{y_k(t)\}$, respectively, whereas $S_{xy}(f)$ is called the *cross-spectral density function* between $\{x_k(t)\}$ and $\{y_k(t)\}$.

Inverse Fourier transforms of Equation (5.27) yield

$$R_{xx}(\tau) = \int_{-\infty}^{\infty} S_{xx}(f) e^{j2\pi f\tau} \, df$$

$$R_{yy}(\tau) = \int_{-\infty}^{\infty} S_{yy}(f) e^{j2\pi f\tau} \, df \qquad (5.28)$$

$$R_{xy}(\tau) = \int_{-\infty}^{\infty} S_{xy}(f) e^{j2\pi f\tau} \, df$$

To handle practical problems, both $R(\tau)$ and $S(f)$ are permitted to include delta functions. The results in Equations (5.27) and (5.28) are often called the *Wiener–Khinchine relations* in honor of the two mathematicians, N. Wiener in the United States and A. I. Khinchine in the USSR, who independently proved the Fourier transform relationship between correlation functions and spectral density functions in the early 1930s.

From the symmetry properties of stationary correlation functions given in Equations (5.9) and (5.10), it follows that

$$S_{xx}(-f) = S_{xx}^*(f) = S_{xx}(f)$$
$$S_{yy}(-f) = S_{yy}^*(f) = S_{yy}(f) \qquad (5.29)$$

$$S_{xy}(-f) = S_{xy}^*(f) = S_{yx}(f) \qquad (5.30)$$

Thus the autospectral density functions $S_{xx}(f)$ and $S_{yy}(f)$ are real-valued even functions of f, whereas the cross-spectral density function is a complex-valued function of f. It is also true that both $S_{xx}(f)$ and $S_{yy}(f)$ are nonnegative for all f, to be proved later.

Equation (5.30) can be proved as follows. By definition,

$$S_{xy}(-f) = \int_{-\infty}^{\infty} R_{xy}(\tau) e^{j2\pi f\tau} \, d\tau$$

$$S_{xy}^*(f) = \int_{-\infty}^{\infty} R_{xy}(\tau) e^{j2\pi f\tau} \, d\tau$$

$$S_{xy}(f) = \int_{-\infty}^{\infty} R_{yx}(\tau) e^{-j2\pi f\tau} \, d\tau$$

It is immediately obvious that $S_{xy}(-f) = S_{xy}^*(f)$. Now make a change of variable in the first integral by letting $\tau = -u$, $d\tau = -du$. Then

$$S_{xy}(-f) = \int_{-\infty}^{\infty} R_{xy}(-u) e^{-j2\pi fu} \, du$$

But $R_{xy}(-u) = R_{yx}(u)$ from Equation (5.10). Hence

$$S_{xy}(-f) = \int_{-\infty}^{\infty} R_{yx}(u) e^{-j2\pi fu} \, du = S_{yx}(f)$$

This completes the proof. Results in Equation (5.29) are special cases of Equation (5.30) when $x(t) = y(t)$.

The autospectral relations in Equation (5.27) may be simplified to

$$S_{xx}(f) = \int_{-\infty}^{\infty} R_{xx}(\tau)\cos 2\pi f\tau \, d\tau = 2\int_{0}^{\infty} R_{xx}(\tau)\cos 2\pi f\tau \, d\tau$$

$$S_{yy}(f) = \int_{-\infty}^{\infty} R_{yy}(\tau)\cos 2\pi f\tau \, d\tau = 2\int_{0}^{\infty} R_{yy}(\tau)\cos 2\pi f\tau \, d\tau$$

$$(5.31)$$

Conversely,

$$R_{xx}(\tau) = 2\int_{0}^{\infty} S_{xx}(f)\cos 2\pi f\tau \, df$$

$$R_{yy}(\tau) = 2\int_{0}^{\infty} S_{yy}(f)\cos 2\pi f\tau \, df$$

$$(5.32)$$

The *one-sided* autospectral density functions, $G_{xx}(f)$ and $G_{yy}(f)$, where f varies only over $(0, \infty)$, are defined by

$$G_{xx}(f) = 2S_{xx}(f) \qquad 0 \le f < \infty \qquad \text{otherwise zero}$$

$$G_{yy}(f) = 2S_{yy}(f) \qquad 0 \le f < \infty \qquad \text{otherwise zero}$$

$$(5.33)$$

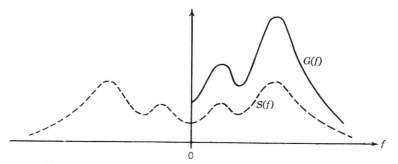

Figure 5.3   One-sided and two-sided autospectral density functions.

These are the quantities measured by direct filtering procedures in practice. For mathematical calculations, however, the use of $S_{xx}(f)$ and $S_{yy}(f)$ defined over $(-\infty, \infty)$ and exponentials with imaginary exponents often simplifies the analysis. It is important to be able to deal properly with both of these representations, and both will be used in this book. See Figure 5.3 for a graphical illustration of the relationship.

In terms of the one-sided autospectral density functions $G_{xx}(f)$ and $G_{yy}(f)$, the correspondence with the stationary correlation functions $R_{xx}(\tau)$ and $R_{yy}(\tau)$ becomes

$$G_{xx}(f) = 4\int_0^\infty R_{xx}(\tau)\cos 2\pi f\tau \, d\tau \qquad 0 \le f < \infty$$

$$G_{yy}(f) = 4\int_0^\infty R_{yy}(\tau)\cos 2\pi f\tau \, d\tau \qquad 0 \le f < \infty$$

$$(5.34)$$

Conversely,

$$R_{xx}(\tau) = \int_0^\infty G_{xx}(f)\cos 2\pi f\tau \, df$$

$$R_{yy}(\tau) = \int_0^\infty G_{yy}(f)\cos 2\pi f\tau \, df$$

$$(5.35)$$

In particular, at $\tau = 0$, one obtains

$$R_{xx}(0) = E[x^2(t)] = \psi_x^2 = \int_0^\infty G_{xx}(f) \, df$$

$$R_{yy}(0) = E[y^2(t)] = \psi_y^2 = \int_0^\infty G_{yy}(f) \, df$$

$$(5.36)$$

The *one-sided* cross-spectral density function $G_{xy}(f)$, where f varies only over $(0, \infty)$, is defined by

$$G_{xy}(f) = 2S_{xy}(f) \qquad 0 \le f < \infty \qquad \text{otherwise zero} \qquad (5.37)$$

From Equation (5.27)

$$G_{xy}(f) = 2\int_{-\infty}^{\infty} R_{xy}(\tau)e^{-j2\pi f\tau}\,d\tau = C_{xy}(f) - jQ_{xy}(f) \qquad (5.38)$$

where $C_{xy}(f)$ is called the *coincident spectral density function* (*co-spectrum*), and $Q_{xy}(f)$ is called the *quadrature spectral density function* (*quad-spectrum*). In terms of $C_{xy}(f)$ and $Q_{xy}(f)$, the cross-correlation function

$$R_{xy}(\tau) = \int_0^{\infty} \left[C_{xy}(f)\cos 2\pi f\tau + Q_{xy}(f)\sin 2\pi f\tau \right] df \qquad (5.39)$$

Observe that $C_{xy}(f)$ and $Q_{xy}(f)$ are defined in terms of $G_{xy}(f)$ rather than $S_{xy}(f)$. Note also that $\tau = 0$ yields the result

$$R_{xy}(0) = E[x(t)y(t)] = \int_0^{\infty} C_{xy}(f)\,df \qquad (5.40)$$

$R_{xy}(0)$ can be determined solely from $C_{xy}(f) = \text{Re}[G_{xy}(f)]$.

The one-sided cross-spectral density function may be presented in complex polar notation as

$$G_{xy}(f) = |G_{xy}(f)|e^{-j\theta_{xy}(f)} \qquad 0 \le f < \infty \qquad (5.41)$$

where the *absolute value* and *phase angle* are determined by

$$|G_{xy}(f)| = \sqrt{C_{xy}^2(f) + Q_{xy}^2(f)} \qquad (5.42)$$

$$\theta_{xy}(f) = \tan^{-1}\frac{Q_{xy}(f)}{C_{xy}(f)} \qquad (5.43)$$

The signs of the terms $C_{xy}(f)$ and $Q_{xy}(f)$ may be positive or negative and give the quadrant for the phase angle. These signs determine also at each frequency f whether $y(t)$ leads $x(t)$ or $x(t)$ leads $y(t)$. When the record $y(t)$ leads $x(t)$, this means $y(t) = x(t - \tau_0)$ where $\tau_0 > 0$ and $\theta_{xy}(f) = 2\pi f\tau_0$. The relation of the phase angle to $C_{xy}(f)$ and $Q_{xy}(f)$ is illustrated in Figure 5.4. Similarly, the two-sided cross-spectral density function in complex polar notation is

$$S_{xy}(f) = |S_{xy}(f)|e^{-j\theta_{xy}(f)} \qquad (5.44)$$

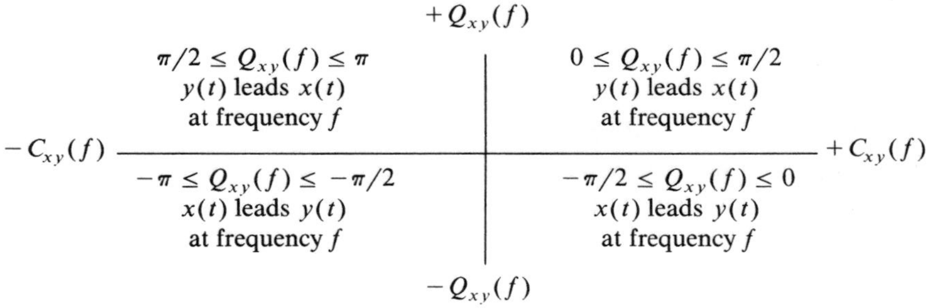

$$+Q_{xy}(f)$$

$\pi/2 \le Q_{xy}(f) \le \pi$ $y(t)$ leads $x(t)$ at frequency f	$0 \le Q_{xy}(f) \le \pi/2$ $y(t)$ leads $x(t)$ at frequency f
$-\pi \le Q_{xy}(f) \le -\pi/2$ $x(t)$ leads $y(t)$ at frequency f	$-\pi/2 \le Q_{xy}(f) \le 0$ $x(t)$ leads $y(t)$ at frequency f

$-C_{xy}(f)$ ⟷ $+C_{xy}(f)$

$$-Q_{xy}(f)$$

Figure 5.4 Relation of phase angle to cross-spectral terms.

where $|S_{xy}(f)| = \frac{1}{2}|G_{xy}(f)|$ and $\theta_{xy}(f)$ is the same as in Equations (5.41) and (5.43).

Referring now to Equation (5.38), it follows that

$$C_{xy}(f) = 2\int_0^\infty \left[R_{xy}(\tau) + R_{yx}(\tau) \right] \cos 2\pi f\tau \, d\tau = C_{xy}(-f)$$

$$(5.45)$$

$$Q_{xy}(f) = 2\int_0^\infty \left[R_{xy}(\tau) - R_{yx}(\tau) \right] \sin 2\pi f\tau \, d\tau = -Q_{xy}(-f)$$

Thus $C_{xy}(f)$ is a real-valued even function of f, whereas $Q_{xy}(f)$ is a real-valued odd function of f. Also,

$$C_{xy}(f) = \frac{1}{2}\left[G_{xy}(f) + G_{yx}(f) \right] = |G_{xy}(f)|\cos\theta_{xy}(f)$$

$$(5.46)$$

$$Q_{xy}(f) = (j/2)\left[G_{xy}(f) - G_{yx}(f) \right] = |G_{xy}(f)|\sin\theta_{xy}(f)$$

The spectral properties of stationary random processes $\{x_k(t)\}$ and $\{y_k(t)\}$, which are described by the three functions $S_{xx}(f)$, $S_{yy}(f)$, and $S_{xy}(f)$, or by the four functions $S_{xx}(f)$, $S_{yy}(f)$, $C_{xy}(f)$, and $Q_{xy}(f)$, need be calculated only for values of $f \ge 0$, since the relations of Equations (5.29), (5.30), and (5.45) yield results for $f < 0$. Of course, corresponding G functions should be calculated only for $f \ge 0$.

When dealing with spectral functions involving delta functions at $f = 0$, it is convenient to let the lower limit of integration, zero, be approached from below. In particular, for $R(\tau) = c^2$, this allows the corresponding $G(f)$ to be $G(f) = c^2\delta(f)$. For this situation, $S(f)$ is also given by $S(f) = c^2\delta(f)$, showing that the factor of 2 in Equation (5.37) should not be applied to delta functions at $f = 0$. This consideration does not exist for correlation functions involving delta functions at $\tau = 0$, since correlation functions are defined for all τ. Thus $R(\tau) = a\delta(\tau)$ corresponds to $S(f) = a$ for all f and $G(f) = 2a$ for $f \geq 0$.

Examples of special autospectral density functions that are useful in theoretical studies are given in Table 5.2.

BANDWIDTH LIMITED WHITE NOISE. By definition, bandwidth limited white noise is a stationary random process with a constant autospectral density function as follows.

$$G_{xx}(f) = \begin{cases} a & 0 \leq f_0 - (B/2) \leq f \leq f_0 + (B/2) \\ 0 & \text{otherwise} \end{cases} \quad (5.47)$$

Here, f_0 is the center of a rectangular filter of bandwidth B. This case is also called *band-pass white noise*. From Equation (5.35), it follows that the associated autocorrelation function is

$$R_{xx}(\tau) = \int_{f_0-(B/2)}^{f_0+(B/2)} a \cos 2\pi f \tau \, d\tau = aB \left(\frac{\sin \pi B \tau}{\pi B \tau} \right) \cos 2\pi f_0 \tau \quad (5.48)$$

For the special case where $f_0 = (B/2)$, this case becomes *low-pass white noise* defined as follows:

$$G_{xx}(f) = \begin{cases} a & 0 \leq f \leq B \\ 0 & \text{otherwise} \end{cases} \quad (5.49)$$

with

$$R_{xx}(\tau) = aB \left(\frac{\sin 2\pi B \tau}{2\pi B \tau} \right) \quad (5.50)$$

These results are pictured in Tables 5.1 and 5.2. For either band-pass white noise or low-pass white noise, the data have a finite mean square value given by

$$\int_0^\infty G_{xx}(f) \, df = aB = R_{xx}(0) \quad (5.51)$$

It is theoretically possible to approximate such cases with real data.

An extreme version of low-pass white noise, called *white noise*, is defined by a $G_{xx}(f)$ that is assumed to be a constant over all frequencies. This case never occurs for real data. Specifically, for white noise,

$$G_{xx}(f) = a \qquad f \geq 0 \text{ only} \quad (5.52)$$

Table 5.2

Special Autospectral Density Functions

Type	(One-Sided) Autospectral Density Function
Constant	$G_{xx}(f) = c^2 \delta(f)$
Sine wave	$G_{xx}(f) = \dfrac{X^2}{2} \delta(f - f_0)$
White noise	$G_{xx}(f) = 2a,\ f \geq 0;$ otherwise zero
Low-pass white noise	$G_{xx}(f) = a,\ 0 \leq f \leq B;$ otherwise zero
Band-pass white noise	$G_{xx}(f) = a,\ \ 0 \leq f_0 - (B/2) \leq f \leq f_0 + (B/2);$ otherwise zero
Exponential	$G_{xx}(f) = \dfrac{4a}{a^2 + 4\pi^2 f^2}$
Exponential cosine	$G_{xx}(f) = 2a\left[\dfrac{1}{a^2 + 4\pi^2(f + f_0)^2} + \dfrac{1}{a^2 + 4\pi^2(f - f_0)^2}\right]$
Exponential cosine, exponential sine	$G_{xx}(f) = \dfrac{2ab + 4\pi c(f + f_0)}{a^2 + 4\pi^2(f + f_0)^2} + \dfrac{2ab - 4\pi c(f - f_0)}{a^2 + 4\pi^2(f - f_0)^2}$

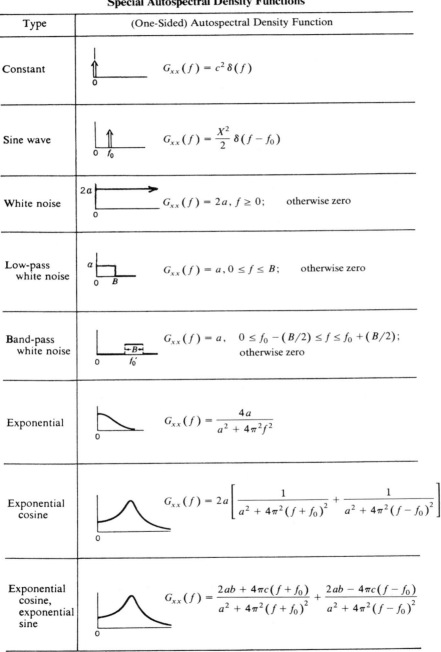

Hence

$$S_{xx}(f) = (a/2) \qquad \text{all } f \qquad\qquad (5.53)$$

$$R_{xx}(\tau) = (a/2)\delta(\tau) \qquad\qquad (5.54)$$

$$\int_0^\infty G_{xx}(f)\, df = \infty = R_{xx}(0) \qquad\qquad (5.55)$$

showing that white noise has an infinite mean square value. Such theoretical white noise *cannot* be Gaussian since a Gaussian process must have a finite mean square value in order to be well defined.

Example 5.5. **Autospectral Density Function of Sine Wave Process.**   The sine wave process described in Example 5.1 has an autocorrelation function given by

$$R_{xx}(\tau) = \frac{X^2}{2}\cos 2\pi f_0 \tau$$

Substitution into Equation (5.27) yields the two-sided result

$$S_{xx}(f) = \frac{X^2}{4}[\delta(f - f_0) + \delta(f + f_0)]$$

which consists of two delta functions at $f = f_0$ and $f = -f_0$. Then, the one-sided autospectral density function is

$$G_{xx}(f) = \frac{X^2}{2}\delta(f - f_0)$$

as pictured in Table 5.2. Note that

$$\int_0^\infty G_{xx}(f)\, df = \frac{X^2}{2} = R_{xx}(0)$$

Example 5.6. **Autospectral Density Function of Rectangular Wave Process.** The rectangular wave process described in Example 5.2 has an exponential autocorrelation function given by

$$R_{xx}(\tau) = c^2 e^{-2\lambda|\tau|}$$

Substitution into Equation (5.27) yields the two-sided result

$$S_{xx}(f) = \frac{\lambda c^2}{\lambda^2 + \pi^2 f^2}$$

Then, the one-sided autospectral density function is

$$G_{xx}(f) = \frac{2\lambda c^2}{\lambda^2 + \pi^2 f^2}$$

as pictured in Table 5.2, where $a = 2\lambda$ and $c^2 = 1$. Note that

$$\int_0^\infty G_{xx}(f)\, df = c^2 = R_{xx}(0)$$

Example 5.7. **Autospectral Density Function of Sum of Two Processes.** The autocorrelation function of the sum of two stationary random processes described in Example 5.3 is

$$R_{yy}(\tau) = a_1^2 R_{x_1 x_1}(\tau) + a_1 a_2 \left[R_{x_1 x_2}(\tau) + R_{x_2 x_1}(\tau) \right] + a_2^2 R_{x_2 x_2(\tau)}$$

Substitution into Equation (5.27) yields the two-sided result

$$S_{yy}(f) = a_1^2 S_{x_1 x_1}(f) + a_1 a_2 \left[S_{x_1 x_2}(f) + S_{x_2 x_1}(f) \right] + a_2^2 S_{x_2 x_2}(f)$$

But

$$S_{x_2 x_1}(f) = S_{x_1 x_2}^*(f)$$

Hence

$$S_{x_1 x_2}(f) + S_{x_2 x_1}(f) = 2\,\text{Re}\left[S_{x_1 x_2}(f) \right] = C_{x_1 x_2}(f)$$

Thus $S_{yy}(f)$ is real valued and may be expressed as

$$S_{yy}(f) = a_1^2 S_{x_1 x_1}(f) + a_1 a_2 C_{x_1 x_2}(f) + a_2^2 S_{x_2 x_2}(f)$$

The corresponding one-sided result is

$$G_{yy}(f) = a_1^2 G_{x_1 x_1}(f) + 2 a_1 a_2 C_{x_1 x_2}(f) + a_2^2 G_{x_2 x_2}(f)$$

5.2.2   *Spectra via Finite Fourier Transforms*

The second method to define spectral density functions is also mathematical. It is based on finite Fourier transforms of the original data records and represents the procedure that is followed in present spectral density calculations.

Consider a pair of associated sample records $x_k(t)$ and $y_k(t)$ from stationary random processes $\{x_k(t)\}$ and $\{y_k(t)\}$. For a finite time interval $0 \le t \le T$, define

$$S_{xy}(f, T, k) = \frac{1}{T} X_k^*(f, T) Y_k(f, T) \tag{5.56}$$

where

$$X_k(f, T) = \int_0^T x_k(t) e^{-j2\pi ft} \, dt$$

$$\tag{5.57}$$

$$Y_k(f, T) = \int_0^T y_k(t) e^{-j2\pi ft} \, dt$$

The quantities $X_k(f, T)$ and $Y_k(f, T)$ represent finite Fourier transforms of $x_k(t)$ and $y_k(t)$ respectively, and $X_k^*(f, T)$ is the complex conjugate of $X_k(f, T)$. These finite-range Fourier transforms will exist for general stationary records, whereas their infinite-range Fourier transforms would not exist, since the stationary data theoretically persist forever.

A common mistake made by many people is now to use analogies from periodic data to define the cross-spectral density function by

$$S_{xy}(f, k) = \lim_{T \to \infty} S_{xy}(f, T, k) \tag{5.58}$$

This is an *unsatisfactory* definition for general stationary random data because the estimate of $S_{xy}(f, k)$ by $S_{xy}(f, T, k)$ does not improve in the statistical sense of consistency (defined in Section 4.1) as T tends to infinity. Also, observe that the left-hand side is still a function of the index k. The correct way to define $S_{xy}(f)$ is by the expression

$$S_{xy}(f) = \lim_{T \to \infty} E\left[S_{xy}(f, T, k)\right] \tag{5.59}$$

where $E[S_{xy}(f, T, k)]$ is, of course, the expected value operation over the ensemble index k in question. The autospectral density functions $S_{xx}(f)$ and $S_{yy}(f)$ are merely special cases of Equation (5.59). The equivalence of the result in Equation (5.59) with that previously given in Equation (5.27) will now be proved.

Using different variables of integration to avoid confusion, Equation (5.56) becomes

$$S_{xy}(f, T, k) = \frac{1}{T} \int_0^T x_k(\alpha) e^{j2\pi f\alpha} \, d\alpha \int_0^T y_k(\beta) e^{-j2\pi f\beta} \, d\beta$$

$$= \frac{1}{T} \int_0^T \int_0^T x_k(\alpha) y_k(\beta) e^{-j2\pi f(\beta - \alpha)} \, d\alpha \, d\beta \qquad (5.60)$$

Now, change the region of integration from (α, β) to (α, τ), where $\tau = \beta - \alpha$, $d\tau = d\beta$. This changes the limits of integration as shown in the sketch below.

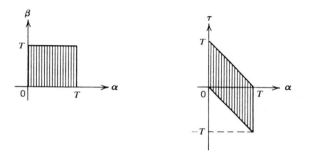

Proceed by integrating in the order (α, τ) instead of the order (α, β). This leads to

$$\int_0^T \int_0^T d\alpha \, d\beta = \int_{-T}^0 \int_{-\tau}^T d\alpha \, d\tau + \int_0^T \int_0^{T-\tau} d\alpha \, d\tau \qquad (5.61)$$

As a check, it is readily verified that both sides of Equation (5.61) yield the value T^2. Hence this change of the region of integration allows Equation (5.60) to take the form

$$S_{xy}(f, T, k) = \int_{-T}^0 \left[\frac{1}{T} \int_{-\tau}^T x_k(\alpha) y_k(\alpha + \tau) \, d\alpha \right] e^{-j2\pi f\tau} \, d\tau$$

$$+ \int_0^T \left[\frac{1}{T} \int_0^{T-\tau} x_k(\alpha) y_k(\alpha + \tau) \, d\alpha \right] e^{-j2\pi f\tau} \, d\tau \qquad (5.62)$$

By definition, the cross-correlation function $R_{xy}(\tau)$ is given by the expected value

$$R_{xy}(\tau) = E[x_k(\alpha) y_k(\alpha + \tau)] \qquad (5.63)$$

The expected value of both sides of Equation (5.62) then gives

$$E\left[S_{xy}(f, T, k)\right]$$

$$= \int_{-T}^{0}\left[\frac{1}{T}\int_{-\tau}^{T}R_{xy}(\tau)\,d\alpha\right]e^{-j2\pi f\tau}\,d\tau + \int_{0}^{T}\left[\frac{1}{T}\int_{0}^{T-\tau}R_{xy}(\tau)\,d\alpha\right]e^{-j2\pi f\tau}\,d\tau$$

$$= \int_{-T}^{T}\left(1 - \frac{|\tau|}{T}\right)R_{xy}(\tau)e^{-j2\pi f\tau}\,d\tau \tag{5.64}$$

In the limit as T tends to infinity, it follows that

$$\lim_{T\to\infty} E\left[S_{xy}(f, T, k)\right] = \int_{-\infty}^{\infty} R_{xy}(\tau)e^{-j2\pi f\tau}\,d\tau \tag{5.65}$$

This is the desired result of Equation (5.59) since the right-hand side of Equation (5.65) is $S_{xy}(f)$, as previously defined in Equation (5.27).

Observe that when $S(f)$ is replaced by its corresponding $G(f)$, the following formulas are obtained:

$$G_{xy}(\overset{\cdot}{f}) = 2\lim_{T\to\infty}\frac{1}{T}E\left[X_k^*(f, T)Y_k(f, T)\right] \tag{5.66}$$

$$G_{xx}(f) = 2\lim_{T\to\infty}\frac{1}{T}E\left[|X_k(f, T)|^2\right]$$

$$G_{yy}(f) = 2\lim_{T\to\infty}\frac{1}{T}E\left[|Y_k(f, T)|^2\right] \tag{5.67}$$

These formulas are estimated in fast finite Fourier transform digital computer procedures discussed later in Chapter 11. In practice, the record length T will always be finite since the limiting operation $T \to \infty$ can never be performed. The expected value operation $E[\ ]$ will also always be taken over only a finite number of ensemble elements since an infinite ensemble is impossible to obtain with real data.

5.2.3   Spectra via Filtering-Squaring-Averaging

The third analog based physical way to compute autospectral density functions consists of the following operations, as pictured in Figure 5.5.

1. Frequency filtering of the signal $x(t)$ by a narrow bandpass filter of bandwidth Δf and center frequency f_0 to obtain $x(f_0, \Delta f, t)$

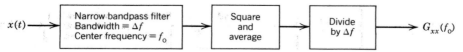

Figure 5.5 Autospectrum measurement by analog operations.

2. Squaring of the instantaneous value of the filtered signal
3. Averaging of the squared instantaneous value over the record length T to obtain a mean square value estimate of the filtered signal
4. Division by the filter bandwidth Δf to obtain an estimate of the rate of change of mean square value with frequency at the center frequency f_0

The autospectral density function estimate is then

$$\hat{G}_{xx}(f) = \frac{1}{(\Delta f)T} \int_0^T x^2(f_0, \Delta f, t)\, dt \qquad (5.68)$$

Computation of the cross-spectral density function by analog methods is a direct extension of this procedure using two different signals $x(t)$ and $y(t)$ with the following operations:

1. Individual frequency filtering of the two signals $x(t)$ and $y(t)$ by narrow bandpass filters having identical bandwidths Δf and the same center frequency f_0 to obtain $x(f_0, \Delta f, t)$ and $y(f_0, \Delta f, t)$
2. Multiplying the instantaneous values of the two filtered signals with no phase shift to obtain the in-phase terms, needed for the co-spectrum
3. Multiplying the instantaneous values of the two filtered signals with $y(f_0, \Delta f, t)$ shifted 90 degrees out of phase compared with $x(f_0, \Delta f, t)$ to obtain the out-of-phase terms, needed for the quad-spectrum
4. Averaging each of the above instantaneous product values over the record length T to obtain mean product value estimates of the in-phase and out-of-phase terms
5. Division of each of the two mean product value estimates by the filter bandwidth Δf to obtain estimates of $\hat{C}_{xy}(f_0)$ and $\hat{Q}_{xy}(f_0)$

The cross-spectral density function estimate is then

$$\hat{G}_{xy}(f_0) = \hat{C}_{xy}(f_0) - j\hat{Q}_{xy}(f_0) \qquad (5.69)$$

where

$$\hat{C}_{xy}(f_0) = \frac{1}{(\Delta f)T} \int_0^T x(f_0, \Delta f, t)\, y(f_0, \Delta f, t)\, dt$$

$$\hat{Q}_{xy}(f_0) = \frac{1}{(\Delta f)T} \int_0^T x(f_0, \Delta f, t)\, y^0(f_0, \Delta f, t)\, dt \qquad (5.70)$$

The symbol $y^0(f_0, \Delta f, t)$ denotes a 90 degree phase shift from the filtered signal $y(f_0, \Delta f, t)$.

The equivalence of this third analog definition of spectral density functions with the previous two mathematical definitions in Sections 5.2.1 and 5.2.2 is not obvious and requires a proof. This will now be done for the autospectral computation. A similar proof applies for the cross-spectral computation.

Consider the autospectral formula of Equation (5.67) where

$$G_{xx}(f) = 2 \lim_{T \to \infty} \frac{1}{T} E\left[|X_k(f, T)|^2\right] \tag{5.71}$$

Although not obvious, the term

$$|X_k(f, T)|^2 = \left[\int_0^T x_k(t) \cos 2\pi f t \, dt\right]^2 + \left[\int_0^T x_k(t) \sin 2\pi f t \, dt\right]^2 \tag{5.72}$$

acts as a filter to help obtain the mean square value in any $x_k(t)$ associated with a narrow frequency band around the center frequency f. To derive this result, define for each $x_k(t)$ in $\{x_k(t)\}$

$$x_k(t, T) = \begin{cases} x_k(t) & 0 \le t \le T \\ 0 & \text{otherwise} \end{cases} \tag{5.73}$$

Then the mean square value of a particular $x_k(t)$ may be determined by

$$\psi_x^2(k) = \lim_{T \to \infty} \frac{1}{T} \int_0^T x_k^2(t) \, dt = \lim_{T \to \infty} \frac{1}{T} \int_{-\infty}^{\infty} x_k^2(t, T) \, dt \tag{5.74}$$

By Parseval's theorem, if $F(f)$ is the Fourier transform of $f(t)$, then

$$\int_{-\infty}^{\infty} f^2(t) \, dt = \int_{-\infty}^{\infty} |F(f)|^2 \, df \tag{5.75}$$

a relation that is an easy exercise to prove. Hence, since $X_k(f, T)$ is the Fourier transform for $x_k(t, T)$, namely,

$$X_k(f, T) = \int_0^T x_k(t) e^{-j2\pi f t} \, dt = \int_{-\infty}^{\infty} x_k(t, T) e^{-j2\pi f t} \, dt \tag{5.76}$$

it follows that

$$\psi_x^2(k) = \lim_{T \to \infty} \frac{1}{T} \int_{-\infty}^{\infty} |X_k(f, T)|^2 \, df = 2 \lim_{T \to \infty} \frac{1}{T} \int_{0}^{\infty} |X_k(f, T)|^2 \, df$$

$$(5.77)$$

Now, the expected value of $\psi_x^2(k)$ over all possible records $x_k(t)$ in $\{x_k(t)\}$ yields the familiar formula

$$\psi_x^2 = E\left[\psi_x^2(k)\right] = \int_{0}^{\infty} G_{xx}(f) \, df \qquad (5.78)$$

where $G_{xx}(f)$ is defined by Equation (5.71).

Next, suppose that $x_k(t)$ is passed through a narrow bandpass filter of center frequency f_o and bandwidth Δf such that the frequency response function $H(f)$ of the filter is

$$H(f) = \begin{cases} 1 & 0 \le f_o - (\Delta f/2) \le f \le f_o + (\Delta f/2) \\ 0 & \text{otherwise} \end{cases} \qquad (5.79)$$

Then, the Fourier transform of the filter output is given by $H(f)X_k(f, T)$ instead of by $X_k(f, T)$. Hence in place of Equation (5.77), the mean square value of a particular filtered $x_k(t)$ becomes

$$\psi_x^2(f_o, \Delta f, k) = 2 \lim_{T \to \infty} \frac{1}{T} \int_{0}^{\infty} |H(f)|^2 |X_k(f, T)|^2 \, df \qquad (5.80)$$

The expected value of both sides of Equation (5.80) gives

$$\psi_x^2(f_o, \Delta f) = \int_{0}^{\infty} |H(f)|^2 G_{xx}(f) \, df = \int_{f_o - (\Delta f/2)}^{f_o + (\Delta f/2)} G_{xx}(f) \, df \qquad (5.81)$$

In words, Equation (5.81) states that $G_{xx}(f)$ is *the rate of change of mean square value with frequency*. Furthermore, the term $|X_k(f, T)|^2$ must be acting as a filter on $x_k(t)$, which passes only a narrow band of frequency components in a certain range, and then squares these outputs prior to the final desired averaging operations. This is precisely the basis on which analog spectral density analyzers are designed.

5.2.4 *Coherence Functions*

A simple, direct mathematical proof will now be carried out to show that the cross-spectral density function is bounded by the *cross-spectrum inequality*

$$|G_{xy}(f)|^2 \le G_{xx}(f)G_{yy}(f) \tag{5.82}$$

This result is much more powerful than the corresponding cross-correlation inequality of Equation (5.11), which bounds $|R_{xy}(\tau)|^2$ in terms of the product $R_{xx}(0)R_{yy}(0)$ using values of $R_{xx}(\tau)$ and $R_{yy}(\tau)$ at $\tau = 0$.

For any value of f, the function $G_{xy}(f)$ can be expressed as

$$G_{xy}(f) = |G_{xy}(f)|e^{-j\theta_{xy}(f)}$$

using the magnitude factor $|G_{xy}(f)|$ and the phase factor $\theta_{xy}(f)$. It is known also that

$$G_{yx}(f) = G_{xy}^*(f) = |G_{xy}(f)|e^{j\theta_{xy}(f)}$$

Consider now the quantities $X_k(f)$ and $Y_k(f)e^{j\theta_{xy}(f)}$, where $X_k(f)$ and $Y_k(f)$ are finite Fourier transforms of records $x_k(t)$ and $y_k(t)$, respectively. For any real constants a and b, the absolute value quantity shown below will be greater than or equal to zero, namely,

$$|aX_k(f) + bY_k(f)e^{j\theta_{xy}(f)}|^2 \ge 0$$

This is the same as

$$a^2|X_k(f)|^2 + ab\left[X_k^*(f)Y_k(f)e^{j\theta_{xy}(f)} + X_k(f)Y_k^*(f)e^{-j\theta_{xy}(f)}\right]$$
$$+ b^2|Y_k(f)|^2 \ge 0$$

By taking the expectation of this equation over the index k, multiplying by $(2/T)$, and letting T increase without bound, one obtains

$$a^2G_{xx}(f) + ab\left[G_{xy}(f)e^{j\theta_{xy}(f)} + G_{yx}(f)e^{-j\theta_{xy}(f)}\right] + b^2G_{yy}(f) \ge 0$$

Use has been made here of the autospectra and cross-spectra formulas of Equations (5.66) and (5.67). From Equation (5.41),

$$G_{xy}(f)e^{j\theta_{xy}(f)} + G_{yx}(f)e^{-j\theta_{xy}(f)} = 2|G_{xy}(f)|$$

Hence

$$a^2G_{xx}(f) + 2ab|G_{xy}(f)| + b^2G_{yy}(f) \ge 0$$

It now follows exactly as in the earlier proof for the cross-correlation inequality that Equation (5.82) is true, namely,

$$|G_{xy}(f)|^2 \le G_{xx}(f)G_{yy}(f)$$

This also proves that for any f, the two-sided quantities satisfy

$$|S_{xy}(f)|^2 \le S_{xx}(f)S_{yy}(f) \tag{5.83}$$

The *coherence function* (sometimes called the coherency squared function) may now be defined by

$$\gamma_{xy}^2(f) = \frac{|G_{xy}(f)|^2}{G_{xx}(f)G_{yy}(f)} = \frac{|S_{xy}(f)|^2}{S_{xx}(f)S_{yy}(f)} \tag{5.84}$$

and satisfies for all f,

$$0 \le \gamma_{xy}^2(f) \le 1 \tag{5.85}$$

A *complex coherence function* $\gamma_{xy}(f)$ may be defined by

$$\gamma_{xy}(f) = |\gamma_{xy}(f)|e^{-j\theta_{xy}(f)} \tag{5.86}$$

where

$$|\gamma_{xy}(f)| = +\sqrt{\gamma_{xy}^2(f)} \tag{5.87}$$

and $\theta_{xy}(f)$ is the phase angle of $G_{xy}(f)$. Throughout this book, the coherence function will always stand for the real-valued squared function of Equation (5.84) with $|\gamma_{xy}(f)|$ as the positive square root of $\gamma_{xy}^2(f)$.

5.2.5 *Cross-Spectrum for Time Delay*

From Equation (5.20), the cross-correlation function for the time delay problem illustrated in Figure 5.1 is

$$R_{xy}(\tau) = \alpha R_{xx}(\tau - \tau_0) \tag{5.88}$$

Substitution into Equation (5.27) yields the two-sided cross-spectral density function

$$S_{xy}(f) = \alpha S_{xx}(f)e^{-j2\pi f\tau_0} \tag{5.89}$$

The corresponding one-sided cross-spectral density function is

$$G_{xy}(f) = \alpha G_{xx}(f)e^{-j2\pi f\tau_0} \tag{5.90}$$

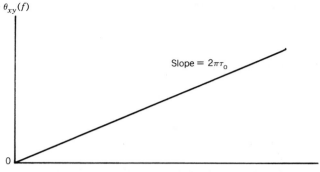

Figure 5.6   Typical phase angle plot for time-delay problem.

Hence, from Equation (5.41),

$$|G_{xy}(f)| = \alpha G_{xx}(f) \tag{5.91}$$

$$\theta_{xy}(f) = 2\pi f \tau_0 \tag{5.92}$$

Thus the time delay τ_0 appears only in the phase angle $\theta_{xy}(f)$. Measurement of $\theta_{xy}(f)$ enables one to determine the time delay by noting that $\theta_{xy}(f)$ is a linear function of f with a slope equal to $2\pi\tau_0$, as illustrated in Figure 5.6. The attenuation factor α is given at any frequency f by

$$\alpha = \left[|G_{xy}(f)| / G_{xx}(f) \right] \tag{5.93}$$

The one-sided autospectral density functions for the transmitted signal $x(t)$ and the received signal $y(t)$ in Figure 5.1 are given by $G_{xx}(f)$ and $G_{yy}(f)$, respectively, where

$$G_{yy}(f) = \alpha^2 G_{xx}(f) + G_{nn}(f) \tag{5.94}$$

This result also follows directly from Equation (5.26). The coherence function at any value of f from Equations (5.84) and (5.91) is given by

$$\gamma_{xy}^2(f) = \frac{|G_{xy}(f)|^2}{G_{xx}(f)G_{yy}(f)} = \alpha^2 \left[\frac{G_{xx}(f)}{G_{yy}(f)} \right] \tag{5.95}$$

Observe also that the two components in $G_{yy}(f)$ are

$$\alpha^2 G_{xx}(f) = \gamma_{xy}^2(f)G_{yy}(f) = \text{spectrum in } y(t) \text{ due to } x(t) \tag{5.96}$$

$$G_{nn}(f) = \left[1 - \gamma_{xy}^2(f) \right] G_{yy}(f) = \text{spectrum in } y(t) \text{ due to } n(t)$$

These results are more significant than their integrated values over all frequencies as shown in Equation (5.25) where there is no frequency discrimination.

The coherence function of Equation (5.95) using Equation (5.94) for $G_{yy}(f)$ takes the form

$$\gamma_{xy}^2(f) = \frac{\alpha^2 G_{xx}(f)}{\alpha^2 G_{xx}(f) + G_{nn}(f)}$$

where the terms on the right-hand side are nonnegative. For all values of f, it is clear that $\gamma_{xy}^2(f) \le 1$ since the denominator is larger than the numerator. This result is equivalent to

$$\gamma_{xy}^2(f) = \frac{\left[\alpha^2 G_{xx}(f) + G_{nn}(f)\right] - G_{nn}(f)}{\alpha^2 G_{xx}(f) + G_{nn}(f)} = \frac{G_{yy}(f) - G_{nn}(f)}{G_{yy}(f)}$$

$$= 1 - \left[G_{nn}(f)/G_{yy}(f)\right] \qquad (5.97)$$

Again, it is clear that $\gamma_{xy}^2(f) \le 1$ for all f since $G_{nn}(f) \le G_{yy}(f)$. When $G_{nn}(f) = 0$, then $\gamma_{xy}^2(f) = 1$. When $G_{nn}(f) = G_{yy}(f)$, then $\gamma_{xy}^2(f) = 0$.

LOCATION OF PEAK VALUE. From Equations (5.21) and (5.28), the peak value of $R_{xy}(\tau)$ is given by

$$R_{xy}(\tau)_{\text{peak}} = R_{xy}(\tau_0) = \int_{-\infty}^{\infty} S_{xy}(f) e^{j2\pi f \tau_0} \, df \qquad (5.98)$$

Let estimates of $R_{xy}(\tau_0)$ and $S_{xy}(f)$ be denoted by $\hat{R}_{xy}(\tau_0)$ and $\hat{S}_{xy}(f)$, respectively. Then

$$\hat{S}_{xy}(f) = |\hat{S}_{xy}(f)| e^{-j\hat{\theta}_{xy}(f)} \qquad (5.99)$$

Equation (5.98) becomes

$$\hat{R}_{xy}(\tau_0) = \int_{-\infty}^{\infty} |\hat{S}_{xy}(f)| e^{j[2\pi f \tau_0 - \hat{\theta}_{xy}(f)]} \, df$$

$$= \int_{-\infty}^{\infty} |\hat{S}_{xy}(f)| \cos\left[2\pi f \tau_0 - \hat{\theta}_{xy}(f)\right] df$$

since $\hat{R}_{xy}(\tau_0)$ is real valued. At the peak location τ_0,

$$\frac{\partial \hat{R}_{xy}(\tau_0)}{\partial \tau_0} = 0 = \int_{-\infty}^{\infty} (-2\pi f) |\hat{S}_{xy}(f)| \sin\left[2\pi f \tau_0 - \hat{\theta}_{xy}(f)\right] df$$

From Equation (5.92),

$$\hat{\theta}_{xy}(f) \simeq 2\pi f \tau_0$$

so that

$$\sin\left[2\pi f \tau_0 - \hat{\theta}_{xy}(f)\right] \simeq 2\pi f \tau_0 - \hat{\theta}_{xy}(f)$$

Hence one obtains the approximate formula

$$\int_{-\infty}^{\infty} (-2\pi f)|\hat{S}_{xy}(f)|\left[2\pi f \tau_0 - \hat{\theta}_{xy}(f)\right] df \simeq 0$$

This can be solved for τ_0 to yield the result

$$\tau_0 \simeq \frac{\int_{-\infty}^{\infty} (2\pi f)|\hat{S}_{xy}(f)|\hat{\theta}_{xy}(f)\, df}{\int_{-\infty}^{\infty} (2\pi f)^2|\hat{S}_{xy}(f)|\, df} \qquad (5.100)$$

The use of one-sided cross-spectral density function estimates $\hat{G}_{xy}(f)$ instead of the two-sided $\hat{S}_{xy}(f)$ gives the equivalent result

$$\tau_0 \simeq \frac{\int_{0}^{\infty} (2\pi f)|\hat{G}_{xy}(f)|\hat{\theta}_{xy}(f)\, df}{\int_{0}^{\infty} (2\pi f)^2|\hat{G}_{xy}(f)|\, df} \qquad (5.101)$$

5.2.6   Uncertainty Relation

Consider a zero mean value stationary random process $\{y(t)\}$ with an autocorrelation function $R_{yy}(\tau)$ and an associated two-sided spectral density function $S_{yy}(f)$ as defined in Equations (5.6) and (5.27), where

$$S_{yy}(f) = \int_{-\infty}^{\infty} R_{yy}(\tau)e^{-j2\pi f\tau}\, d\tau = 2\int_{0}^{\infty} R_{yy}(\tau)\cos 2\pi f\tau\, d\tau$$

$$R_{yy}(\tau) = \int_{-\infty}^{\infty} S_{yy}(f)e^{j2\pi f\tau}\, df = 2\int_{0}^{\infty} S_{yy}(f)\cos 2\pi f\tau\, df$$

$$(5.102)$$

It is known also that

$$R_{yy}(0) \geq R_{yy}(\tau) \qquad \text{for all } \tau$$

$$S_{yy}(f) \geq 0 \qquad\qquad \text{for all } f$$

$$(5.103)$$

In place of $S_{yy}(f)$, a one-sided, spectral density function $G_{yy}(f)$ can be denoted by

$$G_{yy}(f) = \begin{cases} 2S_{yy}(f) & f \geq 0 \\ 0 & f < 0 \end{cases} \tag{5.104}$$

Thus

$$G_{yy}(f) = 4\int_0^\infty R_{yy}(\tau)\cos 2\pi f\tau \, d\tau \qquad R_{yy}(\tau) = \int_0^\infty G_{yy}(f)\cos 2\pi f\tau \, df$$

$$\tag{5.105}$$

Definitions can now be given for equivalent noise spectral bandwidth and equivalent noise correlation duration. A useful uncertainty relation can be proved for the product of these two quantities.

The *noise spectral bandwidth* is defined by

$$B_n = \int_0^\infty G_{yy}(f) \, df \Big/ G_{yy}(f)\big|_{max} = R_{yy}(0) \Big/ G_{yy}(f)\big|_{max} \tag{5.106}$$

The *noise correlation duration* is defined by

$$T_n = \int_{-\infty}^\infty |R_{yy}(\tau)| \, d\tau \Big/ R_{yy}(\tau)\big|_{max} = 2\int_0^\infty |R_{yy}(\tau)| \, d\tau \Big/ R_{yy}(0) \tag{5.107}$$

From these definitions, it follows that

$$B_n T_n = 2\int_0^\infty |R_{yy}(\tau)| \, d\tau \Big/ G_{yy}(f)\big|_{max} \tag{5.108}$$

and estimation of this product leads to the *uncertainty relation*: for an arbitrary $R_{yy}(\tau)$ and the associated $G_{yy}(f)$, the product of B_n and T_n satisfies the inequality

$$B_n T_n \geq \tfrac{1}{2} \tag{5.109}$$

Hence, as B_n becomes small, T_n must become large and, conversely, as T_n becomes small, B_n must become large.

The uncertainty relation can be proved as follows. From equation (5.105), for any f,

$$G_{yy}(f) \leq 4\int_0^\infty |R_{yy}(\tau)\cos 2\pi f\tau| \, d\tau \leq 4\int_0^\infty |R_{yy}(\tau)| \, d\tau \tag{5.110}$$

Hence

$$G_{yy}(f)\big|_{max} \leq 4\int_0^\infty |R_{yy}(\tau)| \, d\tau \tag{5.111}$$

Substitution of Equation (5.111) into Equation (5.108) leads immediately to the result stated in Equation (5.109). Note that this simple proof does *not* require use of Schwartz's inequality.

Example 5.8. **Low-Pass White Noise.** For low-pass white noise, one has

$$S_{yy}(f) = a/2 \quad -B \le f \le B \quad \text{otherwise zero}$$

$$G_{yy}(f) = a \quad 0 \le f \le B \quad \quad \text{otherwise zero}$$

$$R_{yy}(\tau) = aB\{\sin(2\pi B\tau)/(2\pi B\tau)\}$$

The noise spectral bandwidth is given here by

$$B_n = R_{yy}(0)/G_{yy}(f)|_{\max} = B$$

However, the noise correlation duration becomes

$$T_n = \frac{2\int_0^\infty |R_{yy}(\tau)|\, d\tau}{R_{yy}(0)} = \frac{1}{\pi B}\int_0^\infty \left|\frac{\sin u}{u}\right| du = \infty$$

Thus $B_n T_n = \infty$, which clearly satisfies the uncertainty relation of Equation (5.109). Because of the shape of $R_{yy}(\tau)$, it would appear to be appropriate here to define T_n as the width of the main lobe that represents most of the energy, namely,

$$T_n = 1/B$$

This gives $B_n T_n = 1$, which still satisfies the uncertainty relation.

Example 5.9. **Gaussian Spectrum Noise.** For such noise, one has

$$R_{yy}(\tau) = ae^{-2\pi\sigma^2\tau^2}$$

$$S_{yy}(f) = \left(a/\sigma\sqrt{2\pi}\right)e^{-f^2/2\sigma^2} \quad -\infty < f < \infty$$

$$G_{yy}(f) = \left(a\sqrt{2}/\sigma\sqrt{\pi}\right)e^{-f^2/2\sigma^2} \quad f \ge 0 \text{ only}$$

$$\int_0^\infty |R_{yy}(\tau)|\, d\tau = \frac{a}{2\sigma\sqrt{2\pi}} = \frac{G_{yy}(0)}{4}.$$

Here, the noise spectral bandwidth is

$$B_n = \frac{\sigma\sqrt{\pi}}{\sqrt{2}} \approx 1.25\sigma$$

Now

$$G_{yy}(B_n) = G_{yy}(0)e^{-\pi/4} \simeq 0.456 G_{yy}(0)$$

The noise correlation duration is

$$T_n = \frac{1}{\sigma\sqrt{2\pi}} \simeq \frac{0.40}{\sigma}$$

Then

$$R_{yy}(T_n) = R_{yy}(0)e^{-\pi} \simeq 0.043 R_{yy}(0)$$

Note that Gaussian noise achieves the minimum uncertainty value,

$$B_n T_n = \tfrac{1}{2}$$

Example 5.10. **Exponential Autocorrelation Function Noise.** For noise with an exponential autocorrelation function, one has

$$R_{yy}(\tau) = Ae^{-a|\tau|} \qquad a > 0$$

$$S_{yy}(f) = 2Aa/\{a^2 + (2\pi f)^2\} \qquad -\infty < f < \infty$$

$$G_{yy}(f) = 4Aa/\{a^2 + (2\pi f)^2\} \qquad f \geq 0 \text{ only}$$

$$\int_0^\infty |R_{yy}(\tau)|\, d\tau = \frac{A}{a} = \frac{G_{yy}(0)}{4}$$

The noise spectral bandwidth is given by

$$B_n = \frac{a}{4}$$

Then

$$G_{yy}(B_n) = G_{yy}(0)\left[1/\{1 + (\pi/2)^2\}\right] \simeq 0.288 G_{yy}(0)$$

The noise correlation duration is

$$T_n = \frac{2}{a}$$

Thus

$$R_{yy}(T_n) = R_{yy}(0)e^{-2} \simeq 0.10 R_{yy}(0)$$

In this case also the minimum uncertainty value is achieved, that is,

$$B_n T_n = \tfrac{1}{2}$$

Other illustrations of the uncertainty relation can be seen from corresponding $R_{xx}(\tau)$ and $G_{xx}(f)$ functions in Tables 5.1 and 5.2.

5.3   ERGODIC AND GAUSSIAN RANDOM PROCESSES

The most important stationary random processes in practice are those considered to be

   a.   Ergodic with arbitrary probability structure
   b.   Gaussian whether ergodic or not

Various cases will now be examined, including linear transformations of random processes.

5.3.1   *Ergodic Random Processes*

Consider two weakly stationary random processes $\{x_k(t)\}$ and $\{y_k(t)\}$ with two arbitrary sample functions $x_k(t)$ and $y_k(t)$. These stationary random processes are said to be *weakly ergodic* if the mean values and covariance (correlation) functions, which are defined by certain *ensemble averages* in Section 5.1.1, may be calculated by performing corresponding *time averages* on the arbitrary pair of sample functions. In this way the underlying statistical structure of the weakly stationary random processes may be determined quite simply from an available sample pair without the need for collecting a considerable amount of data.

   To be more specific, the mean values of the individual sample functions $x_k(t)$ and $y_k(t)$, when computed by a time average, may be represented by

$$\mu_x(k) = \lim_{T \to \infty} \frac{1}{T} \int_0^T x_k(t)\, dt$$

$$\mu_y(k) = \lim_{T \to \infty} \frac{1}{T} \int_0^T y_k(t)\, dt$$

(5.112)

Observe that the answer is no longer a function of t, since t has been averaged out. In general, however, the answer is a function of the particular sample function chosen, denoted by the index k.

   The cross-covariance function and cross-correlation function between $x_k(t)$ and $y_k(t + \tau)$, when computed by a time average, are defined by the expression

$$C_{xy}(\tau, k) = \lim_{T \to \infty} \frac{1}{T} \int_0^T \left[x_k(t) - \mu_x(k) \right] \left[y_k(t + \tau) - \mu_y(k) \right] dt$$

$$= \lim_{T \to \infty} \frac{1}{T} \int_0^T x_k(t) y_k(t + \tau) \, dt - \mu_x(k) \mu_y(k)$$

$$= R_{xy}(\tau, k) - \mu_x(k) \mu_y(k) \tag{5.113}$$

The autocovariance functions and autocorrelation functions are defined by

$$C_{xx}(\tau, k) = \lim_{T \to \infty} \frac{1}{T} \int_0^T \left[x_k(t) - \mu_x(k) \right] \left[x_k(t + \tau) - \mu_x(k) \right] dt$$

$$= R_{xx}(\tau, k) - \mu_x^2(k)$$

$$\tag{5.114}$$

$$C_{yy}(\tau, k) = \lim_{T \to \infty} \frac{1}{T} \int_0^T \left[y_k(t) - \mu_y(k) \right] \left[y_k(t + \tau) - \mu_y(k) \right] dt$$

$$= R_{yy}(\tau, k) - \mu_y^2(k)$$

These quantities should now be compared with the previously defined ensemble mean values μ_x, μ_y, and ensemble covariance functions $C_{xx}(\tau)$, $C_{yy}(\tau), C_{xy}(\tau)$ for stationary random processes developed in Section 5.1.1. If it turns out that, independent of k,

$$\mu_x(k) = \mu_x$$

$$\mu_y(k) = \mu_y$$

$$C_{xx}(\tau, k) = C_{xx}(\tau) \tag{5.115}$$

$$C_{yy}(\tau, k) = C_{yy}(\tau)$$

$$C_{xy}(\tau, k) = C_{xy}(\tau)$$

then the random processes $\{x_k(t)\}$ and $\{y_k(t)\}$ are said to be *weakly ergodic*. If all ensemble averaged statistical properties of $\{x_k(t)\}$ and $\{y_k(t)\}$, not just the means and covariances, are deducible from corresponding time averages, then the random processes are said to be *strongly ergodic*. Thus strong ergodicity implies weak ergodicity, but not conversely. No distinction between these concepts exists for Gaussian random processes.

For an arbitrary random process to be ergodic, it must first be stationary. Each sample function must then be representative of all the others in the sense described above so that it does not matter which particular sample function is used in the time-averaging calculations. With arbitrary ergodic processes $x(t)$ and $y(t)$, in place of Equation (5.6), their autocorrelation and cross-correlation functions are defined by

$$R_{xx}(\tau) = \lim_{T \to \infty} \frac{1}{T} \int_0^T x(t)x(t + \tau)\, dt$$

$$R_{yy}(\tau) = \lim_{T \to \infty} \frac{1}{T} \int_0^T y(t)y(t + \tau)\, dt \qquad (5.116)$$

$$R_{xy}(\tau) = \lim_{T \to \infty} \frac{1}{T} \int_0^T x(t)y(t + \tau)\, dt$$

***Example 5.11.* Nonergodic Stationary Random Process.**   A simple example of a nonergodic stationary random process follows. Consider a hypothetical random process $\{x_k(t)\}$ composed of sinusoidal sample functions such that

$$\{x_k(t)\} = \{X_k \sin[2\pi ft + \theta_k]\}$$

Let the amplitude X_k and the phase angle θ_k be random variables that take on a different set of values for each sample function, as illustrated in Figure 5.7. If

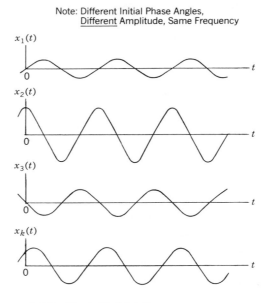

$$\{x(t)\} = \{X_k \sin(2\pi ft + \theta_k)\}$$

Figure 5.7   Illustration of nonergodic stationary sine wave process.

θ_k is uniformly distributed, the properties of the process computed over the ensemble at specific times will be independent of time; hence the process is stationary. The properties computed by time averaging over individual sample functions are not always the same, however. For example, the autocovariance (or autocorrelation) function for each sample function is given here by

$$C_{xx}(\tau, k) = \frac{X_k^2}{2} \sin 2\pi f \tau$$

Since X_k is a function of k, $C_{xx}(\tau, k) \neq C_x(\tau)$. Hence the random process is nonergodic.

Instead of having random amplitudes $\{X_k\}$, suppose each amplitude is the same X independent of k. Now the random process consists of sinusoidal sample functions such that

$$\{x_k(t)\} = \{X \sin(2\pi f t + \theta_k)\}$$

For this case, the random process is ergodic, with each record statistically equivalent to every other record for any time averaging results, as illustrated in Figure 5.8. This concludes Example 5.11.

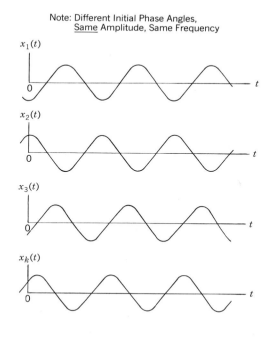

Figure 5.8   Illustration of ergodic sine wave process.

5.3.2 *Sufficient Conditions for Ergodicity*

There are two important classes of random processes that one can state in advance will be ergodic. The first ergodic class is the class of stationary Gaussian random processes whose autospectral density functions are absolutely continuous, that is, no delta functions appear in the autospectra corresponding to infinite mean square densities at discrete frequencies. The second ergodic class (a special case of the first class) is the class of stationary Gaussian Markov processes; a Markov process is one whose relationship to the past does not extend beyond the immediately preceding observation. The autocorrelation function of a stationary Gaussian Markov process may be shown to be of a simple exponential form [Reference 5.3].

Sufficient conditions for a random process to be ergodic are as follows.

I. A sufficient condition for an arbitrary random process to be weakly ergodic is that it be weakly stationary, and that the time averages $\mu_x(k)$ and $C_{xx}(\tau, k)$ be the same for all sample functions k.

The proof of this result is as follows. By definition,

$$\mu_x(k) = \lim_{T \to \infty} \frac{1}{T} \int_0^T x_k(t) \, dt$$

By hypothesis, $\mu_x(k)$ is independent of k. Hence the expected value over k is the same as an individual estimate, namely,

$$E[\mu_x(k)] = \mu_x(k)$$

Also, as will be proved later, expected values commute with linear operations. Hence

$$E[\mu_x(k)] = \lim_{T \to \infty} \frac{1}{T} \int_0^T E[x_k(t)] \, dt$$

$$= \lim_{T \to \infty} \frac{1}{T} \int_0^T \mu_x \, dt = \mu_x$$

The assumption of weak stationarity is used in setting $E[x_k(t)] = \mu_x$. Thus

$$\mu_x(k) = \mu_x$$

Similarly,

$$C_{xx}(\tau, k) = C_{xx}(\tau)$$

since the hypothesis that $C_x(\tau, k)$ is independent of k yields

$$E[C_{xx}(\tau, k)] = C_{xx}(\tau, k)$$

whereas the stationary hypothesis yields

$$E[C_{xx}(\tau, k)] = C_{xx}(\tau)$$

This completes the proof.

II. A sufficient condition for a Gaussian random process to be ergodic is that it be weakly stationary, and the autocovariance function $C_{xx}(\tau)$ has the following four integrable properties.

$$\int_{-\infty}^{\infty} |C_{xx}(\tau)|\, d\tau < \infty \qquad \int_{-\infty}^{\infty} C_{xx}^2(\tau)\, d\tau < \infty$$

$$\int_{-\infty}^{\infty} |\tau C_{xx}(\tau)|\, d\tau < \infty \qquad \int_{-\infty}^{\infty} |\tau| C_{xx}^2(\tau)\, d\tau < \infty \tag{5.117}$$

The four conditions of Equation (5.117) can be replaced by the single requirement that

$$\frac{1}{T} \int_{-T}^{T} |C_{xx}(\tau)|\, d\tau \to 0 \qquad \text{as } T \to \infty \tag{5.118}$$

The proof of this result is in Sections 8.2.1 and 8.2.2, where it is shown that mean value and autocorrelation function estimates produced by time averages are independent of the particular sample record when Equation (5.117) is satisfied. Result II then follows from Result I. In practice, these conditions are frequently satisfied, justifying the assumption of ergodicity.

5.3.3 *Gaussian Random Processes*

The formal definition of a Gaussian random process is as follows. A random process $\{x_k(t)\}$ is said to be a Gaussian random process if, for every set of fixed times $\{t_n\}$, the random variables $x_k(t_n)$ follow a multidimensional normal distribution as defined by Equation (3.62). Gaussian random processes are quite prevalent in physical problems, and often may be mathematically predicted by the multidimensional central limit theorem. Also, it can be shown that if a Gaussian process undergoes a linear transformation, then the output will still be a Gaussian process. This property is quite important in various theoretical and practical applications of random process theory.

Consider a time history $x(t)$, which is a sample function from an ergodic Gaussian random process with zero mean value. Note that the index k is no longer needed, since the properties of any one sample function will be representative of all other sample functions. From the ergodic property, the behavior of $x(t)$ over a long period of time will exhibit the same statistical characteristics as corresponding ensemble averages at various fixed times. As a consequence, it follows that the probability density function associated with the instantaneous values of $x(t)$ that will occur over a long time interval is given by the Gaussian probability density function with zero mean value, as follows:

$$p(x) = \left(\sigma_x\sqrt{2\pi}\right)^{-1}e^{-x^2/2\sigma_x^2} \qquad (5.119)$$

The variance σ_x^2 when $x(t)$ has zero mean is determined by

$$\sigma_x^2 = E\left[x^2(t)\right] = \int_{-\infty}^{\infty} x^2 p(x)\,dx \qquad \text{independent of } t$$

$$\approx \frac{1}{T}\int_0^T x^2(t)\,dt \qquad \text{for large } T$$

$$= \int_{-\infty}^{\infty} S_{xx}(f)\,df = 2\int_0^{\infty} S_{xx}(f)\,df = \int_0^{\infty} G_{xx}(f)\,df \qquad (5.120)$$

Thus the Gaussian probability density function $p(x)$ is completely characterized through knowledge of $S_{xx}(f)$ or $G_{xx}(f)$ since they alone determine σ_x. This important result places knowledge of $S_{xx}(f)$ or $G_{xx}(f)$ at the forefront of much work in the analysis of random records. It should be noted that no restriction is placed on the shape of the autospectral density function or its associated autocorrelation function.

If the mean value of $x(t)$ is not zero, then the underlying probability density function is given by the general Gaussian formula

$$p(x) = \left(\sigma_x\sqrt{2\pi}\right)^{-1}e^{-(x-\mu_x)^2/2\sigma_x^2} \qquad (5.121)$$

where the mean value

$$\mu_x = E\left[x(t)\right] = \int_{-\infty}^{\infty} xp(x)\,dx \qquad \text{independent of } t$$

$$\approx \frac{1}{T}\int_0^T x(t)\,dt \qquad \text{for large } T \qquad (5.122)$$

and the variance

$$\sigma_x^2 = E\left[(x(t) - \mu_x)^2\right] = E\left[x^2(t)\right] - \mu_x^2 \qquad (5.123)$$

Assume $\{x(t)\}$ is a stationary Gaussian random process where the index k is omitted for simplicity in notation. Consider the two random variables $x_1 = x(t)$ and $x_2 = x(t + \tau)$ at an arbitrary pair of fixed times t and $t + \tau$. Assume that x_1 and x_2 follow a two-dimensional (joint) Gaussian distribution with *zero means* and *equal variances* σ_x^2. By definition, then

$$\sigma_x^2 = E[x^2(t)] = E[x^2(t + \tau)] = \int_{-\infty}^{\infty} x^2 p(x)\, dx \tag{5.124}$$

$$R_{xx}(\tau) = E[x(t)x(t + \tau)] = \rho_{xx}(\tau)\sigma_x^2 = \int_{-\infty}^{\infty}\!\!\int x_1 x_2 p(x_1, x_2)\, dx_1\, dx_2 \tag{5.125}$$

The quantity $\rho_{xx}(\tau)$ is the correlation coefficient function of Equation (5.16) for $C_{x_1 x_2}(\tau) = R_{xx}(\tau)$ and $\sigma_{x_1} = \sigma_{x_2} = \sigma_x$, namely,

$$\rho_{xx}(\tau) = \frac{R_{xx}(\tau)}{\sigma_x^2} \tag{5.126}$$

Letting $\rho = \rho_{xx}(\tau)$ and $\mu = 0$, the joint Gaussian probability density function is given by

$$p(x_1, x_2) = \left(2\pi\sigma_x^2\sqrt{1 - \rho^2}\right)^{-1}\exp\left[\frac{-1}{2\sigma_x^2(1 - \rho^2)}(x_1^2 - 2\rho x_1 x_2 + x_2^2)\right] \tag{5.127}$$

All properties developed in Chapter 3 apply to Gaussian random processes at any set of fixed times.

Consider four random variables x_1, x_2, x_3, x_4, with zero mean values, which follow a four-dimensional Gaussian distribution. From Equation (3.72)

$$E[x_1 x_2 x_3 x_4] = E[x_1 x_2]E[x_3 x_4] + E[x_1 x_3]E[x_2 x_4] + E[x_1 x_4]E[x_2 x_3] \tag{5.128}$$

In particular, let $x_1 = x(u)$, $x_2 = y(u + \tau)$, $x_3 = x(v)$, $x_4 = y(v + \tau)$, and let $R_{xy}(\tau)$ be the stationary cross-correlation function given by

$$R_{xy}(\tau) = E[x(t)y(t + \tau)] \tag{5.129}$$

It now follows from Equation (5.128) that

$$E[x(u)y(u + \tau)x(v)y(v + \tau)] = R_{xy}^2(\tau) + R_{xx}(v - u)R_{yy}(v - u)$$
$$+ R_{xy}(v - u + \tau)R_{yx}(v - u - \tau) \tag{5.130}$$

This result will be used later in Chapter 8 in Equation (8.94).

5.3.4 *Linear Transformations of Random Processes*

The dynamic behavior of representative linear physical systems has been discussed in practical terms in Chapter 2. It will be helpful at this time to consider very briefly the mathematical properties of linear transformations of random processes. This background will be assumed in Chapters 6 and 7 to develop important input/output relationships for linear systems subjected to random inputs.

Consider an arbitrary random process $\{x_k(t)\}$. An operator A that transforms a sample function $x_k(t)$ into another function $y_k(v)$ may be written as

$$y_k(v) = A[x_k(t)] \tag{5.131}$$

where A denotes a functional operation on the term inside the brackets [ ]. The argument v may or may not be the same as t. For example, if the operation in question is differentiation, then $v = t$ and $y_k(t)$ will be a sample function from the derivative random process $\{\dot{x}_k(t)\}$, assuming of course that the derivative exists. A different example is when the operation in question is integration between definite limits. Here, $v \neq t$, and $y_k(v)$ will be a random variable over the index k, determined by $x_k(t)$ and the definite limits. The operator A can take many different forms. In the following, the sample space index k will be omitted for simplicity in notation.

The operator A is said to be *linear* if, for any set of admissible values $x_1, x_2, \ldots, x_N$ and constants $a_1, a_2, \ldots, a_N$, if follows that

$$A\left[\sum_{i=1}^{N} a_i x_i\right] = \sum_{i=1}^{N} a_i A[x_i] \tag{5.132}$$

In words, the operation is both additive and homogeneous. The admissible values here may be different sample functions at the same t, or they may be different values from the same sample function at different t.

The operator A is said to be *time-invariant* if any shift t_0 of the input $x(t)$ to $x(t + t_0)$ causes a similar shift of the output $y(t)$ to $y(t + t_0)$. In equation form,

$$y(t + t_0) = A[x(t + t_0)] \qquad \text{for any } t_0 \tag{5.133}$$

Unless stated otherwise, all linear systems will henceforth be assumed to be time-invariant. Such systems are the constant-parameter linear systems of Chapter 2.

For any linear operation where all quantities exist, the procedure of taking expected values of random variables is commutative with the linear operation. That is, for fixed t and v,

$$E[y(v)] = E[A[x(t)]] = A[E[x(t)]] \tag{5.134}$$

This result is proved easily, as follows. Assume $x(t)$ takes on N discrete values $x_1, x_2, \ldots, x_N$, and $y(v)$ takes on N corresponding discrete values $y_1, y_2, \ldots, y_N$, where $y_i = A[x_i]$. Then

$$E[y(v)] = \frac{1}{N} \sum_{i=1}^{N} y_i = \frac{1}{N} \sum_{i=1}^{N} A[x_i] \quad \text{and} \quad E[x(t)] = \frac{1}{N} \sum_{i=1}^{N} x_i$$

Now, since A is a linear operator,

$$\frac{1}{N} \sum_{i=1}^{N} A[x_i] = A\left[\frac{1}{N} \sum_{i=1}^{N} x_i\right] = A[E[x(t)]]$$

Hence

$$E[y(v)] = A[E[x(t)]] \tag{5.135}$$

The continuous case follows by letting N approach infinity and using an appropriate convergence criterion, such as Equation (5.138) to follow. This completes the proof.

A basic result whose proof evolves directly from definitions is as follows. *If $x(t)$ is from a weakly (strongly) stationary random process and if the operator A is linear and time-invariant, then $y(v) = A[x(t)]$ will form a weakly (strongly) stationary random process.* Another result of special significance, proved in Reference 5.4, is as follows. *If $x(t)$ follows a Gaussian distribution and the operator A is linear, then $y(v) = A[x(t)]$ will also follow a Gaussian distribution.*

An integral transformation of any particular sample function $x(t)$ from an arbitrary random process $\{x(t)\}$ is defined by

$$I = \int_a^b x(t)\phi(t)\, dt \tag{5.136}$$

where $\phi(t)$ is an arbitrary given function for which the integral exists. For any given $\phi(t)$ and limits (a, b), the quantity I is a random variable that depends on the particular sample function $x(t)$. In order to investigate statistical properties of the random variable I, it is customary to break up the integration interval (a, b) into subintervals Δt, and consider the approximation linear sum

$$I_N = \sum_{i=1}^{N} x(i\Delta t)\phi(i\Delta t)\, \Delta t \tag{5.137}$$

Convergence of I_N to I may now be defined by different criteria. The sequence $\{I_N\}$ is said to converge to I

1. In the *mean square sense* if

$$\lim_{N \to \infty} E\left[|I_N - I|^2\right] = 0 \tag{5.138}$$

2.   In *probability* if for every $\varepsilon > 0$

$$\lim_{N \to \infty} \text{Prob}\big[|I_N - I| \geq \varepsilon\big] = 0$$

From the Chebyshev inequality of Equation (3.22), it follows directly that convergence in the mean square sense implies convergence in probability. In practice, most integral expressions involving random variables exist by assuming convergence in the mean square sense.

5.4   DERIVATIVE RANDOM PROCESSES

The derivative of any particular sample function $x(t)$ from an arbitrary random process $\{x(t)\}$ is defined by

$$\dot{x}(t) = \frac{dx(t)}{dt} = \lim_{\varepsilon \to 0}\left[\frac{x(t + \varepsilon) - x(t)}{\varepsilon}\right] \qquad (5.139)$$

Existence of this limit may occur in different ways. The derivative $\dot{x}(t)$ is said to exist

1.   In the *usual sense* if the limit exists for all functions $x(t)$ in $\{x(t)\}$.
2.   In the *mean square sense* if

$$\lim_{\varepsilon \to 0} E\left[\left|\frac{x(t + \varepsilon) - x(t)}{\varepsilon} - \dot{x}(t)\right|^2\right] = 0 \qquad (5.140)$$

For a stationary random process, a necessary and sufficient condition for $\dot{x}(t)$ to exist in the mean square sense is that its autocorrelation function $R_{xx}(\tau)$ should have derivatives of order up to 2, that is, $R'_{xx}(\tau)$ and $R''_{xx}(\tau)$ must exist [Reference 5.4].

5.4.1   *Correlation Functions*

Consider the following derivative functions, which are assumed to be well defined:

$$R'_{xx}(\tau) = \frac{dR_{xx}(\tau)}{d\tau} \qquad R''_{xx}(\tau) = \frac{d^2 R_{xx}(\tau)}{d\tau^2}$$

$$\dot{x}(t) = \frac{dx(t)}{dt} \qquad\qquad \ddot{x}(t) = \frac{d^2 x(t)}{dt^2} \qquad (5.141)$$

By definition, for stationary random data,

$$R_{xx}(\tau) = E[x(t)x(t + \tau)] = E[x(t - \tau)x(t)]$$

$$R_{x\dot{x}}(\tau) = E[x(t)\dot{x}(t + \tau)] = E[x(t - \tau)\dot{x}(t)] \qquad (5.142)$$

$$R_{\dot{x}\dot{x}}(\tau) = E[\dot{x}(t)\dot{x}(t + \tau)] = E[\dot{x}(t - \tau)\dot{x}(t)]$$

Now

$$R'_{xx}(\tau) = \frac{d}{d\tau} E[x(t)x(t + \tau)] = E[x(t)\dot{x}(t + \tau)] = R_{x\dot{x}}(\tau) \quad (5.143)$$

Also

$$R'_{xx}(\tau) = \frac{d}{d\tau} E[x(t - \tau)x(t)] = -E[\dot{x}(t - \tau)x(t)] = -R_{\dot{x}x}(\tau)$$

Hence

$$R'_{xx}(0) = R_{x\dot{x}}(0) = -R_{\dot{x}x}(0) = 0 \qquad (5.144)$$

since $R'_{xx}(0)$ equals the positive and negative of the same quantity. The corresponding $R_{xx}(0)$ is a maximum value of $R_{xx}(\tau)$. This proves that for stationary random data

$$E[x(t)\dot{x}(t)] = 0 \qquad (5.145)$$

In words, at any t, Equation (5.145) indicates that the derivative $\{\dot{x}(t)\}$ for stationary random data $\{x(t)\}$ is equally likely to be positive or negative. Equation (5.143) states that the derivative $R'_{xx}(\tau)$ of the autocorrelation function $R_{xx}(\tau)$ with respect to τ is the same as the cross-correlation function between $\{x(t)\}$ and $\{\dot{x}(t)\}$. A maximum value for the autocorrelation function $R_{xx}(\tau)$ corresponds to a zero crossing for its derivative $R'_{xx}(\tau)$, which becomes a zero crossing for the cross-correlation function between $\{x(t)\}$ and $\{\dot{x}(t)\}$. This crossing of zero by $R'_{xx}(\tau)$ will be with *negative* slope, that is

$$R'_{xx}(0 -) > 0 \qquad \text{and} \qquad R'_{xx}(0 +) < 0 \qquad (5.146)$$

as can be seen from the picture in Figure 5.9. In practice, determining the location where zero crossings will occur is usually easier than determining the location of maximum values.

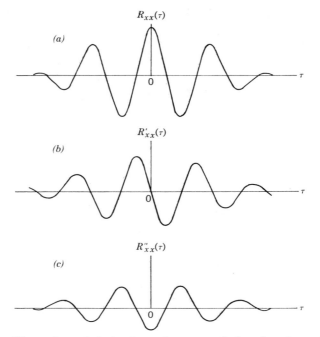

Figure 5.9 Illustration of derivatives of autocorrelation functions. (*a*) Original function. (*b*) First derivative. (*c*) Second derivative.

It will now be shown that $R'_{xx}(\tau)$ is an odd function of τ corresponding to $R_{xx}(\tau)$ being an even function of τ. By definition

$$R_{xx}(-\tau) = E[x(t)x(t - \tau)] = E[x(t + \tau)x(t)] \qquad (5.147)$$

Hence

$$R'_{xx}(-\tau) = \frac{d}{d\tau}E[x(t + \tau)x(t)] = E[\dot{x}(t + \tau)x(t)] = R_{\dot{x}x}(\tau) \qquad (5.148)$$

But, from Equation (5.143), $R_{\dot{x}x}(\tau) = -R'_{xx}(\tau)$. Hence Equation (5.148) becomes

$$R'_{xx}(-\tau) = -R'_{xx}(\tau) \qquad (5.149)$$

This proves that $R'_{xx}(\tau)$ is an odd function of τ.

The second derivative gives

$$R''_{xx}(\tau) = \frac{d}{d\tau}R'_{xx}(\tau) = \frac{d}{d\tau}R_{x\dot{x}}(\tau) = \frac{d}{d\tau}E[x(t-\tau)\dot{x}(t)]$$

$$= -E[\dot{x}(t-\tau)\dot{x}(t)] = -R_{\dot{x}\dot{x}}(\tau) \qquad (5.150)$$

Also

$$R''_{xx}(\tau) = \frac{d}{d\tau}R'_{xx}(\tau) = \frac{d}{d\tau}R_{x\dot{x}}(\tau) = \frac{d}{d\tau}E[x(t)\dot{x}(t+\tau)]$$

$$= E[x(t)\ddot{x}(t+\tau) = R_{x\ddot{x}}(\tau) \qquad (5.151)$$

One can also verify directly that $R''_{xx}(\tau)$ is an even function of τ, namely,

$$R''_{xx}(-\tau) = R''_{xx}(\tau) \qquad (5.152)$$

At $\tau = 0$, one obtains

$$E[\dot{x}^2(t)] = R_{\dot{x}\dot{x}}(0) = -R_{x\ddot{x}}(0) = -R''_{xx}(0) \qquad (5.153)$$

As shown earlier,

$$R_{x\dot{x}}(\tau) = \frac{d}{d\tau}R_{xx}(\tau) = R'_{xx}(\tau) \qquad (5.154)$$

Typical plots for $R_{xx}(\tau)$, $R'_{xx}(\tau)$, and $R''_{xx}(\tau)$ are drawn in Figure 5.9, based on a sine wave process where

$$R_{xx}(\tau) = X\cos 2\pi f_0\tau$$

$$R'_{xx}(\tau) = -X(2\pi f_0)\sin 2\pi f_0\tau \qquad (5.155)$$

$$R''_{xx}(\tau) = -X(2\pi f_0)^2\cos 2\pi f_0\tau$$

The results given above can be extended to higher-order derivatives. For example,

$$R_{\dot{x}\ddot{x}}(\tau) = \frac{d}{d\tau}R_{\dot{x}\dot{x}}(\tau) = -R'''_{xx}(\tau) \qquad (5.156)$$

$$R_{\ddot{x}\ddot{x}}(\tau) = -\frac{d}{d\tau}R_{\dot{x}\ddot{x}}(\tau) = R''''_{xx}(\tau) \qquad (5.157)$$

At $\tau = 0$, one obtains

$$E[\ddot{x}^2(t)] = R_{\ddot{x}\ddot{x}}(0) = R''''_{xx}(0) \qquad (5.158)$$

Thus knowledge of $R_{xx}(\tau)$ and its successive derivatives can enable one to state properties for autocorrelation and cross-correlation functions between $\{x(t)\}$ and its successive derivatives $\{\dot{x}(t)\}$, $\{\ddot{x}(t)\}$, and so on.

5.4.2 *Spectral Density Functions*

It is easy to derive corresponding properties for autospectral and cross-spectral density functions between $\{x(t)\}$ and its successive derivatives $\{\dot{x}(t)\}$ and $\{\ddot{x}(t)\}$. Let

$$X(f) = \mathscr{F}[x(t)] = \text{Fourier transform}[x(t)] \tag{5.159}$$

Then

$$\mathscr{F}[\dot{x}(t)] = (j2\pi f)X(f) \tag{5.160}$$

$$\mathscr{F}[\ddot{x}(t)] = -(2\pi f)^2 X(f) \tag{5.161}$$

From Equations (5.66) and (5.67), it follows directly that

$$G_{x\dot{x}}(f) = j(2\pi f)G_{xx}(f) \tag{5.162}$$

$$G_{\dot{x}\dot{x}}(f) = (2\pi f)^2 G_{xx}(f) \tag{5.163}$$

$$G_{\dot{x}\ddot{x}}(f) = j(2\pi f)^3 G_{xx}(f) \tag{5.164}$$

$$G_{\ddot{x}\ddot{x}}(f) = (2\pi f)^4 G_{xx}(f) \tag{5.165}$$

and so on. These formulas are the same with one-sided G's replaced by the corresponding two-sided S's.

These results can also be derived from the Wiener–Khinchine relations of Equation (5.28). Start with the basic relation

$$R_{xx}(\tau) = \int_{-\infty}^{\infty} S_{xx}(f)e^{j2\pi f\tau}\,df \tag{5.166}$$

Then successive derivatives will be

$$R'_{xx}(\tau) = j\int_{-\infty}^{\infty} (2\pi f)S_{xx}(f)e^{j2\pi f\tau}\,df \tag{5.167}$$

$$R''_{xx}(\tau) = -\int_{-\infty}^{\infty} (2\pi f)^2 S_{xx}(f)e^{j2\pi f\tau}\,df \tag{5.168}$$

$$R'''_{xx}(\tau) = -j\int_{-\infty}^{\infty} (2\pi f)^3 S_{xx}(f)e^{j2\pi f\tau}\,df \tag{5.169}$$

$$R''''_{xx}(\tau) = \int_{-\infty}^{\infty} (2\pi f)^4 S_{xx}(f)e^{j2\pi f\tau}\,df \tag{5.170}$$

The Wiener–Khinchine relations, together with previous formulas in Section 5.4.1, show that these four derivative expressions are the same as

$$R'_{xx}(\tau) = R_{x\dot{x}}(\tau) = \int_{-\infty}^{\infty} S_{x\dot{x}}(f) e^{j2\pi f\tau} \, df \tag{5.171}$$

$$R''_{xx}(\tau) = -R_{\dot{x}\dot{x}}(\tau) = -\int_{-\infty}^{\infty} S_{\dot{x}\dot{x}}(f) e^{j2\pi f\tau} \, df \tag{5.172}$$

$$R'''_{xx}(\tau) = -R_{\dot{x}\ddot{x}}(\tau) = -\int_{-\infty}^{\infty} S_{\dot{x}\ddot{x}}(f) e^{j2\pi f\tau} \, df \tag{5.173}$$

$$R''''_{xx}(\tau) = R_{\ddot{x}\ddot{x}}(\tau) = \int_{-\infty}^{\infty} S_{\ddot{x}\ddot{x}}(f) e^{j2\pi f\tau} \, df \tag{5.174}$$

Corresponding terms in the last eight formulas yield Equations (5.162)–(5.165).

5.4.3 Expected Number of Zero Crossings

Consider a stationary random noise record $x(t)$ that has the time derivative $\dot{x}(t)$. Let $p(\alpha, \beta)$ represent the joint probability density function of $x(t)$ and $\dot{x}(t)$ at $x(t) = \alpha$ and $\dot{x}(t) = \beta$. By definition, for all t,

$$p(\alpha, \beta) \, \Delta\alpha \, \Delta\beta \simeq \text{Prob}[\alpha < x(t) \leq +\alpha + \Delta\alpha \text{ and } \beta < \dot{x}(t) \leq \beta + \Delta\beta] \tag{5.175}$$

In words, $p(\alpha, \beta) \, \Delta\alpha \, \Delta\beta$ estimates the probability over all time that $x(t)$ lies in the interval $[\alpha, \alpha + \Delta\alpha]$ when its derivative $\dot{x}(t)$ is between β and $\beta + \Delta\beta$. For unit total time, this represents the amount of time that $x(t)$ spends in the interval $[\alpha, \alpha + \Delta\beta]$ with a given derivative value between β and $\beta + \Delta\beta$. When $\Delta\beta$ is negligible compared to β, this means that the derivative value is essentially β.

To find the expected number of crossings of $x(t)$ through the interval $[\alpha, \alpha + \Delta\alpha]$, the amount of time that $x(t)$ is inside this interval should be divided by the time required to cross the interval. If t_β is the crossing time for a particular derivative value β, then

$$t_\beta = \frac{\Delta\alpha}{|\beta|} \tag{5.176}$$

where the absolute value of β is used since crossing time must be a positive quantity. Hence, the expected number of passages per unit time of $x(t)$ through the interval $[\alpha, \alpha + \Delta\alpha]$ for a given value of $\dot{x}(t) = \beta$ is

$$\frac{p(\alpha, \beta) \, \Delta\alpha \, \Delta\beta}{t_\beta} \simeq |\beta| p(\alpha, \beta) \, \Delta\beta \tag{5.177}$$

In the limit as $\Delta\beta \rightarrow 0$, the total expected number of passages per unit time of $x(t)$ through the line $x(t) = \alpha$ for all possible values of β is found by

$$\overline{N}_\alpha = \int_{-\infty}^{\infty} |\beta| p(\alpha, \beta) \, d\beta \qquad (5.178)$$

This represents the expected number of crossings of α per unit time with *both* positive and negative slopes. Assuming that $x(t)$ passes the value α half of the time with positive slope and half of the time with negative slope, then $\frac{1}{2}\overline{N}_\alpha$ gives the expected number of times per unit time that $x(t)$ exceeds the value α, that is, crosses the line $x(t) = \alpha$ with positive slope.

The expected number of zeros of $x(t)$ per unit time is found by the number of crossings of the line $x(t) = 0$ with both positive and negative slopes. This is given by $\overline{N}_\alpha$ when $\alpha = 0$, namely,

$$\overline{N}_0 = \int_{-\infty}^{\infty} |\beta| p(0, \beta) \, d\beta \qquad (5.179)$$

The value of $\overline{N}_0$ can be interpreted as twice the "apparent frequency" of the noise record. For example, if the record were a pure sine wave of frequency f_0 Hz, then N_0 would be $2f_0$ zeros per sec (e.g., a 60 Hz sine wave has 120 zeros/sec). For noise, the situation is more complicated but, still, knowledge of $\overline{N}_0$ together with other quantities helps to characterize a particular noise.

For an arbitrary record $x(t)$ and its derivative $\dot{x}(t)$ from a zero mean value stationary random process, it follows from Equations (5.145) and (5.150) that

$$\sigma_x^2 = E[x^2(t)] = R_{xx}(0) \qquad (5.180)$$

$$\sigma_{\dot{x}}^2 = E[\dot{x}^2(t)] = R_{\dot{x}\dot{x}}(0) = -R_{xx}''(0) \qquad (5.181)$$

$$\sigma_{x\dot{x}} = E[x(t)\dot{x}(t)] = 0 \qquad (5.182)$$

From Equations (5.166) and (5.168), it also follows that

$$\sigma_x^2 = \int_{-\infty}^{\infty} S_{xx}(f) \, df = \int_0^{\infty} G_{xx}(f) \, df \qquad (5.183)$$

$$\sigma_{\dot{x}}^2 = \int_{-\infty}^{\infty} (2\pi f)^2 S_{xx}(f) \, df = \int_0^{\infty} (2\pi f)^2 G_{xx}(f) \, df \qquad (5.184)$$

Assume now that $x(t)$ and $\dot{x}(t)$ have zero mean values and form a two-dimensional normal distribution with the above variances and zero covariance. Then

$$p(\alpha, \beta) = p(\alpha) p(\beta) \qquad (5.185)$$

with

$$p(\alpha) = \frac{1}{\sigma_x\sqrt{2\pi}}\exp(-\alpha^2/2\sigma_x^2) \qquad (5.186)$$

$$p(\beta) = \frac{1}{\sigma_{\dot{x}}\sqrt{2\pi}}\exp(-\beta^2/2\sigma_{\dot{x}}^2) \qquad (5.187)$$

Substitution of Equation (5.185) into Equation (5.178) shows that

$$\overline{N}_\alpha = \frac{\exp(-\alpha^2/2\sigma_x^2)}{2\pi\sigma_x\sigma_{\dot{x}}}\int_{-\infty}^{\infty}|\beta|\exp(-\beta^2/2\sigma_{\dot{x}}^2)\,d\beta$$

$$= \frac{1}{\pi}\left(\frac{\sigma_{\dot{x}}}{\sigma_x}\right)\exp(-\alpha^2/2\sigma_x^2) \qquad (5.188)$$

In particular, for $\alpha = 0$, one obtains

$$\overline{N}_0 = \frac{1}{\pi}\left(\frac{\sigma_{\dot{x}}}{\sigma_x}\right) = \frac{1}{\pi}\left[\frac{-R_{xx}''(0)}{R_{xx}(0)}\right]^{1/2}$$

$$= \frac{1}{\pi}\left[\frac{\int_0^\infty (2\pi f)^2 G_{xx}(f)\,df}{\int_0^\infty G_{xx}(f)\,df}\right]^{1/2} \qquad (5.189)$$

In terms of $\overline{N}_0$, one can express

$$\overline{N}_\alpha = \overline{N}_0\exp(-\alpha^2/2\sigma_x^2) \qquad (5.190)$$

These results from Reference 5.1 were derived originally by Rice [Reference 5.5], who used a different method of proof.

Example 5.12. **Zero Crossings of Low-Pass White Noise.**   To illustrate the above formulas, consider the case of low-pass white noise from 0 to B Hz, where

$$G_{xx}(f) = K \qquad 0 \le f \le B \quad \text{otherwise zero}$$

Now

$$\sigma_x^2 = \int_0^B K\,df = KB$$

$$\sigma_{\dot{x}}^2 = \int_0^B (2\pi f)^2 K\,df = \frac{4\pi^2}{3}KB^3$$

From Equation (5.189),

$$\overline{N}_0 = \frac{2}{\sqrt{3}} B \approx 2(0.58B)$$

A pure sine wave of frequency B Hz would have $\overline{N}_0 = 2B$ zeros/sec. Here, the conclusion is that for low-pass white noise cutting off at B Hz, the apparent frequency of the noise is about 0.58 of the cutoff frequency.

PROBLEMS

5.1   Given data with an autocorrelation function defined by $R_{xx}(\tau) = 25e^{-4|\tau|}\cos 4\pi\tau + 16$, determine
   (a)   the mean value and variance.
   (b)   the associated one-sided autospectral density function.

5.2   Which of the following properties are always true of autocorrelation functions of stationary data?
   (a)   must be an even function.
   (b)   must be nonnegative.
   (c)   must be bounded by its value at zero.
   (d)   can determine the mean value of the data.
   (e)   can determine the variance of the data.

5.3   Which of the properties in Problem 5.2 are always true of cross-correlation functions of stationary data?

5.4   Given data with a two-sided autospectral density function defined by

$$S_{xx}(f) = \begin{cases} 16\delta(f) + 20\left(1 - \dfrac{|f|}{10}\right) & f \le 10 \\ 0 & |f| > 10 \end{cases}$$

   determine for the data
   (a)   the mean value and variance.
   (b)   the associated autocorrelation function.

5.5   Which of the properties in Problem 5.2 are always true of the two-sided
   (a)   autospectral density functions?
   (b)   cross-spectral density functions?

5.6   Which of the following properties are always true for two ergodic random processes?
   (a) $R_{xy}(\infty) = \mu_x \mu_y$.
   (b) $R_{xy}(0) = 0$ when $\mu_x = 0$ or $\mu_y = 0$.
   (c) $R_{xy}(\tau) = 0$ when $R_{xx}(\tau) = 0$ or $R_{yy}(\tau) = 0$.

(d) $|R_{xy}(\tau)|^2 \le R_{xx}(\tau)R_{yy}(\tau)$.
(e) $G_{xy}(0) = 0$ when $\mu_x = 0$ or $\mu_y = 0$.
(f) $|G_{xy}(f)|^2 \le G_{xx}(0)G_{yy}(0)$.
(g) $G_{xy}(f) = 0$ when $G_{xx}(f) = 0$ or $G_{yy}(f) = 0$.

5.7   Assume data have a one-sided cross-spectral density function given by $G_{xy}(f) = (6/f^2) + j(8/f^3)$. Determine the two-sided cross-spectral density function $S_{xy}(f)$ for all frequencies in terms of
(a)   real and imaginary functions.
(b)   gain and phase functions.

5.8   If a record $x(t)$ from an ergodic random process has an autocorrelation function given by $R_{xx}(\tau) = e^{-a|\tau|}\cos 2\pi f_0\tau$ with $a > 0$, determine the autocorrelation function for the first time derivative of the data, $\dot{x}(t)$.

5.9   Assume a record $x(t)$ from an ergodic random process has a one-sided autospectral spectral density function given by

$$G_{xx}(f) = \frac{1}{25 + f^2} \qquad 0 \le f \le 25 \qquad \text{otherwise zero}$$

Determine the average number of zero crossings per second in the record $x(t)$.

5.10   Prove the result in Equation (5.30) from Equation (5.59).

REFERENCES

5.1   Bendat, J. S., *Principles and Applications of Random Noise Theory*, Wiley, New York, 1958. Reprinted by Krieger, Melbourne, Florida, 1977.

5.2   Bendat, J. S., and Piersol, A. G., *Engineering Applications of Correlation and Spectral Analysis*, Wiley-Interscience, New York, 1980.

5.3   Doob, J. L., *Stochastic Processes*, Wiley, New York, 1953.

5.4   Papoulis, A., *Probability, Random Variables, and Stochastic Processes*, McGraw-Hill, New York, 1965.

5.5   Rice, S. O., "Mathematical Analysis of Random Noise," in *Selected Papers on Noise and Stochastic Processes*, (N. Wax, ed.) Dover, New York, 1954.

CHAPTER 6

SINGLE-INPUT / OUTPUT
RELATIONSHIPS

This chapter is concerned with the theory and applications of input/output relationships for single-input problems. It is assumed that records are from stationary random processes with zero mean values and that systems are constant-parameter linear systems. Single-input/single-output models and single-input/multiple-output models are discussed. Ordinary coherence functions are defined for these models. Multiple-input problems are covered in Chapter 7.

6.1   SINGLE-INPUT/SINGLE-OUTPUT MODELS

Consider a constant-parameter linear system with a weighting function $h(\tau)$ and frequency response function $H(f)$ as defined and discussed in Chapter 2. Assume the system is subjected to a well-defined single input $x(t)$ from a stationary random process $\{x(t)\}$ and produces a well-defined output $y(t)$, as illustrated in Figure 6.1. This output will belong to a stationary random process $\{y(t)\}$.

6.1.1   *Correlation and Spectral Relations*

Under ideal conditions, the output $y(t)$ for the system in Figure 6.1 is given by the convolution integral

$$y(t) = \int_0^\infty h(\tau)x(t-\tau)\,d\tau \tag{6.1}$$

164

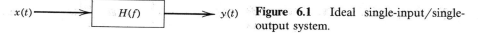

$x(t)$ ⟶ $H(f)$ ⟶ $y(t)$ **Figure 6.1** Ideal single-input/single-output system.

where $h(\tau) = 0$ for $\tau < 0$ when the system is physically realizable. The product $y(t)y(t + \tau)$ is given by

$$y(t)y(t + \tau) = \int_0^\infty h(\alpha)h(\beta)x(t - \beta)x(t + \tau - \alpha)\,d\alpha\,d\beta \quad (6.2)$$

Taking expected values of both sides yields the *input/output autocorrelation relation*

$$R_{yy}(\tau) = \iint_0^\infty h(\alpha)h(\beta)R_{xx}(\tau + \beta - \alpha)\,d\alpha\,d\beta \quad (6.3)$$

Similarly, the product $x(t)y(t + \tau)$ is given by

$$x(t)y(t + \tau) = \int_0^\infty h(\alpha)x(t)x(t + \tau - \alpha)\,d\alpha \quad (6.4)$$

Here, expected values of both sides yields the *input/output cross-correlation relation*

$$R_{xy}(\tau) = \int_0^\infty h(\alpha)R_{xx}(\tau - \alpha)\,d\alpha \quad (6.5)$$

Note that Equation (6.5) is a convolution integral of the same form as Equation (6.1).

Direct Fourier transforms of Equations (6.3) and (6.5) after various algebraic steps yield two-sided spectral density functions $S_{xx}(f)$, $S_{yy}(f)$, and $S_{xy}(f)$, which satisfy the important formulas

$$S_{yy}(f) = |H(f)|^2 S_{xx}(f) \quad (6.6)$$

$$S_{xy}(f) = H(f)S_{xx}(f) \quad (6.7)$$

Here f may be either positive or negative. Note that Equation (6.6) is a real-valued relation containing only the gain factor $|H(f)|$ of the system. Equation (6.7) is a complex-valued relation, which can be broken down into a pair of equations to give both the gain factor $|H(f)|$ and the phase factor $\phi(f)$ of the system. Equation (6.6) is called the *input/output autospectrum relation*, while Equation (6.7) is called the *input/output cross-spectrum relation*. These results apply only to ideal situations where no extraneous noise exists at input or output points, and the systems have no time-varying or nonlinear characteristics. Interpretation of these spectral relations in the frequency domain is much easier than their corresponding correlation relations in the time domain.

In terms of physically measurable one-sided spectral density functions $G_{xx}(f)$, $G_{yy}(f)$, and $G_{xy}(f)$, where $G(f) = 2S(f)$ for $f \geq 0$, otherwise zero, Equations (6.6) and (6.7) become

$$G_{yy}(f) = |H(f)|^2 G_{xx}(f) \tag{6.8}$$

$$G_{xy}(f) = H(f)G_{xx}(f) \tag{6.9}$$

Let

$$G_{xy}(f) = |G_{xy}(f)|e^{-j\theta_{xy}(f)} \tag{6.10}$$

$$H(f) = |H(f)|e^{-j\phi(f)} \tag{6.11}$$

Then Equation (6.9) is equivalent to the pair of relations

$$|G_{xy}(f)| = |H(f)|G_{xx}(f) \tag{6.12}$$

$$\theta_{xy}(f) = \phi(f) \tag{6.13}$$

These results provide the basis for many engineering applications of spectral density functions. See Reference 6.1 for physical illustrations. Figure 6.2 shows how an input spectrum $G_{xx}(f)$ is modified in passing through a linear system described by $H(f)$.

From Equation (6.8), the output mean square value is given by

$$\psi_y^2 = \int_0^\infty G_{yy}(f)\, df = \int_0^\infty |H(f)|^2 G_{xx}(f)\, df \tag{6.14}$$

Equation (6.8) also permits the determination of $G_{xx}(f)$ from a knowledge of $G_{yy}(f)$ and $|H(f)|$, or the determination of $|H(f)|$ from a knowledge of $G_{xx}(f)$ and $G_{yy}(f)$. Equation (6.8) does not yield the compete frequency response function $H(f)$ of the system, however, since it contains no phase information. The complete frequency response function in both gain and phase can only be obtained from Equations (6.9) to (6.13) when both $G_{xy}(f)$ and $G_{xx}(f)$ are known.

An alternative direct transform is available to derive Equations (6.8) and (6.9) without first computing the correlation expressions of Equations (6.3) and (6.5). For any pair of long but finite records of length T, Equation (6.1) is equivalent to

$$Y(f) = H(f)X(f) \tag{6.15}$$

where $X(f)$ and $Y(f)$ are finite Fourier transforms of $x(t)$ and $y(t)$, respec-

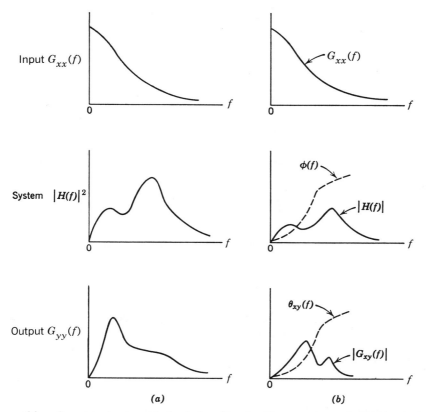

Figure 6.2 Input/output spectral relationships for linear systems. (a) Autospectra. (b) Cross-spectra.

tively. It follows that

$$Y^*(f) = H^*(f)X^*(f)$$

$$|Y(f)|^2 = |H(f)|^2|X(f)|^2$$

$$X^*(f)Y(f) = H(f)|X(f)|^2$$

Taking the expectation of the last two equations over different independent records, multiplying by $(2/T)$, and letting T increase without bound now proves from Equations (5.66) and (5.67) that

$$G_{yy}(f) = |H(f)|^2 G_{xx}(f) \tag{6.16}$$

$$G_{xy}(f) = H(f)G_{xx}(f) \tag{6.17}$$

Note the simplicity of this direct derivation. This method will be used in Section 6.1.4 and in Chapter 7.

Complex conjugation of Equation (6.17) yields the result

$$G_{xy}^*(f) = G_{yx}(f) = H^*(f)G_{xx}(f) \qquad (6.18)$$

where

$$G_{yx}(f) = |G_{xy}(f)|e^{j\theta_{xy}(f)} \qquad (6.19)$$

$$H^*(f) = |H(f)|e^{j\phi(f)} \qquad (6.20)$$

Thus, to determine the phase factor of the system, one can use the formula

$$\frac{G_{xy}(f)}{G_{yx}(f)} = \frac{H(f)}{H^*(f)} = e^{-j2\phi(f)} \qquad (6.21)$$

To determine the complete frequency response function of the system, Equations (6.16) and (6.18) show that

$$G_{yy}(f) = H(f)[H^*(f)G_{xx}(f)] = H(f)G_{yx}(f) \qquad (6.22)$$

Hence, for the ideal single-input/single-output model, one can determine $H(f)$ using Equation (6.17) to yield

$$H(f) = \frac{G_{xy}(f)}{G_{xx}(f)} \qquad (6.23)$$

while Equation (6.22) gives

$$H(f) = \frac{G_{yy}(f)}{G_{yx}(f)} \qquad (6.24)$$

Thus

$$\frac{G_{xy}(f)}{G_{xx}(f)} = \frac{G_{yy}(f)}{G_{yx}(f)} \qquad (6.25)$$

which is equivalent to

$$|G_{xy}(f)|^2 = G_{xx}(f)G_{yy}(f) \qquad (6.26)$$

For transient data, discussed in Chapter 12, "energy" spectral density functions are used instead of "power"-type spectral density functions defined in Chapters 5 and 6. These are related by $\mathscr{G}_{xy}(f) = TG_{xy}(f)$, where $\mathscr{G}_{xy}(f)$ represents the "energy" cross-spectral density function. The transients $x(t)$ and $y(t)$ are assumed to exist only in the range $0 \leq t \leq T$. Input/output formulas derived in this chapter and Chapter 7 apply to such transient data by merely replacing "power" spectral density functions by corresponding "energy" spectral density functions.

***Example 6.1.* Response Properties of Low-Pass Filter to White Noise.** Assume white noise is applied to the input of a low-pass RC filter with a time constant $K = RC$. Determine the output autospectral density function, the output mean square value, and the output autocorrelation function.

The frequency response function of the low-pass filter is

$$H(f) = (1 + j2\pi Kf)^{-1} = |H(f)|e^{-j\phi(f)}$$

corresponding to a weighting function of

$$h(\tau) = \begin{cases} \dfrac{1}{K}e^{-\tau/K} & \tau \geq 0 \\ 0 & \tau < 0 \end{cases}$$

Here

$$|H(f)| = \left[1 + (2\pi Kf)^2\right]^{-1/2}$$

$$\phi(f) = \tan^{-1}(2\pi Kf)$$

If $G_{xx}(f)$ is white noise where $G_{xx}(f) = A$, a constant for all $f \geq 0$, then

$$G_{yy}(f) = |H(f)|^2 G_{xx}(f) = \frac{A}{1 + (2\pi Kf)^2}$$

$$\psi_y^2 = \int_0^\infty G_{yy}(f)\, df = \int_0^\infty \frac{A}{1 + (2\pi Kf)^2}\, df = \frac{A}{4K}$$

$$R_{yy}(\tau) = \int_0^\infty G_{yy}(f)\cos 2\pi f\tau\, df = \frac{A}{4K}e^{-|\tau|/K}$$

***Example 6.2.* Response Properties of Low-Pass Filter to Sine Wave.** Assume a sine wave with an autospectral density function

$$G_{xx}(f) = (X^2/2)\delta(f - f_0)$$

is applied to the low-pass RC filter in Example 6.1. Determine the output autospectral density function, the output mean square value, and the output autocorrelation function.

For this problem,

$$G_{yy}(f) = |H(f)|^2 G_{xx}(f) = \frac{(X^2/2)\delta(f - f_0)}{1 + (2\pi Kf)^2}$$

$$\psi_y^2 = \int_0^\infty G_{yy}(f)\, df = \frac{X^2/2}{1 + (2\pi Kf)^2}$$

$$R_{yy}(\tau) = \int_0^\infty G_{yy}(f)\cos 2\pi f\tau\, df = \frac{X^2/2}{1 + (2\pi Kf_0)^2}\cos 2\pi f_0\tau$$

Example 6.3. **Force-Input / Displacement-Output System.**   Determine the output autospectral density function, the output autocorrelation function, and the output mean square value when the input is white noise, for the force-input/displacement-output system in Figure 2.2. These results apply also to other analogous systems, as discussed in Chapter 2.

Assume $G_{xx}(f) = G$, a constant. Then, from Equation (2.24a) or Table 2.1, when the force is expressed in displacement units, that is $x(t) = F(t)/k$, the output autospectral density function becomes

$$G_{yy}(f) = |H(f)|^2_{f-d}G = \frac{G}{\left[1 - (f/f_n)^2\right]^2 + (2\zeta f/f_n)^2}   \qquad 0 \le f < \infty$$

The corresponding output autocorrelation function is given by

$$R_{yy}(\tau) = \frac{G\pi f_n e^{-2\pi f_n \zeta |\tau|}}{4\zeta}\cos\!\left(2\pi f\sqrt{1 - \zeta^2}\,|\tau|\right)$$

$$+ \frac{\zeta}{\sqrt{1 - \zeta^2}}\sin\!\left(2\pi f_n\sqrt{1 - \zeta^2}\,|\tau|\right)$$

The output mean square value is

$$\psi_y^2 = \int_0^\infty G_{yy}(f)\, df = R_{yy}(0) = \frac{G\pi f_n}{4\zeta}$$

Thus, for a noise input, the output mean square value ψ_y^2 is inversely proportional to ζ.

For a sine wave input, it will now be shown that the maximum value of ψ_y^2 is inversely proportional to ζ^2. Consider

$$F(t) = kx(t) = kX\cos 2\pi f_0 t   \qquad f_0 = (1/T)$$

passing through a force-input/displacement-output system specified by the $H(f)$ in Figure 2.2. Here, the output $y(t)$ becomes

$$y(t) = kX|H(f_0)|\cos[2\pi f_0 t - \phi(f_0)]$$

with

$$\psi_y^2 = \frac{1}{T}\int_0^T y^2(t)\, dt = k^2(X^2/2)|H(f_0)|^2$$

From Equation (2.26), for small ζ,

$$\max|H(f_0)| = |H(f_r)| \approx \frac{1}{2k\zeta}$$

Hence

$$\max \psi_y^2 \approx \frac{X^2}{8\zeta^2}$$

proving the stated result.

***Example 6.4.* Displacement-Input / Displacement-Output System.** Determine the output autospectral density function, the output autocorrelation function, and the output mean square value when the input is white noise, for the displacement-input/displacement-output system in Figure 2.4. These results apply also to other analogous systems, as discussed in Chapter 2.

Assume $G_x(f) = G$, a constant. Then, from Equation (2.38a) or Table 2.1, the output autospectral density function becomes

$$G_{yy}(f) = |H(f)|_{d-d}^2 G = \frac{G\left[1 + (2\zeta f/f_n)^2\right]}{\left[1 - (f/f_n)^2\right]^2 + (2\zeta f/f_n)^2}$$

The corresponding output autocorrelation function is given by

$$R_{yy}(\tau) = \frac{G\pi f_n(1 + 4\zeta^2)}{4\zeta} e^{-2f_n\zeta|\tau|}\left[\cos\left(2\pi f_n\sqrt{1-\zeta^2}\,|\tau|\right)\right.$$

$$\left. + \frac{\zeta(1 - 4\zeta^2)}{\sqrt{1-\zeta^2}\,(1 + 4\zeta^2)}\sin\left(2\pi f_n\sqrt{1-\zeta^2}\,|\tau|\right)\right]$$

The output mean square value is

$$\psi_y^2 = \int_0^\infty G_{yy}(f)\, df = R_{yy}(0) = \frac{G\pi f_n(1 + 4\zeta^2)}{4\zeta}$$

The importance of exponential-cosine and exponential-sine autocorrelation functions for many physical problems is apparent from the last two examples. In cases where $\zeta \ll 1$, results can be approximated using only exponential-cosine functions.

6.1.2   *Ordinary Coherence Functions*

Assuming $G_{xx}(f)$ and $G_{yy}(f)$ are both different from zero and do not contain delta functions, the coherence function between the input $x(t)$ and the output $y(t)$ is a real-valued quantity defined by

$$\gamma_{xy}^2(f) = \frac{|G_{xy}(f)|^2}{G_{xx}(f)G_{yy}(f)} = \frac{|S_{xy}(f)|^2}{S_{xx}(f)S_{yy}(f)} \tag{6.27}$$

where the G's are the one-sided spectra and the S's are the two-sided theoretical spectra defined previously. From Equation (5.85), it follows that the coherence function satisfies for all f

$$0 \le \gamma_{xy}^2(f) \le 1 \tag{6.28}$$

To eliminate delta functions at the origin, mean values different from zero should be removed from the data before applying these last two results. Note that the coherence function is analogous to the square of the correlation function coefficient $\rho_{xy}^2(\tau)$ defined by Equation (5.16).

For a constant-parameter linear system, Equations (6.16) and (6.17) apply and may be substituted into Equations (6.27) to obtain

$$\gamma_{xy}^2(f) = \frac{|H(f)|^2 G_{xx}^2(f)}{G_{xx}(f)|H(f)|^2 G_{xx}(f)} = 1 \tag{6.29}$$

Hence for the ideal case of a constant-parameter linear system with a single clearly defined input and output, the coherence function will be unity. If $x(t)$ and $y(t)$ are completely unrelated, the coherence function will be zero. If the coherence function is greater than zero but less than unity, one or more of three possible physical situations exist.

    a. Extraneous noise is present in the measurements.
    b. The system relating $x(t)$ and $y(t)$ is not linear.
    c. $y(t)$ is an output due to an input $x(t)$ as well as to other inputs.

For linear systems, the coherence function $\gamma_{xy}^2(f)$ can be interpreted as the fractional portion of the mean square value at the output $y(t)$ that is contributed by the input $x(t)$ at frequency f. Conversely, the quantity $[1 - \gamma_{xy}^2(f)]$ is a measure of the mean square value of $y(t)$ not accounted for by $x(t)$ at frequency f.

Output $y(t)$ = Vertical C.G. acceleration

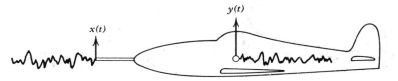

$y(t)$

$x(t)$

Input $x(t)$ = Vertical gust velocity

Figure 6.3   Airplane flying through atmospheric turbulence.

Example 6.5. **Physical Illustration of Coherence Measurement.** Consider an airplane flying through a patch of atmospheric turbulence, as illustrated in Figure 6.3. Let the input $x(t)$ be vertical gust velocity in feet/second as measured with a probe extending forward of the airplane, and the output $y(t)$ be vertical acceleration in g's measured with an accelerometer at the center of gravity of the airplane. The resulting coherence function and autospectra for actual data of this type are presented in Figures 6.4 and 6.5. In this problem, the spectral data were computed over a frequency range from 0.1 to 4.0 Hz with a resolution bandwidth of 0.05 Hz and a record length of about 10 min.

From Figure 6.4, it is seen that the input gust velocity and output airplane acceleration display a relatively strong coherence of 0.8 to 0.9 over the frequency range from about 0.3 to about 2.0 Hz. Below and above this range, however, the coherence function diminishes. At the lower frequencies, the vertical acceleration of the airplane is increasingly due to maneuver loads induced through the control system by the pilot, rather than due to atmospheric turbulence loads. Hence, the loss of coherence at the lower frequencies probably reflects contributions to the output $y(t)$ from inputs other than the measured input $x(t)$. At the higher frequencies, the low pass filtering characteristics of the airplane response plus the decaying nature of the input autospectrum cause the output autospectrum to fall off sharply, as indicated in Figure 6.5. On the other hand, the noise floor for the data acquisition and recording equipment generally does not fall off with increasing frequency. Hence, the diminishing coherence at the higher frequencies probably results from the contributions of extraneous measurement noise. This concludes Example 6.5.

For any two arbitrary records $x(t)$ and $y(t)$, one can always compute the ordinary coherence function from $G_{xx}(f)$, $G_{yy}(f)$, and $G_{xy}(f)$. The value of this coherence function indicates how much of one record is linearly related to the other record. It does not necessarily indicate a cause-and-effect relationship between the two records. The issue of causality is addressed in Section 6.2.1.

For applications of coherence functions to problems of estimating linear frequency response functions, the coherence function may be considered to be the ratio of two different measures of the square of the system gain factor.

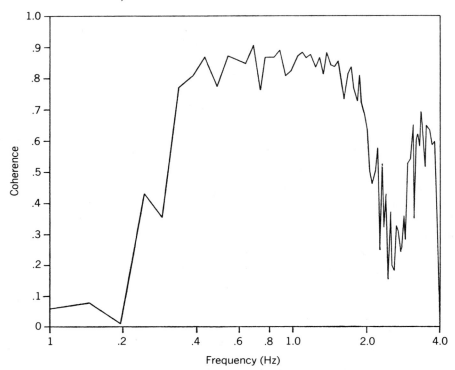

Figure 6.4 Coherence function between gust velocity and response acceleration. These data resulted from studies funded by the NASA Langley Research Center, Hampton, Virginia, under Contract NAS 1-8538.

From Equation (6.16), one measure is given by

$$|H(f)|_1^2 = \frac{G_{yy}(f)}{G_{xx}(f)} \tag{6.30}$$

From Equation (6.17), the second measure is given by

$$|H(f)|_2^2 = \frac{|G_{xy}(f)|^2}{G_{xx}^2(f)} \tag{6.31}$$

Now, their ratio gives the coherence function

$$\frac{|H(f)|_2^2}{|H(f)|_1^2} = \frac{|G_{xy}(f)|^2}{G_{xx}(f)G_{yy}(f)} = \gamma_{xy}^2(f) \tag{6.32}$$

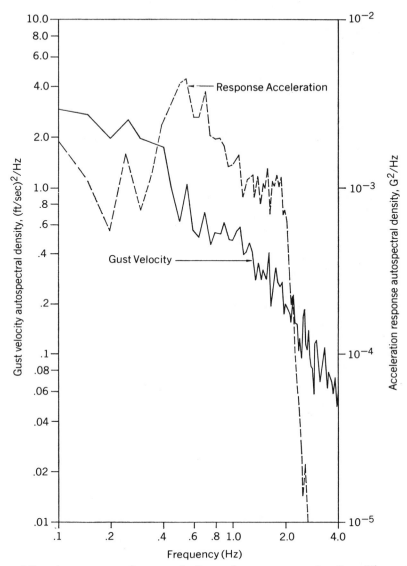

Figure 6.5 Autospectra of gust velocity and response acceleration. These data resulted from studies funded by the NASA Langley Research Center, Hampton, Virginia, under Contract NAS 1-8538.

175

In practice, measured values of Equation (6.32) will be between zero and unity. The gain factor estimate of Equation (6.30), based on autospectra calculations of input and output, will be a biased estimate for all cases except when the coherence function equals unity. The gain factor estimate of Equation (6.31), however, based on the input autospectrum and the cross-spectrum between input and output, will be a biased estimate for cases where extraneous noise is present at the input, but will be an unbiased estimate for cases where extraneous noise is present at the output only. In particular, Equation (6.31) provides an unbiased estimate of the frequency response function gain factors in multiple-input problems when the inputs are uncorrelated. These matters are discussed further in Chapter 9, where it is shown how the accuracy of frequency response function estimates increases as the coherence function approaches unity.

Coherence functions are preserved under linear transformations. To be specific, suppose one desires the coherence function between $x(t)$ and $y(t)$ where these two quantities cannot be measured easily. Assume, however, that one can measure two other quantities $x_1(t)$ and $y_1(t)$, where, from physical considerations, it can be stated that $x_1(t)$ is perfectly linearly related to $x(t)$ and $y_1(t)$ is perfectly linearly related to $y(t)$. Then the coherence function between $x_1(t)$ and $y_1(t)$ will give the desired coherence function between $x(t)$ and $y(t)$. The proof is as follows.

Perfect linear relationships mean that there exist hypothetical frequency response functions $A(f)$ and $B(f)$, which do not have to be computed, such that

$$X_1(f) = A(f)X(f) \qquad Y_1(f) = B(f)Y(f)$$

Then, at any value of f,

$$G_{x_1 x_1} = |A|^2 G_{xx} \qquad G_{y_1 y_1} = |B|^2 G_{yy} \qquad G_{x_1 x_1} = A^* B G_{xy}$$

Hence,

$$\gamma_{x_1 y_1}^2 = \frac{|G_{x_1 y_1}|^2}{G_{x_1 x_1} G_{y_1 y_1}} = \frac{|A^* B|^2 |G_{xy}|^2}{|A|^2 G_{xx} |B|^2 G_y} = \frac{|G_{xy}|^2}{G_{xx} G_{yy}} = \gamma_{xy}^2$$

This result is important for many applications discussed in Reference 6.1.

6.1.3 Models with Extraneous Noise

Consider models where extraneous noise is measured at the input and output points to a linear system $H(f)$. Let the true signals be $u(t)$ and $v(t)$ and the extraneous noise be $m(t)$ and $n(t)$, respectively, as shown in Figure 6 6. Assume that only $u(t)$ passes through the system to produce the true output

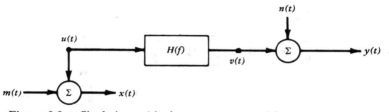

Figure 6.6 Single-input/single-output system with extraneous noise.

$v(t)$, but that the measured input and output records are

$$x(t) = u(t) + m(t)$$
$$y(t) = v(t) + n(t)$$
$$(6.33)$$

For arbitrary correlations between the signal and noise terms, autospectral and cross-spectral density functions for $x(t)$ and $y(t)$ will be

$$G_{xx}(f) = G_{uu}(f) + G_{nn}(f) + G_{um}(f) + G_{mu}(f)$$

$$G_{yy}(f) = G_{vv}(f) + G_{nn}(f) + G_{vn}(f) + G_{nv}(f) \qquad (6.34)$$

$$G_{xy}(f) = G_{uv}(f) + G_{un}(f) + G_{mv}(f) + G_{mn}(f)$$

where

$$G_{vv}(f) = |H(f)|^2 G_{uu}(f) \qquad (6.35)$$

$$G_{uv}(f) = H(f) G_{uu}(f) \qquad (6.36)$$

Various cases occur, depending on the correlation properties of $m(t)$ and $n(t)$ to each other and to the signals. Three cases of interest are

Case 1. No input noise; uncorrelated output noise
Case 2. No output noise; uncorrelated input noise
Case 3. Both noises present; uncorrelated with each other and with the signals

CASE 1

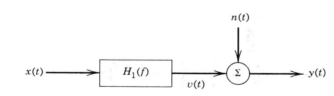

CASE 1. *No Input Noise; Uncorrelated Output Noise*

$$x(t) = u(t)$$

$$y(t) = v(t) + n(t) \qquad G_{vn}(f) = 0$$

$$G_{xx}(f) = G_{uu}(f)$$

$$G_{yy}(f) = G_{vv}(f) + G_{nn}(f)$$

$$G_{xy}(f) = G_{uv}(f) = H_1(f)G_{xx}(f)$$

$$H_1(f) = \frac{G_{xy}(f)}{G_{xx}(f)} \tag{6.37}$$

$$G_{vv}(f) = |H_1(f)|^2 G_{uu}(f) = \frac{|G_{xy}(f)|^2}{G_{xx}(f)} \tag{6.38}$$

Note that $G_{vv}(f)$ can be calculated from $x(t)$ and $y(t)$ even though $v(t)$ cannot be measured. Also, $G_{nn}(f)$ can be calculated without measuring $n(t)$. For applications in practice, this is by far the most important case because one can often define inputs and minimize input noise. However, one will have no control over the output noise which is due to nonlinear operations or the contributions from other unmeasured inputs.

For Case 1, the ordinary coherence function is

$$\gamma_{xy}^2(f) = \frac{|G_{xy}(f)|^2}{G_{xx}(f)G_{yy}(f)} = \frac{|G_{uv}(f)|^2}{G_{uu}(f)[G_{vv}(f) + G_{nn}(f)]}$$

$$= \frac{1}{1 + G_{nn}(f)/G_{vv}(f)} \tag{6.39}$$

since

$$\gamma_{uv}^2(f) = \frac{|G_{uv}(f)|^2}{G_{uu}(f)G_{vv}(f)} = 1$$

Note that $\gamma_{xy}^2(f) < 1$ when $G_{nn}(f) > 0$ with

$$G_{vv}(f) = \gamma_{xy}^2(f)G_{yy}(f) \tag{6.40}$$

This product of $\gamma_{xy}^2(f)$ with $G_{yy}(f)$ is called the *coherent output spectrum*.

Note also that the *noise output spectrum* is

$$G_{nn}(f) = \left[1 - \gamma_{xy}^2(f)\right] G_{yy}(f) \qquad (6.41)$$

Thus, $\gamma_{xy}^2(f)$ can be interpreted here as the portion of $G_{yy}(f)$ that is due to $x(t)$ at frequency f, while $[1 - \gamma_{xy}^2(f)]$ is a measure of the portion of $G_{yy}(f)$ not due to $x(t)$ at frequency f. Here, the ordinary coherence function decomposes the measured *output* spectrum into its uncorrelated components due to the input signal and extraneous noise. Equation (6.40) is the basis for solving many source identification problems using ordinary coherence functions, as illustrated in Reference 6.1.

CASE 2. *No Output Noise; Uncorrelated Input Noise*

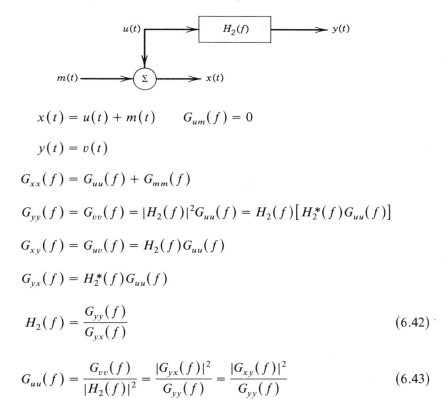

$$x(t) = u(t) + m(t) \qquad G_{um}(f) = 0$$

$$y(t) = v(t)$$

$$G_{xx}(f) = G_{uu}(f) + G_{mm}(f)$$

$$G_{yy}(f) = G_{vv}(f) = |H_2(f)|^2 G_{uu}(f) = H_2(f)\left[H_2^*(f)G_{uu}(f)\right]$$

$$G_{xy}(f) = G_{uv}(f) = H_2(f)G_{uu}(f)$$

$$G_{yx}(f) = H_2^*(f)G_{uu}(f)$$

$$H_2(f) = \frac{G_{yy}(f)}{G_{yx}(f)} \qquad (6.42)$$

$$G_{uu}(f) = \frac{G_{vv}(f)}{|H_2(f)|^2} = \frac{|G_{yx}(f)|^2}{G_{yy}(f)} = \frac{|G_{xy}(f)|^2}{G_{yy}(f)} \qquad (6.43)$$

Here $G_{uu}(f)$ and $G_{mm}(f)$ can be determined from $x(t)$ and $y(t)$. This case is useful for special applications as discussed in Reference 6.2. It should not be applied when output noise is expected because of possible nonlinear operations, the contributions from other inputs, and/or output measurement instrumentation noise. Case 1 is always preferred in these situations.

For Case 2, the ordinary coherence function is

$$\gamma_{xy}^2(f) = \frac{|G_{xy}(f)|^2}{G_{xx}(f)G_{yy}(f)} = \frac{|G_{uv}(f)|^2}{[G_{uu}(f) + G_{mm}(f)]G_{vv}(f)}$$

$$= \frac{1}{1 + [G_{mm}(f)/G_{uu}(f)]} \tag{6.44}$$

It follows that $\gamma_{xy}^2(f) < 1$ when $G_{mm}(f) > 0$ with

$$G_{uu}(f) = \gamma_{xy}^2(f)G_{xx}(f) \tag{6.45}$$

$$G_{mm}(f) = [1 - \gamma_{xy}^2(f)]G_{xx}(f) \tag{6.46}$$

Thus, for Case 2, the ordinary coherence function can decompose a measured *input* spectrum into its uncorrelated signal and noise components.

If one divides the frequency response function $H_1(f)$ for Case 1 as given by Equation (6.37) by the frequency response function $H_2(f)$ for Case 2 as given by Equation (6.42), the result is the coherence function, namely,

$$\frac{H_1(f)}{H_2(f)} = \frac{G_{xy}(f)/G_{xx}(f)}{G_{yy}(f)/G_{yx}(f)} = \gamma_{xy}^2(f) \tag{6.47}$$

CASE 3. *Both Noises Present; Uncorrelated with Each Other and with the Signals*

$$x(t) = u(t) + m(t) \qquad G_{um}(f) = G_{vn}(f) = 0$$

$$y(t) = v(t) + n(t) \qquad G_{mn}(f) = 0$$

$$G_{xx}(f) = G_{uu}(f) + G_{mm}(f)$$

$$G_{yy}(f) = G_{vv}(f) + G_{nn}(f)$$

$$G_{xy}(f) = G_{uv}(f) = H(f)G_{uu}(f)$$

$$G_{vv}(f) = |H(f)|^2 G_{uu}(f) \tag{6.48}$$

Here, $H(f)$ cannot be determined from the measured $x(t)$ and $y(t)$ without a knowledge or measurement of the input noise. Specifically,

$$H(f) = \frac{G_{xy}(f)}{G_{uu}(f)} = \frac{G_{xy}(f)}{G_{xx}(f) - G_{mm}(f)} \tag{6.49}$$

$$|H(f)|^2 = \frac{G_{vv}(f)}{G_{uu}(f)} = \frac{G_{yy}(f) - G_{nn}(f)}{G_{xx}(f) - G_{mm}(f)} \tag{6.50}$$

Note that $H(f)$ is a function of $G_{mm}(f)$, but is independent of $G_{nn}(f)$ by Equation (6.49). Using Equation (6.50) to calculate $|H(f)|^2$ shows that $|H(f)|^2$ is a function of both $G_{mm}(f)$ and $G_{nn}(f)$.

For Case 3, the ordinary coherence function is

$$\gamma_{xy}^2(f) = \frac{|G_{xy}(f)|^2}{G_{xx}(f)G_{yy}(f)} = \frac{|G_{uv}(f)|^2}{[G_{uu}(f) + G_{mm}(f)][G_{vv}(f) + G_{nn}(f)]}$$

$$= \frac{1}{1 + \alpha(f) + \beta(f) + \alpha(f)\beta(f)} \tag{6.51}$$

where $\alpha(f)$ and $\beta(f)$ are the noise-to-signal ratios given by

$$\alpha(f) = [G_{mm}(f)/G_{uu}(f)]$$
$$\beta(f) = [G_{nn}(f)/G_{vv}(f)] \tag{6.52}$$

Clearly, $\gamma_{xy}^2(f) < 1$ whenever $\alpha(f) > 0$ or $\beta(f) > 0$. Here, using only $x(t)$ and $y(t)$, it is *not* possible to decompose $G_{xx}(f)$ or $G_{yy}(f)$ into their separate signal and noise components.

6.1.4   *Optimum Frequency Response Functions*

Return now to the basic single-input/single-output system with output noise only (Case 1), as illustrated in Figure 6.7. Without assuming that $n(t)$ is uncorrelated with $v(t)$, let $H(f)$ be *any* linear frequency response function acting on $x(t)$. Of interest is the specific $H(f)$ that will minimize the noise at the output; that is, the optimum estimate of $H(f)$ in the least squares sense. Note that if there is any correlation between the output signal and noise, as would occur if the output included signal-dependent instrumentation noise or the contributions of other inputs that are correlated with the measured $x(t)$, then the resulting optimum $H(f)$ will not represent the physical direct path between the points where $x(t)$ and $y(t)$ are measured. Also, there may be nonlinear operations between the measured input and output data. In any case, the optimum $H(f)$ will simply constitute a mathematical function that defines the best linear relationship between $x(t)$ and $y(t)$ in the least squares sense.

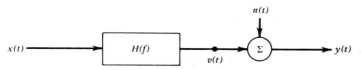

Figure 6.7   Single-input/single-output system with output noise.

For any set of long records of finite length T in Figure 6.7, the governing relation is

$$Y(f) = H(f)X(f) + N(f) \qquad (6.53)$$

where capital letters are finite Fourier transforms of associated time domain records. Solving for $N(f)$ and $N*(f)$ gives

$$N(f) = Y(f) - H(f)X(f) \qquad N*(f) = Y*(f) - H*(f)X*(f)$$

It follows that

$$|N(f)|^2 = |Y(f)|^2 - H(f)Y*(f)X(f)$$

$$- H*(f)X*(f)Y(f) + H(f)H*(f)|X(f)|^2 \qquad (6.54)$$

Taking the expectation of Equation (6.54), multiplying by $(2/T)$, and letting T increase to infinity yields

$$G_{nn}(f) = G_{yy}(f) - H(f)G_{yx}(f) - H*(f)G_{xy}(f) + H(f)H*(f)G_{xx}(f)$$

$$(6.55)$$

This is the form of $G_{nn}(f)$ for *any* $H(f)$. By definition, the *optimum* $H(f)$ will now be defined as that $H(f)$ that minimizes $G_{nn}(f)$ over all possible choices of $H(f)$. This is called the *least-squares estimate*, defined in Section 4.8.2.

The minimization of $G_{nn}(f)$ as a function of $H(f)$ will now be carried out. To simplify the derivation, the dependence on f will be omitted. Thus

$$G_{nn} = G_{yy} - HG_{yx} - H*G_{xy} + HH*G_{xx} \qquad (6.56)$$

Now let the complex numbers be expressed in terms of their real and imaginary parts as follows:

$$H = H_R - jH_I \qquad H* = H_R + jH_I$$

$$G_{xy} = G_R - jG_I \qquad G_{yx} = G_R + jG_I$$

Then

$$G_{nn} = G_{yy} - (H_R - jH_I)G_{yx} - (H_R + jH_I)G_{xy} + (H_R^2 + H_I^2)G_{xx}$$

To find the form of H that will minimize G_{nn}, one should now set the partial derivatives of G_{nn} with respect to H_R and H_I equal to zero and solve the resulting pair of equations. This gives

$$\frac{\partial G_{nn}}{\partial H_R} = -G_{yx} - G_{xy} + 2H_R G_{xx} = 0$$

$$\frac{\partial G_{nn}}{\partial H_I} = jG_{yx} - jG_{xy} + 2H_I G_{xx} = 0$$

which leads to

$$H_R = \frac{G_{xy} + G_{yx}}{2G_{xx}} = \frac{G_R}{G_{xx}}$$

$$H_I = \frac{j(G_{xy} - G_{yx})}{2G_{xx}} = \frac{G_I}{G_{xx}}$$

Hence the optimum H is

$$H = H_R - jH_I = \frac{G_R - jG_I}{G_{xx}} = \frac{G_{xy}}{G_{xx}} \tag{6.57}$$

Again, the optimum H calculated by Equation (6.57), using arbitrary measured records, does not have to be physically realizable; it may be only a theoretically computed result.

Another important property satisfied by the optimum $H(f)$ is revealed by substitution of the optimum system satisfying Equation (6.57) into Equation (6.55). This gives the noise output spectrum

$$G_{nn}(f) = \left[1 - \gamma_{xy}^2(f)\right]G_{yy}(f) \tag{6.58}$$

which leads to the coherent output spectrum

$$G_{vv}(f) = G_{yy}(f) - G_{nn}(f) = \gamma_{xy}^2(f)G_{yy}(f) \tag{6.59}$$

Moreover, using the optimum $H(f)$ shows that

$$G_{xv}(f) = H(f)G_{xx}(f)$$

and

$$G_{xn}(f) = G_{xy}(f) - H(f)G_{xx}(f) = 0 \tag{6.60}$$

It follows that

$$G_{vn}(f) = H^*(f)G_{xn}(f) = 0 \tag{6.61}$$

Thus $n(t)$ and $v(t)$ will *automatically be uncorrelated* when the optimum $H(f)$ is used to estimate the linear system in Figure 6.7.

It should also be noted here that because of the special form of Equation (6.56), a simple way to derive the same optimum H can be obtained by setting either the partial derivative of G_{nn} with respect to H equal to zero (holding H^* fixed) or setting the partial derivative of G_{nn} with respect to H^* equal to zero (holding H fixed). By this alternative method

$$\frac{\partial G_{nn}}{\partial H} = -G_{yx} + H^* G_{xx} = 0$$

$$H^* = \frac{G_{yx}}{G_{xx}} \qquad H = \frac{G_{xy}}{G_{xx}} \tag{6.62}$$

The following steps justify this method. Equation (6.56) shows that G_{nn} is a real-valued function of (H_R, H_I) or of (H, H^*) denoted by

$$G_{nn} = f(H_R, H_I) = g(H, H^*)$$

which has the structure

$$g(H, H^*) = AH + A^* H^* + BHH^* + C \tag{6.63}$$

where A is complex-valued and B, C are real-valued. The quantity $H_R(f)$ is the real part of $H(f)$ while $H_I(f)$ is the imaginary part of $H(f)$ satisfying $H = H_R - jH_I$, $H^* = H_R + jH_I$. Now

$$\frac{\partial G_{nn}}{\partial H_R} = \frac{\partial g}{\partial H}\frac{\partial H}{\partial H_R} + \frac{\partial g}{\partial H^*}\frac{\partial H^*}{\partial H_R} = \frac{\partial g}{\partial H} + \frac{\partial g}{\partial H^*}$$

$$\frac{\partial G_{nn}}{\partial H_I} = \frac{\partial g}{\partial H}\frac{\partial H}{\partial H_I} + \frac{\partial g}{\partial H^*}\frac{\partial H^*}{\partial H_I} = -j\left(\frac{\partial g}{\partial H} - \frac{\partial g}{\partial H^*}\right) \tag{6.64}$$

Hence minimization requirements that both

$$\frac{\partial G_{nn}}{\partial H_R} = 0 \qquad \text{and} \qquad \frac{\partial G_{nn}}{\partial H_I} = 0 \tag{6.65}$$

are equivalent to setting both

$$\frac{\partial g}{\partial H} = A + BH^* = 0 \qquad \text{and} \qquad \frac{\partial g}{\partial H^*} = 0 \tag{6.66}$$

From Equation (6.63), this will occur when

$$\frac{\partial g}{\partial H} = A + BH^* = 0 \qquad \text{giving } H^* = -\frac{A}{B}$$

$$\frac{\partial g}{\partial H^*} = A^* + BH = 0 \qquad \text{giving } H = -\frac{A^*}{B} \tag{6.67}$$

Thus both of the conditions of Equation (6.66) hold when

$$H = -\frac{A^*}{B} \tag{6.68}$$

Note that this solution is obtained by setting either $\partial g/\partial H = 0$ (holding H^* fixed) or setting $\partial g/\partial H^* = 0$ (holding H fixed) without going back to H_R and H_I. From Equation (6.56), $A = -G_{yx}(f)$ and $B = G_{xx}(f)$. Hence

$$H(f) = \frac{G_{xy}(f)}{G_{xx}(f)} \tag{6.69}$$

which is the same as Equation (6.57).

6.2   SINGLE-INPUT/MULTIPLE-OUTPUT MODELS

Models will now be formulated that are appropriate for studying the properties of multiple transmission paths between a single source and different output points. It will be assumed that constant-parameter linear systems can be used to describe these different possible paths and that all unknown deviations from ideal cases can be included in uncorrelated extraneous output noise terms.

6.2.1   Single - Input / Two - Output Model

Consider the special case of a single-input/two-output system, as pictured in Figure 6.8. The following frequency domain equations apply to this situation assuming that the noise terms $n_1(t)$ and $n_2(t)$ are incoherent (uncorrelated) with each other and with the input signal $x(t)$. The dependence on frequency f will be omitted here to simplify the notation.

$$G_{xn_1} = G_{v_1n_1} = G_{xn_2} = G_{v_2n_2} = G_{n_1n_2} = 0$$

$$G_{y_1y_1} = G_{v_1v_1} + G_{n_1n_1} = |H_1|^2G_{xx} + G_{n_1n_1} \tag{6.70}$$

$$G_{y_2y_2} = G_{v_2v_2} + G_{n_2n_2} = |H_2|^2G_{xx} + G_{n_2n_2}$$

$$G_{xy_1} = G_{xv_1} = H_1G_{xx} \qquad G_{xy_2} = G_{xv_2} = H_2G_{xx} \tag{6.71}$$

$$G_{y_1y_2} = G_{v_1v_2} = H_1^*H_2G_{xx} \tag{6.72}$$

For this model, the coherence function between the output records is given by

$$\gamma_{y_1y_2}^2 = \frac{|G_{y_1y_2}|^2}{G_{y_1y_1}G_{y_2y_2}} = \frac{|G_{v_1v_2}|^2}{G_{y_1y_1}G_{y_2y_2}} = \gamma_{xy_1}^2\gamma_{xy_2}^2 \tag{6.73}$$

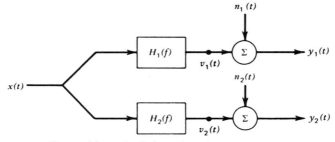

Figure 6.8 Single-input/two-output system.

where the last equality occurs because

$$|G_{v_1 v_2}|^2 = |H_1^* H_2 G_{xx}|^2 = \left(|H_1|^2 G_{xx}\right)\left(|H_2|^2 G_{xx}\right) = G_{v_1 v_1} G_{v_2 v_2}$$

$$\gamma_{x y_1}^2 = \frac{|G_{x y_1}|^2}{G_{xx} G_{y_1 y_1}} = \frac{|H_1 G_{xx}|^2}{G_{xx} G_{y_1 y_1}} = \frac{G_{v_1 v_1}}{G_{y_1 y_1}}$$

$$\gamma_{x y_2}^2 = \frac{|G_{x y_2}|^2}{G_{xx} G_{y_2 y_2}} = \frac{|H_2 G_{xx}|^2}{G_{xx} G_{y_2 y_2}} = \frac{G_{v_2 v_2}}{G_{y_2 y_2}}$$

(6.74)

A number of applications exist for these results, depending on whether or not the input signal $x(t)$ can be measured along with the two output records $y_1(t)$ and $y_2(t)$.

CASE 1. *x(t), y_1(t), and y_2(t) Can Be Measured Simultaneously*

For this case, Equation (6.71) yields H_1 and H_2 by

$$H_1 = \frac{G_{x y_1}}{G_{xx}} \qquad H_2 = \frac{G_{x y_2}}{G_{xx}}$$

Then one can determine from Equations (6.70) and (6.74),

$$G_{v_1 v_1} = |H_1|^2 G_{xx} = \gamma_{x y_1}^2 G_{y_1 y_1} \qquad G_{n_1 n_1} = \left(1 - \gamma_{x y_1}^2\right) G_{y_1 y_1}$$

$$G_{v_2 v_2} = |H_2|^2 G_{xx} = \gamma_{x y_2}^2 G_{y_2 y_2} \qquad G_{n_2 n_2} = \left(1 - \gamma_{x y_2}^2\right) G_{y_2 y_2}$$

Thus all quantities in Equations (6.70)–(6.74) can be found.

CASE 2. *Only $y_1(t)$ and $y_2(t)$ Can Be Measured Simultaneously*

For this case, H_1 and H_2 *cannot* be determined using Equation (6.71). If, however, H_1 and H_2 are known from other considerations, such as theoretical ideas, then Equation (6.72) can be used to obtain G_{xx} by

$$ G_{xx} = \frac{G_{y_1 y_2}}{H_1^* H_2} $$

The assumed H_1 and H_2 together with the computed G_{xx} will give all the remaining quantities in Equations (6.70)–(6.74).

If H_1 and H_2 are not known from other considerations, one can still compute $G_{y_1 y_1}$, $G_{y_2 y_2}$, and $G_{y_1 y_2}$. One can then use Equation (6.73) to determine the coherence function $\gamma_{y_1 y_2}^2$. A high value for $\gamma_{y_1 y_2}^2$ indicates that $y_1(t)$ and $y_2(t)$ can come from an unmeasured common source $x(t)$ via unknown linear transformations, and that extraneous output noise is small compared to the signal terms. It should be noted here that *a high coherence value does not indicate a causal relationship between $y_1(t)$ and $y_2(t)$*. In fact, there is no causal relationship between them in this model. Applications of two output models in energy source location problems are detailed in Reference 6.1.

6.2.2 *Single-Input/Multiple-Output Model*

Consider the system shown in Figure 6.9, consisting of a single stationary random input $x(t)$ and r measured outputs $y_i(t)$, $i = 1, 2, \ldots, r$. Here for $i = 1, 2, \ldots, r$, frequency domain equations are

$$ Y_i(f) = H_i(f) X(f) + N_i(f) \tag{6.75} $$

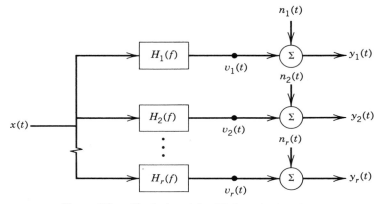

Figure 6.9 Single-input/multiple-output system.

where the capital letters represent finite Fourier transforms of corresponding time domain records in lowercase letters. For simplicity and without loss of generality, assume that all records have zero mean values. Assume also that each of the output noise terms $n_i(t)$ are uncorrelated with $x(t)$ and that they are mutually uncorrelated with each other.

CASE 1.   *Input Plus Output Measurements*

From simultaneous measurements of $x(t)$ with each of the outputs $y_i(t)$, $i = 1, 2, \ldots, r$, one can compute

$$G_{xy_i}(f) = H_i(f)G_{xx}(f) \tag{6.76}$$

These equations are the same as for earlier single-input/single-output models. Each $H_i(f)$ is then given by the formula

$$H_i(f) = \frac{G_{xy_i}(f)}{G_{xx}(f)} \tag{6.77}$$

In words, $H_i(f)$ is the ratio of the cross-spectral density function between the input $x(t)$ and the particular output $y_i(t)$, divided by the autospectral density function of the input. The ordinary coherence function between $x(t)$ and $y_i(t)$ is

$$\gamma_{xy_i}^2(f) = \frac{|G_{xy_i}(f)|^2}{G_{xx}(f)G_{y_iy_i}(f)} \tag{6.78}$$

Each output can be easily decomposed into its separate signal and noise components.

In the time domain, the cross-correlation function between $x(t)$ and $y_i(t)$ is given by

$$R_{xy_i}(\tau) = \int_0^\infty h_i(\tau)R_{xx}(\tau - \alpha)\,d\alpha \tag{6.79}$$

where $h_i(\tau)$ is the inverse Fourier transform of $H_i(f)$; that is, the unit impulse response function of the ith path.

CASE 2.   *Output Measurements Only*

Consider cases where only the outputs $y_i(t)$ can be measured. For each $i = 1, 2, \ldots, r$, one can compute the autospectral density functions

$$G_{y_iy_i}(f) = |H_i(f)|^2 G_{xx}(f) + G_{n_in_i}(f) \tag{6.80}$$

The associated autocorrelation functions are

$$R_{y_i y_i}(\tau) = \iint\limits_0^\infty h_i(\alpha) h_i(\beta) R_{xx}(\tau + \alpha - \beta) \, d\alpha \, d\beta + R_{n_i n_i}(\tau) \quad (6.81)$$

It is also possible here between all pairs of different output records to compute their cross-spectral density functions and cross-correlation functions. For $i \neq j$, since $R_{n_i n_j}(\tau) = 0$ and $G_{n_i n_j}(f) = 0$, the cross-correlation function is given by

$$R_{y_i y_j}(\tau) = E\big[y_i(t) y_j(t + \tau) \big]$$

$$= \iint\limits_0^\infty h_i(\alpha) h_j(\beta) E\big[x(t - \alpha) x(t + \tau - \beta) \big] \, d\alpha \, d\beta$$

$$= \iint\limits_0^\infty h_i(\alpha) h_j(\beta) R_{xx}(\tau + \alpha - \beta) \, d\alpha \, d\beta \quad (6.82)$$

with other terms that average to zero. On the other hand, the cross-spectral density function between any two of the output records is the simple expression

$$G_{y_i y_j}(f) = H_i^*(f) H_j(f) G_{xx}(f) \quad (6.83)$$

Algebraic operations are now involved to interpret this result easily. Measurement of $G_{y_i y_j}(f)$ plus separate knowledge of $H_i(f)$ and $H_j(f)$ enables one to estimate $G_{xx}(f)$ when $G_{xx}(f)$ cannot be measured directly.

From Equations (6.80) and (6.83), the coherence function between any two of the output records is given by

$$\gamma_{y_i y_j}^2(f) = \frac{|G_{y_i y_j}(f)|^2}{G_{y_i y_i}(f) G_{y_j y_j}(f)}$$

$$= \frac{|H_i(f)|^2 |H_j(f)|^2 G_{xx}(f)}{\big[|H_i(f)|^2 G_{xx}(f) + G_{n_i n_i}(f) \big]\big[|H_j(f)|^2 G_{xx}(f) + G_{n_j n_j}(f) \big]}$$

$$(6.84)$$

Various special cases of Equation (6.84) of physical interest are discussed in Chapter 7 of Reference 6.1.

6.2.3 *Removal of Extraneous Noise*

Assume that three or more output measurements, say, $y_1(t)$, $y_2(t)$, and $y_3(t)$, are noisy measurements of the desired output signals $v_1(t)$, $v_2(t)$, and $v_3(t)$, as shown in Figure 6.9. The noise terms $n_1(t)$, $n_2(t)$, and $n_3(t)$ are assumed to be mutually uncorrelated with each other and with $v_1(t)$, $v_2(t)$, and $v_3(t)$. It will now be shown how to determine the autospectral density function properties of the true output signals $v_i(t)$ from an analysis of the measured $y_i(t)$, based on ideas from Reference 6.3.

The following equations are all functions of frequency f, which is omitted to simplify the notation. The measured output autospectra are

$$G_{11} = G_{y_1 y_1} = G_{v_1 v_1} + G_{n_1 n_1} = G_{v_1 v_1}(1 + c_1)$$

$$G_{22} = G_{y_2 y_2} = G_{v_2 v_2} + G_{n_2 n_2} = G_{v_2 v_2}(1 + c_2) \qquad (6.85)$$

$$G_{33} = G_{y_3 y_3} = G_{v_3 v_3} + G_{n_3 n_3} = G_{v_3 v_3}(1 + c_3)$$

where the $c_i \geq 0$ represent

$$c_1 = G_{n_1 n_1}/G_{v_1 v_1} = \text{noise-to-signal ratio at output 1}$$

$$c_2 = G_{n_2 n_2}/G_{v_2 v_2} = \text{noise-to-signal ratio at output 2} \qquad (6.86)$$

$$c_3 = G_{n_3 n_3}/G_{v_3 v_3} = \text{noise-to-signal ratio at output 3}$$

The measured cross-spectra between pairs of outputs are

$$G_{12} = G_{y_1 y_2} = G_{v_1 v_2}$$

$$G_{13} = G_{y_1 y_3} = G_{v_1 v_3} \qquad (6.87)$$

$$G_{23} = G_{y_2 y_3} = G_{v_2 v_3}$$

In Figure 6.9, it is assumed that $v_1(t)$, $v_2(t)$, and $v_3(t)$ are due to a common single unmeasured source $x(t)$. Hence, the ordinary coherence functions between pairs of $v_i(t)$ must be unity, namely,

$$\gamma_{v_1 v_2}^2 = \frac{|G_{v_1 v_2}|^2}{G_{v_1 v_1} G_{v_2 v_2}} = 1$$

$$\gamma_{v_1 v_3}^2 = \frac{|G_{v_1 v_3}|^2}{G_{v_1 v_1} G_{v_3 v_3}} = 1 \qquad (6.88)$$

$$\gamma_{v_2 v_3}^2 = \frac{|G_{v_2 v_3}|^2}{G_{v_2 v_2} G_{v_3 v_3}} = 1$$

Thus

$$|G_{12}|^2 = |G_{v_1 v_2}|^2 = G_{v_1 v_1} G_{v_2 v_2}$$

$$|G_{13}|^2 = |G_{v_1 v_3}|^2 = G_{v_1 v_1} G_{v_3 v_3} \qquad (6.89)$$

$$|G_{23}|^2 = |G_{v_2 v_3}|^2 = G_{v_2 v_2} G_{v_3 v_3}$$

From Equations (6.85) and (6.89), it follows that

$$\gamma_{12}^2 = \frac{|G_{12}|^2}{G_{11} G_{22}} = \frac{1}{(1 + c_1)(1 + c_2)}$$

$$\gamma_{13}^2 = \frac{|G_{13}|^2}{G_{11} G_{33}} = \frac{1}{(1 + c_1)(1 + c_3)} \qquad (6.90)$$

$$\gamma_{23}^2 = \frac{|G_{23}|^2}{G_{22} G_{33}} = \frac{1}{(1 + c_2)(1 + c_3)}$$

It is reasonable to assume that none of the measured coherence functions in Equation (6.90) will be zero, to obtain

$$(1 + c_1) = \frac{1}{\gamma_{12}^2 (1 + c_2)}$$

$$(1 + c_2) = \frac{1}{\gamma_{23}^2 (1 + c_3)} \qquad (6.91)$$

$$(1 + c_3) = \frac{1}{\gamma_{13}^2 (1 + c_1)}$$

This yields

$$(1 + c_1) = \frac{\gamma_{23}^2 (1 + c_3)}{\gamma_{12}^2} = \frac{\gamma_{23}^2}{\gamma_{12}^2 \gamma_{13}^2 (1 + c_1)} \qquad (6.92)$$

with similar formulas for $(1 + c_2)$ and $(1 + c_3)$. Hence

$$(1 + c_1) = \frac{|\gamma_{23}|}{|\gamma_{12}| \, |\gamma_{13}|}$$

$$(1 + c_2) = \frac{|\gamma_{13}|}{|\gamma_{12}| \, |\gamma_{23}|} \qquad (6.93)$$

$$(1 + c_3) = \frac{|\gamma_{12}|}{|\gamma_{13}| \, |\gamma_{23}|}$$

From Equation (6.85), one now obtains the desired results

$$G_{v_1 v_1} = \frac{G_{11}}{(1 + c_1)} = \frac{|\gamma_{12}| \, |\gamma_{13}|}{|\gamma_{23}|} G_{11} = \frac{|G_{12}| \, |G_{13}|}{|G_{23}|}$$

$$G_{v_2 v_2} = \frac{G_{22}}{(1 + c_2)} = \frac{|\gamma_{12}| \, |\gamma_{23}|}{|\gamma_{13}|} G_{22} = \frac{|G_{12}| \, |G_{23}|}{|G_{13}|} \qquad (6.94)$$

$$G_{v_3 v_3} = \frac{G_{33}}{(1 + c_3)} = \frac{|\gamma_{13}| \, |\gamma_{23}|}{|\gamma_{12}|} G_{33} = \frac{|G_{13}| \, |G_{23}|}{|G_{12}|}$$

Since $G_{v_1 v_1} \le G_{11}$, $G_{v_2 v_2} \le G_{22}$, and $G_{v_3 v_3} \le G_{33}$, the coherence functions must satisfy

$$\gamma_{12}^2 \gamma_{13}^2 \le \gamma_{23}^2$$

$$\gamma_{12}^2 \gamma_{23}^2 \le \gamma_{13}^2 \qquad (6.95)$$

$$\gamma_{13}^2 \gamma_{23}^2 \le \gamma_{12}^2$$

One can also solve for the output autospectra $G_{n_1 n_1}$, $G_{n_2 n_2}$, and $G_{n_3 n_3}$ using Equations (6.85) and (6.94).

Example 6.6. **Illustration of Noisy Measurements.**   The spectral densities of three output measurements are computed to be $G_{11}(f) = G_{22}(f) = G_{33}(f) = G$, a constant, and the coherence among the output measurements are computed to be $\gamma_{12}^2 = 0.5$, $\gamma_{13}^2 = 0.5$, and $\gamma_{23}^2 = 0.4$. Determine the signal-to-noise ratios and the spectral densities of the signals in the three output measurements.

From Equation (6.93), the noise-to-signal ratios are given by

$$c_1 = \frac{0.4}{(0.5)(0.5)} - 1 = 0.60$$

$$c_2 = \frac{0.5}{(0.5)(0.4)} - 1 = 1.50$$

$$c_3 = \frac{0.5}{(0.5)(0.4)} - 1 = 1.50$$

Hence the signal-to-noise ratios at the output measurements are

$$(S/N)_1 = 1.67 \qquad (S/N)_2 = 0.67 \qquad (S/N)_3 = 0.67$$

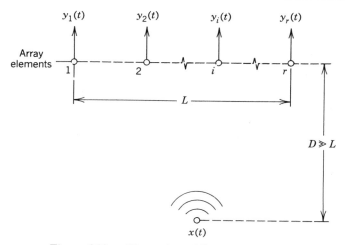

Figure 6.10   Illustration of linear point array.

and the spectral densities of the signals in the output measurements are

$$G_{v_1 v_1} = 0.625G \qquad G_{v_2 v_2} = 0.40G \qquad G_{v_3 v_3} = 0.40G$$

6.3   MULTIPLE-OUTPUT ARRAY MODELS

A single-input/multiple-output problem of special interest involves the model shown in Figure 6.9, where the input $x(t)$ is a point source of radiating energy (e.g., acoustic noise) in a homogeneous medium (e.g., air or water), and the outputs $y_i(t)$, $i = 1, 2, \ldots, r$, are point measurements in the medium. The simplest version of this model is that in which the output measurement locations are on a line that is distant from the source, as illustrated in Figure 6.10. The configuration of output measurements in Figure 6.10 is commonly called a *linear point array,* and the measurement locations are called *array elements.* The underlying assumption in this simple array model is that the frequency response functions $H_i(f)$, $i = 1, 2, \ldots, r$, in Figure 6.9 are essentially identical except for a time delay. A summing of the output measurements can then yield two benefits, (a) an increase in the output signal-to-noise ratio (gain), and (b) a directivity that will isolate individual sources with different angles of incidence θ to the array ($\theta = 0$ degrees when the source is on the normal to the array axis).

6.3.1   Array Gain

Referring to the multiple-output model in Figure 6.9, assume the unit impulse functions between the input $x(t)$ and the various outputs $y_i(t)$, $i = 1, 2, \ldots, r$, are equal, that is

$$h_i(\tau) = h(\tau) \qquad i = 1, 2, \ldots, r \tag{6.96}$$

Such a situation would be closely approximated by the linear point array in Figure 6.10 if (a) the source was sufficiently distant from the array to appear as a plane wave, (b) the source was on a line normal to the axis of the array ($\theta = 0$ degrees), and (c) the medium was homogeneous. Also assume the noise signals in the output measurements have zero mean values and equal variances, and further are statistically independent of the input and of one another such that

$$E[n_i(t)] = 0$$

$$E[n_i^2(t)] = \sigma_n^2 \qquad i = 1, 2, \ldots, r$$

$$E[n_i(t)x(t)] = \sigma_{n_i x} = 0 \tag{6.97}$$

$$E[n_i(t)n_j(t)] = \sigma_{n_i n_j} = 0 \qquad i, j = 1, 2, \ldots, r \quad i \neq j$$

It follows that the output measurements are given by

$$y_i(t) = \int_0^\infty h(\tau)x(t - \tau)\, d\tau + n_i(t)$$

$$= v(t) + n_i(t) \qquad i = 1, 2, \ldots, r \tag{6.98}$$

The average of the output measurements is then

$$\bar{y} = \frac{1}{r} \sum_{i=1}^r [v(t) + n_i(t)] = v(t) + \frac{1}{r} \sum_{i=1}^r n_i(t) \tag{6.99}$$

Using Equation (4.33), the variance of the total output is computed as

$$\sigma_{\bar{y}}^2 = \sigma_v^2 + \left(\sigma_n^2/r\right) \tag{6.100}$$

Hence, if the input signal-to-noise ratio is defined as

$$(S/N)_{\text{in}} = \frac{\sigma_v^2}{\sigma_n^2} \tag{6.101}$$

the output signal-to-noise ratio of the array becomes

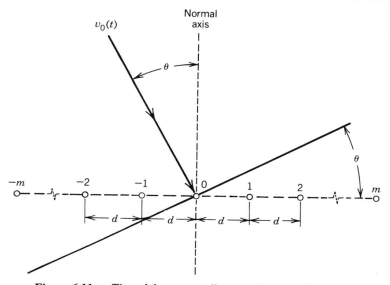

Figure 6.11   Time delay among linear point array elements.

$$(S/N)_{\text{out}} = \frac{\sigma_v^2}{(\sigma_n^2/r)} = r(S/N)_{\text{in}} \qquad (6.102)$$

It follows from Equation (6.102) that the signal-to-noise ratio of the array output increases in direct proportion to the number of individual output measurements (the number of elements) in the array. This is often referred to as the (power) *gain of the array*. Array gains are often expressed in decibels where dB gain = $10 \log_{10} r$.

6.3.2   *Array Directivity*

As the source in Figure 6.10 moves off a line normal to the array axis, the signals $v_i(t)$ in the output measurements $y_i(t)$, $i = 1, 2, \ldots, r$, will no longer be in phase and thus the sensitivity of the array to the source signal will diminish as the angle of incidence increases; that is, the array has directivity. To evaluate this directivity, let the propagation velocity of the source energy in the medium be c, the separation distance between the array elements be d, and the angle of incidence of the source energy be θ, as shown in Figure 6.11. Note that an odd number of array elements is assumed for convenience, and the locations of the elements are identified relative to the center of the array by nd, $n = 0, \pm 1, \pm 2, \ldots, \pm m$, so that $r = 2m + 1$. Also, all assumptions stated previously in Equation (6.97) still apply.

Let the source signal reaching the center element in the array be a sine wave with unit magnitude, that is

$$v_0(t) = \sin 2\pi ft \qquad (6.103)$$

It is clear from Figure 6.11 that the source signals reaching other elements in the array will be displaced in time by $\tau_n = (nd/c)\sin\theta$, $n = 0, \pm 1, \pm 2, \ldots, \pm m$. Hence,

$$v_n(t) = \sin(2\pi ft + Knd) \tag{6.104}$$

where the term K, called the *trace wave number*, is given by

$$K = (2\pi f/c)\sin\theta \tag{6.105}$$

The total output of the array is then

$$T = \sum_{n=-m}^{m} \sin(2\pi ft + Knd)$$

$$= \left[\sin 2\pi ft \sum_{n=-m}^{m} \cos Knd\right] + \left[\cos 2\pi ft \sum_{n=-m}^{m} \sin Knd\right] \tag{6.106}$$

Since the phase of the total output is of no great interest, the modulus of Equation (6.106) can be computed to obtain

$$|T| = D(K) = \left\{\left[\sum_{n=-m}^{m} \cos Knd\right]^2 + \left[\sum_{n=-m}^{m} \sin Knd\right]^2\right\}^{1/2} \tag{6.107}$$

However, because of the symmetry around the midpoint of the array, the second term in Equation (6.107) is zero so

$$D(K) = \sum_{n=-m}^{m} \cos Knd = \frac{\sin(rKd/2)}{\sin(Kd/2)} \qquad r = 2m + 1 \tag{6.108}$$

The last equality is obtained using the finite sum relationship from Reference 6.4,

$$\sum_{n=0}^{m} \cos nx = \frac{\cos[mx/2]\sin[(m+1)x/2]}{\sin(x/2)} \tag{6.109}$$

along with the trigonometric identity

$$2\cos a \sin b = \sin(a + b) + \sin(b - a) \tag{6.110}$$

The term $D(K)$ in Equation (6.108) is called the *directional pattern* and defines the gain of the array as a function of the number of elements r and the trace wave number K, which in turn is a function of the frequency f and the angle of incidence θ. The normalized directional pattern, which has a maxi-

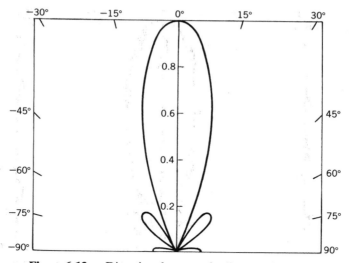

Figure 6.12   Directional pattern for linear point array.

mum value of unity, is defined by

$$D_0(K) = \frac{\sin(rKd/2)}{r\sin(Kd/2)} \qquad r = 2m + 1 \qquad (6.111)$$

The quantity $20\log_{10}D_0(K)$ defines the directional pattern in dB.

A polar plot of $D_0(K)$ for a typical linear point array is shown in Figure 6.12. The main lobe in the directional pattern is called the *beam of the array*. Note that there are various side lobes in the directional pattern that are closely related to the side lobes inherent in the spectral windows associated with finite Fourier transforms of time history data, to be detailed in Chapter 11. In fact, Equation (6.107) represents the modulus of a discrete Fourier transform in wave number space. Hence, many of the procedures and problems addressed in Chapters 10 and 11, including aliasing effects and the use of tapering to suppress side lobes, are directly applicable to array problems.

Example 6.7. Half-Power Point Array Beam Width.   The width of array beams are sometimes described by the angle subtended from the half-power points of the array directional pattern; that is, those points on the directional pattern where the array gain is 3dB less than the maximum gain. For any given source distance, this angular beam width can then be specified in distance units. As an illustration, consider an acoustic linear point array in air ($c = 1120$ ft/sec) with $r = 50$ elements separated by $d = 0.1$ ft. Determine the half-power point width of the array beam for an acoustic source at 1000 Hz located 1000 ft from the array.

Given the above stated array parameters, $Kd/2 = 0.280 \sin \theta$. Hence, using Equation (6.111), the angular beam width of the array at the half-power points is given by twice the angle, which satisfies

$$D_0^2(\theta) = 0.5 = \left[\frac{\sin(14.0 \sin \theta)}{50 \sin(0.280 \sin \theta)} \right]^2$$

An iterative solution to the above equation yields $\theta = 5.70$ degrees, so the angular beam width at the half-power points is $2\theta = 11.4$ degrees. Then for a source located 1000 ft from the array, the beam width in feet at the half-power points is

$$\text{Width} = 2(1000)\tan(\theta) = 273 \text{ ft}$$

PROBLEMS

6.1   Consider the simple mechanical system shown in Figure 2.2, where the frequency response function between the input force in displacement units $x(t) = F(t)/k$ and the output displacement $y(t)$ is given by

$$H(f) = \frac{1}{1 - (f/f_n)^2 + j2\zeta f/f_n}$$

Assuming $f_n = 10$ Hz and $\zeta = 0.01$, determine the rms value of the response displacement to a white noise excitation with a mean value of zero and a spectral density of $G_{xx}(f) = 0.1$ at all frequencies.

6.2   For the mechanical system in Problem 6.1, determine the maximum rms value of the response displacement to a sinusoidal excitation given by $x(t) = 2 \sin 20\pi t$.

6.3   In Problem 6.1, assume there is extraneous noise in the output measurement with a spectral density of $G_{nn}(f) = 0.1$ at all frequencies. Determine the coherence function between $x(t)$ and $y(t)$ at
(a) $f = 0$ Hz.
(b) $f = 10$ Hz.
(c) $f = 100$ Hz.

6.4   In Problem 6.1, assume also there is extraneous noise in the output measurement with a spectral density of $G_{nn}(f) = 0.1$ at all frequencies. Determine the frequency response function that would be computed (ignoring estimation errors) using the equations
(a) $H_1(f) = G_{xy}(f)/G_{xx}(f)$.
(b) $H_2(f) = G_{yy}(f)/G_{yx}(f)$.

6.5   The input to a physical system has an autospectral density function given
by $G_{xx}(f) = 2/f^2$. The cross-spectral density function between the input
and output is $G_{xy}(f) = (4/f^3) - j(4/f^2)$. Determine
(a)   the gain factor of the system.
(b)   the phase factor of the system.
(c)   the time delay through the system.

6.6   Assume three identical physical systems with uniform gain factors of
$|H(f)| = 10$ have statistically independent white noise inputs with spectral densities of $G_{x_1 x_1}(f) = 1$, $G_{x_2 x_2}(f) = 2$, and $G_{x_3 x_3}(f) = 3$. Further
assume the systems produce a common output with extraneous noise
having a spectral density of $G_{nn}(f) = 100$. Determine the coherence
function between each input and the output.

6.7   Consider two physical systems with frequency response functions given
by

$$H_1(f) = (1 + j4f)^{-1}   H_2(f) = (1 + j8f)^{-1}$$

Assume the two systems have a common input with a spectral density of
$G_{xx}(f) = 10$ and that the outputs of the two systems are contaminated
by extraneous noise signals with identical spectral densities of $G_{n_1 n_1}(f) =$
$G_{n_2 n_2}(f) = 0.1$. At a frequency of 1 Hz, determine
(a)   the coherence function between the input and each output.
(b)   the coherence function between the two outputs.

6.8   An array is to be constructed which will detect distant acoustic sources. If
a signal-to-noise ratio of $(S/N)_{out} = 1.5$ is needed to make confident
detections, and the anticipated signal-to-noise ratio in an individual
measurement is $(S/N)_{in} = 0.1$, determine
(a)   the number of elements needed in the array.
(b)   the coherence function between the outputs of any two elements.

6.9   An underwater linear point array consists of $r = 100$ elements with a
separation distance of $d = 1$ ft. What is the gain of the array for a distant
acoustic source that is 10 degrees off the normal axis to the array and has
a frequency of $f = 1000$ Hz? (Assume the speed of sound in water is
$c = 5000$ ft/sec.)

6.10 An underwater linear point array is desired which will have a half-power
point beam width of 200 ft for a 5000-Hz source that is 1000 ft away. If
the separation distance between elements is $d = 0.2$ ft, how many elements must the array include?

REFERENCES

6.1 Bendat, J. S., and Piersol, A. G., *Engineering Applications of Correlation and Spectral Analysis*, Wiley-Interscience, New York, 1980.

6.2 Upton, R., "Innovative Functions for Two-Channel FFT Analyzers," *Sound and Vibration*, Vol. 18, p. 18, March 1984.

6.3 Chung, J. Y., "Rejection of Flow Noise Using a Coherence Function Method," *Journal of the Acoustical Society of America*, Vol. 62, p. 388, 1977.

6.4 Adams, E. P., *Smithsonian Mathematical Formulae and Tables of Elliptic Functions*, Smithsonian Institution, Washington, D.C., 1947.

CHAPTER 7

MULTIPLE-INPUT/OUTPUT RELATIONSHIPS

Material contained in Chapter 6 is now extended to multiple-input problems. As before, records are assumed to be from stationary random processes with zero mean values, and systems are constant-parameter linear systems. The discussion includes multiple-input/single-output models and multiple-input/multiple-output models. Partial and multiple coherence functions are defined for these models. Iterative computational procedures are explained that provide physical insight to decompose the models, and are also more efficient than other matrix techniques. The various models considered in Chapters 6 and 7 represent useful building blocks to analyze more complicated physical systems that can be broken down into such combinations of basic subsystems.

7.1 MULTIPLE-INPUT/SINGLE-OUTPUT MODELS

This section describes multiple-input/single-output analysis problems as previously discussed and solved in References 7.1–7.3. Later material gives recommended procedures and interpretation of the various results. Consider q constant-parameter linear systems $H_i(f)$, $i = 1, 2, \ldots, q$, with q clearly defined measurable inputs $x_i(t)$, $i = 1, 2, \ldots, q$, and one measured output $y(t)$, as illustrated in Figure 7.1. There is *no* requirement that the inputs be mutually uncorrelated. The output noise term $n(t)$ accounts for all deviations from the ideal model, which may be due to unmeasured inputs, nonlinear operations, nonstationary effects, and instrument noise. Measured records are assumed to be realizations of stationary (ergodic) random processes, where nonzero mean values and any periodicities have been removed prior to this analysis. Situations of multiple-input/multiple-output models are direct extensions of these techniques by merely considering such models to be combinations of the multiple-input/single-output cases of Figure 7.1.

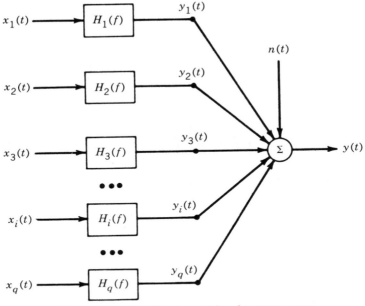

Figure 7.1 Multiple-input/signal-output system.

7.1.1 General Relationships

Referring to the multiple-input model in Figure 7.1, four conditions are required for this model to be well defined.

1. None of the ordinary coherence functions between any pair of input records should equal unity. If this occurs, these two inputs contain redundant information and one of the inputs should be eliminated from the model. This consideration allows distributed input systems to be studied as discrete inputs.
2. None of the ordinary coherence functions between any input and the total output should equal unity. If this occurs, then the other inputs are not contributing to this output and the model should be considered as simply a single-input/single-output model.
3. The multiple coherence function between any input and the other inputs, excluding the given input, should not equal unity. If this occurs, then this input can be obtained by linear operations from the other inputs. This input is not providing any new information to the output and should be eliminated from the model.
4. The multiple coherence function between the output and the given inputs, in a practical situation, should be sufficiently high, say above 0.50, for the theoretical assumptions and later conclusions to be

reasonable. Otherwise, some important inputs are probably being omitted, or nonlinear effects should be considered. This value of 0.50 is not precise, but is a matter of engineering and statistical judgment, based on the physical environment and the amount of data available for analysis.

It is assumed that one can make simultaneous measurements of the input and output time-history records in any specified multiple-input/output model. It is assumed also that possible system errors and statistical errors in computed quantities have been minimized by careful calibration and choice of data processing parameters. In particular, it is required that one obtain good estimates of the real-valued autospectral density functions of each record and the complex-valued cross-spectral density functions between every pair of records. From these stored spectral quantities, the following results are desired:

1. Decomposition of the output spectra into physically meaningful components due to the measured input records
2. Determination of optimum constant-parameter linear systems between each input and the output to minimize any noise spectra at the output that is not due to linear operations from the measured input records

The output $y(t)$ may be considered to be the sum of the unmeasured q outputs $y_i(t)$, $i = 1, 2, \ldots, q$, plus the noise term $n(t)$, namely,

$$y(t) = \sum_{i=1}^{q} y_i(t) + n(t) \tag{7.1}$$

with corresponding finite Fourier transforms

$$Y(f) = \sum_{i=1}^{q} Y_i(f) + N(f) \tag{7.2}$$

Each output term $Y_i(f)$ for $i = 1, 2, \ldots, q$ satisfies

$$Y_i(f) = H_i(f) X_i(f) \tag{7.3}$$

Thus, the basic frequency domain equation for Figure 7.1 is

$$Y(f) = \sum_{i=1}^{q} H_i(f) X_i(f) + N(f) \tag{7.4}$$

where each $X_i(f)$ and $Y(f)$ can be computed from the measured $x_i(t)$ and $y(t)$. From this information it is required to determine the systems $H_i(f)$ and other quantities of interest where all results are functions of frequency.

Finite Fourier transforms $X_i(f)$ and $Y(f)$ for single records $x_i(t)$ and $y(t)$ of length T are

$$X_i(f) = \int_0^T x_i(t) e^{-j2\pi ft} \, dt \qquad Y(f) = \int_0^T y(t) e^{-j2\pi ft} \, dt \qquad (7.5)$$

As per Equations (5.66) and (5.67), one-sided autospectral and cross-spectral density functions are defined by

$$G_{ii}(f) = G_{x_i x_i}(f) = \lim_{T \to \infty} \frac{2}{T} E\left[|X_i(f)|^2\right]$$

$$G_{ij}(f) = G_{x_i x_j}(f) = \lim_{T \to \infty} \frac{2}{T} E\left[X_i^*(f) X_j(f)\right] \qquad (7.6)$$

$$G_{yy}(f) = \lim_{T \to \infty} \frac{2}{T} E\left[|Y(f)|^2\right]$$

$$G_{iy}(f) = G_{x_i y}(f) = \lim_{T \to \infty} \frac{2}{T} E\left[X_i^*(f) Y(f)\right] \qquad (7.7)$$

These one-sided functions where $G(f) = 0$ for $f < 0$ can be replaced by theoretical two-sided functions $S(f)$, as discussed in Section 5.2.1. All formulas in this chapter can be stated in terms of either $G(f)$ or $S(f)$ quantities on both sides of the equations. The ratio of two $G(f)$ quantities is the same as the ratio of the corresponding two $S(f)$ quantities. Further work here, as in Chapter 6, will use the one-sided $G(f)$ quantities.

In practice, one obtains only estimates of Equations (7.6) and (7.7) since T will be finite and the expected value operation $E[\ ]$ can be taken only over a finite number of sample records. Also, when records are digitized, as discussed in Chapters 10 and 11, results will be obtained only at selected discrete frequencies. At any such frequency f, an estimate of $G_{xy}(f)$, denoted $\hat{G}_{xy}(f)$, is usually obtained by

$$\hat{G}_{xy}(f) = \frac{2}{n_d T} \sum_{k=1}^{n_d} X_k^*(f) Y_k(f) \qquad (7.8)$$

where n_d is the number of different (disjoint) sample records of $x(t)$ and $y(t)$, each of length T, so that the total record length $T_r = n_d T$. It is proved in Chapter 8 that to reduce bias errors, T should be made as large as possible, while to reduce random errors, n_d should be as large as possible. Hence compromise choices are necessary when T_r is fixed.

If desired, corresponding time domain equations can be written involving convolution integrals of the respective weighting functions $h_i(\tau)$, $i = 1, 2, \ldots, q$, associated with the $H_i(f)$. In place of Equation (7.3), one would have

$$y_i(t) = \int_0^\infty h_i(\tau) x_i(t - \tau) \, d\tau \tag{7.9}$$

where the lower limit is zero only when the systems are physically realizable. Such convolution integrals and their extensions to correlation functions are much more complicated than associated Fourier transform spectral relations and will not be used in the further development here. These correlation relations do not show how results vary as a function of frequency. That such information is hidden within the correlation functions is no justification for computing them when it is unnecessary and more efficient to compute spectral quantities directly.

7.1.2   *General Case of Arbitrary Inputs*

Consider the general case of arbitrary inputs. From Equation (7.4), with a different index of summation j instead of i,

$$Y(f) = \sum_{j=1}^{q} H_j(f) X_j(f) + N(f) \tag{7.10}$$

Multiplication of both sides by $X_i^*(f)$ for any fixed $i = 1, 2, \ldots, q$ yields

$$X_i^*(f) Y(f) = \sum_{j=1}^{q} H_j(f) X_i^*(f) X_j(f) + X_i^*(f) N(f)$$

Expected values of both sides shows that

$$E[X_i^*(f) Y(f)] = \sum_{j=1}^{q} H_j(f) E[X_i^*(f) X_j(f)] + E[X_i^*(f) N(f)]$$

Equations (7.6) and (7.7), with a scale factor of $(2/T)$ to obtain one-sided results, now proves at any f that

$$G_{iy}(f) = \sum_{j=1}^{q} H_j(f) G_{ij}(f) + G_{in}(f) \qquad i = 1, 2, \ldots, q$$

where the cross-spectral terms $G_{in}(f)$ will be zero if $n(t)$ is uncorrelated with each $x_i(t)$. Making this assumption gives the set of equations

$$G_{iy}(f) = \sum_{j=1}^{q} H_j(f) G_{ij}(f) \qquad i = 1, 2, \ldots, q \tag{7.11}$$

This is a set of q equations in q unknowns, the $H_i(f)$ for $i = 1, 2, \ldots, q$, where all the spectral terms shown are computed from the measured input and output records. If the model is well defined, one can solve for the $H_i(f)$ by matrix techniques using Cramer's rule or the equivalent.

In terms of the computed $H_i(f)$, as found by solving Equation (7.11), the total output autospectral density function $G_{yy}(f)$ is

$$G_{yy}(f) = \sum_{i=1}^{q} \sum_{j=1}^{q} H_i^*(f)H_j(f)G_{ij}(f) + G_{nn}(f) \qquad (7.12)$$

assuming, as before, that $n(t)$ is uncorrelated with each $x_i(t)$. Equation (7.12) is derived from

$$Y^*(f) = \sum_{i=1}^{q} H_i^*(f)X_i^*(f) + N^*(f)$$

$$Y^*(f)Y(f) = \left[\sum_{i=1}^{q} H_i^*(f)X_i^*(f) + N^*(f) \right]\left[\sum_{j=1}^{q} H_j(f)X_j(f) + N(f) \right]$$

$$= \sum_{i=1}^{q} \sum_{j=1}^{q} H_i^*(f)H_j(f)X_i^*(f)X_j(f) + N^*(f)N(f)$$

$$+ \sum_{i=1}^{q} H_i^*(f)X_i^*(f)N(f) + \sum_{j=1}^{q} H_j(f)N^*(f)X_j(f)$$

Expected values of both sides, multiplication of both sides by the scale factor $(2/T)$, and passage to the limit gives the result

$$G_{yy}(f) = \sum_{i=1}^{q} \sum_{j=1}^{q} H_i^*(f)H_j(f)G_{ij}(f) + G_{nn}(f)$$

$$+ \sum_{i=1}^{q} H_i^*(f)G_{in}(f) + \sum_{j=1}^{q} H_j(f)G_{nj}(f)$$

This reduces to Equation (7.12) when the cross-spectral terms $G_{in}(f) = 0$ for all f and all i. For q different inputs and for terms $G_{ij}(f) \neq 0$ when $i \neq j$, the output $G_{yy}(f)$ of Equation (7.12) contains $(q^2 + 1)$ parts. Decomposing $G_{yy}(f)$ into each of these parts and interpreting their meaning can be a very involved exercise.

7.1.3 Special Case of Mutually Uncorrelated Inputs

Consider the important special case when not only is $n(t)$ uncorrelated with each of the $x_i(t)$, but the inputs are mutually uncorrelated with each other. For this special case, Equations (7.11) and (7.12) become

$$G_{iy}(f) = H_i(f)G_{ii}(f) \qquad i = 1, 2, \ldots, q \tag{7.13}$$

$$G_{yy}(f) = \sum_{i=1}^{q} |H_i(f)|^2 G_{ii}(f) + G_{nn}(f) \tag{7.14}$$

These relations have a very simple interpretation as a collection of distinct single-input/single-output models. No system of equations is required to solve for the $H_i(f)$. From Equation (7.13), each $H_i(f)$ is given by the ratio

$$H_i(f) = \frac{G_{iy}(f)}{G_{ii}(f)} \qquad i = 1, 2, \ldots, q \tag{7.15}$$

The output $G_{yy}(f)$ of Equation (7.14), unlike the output of Equation (7.12), now contains only $(q + 1)$ parts. Each of the q parts is the coherent output spectrum result of single-input/single-output models, namely,

$$|H_i(f)|^2 G_{ii}(f) = \gamma_{iy}^2(f)G_{yy}(f) \tag{7.16}$$

These quantities represent the input record $x_i(t)$ passing only through its particular $H_i(f)$ to reach $y(t)$. No leakage of $x_i(t)$ occurs through any other system $H_j(f)$ since $x_i(t)$ is uncorrelated with $x_j(t)$ for $i \neq j$.

For general cases of multiple input records, with arbitrary correlations between the inputs, any input record $x_i(t)$ can reach $y(t)$ by passage through any of the systems $H_i(f)$ for all $i = 1, 2, \ldots, q$. It can now be quite difficult to decompose $G_{yy}(f)$ into its respective contributions from each of the input records if one attempts to solve the general multiple-input/output problem by the usual "brute force" matrix techniques. These matrix techniques also provide no physical insight into how models should be formulated to provide good agreements between mathematical results and physical situations. Better methods are required to solve many problems, as explained in this chapter.

7.2 TWO-INPUT/ONE-OUTPUT MODELS

For an understanding of the methodology to solve general multiple-input/output problems, one should consider special cases of two-input/one-output models and three-input/one-output models. The general case is then merely more of the same features encountered in these special cases. This will now be done in detail for the two-input/one-output model. Equations for the three-input/one-output model can be developed from the material in Section 7.3.

7.2.1 Basic Relationships

Consider a two-input/one-output model as pictured in Figure 7.2, where the two inputs may be correlated. Assume that $x_1(t)$, $x_2(t)$, and $y(t)$ can be

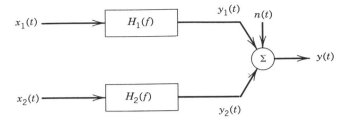

Figure 7.2 Two-input/one-output model.

measured. This system is defined by the basic transform relation

$$Y(f) = H_1(f)X_1(f) + H_2(f)X_2(f) + N(f) \qquad (7.17)$$

For arbitrary $n(t)$, which may be correlated with $x_1(t)$ and/or $x_2(t)$, the one-sided autospectral density function of $y(t)$ from Equations (7.5)–(7.8), for a very long but finite record length T, is given by

$$G_{yy}(f) = G_{yy} = \frac{2}{T}E[Y^*Y]$$

$$= \frac{2}{T}E[\{H_1^*X_1^* + H_2^*X_2^* + N^*\}\{H_1X_1 + H_2X_2 + N\}] \qquad (7.18)$$

where the dependence on f and the limiting operation on T have been omitted to simplify the notation. Equation (7.18) becomes

$$G_{yy} = |H_1|^2G_{11} + H_1^*H_2G_{12} + H_1H_2^*G_{21} + |H_2|^2G_{22} + G_{nn}$$

$$+ H_1^*G_{1n} + H_1G_{n1} + H_2^*G_{2n} + H_2G_{n2} \qquad (7.19)$$

where the last four terms occur only when $n(t)$ is correlated with $x_1(t)$ and $x_2(t)$.

For arbitrary $n(t)$, the cross-spectral density functions between $x_1(t)$ and $y(t)$, and between $x_2(t)$ and $y(t)$, are given in a similar way by

$$G_{1y} = \frac{2}{T}E[X_1^*Y] = \frac{2}{T}E[X_1^*(H_1X_1 + H_2X_2 + N)]$$

$$= H_1G_{11} + H_2G_{12} + G_{1n} \qquad (7.20)$$

$$G_{2y} = \frac{2}{T}E[X_2^*Y] = \frac{2}{T}E[X_2^*(H_1X_1 + H_2X_2 + N)]$$

$$= H_1G_{21} + H_2G_{22} + G_{2n} \qquad (7.21)$$

where, again, the dependence on f and the limiting operation on T have been omitted to simplify the notation. The terms G_{1n} and G_{2n} occur only when $n(t)$ is correlated with $x_1(t)$ and $x_2(t)$.

From knowledge of $x_1(t)$, $x_2(t)$, and $y(t)$, the one-sided spectral quantities will be

$$G_{11} = G_{11}(f) = \text{autospectrum of } x_1(t)$$

$$G_{22} = G_{22}(f) = \text{autospectrum of } x_2(t)$$

$$G_{yy} = G_{yy}(f) = \text{autospectrum of } y(t)$$

$$G_{12} = G_{12}(f) = \text{cross-spectrum between } x_1(t) \text{ and } x_2(t)$$

$$G_{1y} = G_{1y}(f) = \text{cross-spectrum between } x_1(t) \text{ and } y(t)$$

$$G_{2y} = G_{2y}(f) = \text{cross-spectrum between } x_2(t) \text{ and } y(t)$$

$$G_{21} = G_{12}^*(f) = \text{complex conjugate of } G_{12}$$

All spectral quantities involving $n(t)$ will be unknown. In particular, Equations (7.20) and (7.21) cannot be solved for H_1 and H_2 unless the terms $G_{1n} = 0$ and $G_{2n} = 0$, where

$$G_{1n} = G_{1n}(f) = \text{cross-spectrum between } x_1(t) \text{ and } n(t)$$

$$G_{2n} = G_{2n}(f) = \text{cross-spectrum between } x_2(t) \text{ and } n(t)$$

When $n(t)$ is uncorrelated with $x_1(t)$ and $x_2(t)$, Equations (7.20) and (7.21) become

$$G_{1y}(f) = H_1(f)G_{11}(f) + H_2(f)G_{12}(f)$$
$$G_{2y}(f) = H_1(f)G_{21}(f) + H_2(f)G_{22}(f)$$

$$(7.22)$$

The solutions for $H_1(f)$ and $H_2(f)$, assuming $\gamma_{12}^2(f) \neq 1$, are

$$H_1(f) = \frac{G_{1y}(f)\left[1 - \dfrac{G_{12}(f)G_{2y}(f)}{G_{22}(f)G_{1y}(f)}\right]}{G_{11}(f)\left[1 - \gamma_{12}^2(f)\right]}$$

$$(7.23)$$

$$H_2(f) = \frac{G_{2y}(f)\left[1 - \dfrac{G_{21}(f)G_{1y}(f)}{G_{11}(f)G_{2y}(f)}\right]}{G_{22}(f)\left[1 - \gamma_{12}^2(f)\right]}$$

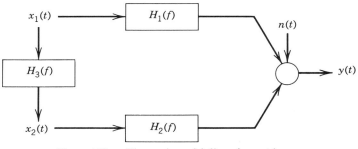

Figure 7.3 Illustration of fully coherent inputs.

where the ordinary coherence function

$$\gamma_{12}^2(f) = \frac{|G_{12}(f)|^2}{G_{11}(f)G_{22}(f)} \tag{7.24}$$

For the special case of uncorrelated inputs when $\gamma_{12}^2(f) = 0$, the terms $G_{12}(f)$ and $G_{21}(f)$ are zero also, and Equation (7.23) reduces to the usual relations for single-input/single-output models

$$H_1(f) = \frac{G_{1y}(f)}{G_{11}(f)} \qquad H_2(f) = \frac{G_{2y}(f)}{G_{22}(f)} \tag{7.25}$$

The case in which $\gamma_{12}^2(f) = 1$ must be handled separately. A coherence function of unity between $x_1(t)$ and $x_2(t)$ implies complete linear dependence. Hence one would consider a linear system existing between them, as illustrated in Figure 7.3. The implication is that the first input $x_1(t)$ is actually taking two different paths to arrive at the output $y(t)$. For this situation, a single frequency response function $H(f)$ will relate $y(t)$ to $x_1(t)$, namely,

$$H(f) = H_1(f) + H_2(f)H_3(f)$$

When $n(t)$ is uncorrelated with $x_1(t)$ and $x_2(t)$, but $G_{12}(f) \neq 0$, Equation (7.19) becomes

$$G_{yy}(f) = |H_1(f)|^2 G_{11}(f) + H_1^*(f)H_2(f)G_{12}(f)$$

$$+ H_2^*(f)H_1(f)G_{21}(f) + |H_2(f)|^2 G_{22}(f) + G_{nn}(f) \tag{7.26}$$

This can be written as

$$G_{yy}(f) = G_{vv}(f) + G_{nn}(f) = G_{y:x}(f) + G_{y:n}(f) \tag{7.27}$$

where $G_{nn}(f) = G_{y:n}(f)$ is the output spectrum due to the noise, and $G_{vv}(f) = G_{y:x}(f)$ represents the ideal output spectrum due to the two inputs, as computed by the first four terms in Equation (7.26). To be explicit, $G_{vv}(f)$ comes from knowledge of the terms $G_{11}(f), G_{12}(f), G_{22}(f)$ plus the computed $H_1(f)$ and $H_2(f)$ by the general Equation (7.23). Finally, even though $n(t)$ is unmeasured, the autospectral density function $G_{nn}(f)$ can be computed by the formula

$$G_{nn}(f) = G_{yy}(f) - G_{vv}(f) \tag{7.28}$$

For the special case of uncorrelated inputs where $G_{12}(f) = 0$, the ideal output spectrum reduces to

$$G_{vv}(f) = |H_1(f)|^2 G_{11}(f) + |H_2(f)|^2 G_{22}(f) \tag{7.29}$$

where $H_1(f)$ and $H_2(f)$ are given by the special Equation (7.25). Here

$$
\begin{aligned}
|H_1(f)|^2 G_{11}(f) &= \gamma_{1y}^2(f) G_{yy}(f) \\
|H_2(f)|^2 G_{22}(f) &= \gamma_{2y}^2(f) G_{yy}(f)
\end{aligned}
\tag{7.30}
$$

where $\gamma_{1y}^2(f)$ and $\gamma_{2y}^2(f)$ are the ordinary coherence functions

$$\gamma_{1y}^2(f) = \frac{|G_{1y}(f)|^2}{G_{11}(f) G_{yy}(f)} \qquad \gamma_{2y}^2(f) = \frac{|G_{2y}(f)|^2}{G_{22}(f) G_{yy}(f)} \tag{7.31}$$

It follows for uncorrelated inputs that

$$G_{vv}(f) = \left[\gamma_{1y}^2(f) + \gamma_{2y}^2(f) \right] G_{yy}(f) \tag{7.32}$$

and

$$G_{nn}(f) = \left[1 - \gamma_{1y}^2(f) - \gamma_{2y}^2(f) \right] G_{yy}(f) \tag{7.33}$$

The quantity $|H_1|^2 G_{11}$ represents the spectral output of x_1 through the H_1 system only. Similarly, $|H_2|^2 G_{22}$ represents the spectral output of x_2 through the H_2 system only. These two results are the ordinary coherent output spectra $\gamma_{1y}^2 G_{yy}$ and $\gamma_{2y}^2 G_{yy}$, respectively, between input x_1 and the output y, and between input x_2 and the output y.

7.2.2 Optimum Frequency Response Functions

Equations (7.22) and (7.26) are based on the assumption that $n(t)$ is uncorrelated with $x_1(t)$ and $x_2(t)$. Without making this assumption, when $x_1(t)$ and

$x_2(t)$ pass through any pair of constant parameter linear systems $H_1(f)$ and $H_2(f)$, respectively, Equation (7.17) states that

$$N(f) = Y(f) - H_1(f)X_1(f) - H_2(f)X_2(f)$$

Hence, for any $H_1(f)$ and $H_2(f)$, the output noise spectrum is

$$G_{nn}(f) = \frac{2}{T}E[N^*(f)N(f)]$$

$$= G_{yy}(f) - H_1(f)G_{y1}(f) - H_2(f)G_{y2}(f)$$

$$- H_1^*(f)G_{1y}(f) + H_1^*(f)H_1(f)G_{11}(f) + H_1^*(f)H_2(f)G_{12}(f)$$

$$- H_2^*(f)G_{2y}(f) + H_1(f)H_2^*(f)G_{21}(f) + H_2(f)H_2^*(f)G_{22}(f)$$

The *optimum frequency response functions* are now defined as the particular $H_1(f)$ and $H_2(f)$ that minimizes $G_{nn}(f)$ at any f over all possible choices of $H_1(f)$ and $H_2(f)$. They yield the optimum linear least squares prediction of $y(t)$ from $x_1(t)$ and $x_2(t)$.

To derive the optimum $H_1(f)$ and $H_2(f)$, as explained previously in Chapter 6, it is sufficient to set the following partial derivatives equal to zero.

$$\frac{\partial G_{nn}(f)}{\partial H_1^*(f)} = 0 \quad \text{holding } H_1(f) \text{ fixed}$$

$$\frac{\partial G_{nn}(f)}{\partial H_2^*(f)} = 0 \quad \text{holding } H_2(f) \text{ fixed}$$

This leads to the pair of equations

$$- G_{1y}(f) + H_1(f)G_{11}(f) + H_2(f)G_{12}(f) = 0$$

$$- G_{2y}(f) + H_1(f)G_{21}(f) + H_2(f)G_{22}(f) = 0$$

which are identical to Equation (7.22).

For any pair, $H_1(f)$ and $H_2(f)$, the cross-spectral density functions $G_{1n}(f)$ and $G_{2n}(f)$ are given by

$$G_{1n}(f) = \frac{2}{T}E[X_1^*(f)N(f)]$$

$$= G_{1y}(f) - H_1(f)G_{11}(f) - H_2(f)G_{12}(f)$$

$$G_{2n}(f) = \frac{2}{T}E[X_2^*(f)N(f)]$$

$$= G_{2y}(f) - H_1(f)G_{21}(f) - H_2(f)G_{22}(f)$$

When $H_1(f)$ and $H_2(f)$ are the optimum results that satisfy the relations of Equation (7.22), these two cross-spectra will be identically zero for all f. Thus, computation of the optimum $H_1(f)$ and $H_2(f)$ to minimize $G_{nn}(f)$ automatically makes $n(t)$ uncorrelated with $x_1(t)$ and $x_2(t)$, namely,

$$G_{1n}(f) = 0 \qquad G_{2n}(f) = 0$$

It should be pointed out here that the optimum computed $H_1(f)$ and $H_2(f)$ satisfying Equation (7.22) do not have to be physically realizable. That is, their associated weighting functions $h_1(\tau)$ and $h_2(\tau)$ may not be zero for τ less than zero. In fact, if the actual systems are nonlinear, these results will only represent optimum linear approximations as done by the computer and could never be the true nonlinear systems. Of course, if the actual systems are physically realizable constant-parameter linear systems where the model includes all the inputs producing the output, then these computed results would represent the true conditions provided the output noise is uncorrelated with all the inputs and there is negligible input noise. Henceforth, it will be assumed that $H_1(f)$ and $H_2(f)$ are computed by Equation (7.23). These results come from either (a) assuming in advance that $n(t)$ is uncorrelated with $x_1(t)$ and $x_2(t)$, or (b) requiring $H_1(f)$ and $H_2(f)$ to be the optimum constant-parameter linear systems to minimize $G_{nn}(f)$, whereupon the $n(t)$ in the model will become uncorrelated with $x_1(t)$ and $x_2(t)$.

7.2.3 *Ordinary and Multiple Coherence Functions*

Return to the general case of correlated inputs where $\gamma_{12}^2(f)$ is any positive value less than unity. The ordinary coherence functions between each input and the output are

$$\gamma_{1y}^2(f) = \frac{|H_1(f)G_{11}(f) + H_2(f)G_{12}(f)|^2}{G_{11}(f)G_{yy}(f)}$$

$$\gamma_{2y}^2(f) = \frac{|H_1(f)G_{21}(f) + H_2(f)G_{22}(f)|^2}{G_{22}(f)G_{yy}(f)}$$

(7.34)

where the numerators represent $G_{1y}(f)$ and $G_{2y}(f)$ as computed using Equation (7.22). The product $\gamma_{1y}^2 G_{yy}$ still represents the ordinary coherent output spectrum between the input x_1 and the output y. However, x_1 does not now get to the output via the H_1 system only. Instead since $\gamma_{12}^2 \neq 0$, part of x_1 also gets to the output y via the H_2 system. Similarly, x_2 gets to the output via both H_1 and H_2 when $\gamma_{12}^2 \neq 0$, and the ordinary coherent output spectrum $\gamma_{2y}^2 G_{yy}$ represents all the ways x_2 can get to the output y. In general, for small $G_{nn}(f)$, the sum of $\gamma_{1y}^2(f)$ with $\gamma_{2y}^2(f)$ will be greater than unity when the inputs are correlated.

The *multiple coherence function* is a relatively simple concept and is a direct extension of the ordinary coherence function. By definition, the multiple coherence function is the ratio of the ideal output spectrum due to the measured inputs in the absence of noise to the total output spectrum, which includes the noise. In equation form, the multiple coherence function

$$\gamma_{y:x}^2(f) = \frac{G_{vv}(f)}{G_{yy}(f)} = 1 - \left[\frac{G_{nn}(f)}{G_{yy}(f)}\right] \tag{7.35}$$

since $G_{vv}(f) = G_{yy}(f) - G_{nn}(f)$. For the general two-input case, $G_{vv}(f)$ is shown in Equation (7.26). The *multiple coherent output spectrum* is defined by the product of the multiple coherence function and the output spectrum, namely,

$$G_{vv}(f) = \gamma_{y:x}^2(f)G_{yy}(f) \tag{7.36}$$

Clearly, for all values of f, Equation (7.35) shows

$$0 \le \gamma_{y:x}^2(f) \le 1 \tag{7.37}$$

The value of unity occurs when $G_{nn}(f) = 0$, indicating a perfect linear model, and the value of zero occurs when $G_{yy}(f) = G_{nn}(f)$, indicating that none of the output record comes from linear operations on the measured input records.

For a single-input/single-output model where $x_2(t) = 0$ and $H_2(f) = 0$, the ideal output spectrum as previously derived in Equation (6.59) is

$$G_{vv}(f) = |H(f)|^2 G_{xx}(f) = \gamma_{xy}^2(f)G_{yy}(f) \tag{7.38}$$

where $x_1(t) = x(t)$ and $H(f) = [G_{xy}(f)/G_{xx}(f)]$. It follows that

$$\gamma_{y:x}^2(f) = \frac{G_{vv}(f)}{G_{yy}(f)} = \gamma_{xy}^2(f) \tag{7.39}$$

In words, the multiple coherence function is the same here as the ordinary coherence function.

For a two-input/one-output model with uncorrelated inputs, the ideal output spectrum as previously derived in Equation (7.32) is

$$G_{vv}(f) = \left[\gamma_{1y}^2(f) + \gamma_{2y}^2(f)\right]G_{yy}(f)$$

Here, the multiple coherence function is

$$\gamma_{y:x}^2(f) = \gamma_{1y}^2(f) + \gamma_{2y}^2(f) \tag{7.40}$$

In words, for uncorrelated inputs, the multiple coherence function is the sum of the ordinary coherence functions between each input and the output. No such simple relation exists for correlated inputs.

Example 7.1. Multiple Coherence for Two Uncorrelated Inputs. Consider the example of an uncorrelated two-input/one-output system with negligible output noise, and assume that the two inputs produce equal output spectra. For this example,

$$\gamma_{y:x}^2(f) = \gamma_{1y}^2(f) + \gamma_{2y}^2(f) = 1.0$$

with

$$\gamma_{1y}^2(f) = \gamma_{2y}^2(f) = 0.50$$

Here there are ideal constant-parameter linear systems $H_1(f)$ and $H_2(f)$ between each input and the output. The fact that $\gamma_{1y}^2(f) = 0.50$ is due to the second input. If this was not known and one assumed a single-input/single-output model, erroneous conclusions would be drawn.

7.2.4 *Conditioned Spectral Density Functions*

When correlation exists between any pair of input records, to make mathematical results agree with the physical situation, wherever possible one should try to determine if one record causes part or all of the second record. If this can be determined, then turning off the first record will remove the correlated parts from the second record and leave only the part of the second record that is not due to the first record. To be precise, if engineering considerations state that any correlation between $x_1(t)$ and $x_2(t)$ comes from $x_1(t)$, then the optimum linear effects of $x_1(t)$ to $x_2(t)$ should be found, as denoted by $x_{2:1}(t)$. This should be subtracted from $x_2(t)$ to yield the *conditioned record* (also called the *residual record*) $x_{2\cdot1}(t)$ representing the part of $x_2(t)$ not due to $x_1(t)$. In equation form, $x_2(t)$ is decomposed into the sum of two uncorrelated terms shown in Figure 7.4, where

$$x_2(t) = x_{2:1}(t) + x_{2\cdot1}(t) \tag{7.41}$$

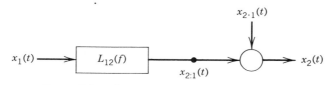

Figure 7.4 Decomposition of $x_2(t)$ from $x_1(t)$.

Fourier transforms yield

$$X_2(f) = X_{2:1}(f) + X_{2\cdot1}(f) \tag{7.42}$$

where

$$X_{2:1}(f) = L_{12}(f)X_1(f) \tag{7.43}$$

This defines the optimum linear least squares prediction of $X_2(t)$ from $X_1(t)$. The Fourier transform of $X_{2\cdot1}(t)$ is

$$X_{2\cdot1}(f) = X_2(f) - L_{12}(f)X_1(f) \tag{7.44}$$

The constant-parameter linear system $L_{12}(f)$ represents the optimum linear system to predict $x_2(t)$ from $x_1(t)$ taken in that order. As proved in Chapter 6, $L_{12}(f)$ is given by the ratio of the cross-spectrum from input to output divided by the autospectrum of the input, namely,

$$L_{12}(f) = \frac{G_{12}(f)}{G_{11}(f)} \tag{7.45}$$

It is also known from Equation (6.60) that this makes $x_{2\cdot1}(t)$ uncorrelated with $x_1(t)$ and decomposes the spectrum of $x_2(t)$ into

$$G_{22}(f) = G_{22:1}(f) + G_{22\cdot1}(f) \tag{7.46}$$

where $G_{22:1}(f)$ is the coherent output spectrum

$$G_{22:1}(f) = |L_{12}(f)|^2 G_{11}(f) = \gamma_{12}^2(f)G_{22}(f) \tag{7.47}$$

and $G_{22\cdot1}(f)$ is the noise output spectrum

$$G_{22\cdot1}(f) = [1 - \gamma_{12}^2(f)]G_{22}(f) \tag{7.48}$$

Note carefully the destinction between the indices $22:1$ and $22\cdot1$.

Example 7.2. Illustration of Erroneous High Coherence. An example of erroneous high ordinary coherence is shown in Figure 7.5. Assume that a coherence function value near unity is computed between the two variables $x_1(t)$ and $y(t)$. One would be inclined to believe that there is a physical linear system relating these two variables as input and output. But suppose there is a third variable $x_2(t)$, which is highly coherent with $x_1(t)$ and also passes through a linear system to make up $y(t)$. In this type of situation, the high coherence computed between $x_1(t)$ and $y(t)$ might be only a reflection of the fact that $x_2(t)$ is highly coherent with $x_1(t)$, and $x_2(t)$ is related via a linear system to $y(t)$. In reality there may be no direct physical system between $x_1(t)$ and $y(t)$ at all. If the partial coherence function (to be defined) were computed between $x_1(t)$ and $y(t)$ in this situation, it would be a very small number near zero. This concludes Example 7.2.

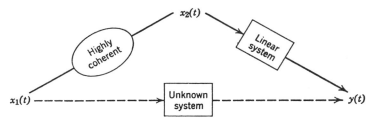

Figure 7.5   Illustration of erroneous high coherence.

For cases where cause-and-effect matters are not clear between the input records, one might compute the cross-correlation function between the two records to see if there is a relative time delay to indicate which record precedes the other record. As noted in Section 6.2.1, a strong cross-correlation and ordinary coherence can exist between any pair of records that come from a common source, even when neither record is the cause of the other.

When no physical basis exists for ordering the records, and when the relative time delay from a cross-correlation function is insignificant, a recommended approach is to compute the ordinary coherence function between each input record and the output record. At special frequencies of interest, such as peaks in the output spectrum where most of the power or energy is being transmitted, the records should be ordered according to their ordinary coherence functions. For example, at a selected frequency f_1, that input $x_1(t)$ or $x_2(t)$ giving the highest ordinary coherence function $\gamma_{1y}^2(f_1)$ or $\gamma_{2y}^2(f_1)$ should be selected as the first record. At a different selected frequency f_2, a similar examination of the ordinary coherence functions $\gamma_{1y}^2(f_2)$ and $\gamma_{2y}^2(f_2)$ may result in choosing a different ordering of the input records. Thus, different models may be appropriate at different frequencies.

For definiteness in this discussion, assume that $x_1(t)$ should precede $x_2(t)$. In place of Figure 7.2, one can now draw the equivalent Figure 7.6, where $H_1 = H_{1y}$ and $H_2 = H_{2y}$. Figure 7.6 shows that the input $x_1(t)$ reaches the output via two parallel paths, whereas the conditioned input $x_{2 \cdot 1}(t)$ goes via only one path. This is drawn explicitly in Figure 7.7.

In the original Figure 7.2, the two measured inputs $x_1(t)$ and $x_2(t)$ are correlated, with linear outputs $y_1(t)$ and $y_2(t)$ that are also correlated. In the equivalent Figure 7.7, the two inputs $x_1(t)$ and $x_{2 \cdot 1}(t)$ will be uncorrelated as accomplished by the data processing. The linear outputs $v_1(t)$ and $v_2(t)$ will also be uncorrelated. Figure 7.7 is equivalent to Figure 7.8 using different systems $L_{1y}(f)$ and $L_{2y}(f)$ where

$$L_{1y}(f) = H_{1y}(f) + L_{12}(f)H_{2y}(f)$$

$$L_{2y}(f) = H_{2y}(f) \tag{7.49}$$

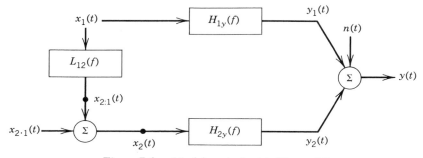

Figure 7.6   Model equivalent to Figure 7.2.

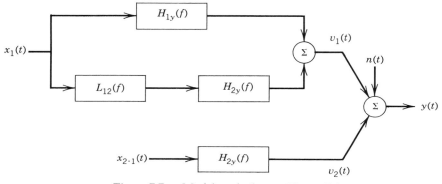

Figure 7.7   Model equivalent to Figure 7.6.

Figure 7.8 represents a two-input/one-output model, where the output $y(t)$ and the noise $n(t)$ are the same as in Figure 7.2. The inputs are now mutually uncorrelated, however, so that Figure 7.8 is equivalent to two separate single-input/single-output models whose nature will now be described. The constant-parameter linear system L_{1y} is the optimum linear system to predict $y(t)$ from $x_1(t)$, whereas the constant-parameter linear system L_{2y} is the optimum linear system to predict $y(t)$ from $x_{2 \cdot 1}(t)$. In equation form, for

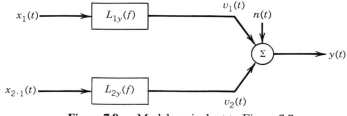

Figure 7.8   Model equivalent to Figure 7.7.

stationary random data, the basic frequency domain relation for Figure 7.8 is

$$Y(f) = L_{1y}(f)X_1(f) + L_{2y}(f)X_{2\cdot1}(f) + N(f) \qquad (7.50)$$

where

$$L_{1y}(f) = \frac{G_{1y}(f)}{G_{11}(f)}$$

$$\qquad\qquad (7.51)$$

$$L_{2y}(f) = \frac{G_{2y\cdot1}(f)}{G_{22\cdot1}(f)}$$

The quantities

$$G_{1y}(f) = \text{cross-spectrum between } x_1(t) \text{ and } y(t)$$

$$G_{11}(f) = \text{autospectrum of } x_1(t)$$

$$G_{2y\cdot1}(f) = \text{cross-spectrum between } x_{2\cdot1}(t) \text{ and } y(t)$$

$$G_{22\cdot1}(f) = \text{autospectrum of } x_{2\cdot1}(t)$$

The quantity $G_{2y\cdot1}(f)$ is called a *conditioned* (*residual*) *cross-spectral density function*, and $G_{22\cdot1}(f)$ is called a *conditioned* (*residual*) *autospectral density function*.

Computation of these conditioned spectral density functions can be done by algebraic operations on previously computed basic spectral density functions of the original measured records. It is *not* necessary to do any averaging of conditioned Fourier transforms except as a way to derive the desired algebraic formula. By definition, if averaging were performed, for finite T,

$$G_{2y\cdot1}(f) = \frac{2}{T}E\big[X_{2\cdot1}^*(f)Y(f)\big]$$

$$G_{22\cdot1}(f) = \frac{2}{T}E\big[X_{2\cdot1}^*(f)X_{2\cdot1}(f)\big]$$

From Equations (7.43) and (7.45),

$$X_{2\cdot1}(f) = X_2(f) - L_{12}(f)X_1(f)$$

where $L_{12}(f) = [G_{12}(f)/G_{11}(f)]$. Hence, for any three records $x_1(t)$, $x_2(t)$, and $y(t)$,

$$G_{2y\cdot 1}(f) = \frac{2}{T}E\left[\left\{X_2^*(f) - L_{12}^*(f)X_1^*(f)\right\}Y(f)\right]$$

$$= \frac{2}{T}E[X_2^*(f)Y(f)] - L_{12}^*(f)\left\{\frac{1}{T}E[X_1^*(f)Y(f)]\right\}$$

$$= G_{2y}(f) - [G_{21}(f)/G_{11}(f)]G_{1y}(f) \tag{7.52}$$

As a special case, replacing $y(t)$ by $x_2(t)$,

$$G_{22\cdot 1}(f) = G_{22}(f) - [G_{21}(f)/G_{11}(f)]G_{12}(f)$$

$$= [1 - \gamma_{12}^2(f)]G_{22}(f) \tag{7.53}$$

showing that $G_{22\cdot 1}(f)$ is the output noise spectrum for a single-input/single-output model with $x_1(t)$ as the input and $x_2(t)$ as the output as in Figure 7.4. Similarly, replacing $x_2(t)$ by $y(t)$ yields

$$G_{yy\cdot 1}(f) = [1 - \gamma_{1y}^2(f)]G_{yy}(f) \tag{7.54}$$

which is the output noise spectrum for a single-input/single-output model with $x_1(t)$ as the input and $y(t)$ as the output. Equations (7.52)–(7.54) are algebraic equations to compute conditioned cross-spectra, and conditioned autospectra from the original computed spectra.

The result in Equation (7.54) applies to the model pictured in Figure 7.9 showing how $y(t)$ can be decomposed into the sum of two uncorrelated terms representing the part of $y(t)$ due to $x_1(t)$ via optimum linear operations and the part of $y(t)$ not due to $x_1(t)$. This model is described in the frequency domain by the Fourier transforms

$$Y(f) = Y_{y:1}(f) + Y_{y\cdot 1}(f) \tag{7.55}$$

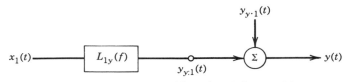

Figure 7.9   Decomposition of $y(t)$ from $x_1(t)$.

where

$$Y_{y:1}(f) = L_{1y}(f) X_1(f) \tag{7.56}$$

$$L_{1y}(f) = [G_{1y}(f)/G_{11}(f)] \tag{7.57}$$

$$Y_{y\cdot1}(f) = Y(f) - L_{1y}(f) X_1(f) \tag{7.58}$$

Note that results here are of the same nature as Equations (7.42)–(7.45).

7.2.5   *Partial Coherence Functions*

Return now to Figure 7.8. Since the output terms $v_1(t)$, $v_2(t)$, and $n(t)$ are mutually uncorrelated, the measured output autospectrum $G_{yy}(f)$ is the sum of three autospectra terms with no cross-spectra terms, namely,

$$G_{yy}(f) = G_{v_1v_1}(f) + G_{v_2v_2}(f) + G_{nn}(f) \tag{7.59}$$

where

$$G_{v_1v_1}(f) = |L_{1y}(f)|^2 G_{11}(f) \tag{7.60}$$

$$G_{v_2v_2}(f) = |L_{2y}(f)|^2 G_{22\cdot1}(f) \tag{7.61}$$

$$G_{nn}(f) = G_{yy\cdot1,2}(f) \tag{7.62}$$

The notation $G_{yy\cdot1,2}(f)$ indicates the autospectrum of $y(t)$ not due to either $x_1(t)$ or $x_2(t)$. Observe also that the first output $v_1(t)$ in Figure 7.8 is the same as the output $y_{y:1}(t)$ in Figure 7.9. The first autospectrum term

$$G_{v_1v_1}(f) = \left| \frac{G_{1y}(f)}{G_{11}(f)} \right|^2 G_{11}(f) = \gamma_{1y}^2(f) G_{yy}(f) \tag{7.63}$$

is the ordinary coherent output spectrum associated with a single-input/ single-output model with $x_1(t)$ as the input and $y(t)$ as the output as in Figure 7.9. Here

$$\gamma_{1y}^2(f) = \frac{|G_{1y}(f)|^2}{G_{11}(f) G_{yy}(f)} \tag{7.64}$$

is the ordinary coherence function between $x_1(t)$ and $y(t)$.

The second autospectrum term

$$G_{v_2v_2}(f) = \left| \frac{G_{2y\cdot1}(f)}{G_{22\cdot1}(f)} \right|^2 G_{22\cdot1}(f) = \gamma_{2y\cdot1}^2(f) G_{yy\cdot1}(f) \tag{7.65}$$

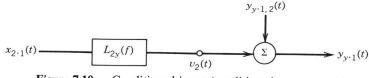

Figure 7.10 Conditioned-input/conditioned-output model.

where

$$\gamma^2_{2y\cdot1}(f) = \frac{|G_{2y\cdot1}(f)|^2}{G_{22\cdot1}(f)G_{yy\cdot1}(f)}$$

(7.66)

is, by definition, the *partial coherence function* between the conditioned records $x_{2\cdot1}(t)$ and $y_{y\cdot1}(t)$ as calculated in Figures 7.4 and 7.9, respectively. It follows that $G_{v_2v_2}(f)$ is the *partial coherent output spectrum* associated with a conditioned-input/conditioned-output model with $x_{2\cdot1}(t)$ as the input and $y_{y\cdot1}(t)$ as the output as in Figure 7.10. The uncorrelated noise term in Figure 7.10, which is denoted $y_{y\cdot1,2}(t)$, is the same as $n(t)$ in Figure 7.8.

Examination of Figure 7.10 shows that it is precisely the same form as Figure 7.9 except that

1. Original records $x_1(t)$ and $y(t)$ have been replaced by conditioned records $x_{2\cdot1}(t)$ and $y_{y\cdot1}(t)$
2. Original spectral quantities $G_{11}(f)$, $G_{yy}(f)$, and $G_{1y}(f)$ have been replaced by conditioned spectral quantities $G_{22\cdot1}(f)$, $G_{yy\cdot1}(f)$, and $G_{2y\cdot1}(f)$
3. The ordinary coherence function $\gamma^2_{1y}(f)$ has been replaced by the partial coherence function $\gamma^2_{2y\cdot1}(f)$

From this viewpoint, it is clear that partial coherence functions play the same role as ordinary coherence functions, except that they apply to conditioned records instead of to the original records. For all values of f, by the cross-spectrum inequality of Equation (5.82), it follows that

$$0 \le \gamma^2_{2y\cdot1}(f) \le 1$$

(7.67)

The two-input/one-output model of Figure 7.8 with mutually uncorrelated inputs $x_1(t)$ and $x_{2\cdot1}(t)$ is equivalent to the two separate single-input/single-output models of Figures 7.9 and 7.10. In particular, note that the input $x_1(t)$ in Figure 7.9 goes to the output $y(t)$, but the conditioned input $x_{2\cdot1}(t)$ in Figure 7.10 goes to the output $y_{y\cdot1}(t)$. *It does not go to the output $y(t)$.* A further observation of importance is that the cross-spectrum between $x_{2\cdot1}(t)$ and $y(t)$ must be the same as the cross-spectrum between $x_{2\cdot1}(t)$ and $y_{y\cdot1}(t)$.

In other words, if correlated effects of $x_1(t)$ are removed from $x_2(t)$ to produce $x_{2\cdot 1}(t)$, it is not necessary to also remove correlated effects of $x_1(t)$ from $y(t)$ to produce $y_{y\cdot 1}(t)$ insofar as the cross-spectrum $G_{2y\cdot 1}(f)$ is concerned. This fact will now be verified by proving that

$$E\left[X_{2\cdot 1}^* Y_{y\cdot 1}\right] = E\left[X_{2\cdot 1}^* Y\right] \tag{7.68}$$

From Equations (7.44) and (7.58)

$$X_{2\cdot 1}^* = X_2^* - \left(G_{21}/G_{11}\right)X_1^*$$

$$Y_{y\cdot 1} = Y - \left(G_{1y}/G_{11}\right)X_1$$

Now

$$E\left[X_{2\cdot 1}^* Y_{y\cdot 1}\right] = E\left[X_{2\cdot 1}^* Y\right] - E\left[X_{2\cdot 1}^*\left(G_{1y}/G_{11}\right)X_1\right]$$

But

$$E\left[X_{2\cdot 1}^* X_1\right] = E\left[\left\{X_2^* - \left(G_{21}/G_{11}\right)X_1^*\right\}X_1\right]$$

$$= E\left[X_2^* X_1\right] - \left(G_{21}/G_{11}\right)E\left[X_1^* X_1\right]$$

$$= 2TG_{21} - \left(G_{21}/G_{11}\right)2TG_{11}$$

$$= 0$$

This proves the desired result of Equation (7.68).

A formula is still needed for the noise output spectrum in Figure 7.10, which is the same as the desired noise output spectrum $G_{nn}(f)$ in Figure 7.8. This is given by

$$G_{yy\cdot 1,2}(f) = G_{yy\cdot 1}(f) - \left|L_{2y}(f)\right|^2 G_{22\cdot 1}(f)$$

$$= \left[1 - \gamma_{2y\cdot 1}^2(f)\right]G_{yy\cdot 1}(f) \tag{7.69}$$

Substitution from Equation (7.54) yields

$$G_{yy\cdot 1,2} = \left[1 - \gamma_{1y}^2(f)\right]\left[1 - \gamma_{2y\cdot 1}^2(f)\right]G_{yy}(f) \tag{7.70}$$

The multiple coherence function $\gamma_{y:x}^2(f)$ from Equation (7.35) for two correlated inputs is now

$$\gamma_{y:x}^2(f) = 1 - \left[\frac{G_{yy\cdot 1,2}(f)}{G_{yy}(f)}\right]$$

$$= 1 - \left[1 - \gamma_{1y}^2(f)\right]\left[1 - \gamma_{2y\cdot 1}^2(f)\right] \tag{7.71}$$

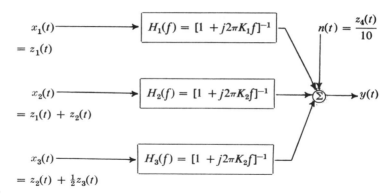

Figure 7.11 Three-input/one-output system with output noise.

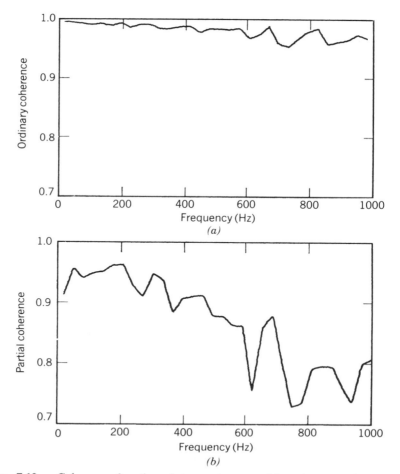

Figure 7.12 Coherence functions between input $x_2(t)$ and output for system in Figure 7.11. (*a*) Ordinary coherence function. (*b*) Partial coherence function.

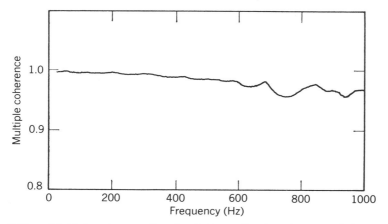

Figure 7.13   Multiple coherence function between the three inputs and the output for the system in Figure 7.11.

This formula shows how the multiple coherence function is related to associated ordinary and partial coherence functions for the prescribed ordering of the two records where $x_1(t)$ precedes $x_2(t)$.

Extensions of these ideas should be followed to solve the three-input/one-output problem for arbitrary correlated inputs, with similar procedures for general multiple-input/output problems. Distinctions must be made between unstructured models using the original collected data and ordered conditioned models created by data processing. Examples 7.3 and 7.4 show results of measurements for a three-input/one-output problem. Formulas for these computations are derived in Section 7.3.

***Example 7.3.* Illustration of Partial Coherence Measurement.**   Consider a three-input/one-output system with output measurement noise and low-pass *RC* filter transmission paths, as detailed in Figure 7.11. Note that input $x_2(t)$ is correlated with both inputs $x_1(t)$ and $x_3(t)$. How coherent is the input $x_2(t)$ with the output $y(t)$ through the indicated transmission path?

To answer this question, the ordinary and partial coherence functions between input $x_2(t)$ and the output $y(t)$ were computed over a frequency range from near zero to about 1000 Hz with a resolution bandwidth of 32 Hz and a record length of 1 sec. The results are presented in Figure 7.12. From Figure 7.12(a), it is seen that the ordinary coherence function between $x_2(t)$ and $y(t)$ is quite strong over the entire frequency range, varying from near unity at low frequencies to about 0.95 at higher frequencies. However, this is due in part to the fact that both $x_1(t)$ and $x_3(t)$ are coherent with $x_2(t)$ and also contribute directly to the output $y(t)$ through separate paths. When the partial coherence function between $x_2(t)$ and $y(t)$ is computed, a somewhat reduced indication of coherence (as low as 0.75) results, as shown in Figure 7.12(b).

Example 7.4. **Illustration of Multiple Coherence Measurement.**   Consider the three-input/one-output system previously discussed in Example 7.3 and detailed in Figure 7.11. The multiple coherence function computed between the output $y(t)$ and the three inputs $x_i(t)$, $i = 1, 2, 3$, is presented in Figure 7.13. Note that the multiple coherence function is quite high (greater than 0.96) over the entire frequency range, but does drop off slightly with increasing frequency. The only reason the multiple coherence is not unity over the entire frequency range is the measurement noise at the output. Since the measurement noise has a uniform spectrum while the inputs are low pass filtered, the relative contribution of measurement noise in the total output signal increases with frequency. This fact is clearly indicated by the decrease in the multiple coherence function with frequency.

7.3   GENERAL AND CONDITIONED MULTIPLE-INPUT MODELS

The general multiple-input/output model for arbitrary inputs is illustrated in Figure 7.14, where the terms $X_i(f)$, $i = 1, 2, \ldots, q$, represent computed finite Fourier transforms of the input records $x_i(t)$. The finite Fourier transform of the computed output record $y(t)$ is represented by $Y(f) = X_{q+1}(f)$. Constant-parameter linear frequency response functions to be determined are represented by $H_{iy}(f)$, $i = 1, 2, \ldots, q$, where the input index *precedes* the output index. All possible deviations from the ideal overall model are accounted for in the finite Fourier transform $N(f)$ of the unknown independent output noise. Similar models can be formulated by interchanging the order of the input records or by selecting different output records. It is assumed that the input and output records are measured simultaneously using a common time base. It is assumed also that nonzero mean values are removed and that possible bias errors due to propagation time delays are corrected prior to computing $X_i(f)$ and $Y(f)$.

An alternative conditioned multiple-input/output model is shown in Figure 7.15 by replacing the original given input records of Figure 7.14 by an ordered set of conditioned input records. No change is made in $Y(f)$ or $N(f)$. One can then compute the finite Fourier transforms $X_{i \cdot (i-1)!}(f)$, $i = 1, 2, \ldots, q$, selected in the order shown in Figure 7.15. For any i, the subscript notation $i \cdot (i - 1)!$ represents the ith record conditioned on the previous $(i - 1)$ records, that is, when the linear effects of $x_1(t)$, $x_2(t)$, up to $x_{i-1}(t)$ have been removed from $x_i(t)$ by optimum linear least squares prediction techniques. These ordered conditioned input records will be mutually uncorrelated, a property not generally satisfied by the original arbitrary records. Constant-parameter linear frequency response functions to be determined are represented by $L_{iy}(f)$, $i = 1, 2, \ldots, q$, where the input index *precedes* the output index.

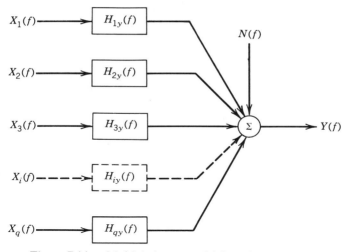

Figure 7.14 Multiple-input model for arbitrary inputs.

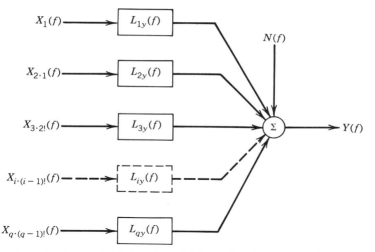

Figure 7.15 Multiple-input model for ordered conditioned inputs.

For the q input records, one can formulate a total of $q!$ different ordered conditioned multiple-input/output models since any of the original q records could be the first record, any of the remaining $(q - 1)$ records could be the second record, any of the remaining $(q - 2)$ records could be the third record, and so on. No one wants to or should analyze $q!$ models when q is a large number. For example, $q = 5$ could conceivably involve 120 different models. Fortunately, in practice, only a few possible orderings make physical sense, as discussed in Section 7.2.4. The suggestions given there should be followed to decide how to order the input records.

Note that for a given set of inputs, the systems $\{H_{iy}\}$ or $\{L_{iy}\}$ apply only to the single output. Different sets of $\{H_{iy}\}$ and $\{L_{iy}\}$ would be obtained for each different output. Thus, matrix notation is helpful to simplify equations for multiple-input/multiple-output problems. However, matrix notation is not needed and will not be used here for multiple-input/single-output problems.

From the previous discussion in Section 7.1, it is clear that solutions for the $\{H_{iy}\}$ systems in Figure 7.14 will be considerably more difficult than solutions for the $\{L_{iy}\}$ systems in Figure 7.15. It is also clear from Equations (7.12) and (7.14) that the output autospectrum $G_{yy}(f)$ in Figure 7.14 will contain $(q^2 + 1)$ terms, whereas this same output autospectrum $G_{yy}(f)$ in Figure 7.15 will contain only $(q + 1)$ terms.

Henceforth, Fourier transforms of original records and of conditional records will be denoted by capital letters where the dependence on frequency f will be omitted. Optimum systems $\{H_{iy}\}$ and optimum systems $\{L_{iy}\}$ will be calculated for both Figures 7.14 and 7.15, as well as the relationship between these optimum systems. To this end, one must know how to compute conditioned Fourier transforms and conditioned spectral density functions, as will be explained in succeeding sections.

7.3.1 Conditioned Fourier Transforms

For Figure 7.15, the defining Fourier transform equation is

$$Y = \sum_{i=1}^{q} L_{iy} X_{i \cdot (i-1)!} + N \tag{7.72}$$

If the output Y is considered to be a $(q + 1)$th record $X_{(q+1)}$ and if the noise N is considered to be this $(q + 1)$th record conditioned on the previous q records, then $N = X_{(q+1) \cdot q!}$ and Equation (7.72) becomes

$$X_{(q+1)} = \sum_{i=1}^{q} L_{i(q+1)} X_{i \cdot (i-1)!} + X_{(q+1) \cdot q!} \tag{7.73}$$

Here, $X_{(q+1) \cdot q!}$ is considered to be the $(q + 1)$th record conditioned on *all* of the preceding q records.

Various subsets are of interest. For the first r conditioned input records only where $r \leq q$, one can write

$$X_{(q+1)} = \sum_{i=1}^{r} L_{i(q+1)} X_{i \cdot (i-1)!} + X_{(q+1) \cdot r!} \tag{7.74}$$

Here, $X_{(q+1) \cdot r!}$ is considered to be the $(q + 1)$th record conditioned on the first r records where $r = 1, 2, \ldots, q$. Substitution of any jth record X_j for $X_{(q+1)}$ where $j > r$ yields the more general relation

$$X_j = \sum_{i=1}^{r} L_{ij} X_{i \cdot (i-1)!} + X_{j \cdot r!} \tag{7.75}$$

Here, the $\{L_{ij}\}$ systems replace the previous $\{L_{i(q+1)}\}$ systems. If r is now changed to $(r - 1)$, then Equation (7.75) becomes

$$X_j = \sum_{i=1}^{r-1} L_{ij} X_{i \cdot (i-1)!} + X_{j \cdot (r-1)!} \tag{7.76}$$

These last two results yield the conditioned Fourier transform algorithm

$$X_{j \cdot r!} = X_{j \cdot (r-1)!} - L_{rj} X_{r \cdot (r-1)!} \tag{7.77}$$

Thus $\{X_{j \cdot r!}\}$ can be computed from knowledge of $\{X_{j \cdot (r-1)!}\}$ and the $\{L_{rj}\}$ systems for all $j > r$. In particular, Equation (7.77) shows that the $\{X_{j \cdot 1}\}$ terms follow from knowledge of the $\{X_j\}$ terms and the $\{L_{1j}\}$ systems. Then the $\{X_{j \cdot 2!}\}$ terms follow from knowledge of the $\{X_{j \cdot 1}\}$ terms and the $\{L_{2j}\}$ systems, and so on.

7.3.2 Conditioned Spectral Density Functions

For the original records $\{X_i\}$, $i = 1, 2, \ldots, q$, and $Y = X_{q+1}$, their original autospectra and cross-spectra can be defined by the expressions

$$G_{ii} = \frac{2}{T} E[X_i^* X_i] \qquad G_{ij} = \frac{2}{T} E[X_i^* X_j] \tag{7.78}$$

$$G_{iy} = \frac{2}{T} E[X_i^* Y] \qquad G_{yy} = \frac{2}{T} E[Y^* Y]$$

Similarly, for conditioned records $X_{j \cdot r!}$ where $j > r$, their *conditioned autospectral density functions* are defined by

$$G_{jj \cdot r!} = \frac{2}{T} E[X_{j \cdot r!}^* X_{j \cdot r!}] \tag{7.79}$$

Conditioned cross-spectral density functions between $X_{i \cdot r!}$ and $X_{j \cdot r!}$ when $i \neq j$ with both $i > r$ and $j > r$ are defined by

$$G_{ij \cdot r!} = \frac{2}{T} E[X_{i \cdot r!}^* X_{j \cdot r!}] \tag{7.80}$$

Results in Equation (7.78) are special cases of Equations (7.79) and (7.80) when $r = 0$. Note also that

$$G_{ij \cdot r!} = \frac{2}{T} E[X_{i \cdot r!}^* X_j] = \frac{2}{T} E[X_i^* X_{j \cdot r!}] \tag{7.81}$$

To compute conditioned spectral density functions from the original spectral density functions, one uses the transform algorithm of Equation (7.77). Multiply both sides by X_i^*, take expected values, and multiply by the scale factor $(2/T)$. This gives the conditioned spectra algorithm

$$G_{ij \cdot r!} = G_{ij \cdot (r-1)!} - L_{rj} G_{ir \cdot (r-1)!} \tag{7.82}$$

Thus $\{G_{ij \cdot r!}\}$ can be computed from knowledge of $\{G_{ij \cdot (r-1)!}\}$ and the $\{L_{rj}\}$ systems for all $i > r$ and $j > r$.

Equation (7.82) is the basic algorithm required to solve multiple-input/output problems. In particular, Equation (7.82) shows that the $\{G_{ij \cdot 1}\}$ terms follow from knowledge of the $\{G_{ij}\}$ terms and the $\{L_{1j}\}$ systems. Then the $\{G_{ij \cdot 2!}\}$ terms follow from knowledge of the $\{G_{ij \cdot 1}\}$ terms and the $\{L_{2j}\}$ systems, and so on. The key to using this algorithm is to determine the $\{L_{rj}\}$ systems for all $r = 1, 2, \ldots, q$, and all $j > r$. This will now be developed.

7.3.3   Optimum Systems for Conditioned Inputs

The systems $\{L_{iy}\}$, $i = 1, 2, \ldots, q$, in Figure 7.15 are optimum linear systems for a collection of single-input/single-output systems since the inputs in Figure 7.15 are mutually uncorrelated. Consequently, as proved in Chapter 6, each system L_{iy} is defined by the ratio of the cross-spectral density function between its input and the output divided by the autospectral density function of its input. Without assuming it, the noise $n(t)$ will be automatically uncorrelated with each input in Figure 7.15 when L_{iy} is computed in this fashion. Thus, since conditioned inputs are present, one obtains the result

$$L_{iy} = \frac{G_{iy \cdot (i-1)!}}{G_{ii \cdot (i-1)!}} \qquad i = 1, 2, \ldots, q \tag{7.83}$$

Equation (7.83) can also be proved directly from Equation (7.72) if one assumes in advance that $n(t)$ is uncorrelated with each input in Figure 7.11. To derive this result, write Equation (7.72) with a different index of summation as

$$Y = \sum_{j=1}^{q} L_{jy} X_{j \cdot (j-1)!} + N$$

Multiply through both sides by $X_{i \cdot (i-1)!}^*$ where $i = 1, 2, \ldots, q$. This gives

$$X_{i \cdot (i-1)!}^* Y = \sum_{j=1}^{q} L_{jy} X_{i \cdot (i-1)!}^* X_{j \cdot (j-1)!} + X_{i \cdot (i-1)!}^* N$$

Now, expected values of both sides and multiplication by the factor $(2/T)$

yields Equation (7.83), namely,

$$G_{iy \cdot (i-1)!} = L_{iy} G_{ii \cdot (i-1)!}$$

since

$$E\left[X^*_{i \cdot (i-1)!} X_{j \cdot (j-1)!}\right] = 0 \qquad \text{for } i \neq j$$

$$E\left[X^*_{i \cdot (i-1)!} N\right] = 0 \qquad \text{for all } i$$

As special cases of Equation (7.83), note that

$$i = 1 \quad L_{1y} = \frac{G_{1y}}{G_{11}}$$

$$i = 2 \quad L_{2y} = \frac{G_{2y \cdot 1}}{G_{22 \cdot 1}} \tag{7.84}$$

$$i = 3 \quad L_{3y} = \frac{G_{3y \cdot 2!}}{G_{33 \cdot 2!}}$$

and so on. The last system where $i = q$ is

$$L_{qy} = \frac{G_{qy \cdot (q-1)!}}{G_{qq \cdot (q-1)!}} \tag{7.85}$$

What are the separate optimum single-input/single-output systems contained in Figure 7.15? To answer this question, note that the output autospectrum from passage of $X_{i \cdot (i-1)!}$ through L_{iy} for any $i = 1, 2, \ldots, q$ is given by

$$|L_{iy}|^2 G_{ii \cdot (i-1)!} = \gamma^2_{iy \cdot (i-1)!} G_{yy \cdot (i-1)!} \tag{7.86}$$

where $G_{ii \cdot (i-1)!}$ represents the input autospectrum and where

$$\gamma^2_{iy \cdot (i-1)!} = \frac{|G_{iy \cdot (i-1)!}|^2}{G_{ii \cdot (i-1)!} G_{yy \cdot (i-1)!}} \tag{7.87}$$

is the *partial coherence function* between $X_{i \cdot (i-1)!}$ and $Y_{y \cdot (i-1)!}$. Also, from Equation (7.82), when $i = j = y$ and $r = i$,

$$G_{yy \cdot i!} = G_{yy \cdot (i-1)!} - L_{iy} G_{yi \cdot (i-1)!}$$

$$= \left[1 - \gamma^2_{iy \cdot (i-1)!}\right] G_{yy \cdot (i-1)!} \tag{7.88}$$

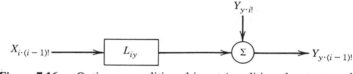

Figure 7.16 Optimum conditioned-input/conditioned-output model.

Here, $G_{yy \cdot i!}$ is the autospectrum of a conditioned record $Y_{y \cdot i!}$. The sum

$$|L_{iy}|^2 G_{ii \cdot (i-1)!} + G_{yy \cdot i!} = G_{yy \cdot (i-1)!} \qquad (7.89)$$

It follows that the single-input/single-output system for any conditioned input $x_{i \cdot (i-1)!}$ must be as shown in Figure 7.16.

The collection of optimum single-input/single-output models equivalent to Figure 7.15 can now be displayed in Figure 7.17. Note that the input X_1 goes to the output Y, the conditioned input $X_{2 \cdot 1}$ goes to the conditioned output

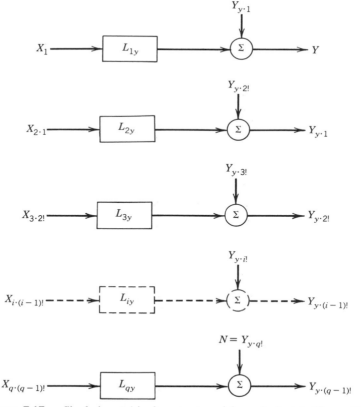

Figure 7.17 Single-input/single-output models equivalent to Figure 7.15.

$Y_{y \cdot 1}$, the conditioned input $X_{3 \cdot 2!}$ goes to the conditioned output $Y_{y \cdot 2!}$, and so on. It would be wrong to assume that any of these conditioned inputs went to the full output Y.

7.3.4   Algorithm for Conditioned Spectra

The algorithm to compute conditioned spectral density functions is contained in Equation (7.82), where the $\{L_{rj}\}$ systems must still be defined. This will now be done by extending the interpretation of the optimum $\{L_{iy}\}$ systems of Equation (7.83) for the inputs X_1, $X_{2 \cdot 1}$, $X_{3 \cdot 2!}$ up to $X_{q \cdot (q-1)!}$ with output Y. In place of Y, consider any output X_j, where $j = 1, 2, \ldots, (q + 1)$. Let the inputs be X_1, $X_{2 \cdot 1}$, $X_{3 \cdot 2!}$ up to $X_{r \cdot (r-1)!}$, where r can be any integer $r < j$, that is, $r = 1, 2, \ldots, (j - 1)$. Conceptually, this creates new conditioned multiple-input/single-output models where the associated optimum linear systems $\{L_{rj}\}$ per the derivation of the optimum $\{L_{iy}\}$ systems must be such that j replaces y and r replaces i to give the result

$$L_{rj} = \frac{G_{rj \cdot (r-1)!}}{G_{rr \cdot (r-1)!}} \qquad \begin{array}{l} r = 1, 2, \ldots, (j - 1) \\ j = 1, 2, \ldots, (q + 1) \end{array} \tag{7.90}$$

Note that L_{rj} involves conditioned records of order $(r - 1)!$. In particular, for $r = 1, 2, 3$, one obtains

$$r = 1 \qquad L_{1j} = \frac{G_{1j}}{G_{11}} \qquad j = 2, 3, \ldots, q, q + 1 \tag{7.91}$$

$$r = 2 \qquad L_{2j} = \frac{G_{2j \cdot 1}}{G_{22 \cdot 1}} \qquad j = 3, 4, \ldots, q, q + 1 \tag{7.92}$$

$$r = 3 \qquad L_{3j} = \frac{G_{3j \cdot 2!}}{G_{33 \cdot 2!}} \qquad j = 4, 5, \ldots, q, q + 1 \tag{7.93}$$

and so on. The case where $r = q$ and $j = (q + 1)$ yields the system $L_{q(q+1)} = L_{qy}$, as in Equation (7.85).

Return now to the previous iterative spectrum algorithm of Equation (7.82) and substitute Equation (7.90) for L_{rj}. This yields the final desired formula

$$G_{ij \cdot r!} = G_{ij \cdot (r-1)!} - \left[\frac{G_{rj \cdot (r-1)!}}{G_{rr \cdot (r-1)!}} \right] G_{ir \cdot (r-1)!} \tag{7.94}$$

to show exactly how conditioned spectral quantities of order $r!$ can be computed from previously known conditioned spectral quantities of order $(r - 1)!$ for any $r = 1, 2, \ldots, q$ and any i, j up to $(q + 1)$, where $i > r$ and $j > r$.

To make this result as explicit as possible, consider special cases where $r = 1, 2, 3$. For $r = 1$, Equation (7.94) states

$$G_{ij \cdot 1} = G_{ij} - \left[\frac{G_{1j}}{G_{11}} \right] G_{i1} \qquad (7.95)$$

All terms on the right-hand side are original autospectra and cross-spectra from the collected input and output records. The conditioned spectral quantities $G_{ij \cdot 1}$ are obtained by the algebraic operations in Equation (7.95) without choosing any bandwidth resolution or performing any averaging. These quantities are real-valued conditioned autospectra $G_{ii \cdot 1}$ when $i = j$, and complex-valued conditioned cross-spectra $G_{ij \cdot 1}$ when $i \neq j$. Indices i and j must be greater than 1.

For $r = 2$, Equation (7.94) states

$$G_{ij \cdot 2!} = G_{ij \cdot 1} - \frac{G_{2j \cdot 1}}{G_{22 \cdot 1}} G_{i2 \cdot 1} \qquad (7.96)$$

Now, all terms on the right-hand side are the computed conditioned spectra $\{G_{ij \cdot 1}\}$ from the previous step. The algebraic operations in Equation (7.96) yield the conditioned spectra $\{G_{ij \cdot 2!}\}$ without choosing any bandwidth resolution or performing any averaging. These quantities are real-valued conditioned autospectra $G_{ii \cdot 2!}$ when $i = j$ and complex-valued conditioned cross-spectra when $i \neq j$. Indices i and j must now be greater than 2.

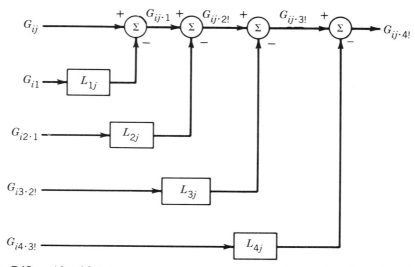

Figure 7.18 Algorithm to compute conditioned spectral density functions. (Results extend to any number of inputs.)

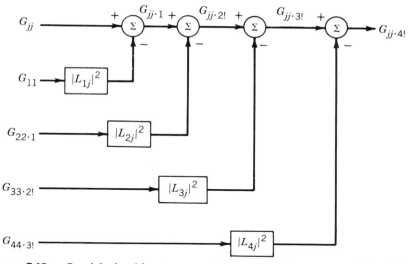

Figure 7.19 Special algorithm to compute conditioned autospectral density functions. (Results extend to any number of inputs.)

For $r = 3$, Equation (7.94) becomes an equation where conditioned spectra $\{G_{ij \cdot 3!}\}$ can be calculated algebraically from the previously computed conditioned spectra $\{G_{ij \cdot 2!}\}$, and so on. This iterative procedure is displayed in Figure 7.18 using the L_{rj} of Equation (7.90), where one goes from the first inner loop to the next inner loop successively. Results in Figure 7.18 apply to both conditioned autospectra where $i = j$ and to conditioned cross-spectra where $i \neq j$. An alternative special algorithm, however, can be used for conditioned autospectra. When $i = j$, Equation (7.94) becomes

$$G_{jj \cdot r!} = G_{jj \cdot (r-1)!} - |L_{rj}|^2 G_{rr \cdot (r-1)!} \tag{7.97}$$

This procedure is displayed in Figure 7.19.

7.3.5 Optimum Systems for Original Inputs

For the original inputs in Figure 7.14, the optimum linear systems $\{H_{iy}\}$, $i = 1, 2, \ldots, q$, are more complicated to compute than the optimum $\{L_{iy}\}$ systems in Figure 7.15. These $\{H_{iy}\}$ systems must satisfy the various q equations in q unknowns specified in Equation (7.11). It is not difficult, however, to derive relations that must exist between the $\{H_{iy}\}$ and $\{L_{iy}\}$ systems, as will now be demonstrated.

The governing equation for Figure 7.14 is

$$Y = \sum_{j=1}^{q} H_{jy} + N \tag{7.98}$$

Multiply through by $X^*_{i \cdot (i-1)!}$, where $i = 1, 2, \ldots, q$, and take expected values of both sides. A scale factor of $(2/T)$ then proves

$$G_{iy \cdot (i-1)!} = \sum_{j=i}^{q} H_{jy} G_{ij \cdot (i-1)!} \tag{7.99}$$

where the index j starts at $j = i$ since $G_{ij \cdot (i-1)!} = 0$ for $j < i$. Divide both sides by $G_{ii \cdot (i-1)!}$ to obtain the result

$$L_{iy} = \sum_{j=i}^{q} H_{jy} L_{ij} \qquad i = 1, 2, \ldots, q \qquad j \geq i \tag{7.100}$$

where, from Equation (7.83)

$$L_{iy} = \frac{G_{iy \cdot (i-1)!}}{G_{ii \cdot (i-1)!}} \qquad L_{ij} = \frac{G_{ij \cdot (i-1)!}}{G_{ii \cdot (i-1)!}} \tag{7.101}$$

Equation (7.100) is the desired relation between the $\{H_{iy}\}$ and $\{L_{iy}\}$ systems.

To understand Equation (7.100), consider some special cases. For $i = j = q$, since $L_{qq} = 1$,

$$L_{qy} = H_{qy} L_{qq} = H_{qy} \tag{7.102}$$

Thus the optimum system H_{qy}, as in Equation (7.85), is

$$H_{qy} = \frac{G_{qy \cdot (q-1)!}}{G_{qq \cdot (q-1)!}} = \frac{G_{qy \cdot 1, 2, \ldots, (q-1)}}{G_{qq \cdot 1, 2, \ldots, (q-1)}} \tag{7.103}$$

The optimum system H_{iy} for any i now follows by merely interchanging X_i with X_q to show that

$$H_{iy} = \frac{G_{iy \cdot 1, 2, \ldots, (i-1), (i+1), \ldots, q}}{G_{ii \cdot 1, 2, \ldots, (i-1), (i+1), \ldots, q}} \tag{7.104}$$

As special cases,

$$H_{1y} = \frac{G_{1y \cdot 2,3,\ldots,q}}{G_{11 \cdot 2,3,\ldots,q}}$$

$$H_{2y} = \frac{G_{2y \cdot 1,3,4,\ldots,q}}{G_{22 \cdot 1,3,4,\ldots,q}} \qquad (7.105)$$

$$H_{3y} = \frac{G_{3y \cdot 1,2,4,5,\ldots,q}}{G_{33 \cdot 1,2,4,5,\ldots,q}}$$

and so on. Comparison of Equation (7.84) with Equation (7.105) shows that the $\{L_{iy}\}$ systems are always simpler than the corresponding $\{H_{iy}\}$ systems for every $i = 1, 2, \ldots, (q-1)$, except for $i = q$, where $L_{qy} = H_{qy}$.

For the two-input/single-output system of Figure 7.2, where H_{1y} and H_{2y} are given by Equation (7.23), one now sees from Equation (7.105) that a shorthand way to describe these answers is

$$H_{1y} = \frac{G_{1y \cdot 2}}{G_{11 \cdot 2}}$$

$$H_{2y} = \frac{G_{2y \cdot 1}}{G_{22 \cdot 1}} \qquad (7.106)$$

Corresponding L_{1y} and L_{2y} systems are given in Equation (7.51). Note that relationships between these L and H systems are shown in Equation (7.49), in agreement with Equation (7.100).

Equation (7.100) provides a general procedure to determine the $\{H_{iy}\}$ systems from the $\{L_{iy}\}$ systems by working backwards, as follows:

$$H_{qy} = L_{qy}$$

$$H_{iy} = L_{iy} - \sum_{j=i+1}^{q} L_{ij} H_{jy} \qquad (7.107)$$

where $i = (q-1), (q-2), \ldots, 2, 1$. For example, if $q = 3$,

$$H_{3y} = L_{3y}$$

$$H_{2y} = L_{2y} - L_{23} H_{3y} \qquad (7.108)$$

$$H_{1y} = L_{1y} - L_{12} H_{2y} - L_{13} H_{3y}$$

In practice, to solve for possible $\{H_{iy}\}$ systems, rather than finding them directly, a simpler two-step method is to compute $\{L_{ij}\}$ systems first and then use Equation (7.107) to obtain the associated $\{H_{iy}\}$ systems.

7.3.6 *Partial and Multiple Coherence Functions*

Ordinary coherence functions between any input x_i for $i = 1, 2, \ldots, q$ and the output y are defined by

$$\gamma_{iy}^2 = \frac{|G_{iy}|^2}{G_{ii}G_{yy}} \tag{7.109}$$

Partial coherence functions between any conditioned input $x_{i \cdot 1}$ for $i = 2, 3, \ldots, q$ and the output y are defined by

$$\gamma_{iy \cdot 1}^2 = \frac{|G_{iy \cdot 1}|^2}{G_{ii \cdot 1}G_{yy \cdot 1}} \tag{7.110}$$

Partial coherence functions between any conditioned input $x_{i \cdot 2!}$ for $i = 3, 4, \ldots, q$ and the output y are defined by

$$\gamma_{iy \cdot 2!}^2 = \frac{|G_{iy \cdot 2!}|^2}{G_{ii \cdot 2!}G_{yy \cdot 2!}} \tag{7.111}$$

and so on up to

$$\gamma_{qy \cdot (q-1)!}^2 = \frac{|G_{qy \cdot (q-1)!}|^2}{G_{qq \cdot (q-1)!}G_{yy \cdot (q-1)!}} \tag{7.112}$$

The noise output spectrum in a single-input/single-output model is defined by

$$G_{yy \cdot 1} = G_{yy}\left(1 - \gamma_{1y}^2\right) \tag{7.113}$$

For a two-input/one-output model

$$G_{yy \cdot 2!} = G_{yy \cdot 1}\left(1 - \gamma_{2y \cdot 1}^2\right) = G_{yy}\left(1 - \gamma_{1y}^2\right)\left(1 - \gamma_{2y \cdot 1}^2\right) \tag{7.114}$$

For a three-input/one-output model

$$G_{yy \cdot 3!} = G_{yy \cdot 2!}\left(1 - \gamma_{3y \cdot 2!}^2\right) = G_{yy}\left(1 - \gamma_{1y}^2\right)\left(1 - \gamma_{2y \cdot 1}^2\right)\left(1 - \gamma_{3y \cdot 2!}^2\right)$$

$$\tag{7.115}$$

and so on. For a q-input/one-output model, the noise output spectrum is

$$G_{yy \cdot q!} = G_{yy} \prod_{i=1}^{q} \left(1 - \gamma_{iy \cdot (i-1)!}^2 \right) \tag{7.116}$$

These results are extensions of Equation (7.88).

It is now simple to define associated multiple coherence functions for any multiple-input/single-output model by using Equation (7.35). For the single-input/single-output model,

$$\gamma_{y:1}^2 = 1 - \left(\frac{G_{yy \cdot 1}}{G_{yy}} \right) = 1 - \left(1 - \gamma_{1y}^2 \right) = \gamma_{1y}^2 \tag{7.117}$$

For the two-input/one-output model,

$$\gamma_{y:2!}^2 = 1 - \left(\frac{G_{yy \cdot 2!}}{G_{yy}} \right) = 1 - \left(1 - \gamma_{1y}^2 \right)\left(1 - \gamma_{2y \cdot 1}^2 \right) \tag{7.118}$$

For the three-input/one-output model,

$$\gamma_{y:3!}^2 = 1 - \left(\frac{G_{yy \cdot 3!}}{G_{yy}} \right) = 1 - \left(1 - \gamma_{1y}^2 \right)\left(1 - \gamma_{2y \cdot 1}^2 \right)\left(1 - \gamma_{3y \cdot 2!}^2 \right) \tag{7.119}$$

and so on. For a q-input/one-output model, the multiple coherence function is given by

$$\gamma_{y:q!}^2 = 1 - \left(\frac{G_{yy \cdot q!}}{G_{yy}} \right) = 1 - \prod_{i=1}^{q} \left(1 - \gamma_{iy \cdot (i-1)!}^2 \right) \tag{7.120}$$

Equations (7.117)–(7.120) are aesthetically satisfying formulas for arbitrary correlated records in a multiple-input/single-output model, which show exactly how multiple coherence functions are related to underlying ordinary and partial coherence functions. Note that these formulas depend on the ordering of the records. For the special case in which the inputs are mutually uncorrelated, Equation (7.120) reduces to the simple known result

$$\gamma_{y:q!}^2 = \sum_{i=1}^{q} \gamma_{iy}^2 \tag{7.121}$$

In words, the multiple coherence is here the sum of all of the ordinary coherence functions between each input and the output.

7.4 MATRIX FORMULAS FOR MULTIPLE-INPUT/MULTIPLE-OUTPUT MODELS

General cases will now be considered as represented in multiple-input/multiple-output models, where the number of output records is the same as the number of input records. It is straightforward to extend these results to cases of uneven numbers of input and output records. For greatest physical insight into these problems, it is recommended that they be broken down into multiple-input/single-output problems and solved by the algebraic procedures outlined in Sections 7.2 and 7.3. If one desires to solve these multiple-input/multiple-output models by employing matrix techniques, however, a consistent set of definitions and matrix formulas are stated in this section. These matrix results provide no useful decomposition of measured output spectra into their components from measured input records. Also, error analysis criteria for these matrix results are considerably more complicated than error analysis criteria for the results computed in Section 7.3, where a multiple-input/single-output model is changed into a set of ordered conditioned single-input/single-output models. Relatively simple error analysis criteria are developed in Chapter 8 for such single-input/single-output models.

7.4.1 *Multiple - Input / Multiple - Output Model*

X is a column vector representing the Fourier transforms of the q input records $X_i = X_i(f)$, $i = 1, 2, \ldots, q$, and Y is a column vector representing the Fourier transforms of the q output records $Y_k = Y_k(f)$, $k = 1, 2, \ldots, q$;

$$\mathbf{X} = \begin{bmatrix} X_1 \\ X_2 \\ \vdots \\ X_q \end{bmatrix} \qquad \mathbf{Y} = \begin{bmatrix} Y_1 \\ Y_2 \\ \vdots \\ Y_q \end{bmatrix} \tag{7.122}$$

$\mathbf{X}^*, \mathbf{Y}^* = $ complex conjugate (column) vectors of $\mathbf{X}, \mathbf{Y}$

$\mathbf{X}', \mathbf{Y}' = $ transpose (row) vectors of $\mathbf{X}, \mathbf{Y}$

$$\mathbf{G}_{xx} = \frac{2}{T} E\{\mathbf{X}^* \mathbf{X}'\} = \text{input spectral density matrix} \tag{7.123}$$

$$\mathbf{G}_{yy} = \frac{2}{T} E\{\mathbf{Y}^* \mathbf{Y}'\} = \text{output spectral density matrix} \tag{7.124}$$

$$\mathbf{G}_{xy} = \frac{2}{T} E\{\mathbf{X}^* \mathbf{Y}'\} = \text{input/output cross-spectral density matrix} \tag{7.125}$$

Ideally, Equations (7.123)–(7.125) involve a limit as $T \to \infty$, but the limit notation, like the frequency dependence notation, is omitted for clarity. In practice, with finite records, the limiting operation is never performed.

The basic matrix terms are defined as follows.

$$G_{ij} = \frac{2}{T} E\left[X_i^* X_j\right] \tag{7.126}$$

$$G_{y_i y_j} = \frac{2}{T} E\left[Y_i^* Y_j\right] \tag{7.127}$$

$$G_{i y_i} = \frac{2}{T} E\left[X_i^* Y_j\right] \tag{7.128}$$

$$\mathbf{G}_{xx} = \frac{2}{T} E\left\{ \begin{bmatrix} X_1^* \\ X_2^* \\ \vdots \\ X_q^* \end{bmatrix} \begin{bmatrix} X_1 & X_2 & \cdots & X_q \end{bmatrix} \right\}$$

$$= \begin{bmatrix} G_{11} & G_{12} & \cdots & G_{1q} \\ G_{21} & G_{22} & \cdots & G_{2q} \\ \vdots & \vdots & & \vdots \\ G_{q1} & G_{q2} & \cdots & G_{qq} \end{bmatrix} \tag{7.129}$$

$$\mathbf{G}_{yy} = \frac{2}{T} E\left\{ \begin{bmatrix} Y_1^* \\ Y_2^* \\ \vdots \\ Y_q^* \end{bmatrix} \begin{bmatrix} Y_1 & Y_2 & \cdots & Y_q \end{bmatrix} \right\}$$

$$= \begin{bmatrix} G_{y_1 y_1} & G_{y_1 y_2} & \cdots & G_{y_1 y_q} \\ G_{y_2 y_1} & G_{y_2 y_2} & \cdots & G_{y_2 y_q} \\ \vdots & & & \vdots \\ G_{y_q y_1} & G_{y_q y_2} & \cdots & G_{y_q y_q} \end{bmatrix} \tag{7.130}$$

Observe that $\mathbf{G}_{xx}$ and $\mathbf{G}_{yy}$ are Hermitian matrices, namely, $G_{ij} = G_{ji}^*$ for all i

and j. For these Hermitian matrices, $\mathbf{G}_{xx}^* = \mathbf{G}_{xx}'$ and $\mathbf{G}_{yy}^* = \mathbf{G}_{yy}'$.

$$
\mathbf{G}_{xy} = \frac{2}{T} E \left\{ \begin{bmatrix} X_1^* \\ X_2^* \\ \vdots \\ X_q^* \end{bmatrix} \begin{bmatrix} Y_1 & Y_2 & \cdots & Y_q \end{bmatrix} \right\}
$$

$$
= \begin{bmatrix} G_{1y_1} & G_{1y_2} & \cdots & G_{1y_q} \\ G_{2y_1} & G_{2y_2} & \cdots & G_{2y_q} \\ \vdots & \vdots & & \vdots \\ G_{qy_1} & G_{qy_2} & \cdots & G_{qy_q} \end{bmatrix} \tag{7.131}
$$

In all these terms, $G_{iy_j} = G_{x_iy_j}(f)$, where the input precedes the output.

Define the system matrix between $\mathbf{X}$ and $\mathbf{Y}$ by $\mathbf{H}_{xy} = \mathbf{H}_{xy}(f)$, where, as above, *input always precedes output*. The matrix terms $H_{iy_k} = H_{x_iy_k}$. Then

$$
\mathbf{H}_{xy} = \begin{bmatrix} H_{1y_1} & H_{1y_2} & \cdots & H_{1y_q} \\ H_{2y_1} & H_{2y_2} & \cdots & H_{2y_q} \\ \vdots & \vdots & & \vdots \\ H_{qy_1} & H_{qy_2} & \cdots & H_{qy_q} \end{bmatrix} \tag{7.132}
$$

From this definition, it follows that

$$
\mathbf{Y} = \mathbf{H}_{xy}' \mathbf{X} \tag{7.133}
$$

where $\mathbf{H}_{xy}'$ is the *transpose* matrix to $\mathbf{H}_{xy}$. Thus

$$
\begin{bmatrix} Y_1 \\ Y_2 \\ \vdots \\ Y_q \end{bmatrix} = \begin{bmatrix} H_{1y_1} & H_{2y_1} & \cdots & H_{qy_1} \\ H_{1y_2} & H_{2y_2} & \cdots & H_{qy_2} \\ \vdots & \vdots & & \vdots \\ H_{1y_q} & H_{2y_q} & \cdots & H_{qy_q} \end{bmatrix} \begin{bmatrix} X_1 \\ X_2 \\ \vdots \\ X_q \end{bmatrix} \tag{7.134}
$$

Note that this gives the algebraic results

$$
Y_k = \sum_{i=1}^{q} H_{iy_k} X_i \qquad k = 1, 2, \ldots, q \tag{7.135}
$$

This is the logical way to relate any Y_k to the inputs $\{X_i\}$, where X_1 passes through H_{1y_k}, X_2 passes through H_{2y_k}, and so on until X_q passes through H_{qy_k}. Schematically,

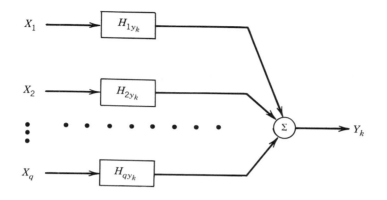

Consequences of these definitions will now be developed which provide matrix solutions for general multiple-input/multiple-output models when the number of outputs is the same as the number of inputs. It is assumed that all inverse matrix operations can be performed.

DETERMINATION OF $\mathbf{G}_{xy}$ AND $\mathbf{H}_{xy}$

$$\mathbf{Y} = \mathbf{H}'_{xy}\mathbf{X} \tag{7.136}$$

$$\mathbf{Y}' = \left(\mathbf{H}'_{xy}\mathbf{X}\right)' = \mathbf{X}'\mathbf{H}_{xy} \tag{7.137}$$

$$\mathbf{X}*\mathbf{Y}' = \mathbf{X}*\mathbf{X}'\mathbf{H}_{xy} \tag{7.138}$$

Taking expected values of both sides and multiplying by $(2/T)$ gives

$$\mathbf{G}_{xy} = \mathbf{G}_{xx}\mathbf{H}_{xy} \tag{7.139}$$

This shows how to obtain $\mathbf{G}_{xy}$ from $\mathbf{G}_{xx}$ and $\mathbf{H}_{xy}$. Multiply both sides of Equation (7.139) by $\mathbf{G}_{xx}^{-1}$ to obtain

$$\mathbf{G}_{xx}^{-1}\mathbf{G}_{xy} = \mathbf{G}_{xx}^{-1}\left(\mathbf{G}_{xx}\mathbf{H}_{xy}\right) \tag{7.140}$$

where

$$\mathbf{G}_{xx}^{-1} \text{ is the } \textit{inverse} \text{ matrix of } \mathbf{G}_{xx} \tag{7.141}$$

Equation (7.140) is the same as

$$\mathbf{H}_{xy} = \mathbf{G}_{xx}^{-1}\mathbf{G}_{xy} \tag{7.142}$$

This shows how to obtain $\mathbf{H}_{xy}$ from $\mathbf{G}_{xx}$ and $\mathbf{G}_{xy}$. The result of Equation (7.139) is easily pictured as the product of two $q \times q$ matrices $\mathbf{G}_{xx}$ and $\mathbf{H}_{xy}$ to give $\mathbf{G}_{xy}$.

DETERMINATION OF $\mathbf{G}_{yy}$ and $\mathbf{G}_{xx}$

$$\mathbf{Y} = \mathbf{H}'_{xy}\mathbf{X} \tag{7.143}$$

$$\mathbf{Y}' = \left(\mathbf{H}'_{xy}\mathbf{X}\right)' = \mathbf{X}'\mathbf{H}_{xy} \tag{7.144}$$

$$\mathbf{Y}^* = \left(\mathbf{H}'_{xy}\mathbf{X}\right)^* = \mathbf{H}'^*_{xy}\mathbf{X}^* \tag{7.145}$$

$$\mathbf{Y}^*\mathbf{Y}' = \left(\mathbf{H}'^*_{xy}\mathbf{X}^*\right)\left(\mathbf{X}'\mathbf{H}_{xy}\right) \tag{7.146}$$

Taking expected values of both sides and multiplying by $(2/T)$ gives

$$\mathbf{G}_{yy} = \mathbf{H}'^*_{xy}\mathbf{G}_{xx}\mathbf{H}_{xy} \tag{7.147}$$

It follows that

$$\mathbf{G}_{xx} = \left(\mathbf{H}'^*_{xy}\right)^{-1}\mathbf{G}_{yy}\left(\mathbf{H}_{xy}\right)^{-1} \tag{7.148}$$

assuming the required inverse matrices exist. The result of Equation (7.147) is easily pictured as the product of three $q \times q$ matrices, $\mathbf{H}'^*_{xy}$, $\mathbf{G}_{xx}$, and $\mathbf{H}_{xy}$, to give $\mathbf{G}_{yy}$. The complex conjugate transpose matrix is as follows.

$$\mathbf{H}'^*_{xy} = \begin{bmatrix} H^*_{1y_1} & H^*_{2y_1} & \cdots & H^*_{qy_1} \\ H^*_{1y_2} & H^*_{2y_2} & \cdots & H^*_{qy_2} \\ \vdots & \vdots & & \vdots \\ H^*_{1y_q} & H^*_{2y_q} & \cdots & H^*_{qy_q} \end{bmatrix} \tag{7.149}$$

Equations (7.139), (7.142), (7.147), and (7.148) are the main matrix formulas to solve multiple-input/multiple-output problems.

Matrix formulas will next be stated for a general multiple-input/single-output model. As before, the input records are not ordered or conditioned in any way.

7.4.2 Multiple-Input/Single-Output Model

$$\mathbf{X} = \begin{bmatrix} X_1 \\ X_2 \\ \vdots \\ X_q \end{bmatrix} \qquad \mathbf{H}_{xy} = \begin{bmatrix} H_{1y} \\ H_{2y} \\ \vdots \\ H_{qy} \end{bmatrix} \qquad \mathbf{Y} = Y \tag{7.150}$$

$$Y = \mathbf{H}'_{xy}\mathbf{X} = \begin{bmatrix} H_{1y} & H_{2y} & \cdots & H_{qy} \end{bmatrix} \begin{bmatrix} X_1 \\ X_2 \\ \vdots \\ X_q \end{bmatrix} \tag{7.151}$$

$$Y' = Y = \mathbf{X}'\mathbf{H}_{xy} = \begin{bmatrix} X_1 & X_2 & \cdots & X_q \end{bmatrix} \begin{bmatrix} H_{1y} \\ H_{2y} \\ \vdots \\ H_{qy} \end{bmatrix} \tag{7.152}$$

$$\mathbf{X}^* = \begin{bmatrix} X_1^* \\ X_2^* \\ \vdots \\ X_q^* \end{bmatrix} \qquad \mathbf{X}' = \begin{bmatrix} X_1 & X_2 & \cdots & X_q \end{bmatrix} \tag{7.153}$$

The input spectral density matrix is here

$$\mathbf{G}_{xx} = \frac{2}{T}E\{\mathbf{X}^*\mathbf{X}'\} = \begin{bmatrix} G_{11} & G_{12} & \cdots & G_{1q} \\ G_{21} & G_{22} & \cdots & G_{2q} \\ \vdots & \vdots & & \vdots \\ G_{q1} & G_{q2} & \cdots & G_{qq} \end{bmatrix} \tag{7.154}$$

The input/output column vector is

$$\mathbf{G}_{xy} = \frac{2}{T}E\{\mathbf{X}^*\mathbf{Y}\} = \begin{bmatrix} G_{1y} \\ G_{2y} \\ \vdots \\ G_{qy} \end{bmatrix} \tag{7.155}$$

The output spectral density function is

$$G_{yy} = \frac{2}{T}E\{Y^*Y'\} = \frac{2}{T}E\{\mathbf{H}'^*_{xy}\mathbf{X}^*\mathbf{X}'\mathbf{H}_{xy}\} = \mathbf{H}'^*_{xy}\mathbf{G}_{xx}\mathbf{H}_{xy} \tag{7.156}$$

Equation (7.156) is equivalent to writing

$$G_{yy} = \begin{bmatrix} H_{1y}^* & H_{2y}^* & \cdots & H_{qy}^* \end{bmatrix} \begin{bmatrix} G_{11} & G_{12} & \cdots & G_{1q} \\ G_{21} & G_{22} & \cdots & G_{2q} \\ \vdots & \vdots & & \vdots \\ G_{q1} & G_{q2} & \cdots & G_{qq} \end{bmatrix} \begin{bmatrix} H_{1y} \\ H_{2y} \\ \vdots \\ H_{qy} \end{bmatrix}$$

$$\tag{7.157}$$

Also

$$G_{yy}^* = \mathbf{H}_{xy}' \mathbf{G}_{xx}^* \mathbf{H}_{xy}^* \tag{7.158}$$

$$\mathbf{G}_{xx}^* = \mathbf{G}_{xx}' \tag{7.159}$$

For the multiple-input/single-output model,

$$G_{yy} = G_{yy}^* = G_{yy}' \tag{7.160}$$

$$\mathbf{G}_{xx}^* = \left(\mathbf{H}_{xy}'\right)^{-1} G_{yy} \left(\mathbf{H}_{xy}^*\right)^{-1} \tag{7.161}$$

$$\mathbf{G}_{xx} = \left(\mathbf{H}_{xy}'^*\right)^{-1} G_{yy} \left(\mathbf{H}_{xy}\right)^{-1} \tag{7.162}$$

$$\mathbf{G}_{xy} = \mathbf{G}_{xx} \mathbf{H}_{xy} \tag{7.163}$$

Equation (7.163) is equivalent to writing

$$\begin{bmatrix} G_{1y} \\ G_{2y} \\ \vdots \\ G_{qy} \end{bmatrix} = \begin{bmatrix} G_{11} & G_{12} & \cdots & G_{1q} \\ G_{21} & G_{22} & \cdots & G_{2q} \\ \vdots & \vdots & & \vdots \\ G_{q1} & G_{q2} & \cdots & G_{qq} \end{bmatrix} \begin{bmatrix} H_{1y} \\ H_{2y} \\ \vdots \\ H_{qy} \end{bmatrix} \tag{7.164}$$

Note also that

$$\mathbf{H}_{xy} = \mathbf{G}_{xx}^{-1} \mathbf{G}_{xy} \tag{7.165}$$

follows directly from Equation (7.163), where $\mathbf{G}_{xx}^{-1}$ is the inverse matrix of $\mathbf{G}_{xx}$. It is assumed here that the model is well defined and that all inverse matrices exist.

7.4.3 *Model with Output Noise*

Consider now a more realistic multiple-input/single-output model where extraneous uncorrelated output noise can occur. Instead of the ideal Equation (7.151), Y is given by

$$Y = \mathbf{H}_{xy}' \mathbf{X} + N \tag{7.166}$$

where $N = N(f)$ is the Fourier transform of the output noise $n(t)$. In place of Equation (7.156), one now obtains

$$G_{yy} = \mathbf{H}_{xy}'^* \mathbf{G}_{xx} \mathbf{H}_{xy} + G_{nn} \tag{7.167}$$

No change occurs in Equations (7.163) so that $\mathbf{G}_{xy}$ with extraneous noise present is still

$$\mathbf{G}_{xy} = \mathbf{G}_{xx}\mathbf{H}_{xy} \qquad (7.168)$$

Thus, G_{yy} can be expressed as

$$G_{yy} = \mathbf{H}'^*_{xy}\mathbf{G}_{xy} + G_{nn} \qquad (7.169)$$

The input spectral density matrix $\mathbf{G}_{xx}$ of Equation (7.154) is a $q \times q$ Hermitian matrix. Define an augmented spectral density matrix of the output $y(t)$ with the inputs $x_i(t)$ by the $(q + 1) \times (q + 1)$ Hermitian matrix

$$\mathbf{G}_{yxx} = \begin{bmatrix} G_{yy} & G_{y1} & G_{y2} & \cdots & G_{yq} \\ G_{1y} & G_{11} & G_{12} & \cdots & G_{1q} \\ G_{2y} & G_{21} & G_{22} & \cdots & G_{2q} \\ \vdots & \vdots & \vdots & & \vdots \\ G_{qy} & G_{q1} & G_{q2} & \cdots & G_{qq} \end{bmatrix} \qquad (7.170)$$

The determinant $|\mathbf{G}_{yxx}|$ of this augmented matrix will now be shown to be zero for all f in an ideal noise-free situation, where $G_{nn} = 0$ in Equation (7.169).

The G_{iy} terms for $i = 1, 2, \ldots, q$ in the first column of $\mathbf{G}_{yxx}$ (below the G_{yy} term) are linear combinations of the G_{ij} terms that appear in the columns of its row. That is, from Equation (7.164),

$$G_{iy} = \sum_{j=1}^{q} H_{jy}G_{ij} \qquad i = 1, 2, \ldots, q \qquad (7.171)$$

For the ideal noise-free case, the output spectrum is

$$G_{yy} = \mathbf{H}'^*_{xy}\mathbf{G}_{xy} = \mathbf{H}'_{xy}\mathbf{G}_{yx} \qquad (7.172)$$

where the last relation occurs because $G_{yy} = G^*_{yy}$ and $\mathbf{G}^*_{xy} = \mathbf{G}_{yx}$. Equation (7.172) is the same as

$$G_{yy} = \sum_{i=1}^{q} H_{iy}G_{yi} \qquad (7.173)$$

Thus, G_{yy} is a linear combination of the G_{yi} terms that appear in the columns of the first row of $\mathbf{G}_{yxx}$. It follows that the matrix $\mathbf{G}_{yxx}$ is such that its entire first column is the result of linear combinations from corresponding terms in the other columns. By a theorem proved in Reference 7.4, the determinant of this matrix must then be zero, that is

$$|\mathbf{G}_{yxx}| = 0 \quad \text{when } G_{nn} = 0 \qquad (7.174)$$

Return to the noise model where G_{yy} is given by Equation (7.169) with $G_{nn} \neq 0$. As noted earlier, no change occurs in Equation (7.163) whether or not extraneous output noise is present. Hence, by using the results derived for the noise-free case, the determinant $|\mathbf{G}_{yxx}|$ of the matrix $\mathbf{G}_{yxx}$ can be computed at any f by the formula

$$|\mathbf{G}_{yxx}| = G_{nn}|\mathbf{G}_{xx}| \qquad (7.175)$$

where $|\mathbf{G}_{xx}|$ is the determinant of the matrix G_{xx}. Note that this equation gives $|\mathbf{G}_{yxx}| = 0$ when $G_{nn} = 0$.

From Equation (7.35), it follows that the multiple coherence function is given by the determinant formula

$$\gamma_{y:x}^2 = 1 - \left(\frac{|\mathbf{G}_{yxx}|}{G_{yy}|\mathbf{G}_{xx}|} \right) \qquad (7.176)$$

As a check on this formula, when $q = 1$, corresponding to a single-input/single-output model,

$$\mathbf{G}_{yxx} = \begin{bmatrix} G_{yy} & G_{yx} \\ G_{xy} & G_{xx} \end{bmatrix} \qquad (7.177)$$

Here, the determinants

$$|\mathbf{G}_{yxx}| = G_{xx}G_{yy} - |G_{xy}|^2 \qquad |\mathbf{G}_{xx}| = G_{xx} \qquad (7.178)$$

Substitution into Equation (7.176) yields

$$\gamma_{y:x}^2 = 1 - \left(\frac{G_{xx}G_{yy} - |G_{xy}|^2}{G_{xx}G_{yy}} \right) = \frac{|G_{xy}|^2}{G_{xx}G_{yy}} \qquad (7.179)$$

which is the required ordinary coherence function for this situation.

7.4.4   Single-Input/Single-Output Model

$$\mathbf{X} = \mathbf{X}' = X \qquad \mathbf{Y} = \mathbf{Y}' = Y \qquad (7.180)$$

$$\mathbf{H}_{xy} = \mathbf{H}'_{xy} = H \qquad (7.181)$$

$$Y = HX \qquad (7.182)$$

$$G_{xx} = \frac{2}{T} E[X^*X] = G_{xx}^* \tag{7.183}$$

$$G_{yy} = \frac{2}{T} E[Y^*Y] = G_{yy}^* \tag{7.184}$$

$$G_{xy} = \frac{2}{T} E[X^*Y] \qquad G_{xy}^* = G_{yx} \tag{7.185}$$

From Equations (7.182)–(7.185), it follows that

$$G_{xy} = HG_{xx} \tag{7.186}$$

with

$$H = \frac{G_{xy}}{G_{xx}} \tag{7.187}$$

$$G_{yy} = H^*G_{xx}H = |H|^2 G_{xx} \tag{7.188}$$

with

$$G_{xx} = \frac{G_{yy}}{|H|^2} \tag{7.189}$$

The relations given above are well known results of Chapter 6 for single-input/single-output problems. They are special cases of previously derived matrix formulas in Sections 7.4.1 and 7.4.2, and show that definitions used there are appropriate for multiple-input/multiple-output problems and multiple-input/single-output problems.

PROBLEMS

7.1 Consider a two-input/single-output system with uncorrelated output noise as shown below.

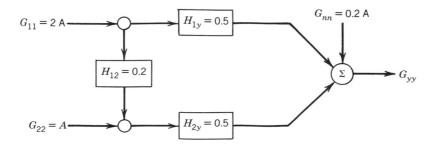

Determine the output spectral density G_{yy}. What would be the output spectral density if the two inputs were uncorrelated ($H_{12} = 0$)?

7.2 In Problem 7.1, determine the coherence γ_{12}^2 between the two inputs as well as the coherence γ_{1y}^2 and γ_{2y}^2 between each input and the output.

7.3 In Problem 7.1, determine the multiple coherence function between the two inputs and the output.

7.4 Consider a two-input/single-output system where the following spectral density functions are measured at a frequency of interest

$$G_{11} = 3 \qquad G_{22} = 2 \qquad G_{yy} = 10$$

$$G_{12} = 1 + j1 \qquad G_{1y} = 4 + j1 \qquad G_{2y} = 3 - j1$$

Determine the conditioned spectral density functions
(a) $G_{22 \cdot 1}$
(b) $G_{2y \cdot 1}$
(c) $G_{yy \cdot 1}$
Also determine the estimated frequency response function H_2.

7.5 Using the data in Problem 7.4, determine the multiple coherence function between the two inputs and the output.

7.6 Consider an ideal two-input/single-output system where the frequency response functions are defined by

$$H_1(f) = \frac{2}{5 + jf} \qquad H_2(f) = \frac{3}{5 + jf}$$

Assume the inputs satisfy

$$R_{11}(\tau) = 3\delta(\tau) \qquad G_{22}(f) = 12 \qquad G_{12}(f) = 8$$

Determine the following quantities:
(a) $\gamma_{12}^2(f)$
(b) $R_{yy}(\tau)$ and $G_{yy}(f)$
(c) $R_{1y}(\tau)$ and $G_{1y}(f)$
(d) $\gamma_{1y}^2(f)$

7.7 Consider a multiple-input linear system where the q inputs are mutually uncorrelated except for nonzero mean values. Determine the equations for
(a) the cross-spectrum between the ith input and the output that replaces Equation (7.13).
(b) the autospectrum of the output that replaces Equation (7.14).

7.8 Using the data in Problem 7.1, determine the systems L_{1y} and L_{2y} that relate the uncorrelated inputs with autospectra G_{11} and $G_{22 \cdot 1}$ to the output, as defined in Equation (7.51).

7.9 In a three-input/single-output model, the following input quantities are measured at a frequency of interest:

$$G_{11} = 10 \qquad G_{22} = 8 \qquad G_{33} = 6$$

$$G_{12} = 2 + j1 \qquad G_{32} = 1 - j2 \qquad G_{13} = 3 + j3$$

Determine an appropriate order for the three input records that would provide the most physically meaningful results from a conditioned analysis using the model in Figure 7.15.

7.10 In Problem 7.1, assume $H_{12} = 0.2 + j0.2$ at the frequency of interest. Determine the output spectral density G_{yy} that would be measured if the second input $G_{22} = 0$.

REFERENCES

7.1 Dodds, C. J., and Robson, J. D., "Partial Coherence in Multivariate Random Processes," *Journal of Sound and Vibration*, Vol. 42, p. 243, 1975.

7.2 Bendat, J. S., and Piersol, A. G., *Engineering Applications of Correlation and Spectral Analysis*, Wiley-Interscience, New York, 1980.

7.3 Bendat, J. S., "Modern Analysis Procedures for Multiple Input/Output Problems," *Journal of the Acoustical Society of America*, Vol. 68, p. 498, 1980.

7.4 Liebeck, H., *Algebra for Scientists and Engineers*, Wiley, New York, 1969.

CHAPTER 8

STATISTICAL ERRORS IN BASIC ESTIMATES

As noted in Chapter 4, the descriptive properties of a random variable cannot be precisely determined from sample data. Only estimates of the parameters of interest can be obtained from a finite sample of observations. The accuracy of certain basic parameter estimates is discussed in Chapter 4 for the case of data in the form of discrete independent observations of sample size N. In this chapter, the accuracy of parameter estimates is developed for data in the form of continuous time-history records of record length T. It is assumed that the data are single sample records from continuous stationary (ergodic) random processes with arbitrary mean values. Statistical error formulas are developed for

Mean value estimates
Mean square value estimates
Probability density function estimates
Correlation function estimates
Autospectral density function estimates

Attention in this chapter and the next chapter is restricted to those errors that are due solely to statistical considerations. Other errors associated with data acquisition and processing are covered in Chapter 10.

8.1 DEFINITION OF ERRORS

Referring to Section 4.1, the accuracy of parameter estimates based on sample values can be described by a mean square error defined as

$$\text{mean square error} = E\left[(\hat{\phi} - \phi)^2\right] \tag{8.1}$$

where $\hat{\phi}$ is an estimator for ϕ. Expanding Equation (8.1) yields

$$E\left[(\hat{\phi} - \phi)^2\right] = E\left[(\hat{\phi} - E[\hat{\phi}] + E[\hat{\phi}] - \phi)^2\right]$$

$$= E\left[(\hat{\phi} - E[\hat{\phi}])^2\right] + 2E\left[(\hat{\phi} - E[\hat{\phi}])(E[\hat{\phi}] - \phi)\right]$$

$$+ E\left[(E[\hat{\phi}] - \phi)^2\right]$$

Note that the middle term in the above expression has a factor equal to zero, namely,

$$E[\hat{\phi} - E[\hat{\phi}]] = E[\hat{\phi}] - E[\hat{\phi}] = 0$$

Hence the mean square error reduces to

$$\text{mean square error} = E\left[(\hat{\phi} - E[\hat{\phi}])^2\right] + E\left[(E[\hat{\phi}] - \phi)^2\right] \qquad (8.2)$$

In words, the mean square error is the sum of two parts. The first part is a variance term that describes the random portion of the error,

$$\text{Var}[\hat{\phi}] = E\left[(\hat{\phi} - E[\hat{\phi}])^2\right] = E[\hat{\phi}^2] - E^2[\hat{\phi}] \qquad (8.3)$$

and the second part is the square of a bias term that describes the systematic portion of the error,

$$b^2[\hat{\phi}] = E[b^2[\hat{\phi}]] = E\left[(E[\hat{\phi}] - \phi)^2\right] \qquad (8.4)$$

Thus the mean square error is the sum of the variance of the estimate plus the square of the bias of the estimate, that is,

$$E\left[(\hat{\phi} - \phi)^2\right] = \text{Var}[\hat{\phi}] + b^2[\hat{\phi}] \qquad (8.5)$$

It is generally more convenient to describe the error of an estimate in terms that have the same engineering units as the parameter being estimated. This can be achieved by taking the positive square roots of the error terms in Equations (8.3)–(8.5). The square root of Equation (8.3) yields the standard deviation for the estimate, called the *standard error* or *random error*, as follows

$$\text{random error} = \sigma[\hat{\phi}] = \sqrt{E[\hat{\phi}^2] - E^2[\hat{\phi}]} \qquad (8.6)$$

The square root of Equation (8.4) gives the *bias error* directly as

$$\text{bias error} = b[\hat{\phi}] = E[\hat{\phi}] - \phi \qquad (8.7)$$

The square root of the sum of the squared errors, as given by Equation (8.5), defines the *root mean square* (rms) error as

$$\text{rms error} = \sqrt{E\left[(\hat{\phi} - \phi)^2\right]} = \sqrt{\sigma^2[\hat{\phi}] + b^2[\hat{\phi}]} \tag{8.8}$$

As a further convenience, it is often desirable to define the error of an estimate in terms of a fractional portion of the quantity being estimated. This is done by dividing the error by the quantity being estimated to obtain a *normalized error*. For $\phi \neq 0$, the normalized random, bias, and rms errors are given by

$$\text{normalized random error} = \varepsilon_r = \frac{\sigma[\hat{\phi}]}{\phi} = \frac{\sqrt{E[\hat{\phi}^2] - E^2[\hat{\phi}]}}{\phi} \tag{8.9a}$$

$$\text{normalized bias error} = \varepsilon_b = \frac{b[\hat{\phi}]}{\phi} = \frac{E[\hat{\phi}]}{\phi} - 1 \tag{8.9b}$$

$$\text{normalized rms error} = \varepsilon = \frac{\sqrt{\sigma^2[\hat{\phi}] + b^2[\hat{\phi}]}}{\phi} = \frac{\sqrt{E\left[(\hat{\phi} - \phi)^2\right]}}{\phi} \tag{8.9c}$$

Note that the normalized random error ε_r is often called the *coefficient of variation*.

For situations where ε_r is small, if one sets

$$\hat{\phi}^2 = \phi^2(1 \pm \varepsilon_r)$$

then

$$\hat{\phi} = \phi(1 \pm \varepsilon_r)^{1/2} \approx \phi\left(1 \pm \frac{\varepsilon_r}{2}\right)$$

Thus

$$\varepsilon_r[\hat{\phi}^2] \approx 2\varepsilon_r[\hat{\phi}] \tag{8.10}$$

In words, when ε_r is small, the normalized random error for squared estimates $\hat{\phi}^2$ is approximately twice the normalized random error for unsquared estimates $\hat{\phi}$.

When estimates $\hat{\phi}$ have a negligible bias error $b[\hat{\phi}] \approx 0$ and a small normalized rms error $\varepsilon = \varepsilon[\hat{\phi}] = \sigma[\hat{\phi}]/\phi$, say $\varepsilon \leq 0.20$, then the probability density function $p(\hat{\phi})$ for these estimates can be approximated by a Gaussian distribution, where the mean value $E[\hat{\phi}] \approx \phi$ and the standard deviation

$\sigma[\hat{\phi}] = \varepsilon\phi$, as follows:

$$p(\hat{\phi}) = \frac{1}{\varepsilon\phi\sqrt{2\pi}} \exp\left[\frac{-(\hat{\phi} - \phi)^2}{2(\varepsilon\phi)^2}\right] \tag{8.11}$$

This gives probability statements for future values of the estimates $\hat{\phi}$, such as

$$\text{Prob}[\phi(1 - \varepsilon) \le \hat{\phi} \le \phi(1 + \varepsilon)] \approx 0.68$$

$$\text{Prob}[\phi(1 - 2\varepsilon) \le \hat{\phi} \le \phi(1 + 2\varepsilon)] \approx 0.95 \tag{8.12}$$

A confidence interval for the unknown true value ϕ based on any single estimate $\hat{\phi}$ then becomes

$$\left[\frac{\hat{\phi}}{1 + \varepsilon} \le \phi \le \frac{\hat{\phi}}{1 - \varepsilon}\right] \quad \text{with 68\% confidence}$$

$$\left[\frac{\hat{\phi}}{1 + 2\varepsilon} \le \phi \le \frac{\hat{\phi}}{1 - 2\varepsilon}\right] \quad \text{with 95\% confidence} \tag{8.13}$$

For small ε, say $\varepsilon \le 0.10$, this simplifies to

$$[\hat{\phi}(1 - \varepsilon) \le \phi \le \hat{\phi}(1 + \varepsilon)] \quad \text{with 68\% confidence}$$

$$[\hat{\phi}(1 - 2\varepsilon) \le \phi \le \hat{\phi}(1 + 2\varepsilon)] \quad \text{with 95\% confidence} \tag{8.14}$$

These confidence claims can be made when ε is small even though the actual unknown sampling distribution for $\hat{\phi}$ is theoretically chi-square, F, or some other more complicated distribution detailed in Chapter 4.

For cases where normalized rms errors are not small, confidence interval statements can still be made, as described in Section 4.4. Such matters will be discussed for spectral density estimates later in this chapter. Along with the derivation of errors, the consistency of various estimates will also be noted, as defined by Equation (4.7).

Example 8.1. **Approximate 95% Confidence Intervals for Mean Square and rms Value Estimates.** Assume the mean square value of a signal $x(t)$ is estimated to be $\hat{\psi}_x^2 = 4$ with a normalized random error of $\varepsilon_r = 0.05$. Determine the approximate 95% confidence intervals for the mean square value ψ_x^2 and the rms value ψ_x of the signal.

From Equation (8.14), the approximate 95% confidence interval for ψ_x^2 is

$$[4(0.90) \le \psi_x^2 \le 4(1.10)] = [3.6 \le \psi_x^2 \le 4.4]$$

Using the result in Equation (8.10), $\varepsilon_r = 0.025$ for $\hat{\psi}_x = 2$, so the approximate

95% confidence interval for ψ_x is

$$\left[2(0.95) \leq \psi_x \leq 2(1.05)\right] = \left[1.9 \leq \psi_x \leq 2.1\right]$$

Note that the confidence interval for ψ_x is approximately the square root of the interval for ψ_x^2.

8.2 MEAN AND MEAN SQUARE VALUE ESTIMATES

8.2.1 Mean Value Estimates

Suppose that a single sample time history record $x(t)$ from a stationary (ergodic) random process $\{x(t)\}$ exists over a finite time T. The mean value of $\{x(t)\}$ can be estimated by

$$\hat{\mu}_x = \frac{1}{T} \int_0^T x(t) \, dt \tag{8.15}$$

The true mean value is

$$\mu_x = E[x(t)] \tag{8.16}$$

and is independent of t since $\{x(t)\}$ is stationary. The expected value of the estimate $\hat{\mu}_x$ is

$$E[\hat{\mu}_x] = E\left[\frac{1}{T} \int_0^T x(t) \, dt\right] = \frac{1}{T} \int_0^T E[x(t)] \, dt = \frac{1}{T} \int_0^T \mu_x \, dt = \mu_x \tag{8.17}$$

since expected values commute with linear operations. Hence $\hat{\mu}_x$ is an *unbiased* estimate of μ_x, independent of T. Since $\hat{\mu}_x$ is unbiased, the mean square error of the estimate $\hat{\mu}_x$ is equal to the variance as follows:

$$\text{Var}[\hat{\mu}_x] = E\left[(\hat{\mu}_x - \mu_x)^2\right] = E[\hat{\mu}_x^2] - \mu_x^2 \tag{8.18}$$

where from Equation (8.15),

$$E[\hat{\mu}_x^2] = \frac{1}{T^2} \int_0^T \int_0^T E[x(\xi)x(\eta)] \, d\eta \, d\xi \tag{8.19}$$

Now the autocorrelation function $R_{xx}(\tau)$ of a stationary random process $\{x(t)\}$ is defined by Equation (5.6) as

$$R_{xx}(\tau) = E[x(t)x(t + \tau)] \tag{8.20}$$

From the stationary hypothesis, $R_{xx}(\tau)$ is independent of t, and an even function of τ with a maximum at $\tau = 0$. It will be assumed that $R_{xx}(\tau)$ is continuous and finite for all values of τ, and that all periodic components in $R_{xx}(\tau)$ have been removed at the onset. The autocovariance function $C_{xx}(\tau)$ is defined by Equation (5.8) as

$$C_{xx}(\tau) = R_{xx}(\tau) - \mu_x^2 \tag{8.21}$$

It turns out that whenever $\mu_x \neq 0$, it is more convenient to work with $C_{xx}(\tau)$ than with $R_{xx}(\tau)$. It will be assumed that $C_{xx}(\tau)$ satisfies the integrable properties of Equation (5.117) so as to make $\{x(t)\}$ ergodic.

In terms of the autocovariance function $C_{xx}(\tau)$, the variance (mean square error) from Equations (8.18) and (8.19) becomes

$$\mathrm{Var}[\hat{\mu}_x] = \frac{1}{T^2} \int_0^T \int_0^T C_{xx}(\eta - \xi) \, d\eta \, d\xi = \frac{1}{T^2} \int_0^T \int_{-\xi}^{T-\xi} C_{xx}(\tau) \, d\tau \, d\xi$$

$$= \frac{1}{T} \int_{-T}^T \left(1 - \frac{|\tau|}{T} \right) C_{xx}(\tau) \, d\tau \tag{8.22}$$

The last expression occurs by reversing the orders of integration between τ and ξ and carrying out the ξ integration. This changes the limits of integration for τ and ξ as shown in the sketch below so that

$$\int_0^T \int_{-\xi}^{T-\xi} C_{xx}(\tau) \, d\tau \, d\xi = \int_{-T}^0 \int_{-\tau}^T C_{xx}(\tau) \, d\xi \, d\tau + \int_0^T \int_0^{T-\tau} C_{xx}(\tau) \, d\xi \, d\tau$$

$$= \int_{-T}^0 (T + \tau) C_{xx}(\tau) \, d\tau + \int_0^T (T - \tau) C_{xx}(\tau) \, d\tau$$

$$= \int_{-T}^T (T - |\tau|) C_{xx}(\tau) \, d\tau$$

Now on letting T tend to infinity, Equation (8.22) becomes

$$\lim_{T \to \infty} T \, \mathrm{Var}[\hat{\mu}_x] = \int_{-\infty}^{\infty} C_{xx}(\tau) \, d\tau < \infty \tag{8.23}$$

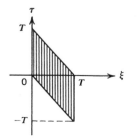

where Equation (5.117) is used, which provides that $C_{xx}(\tau)$ and $\tau C_{xx}(\tau)$ are absolutely integrable over $(-\infty, \infty)$ to justify passage to the limit inside the integral sign. In particular, Equation (8.23) shows that for large T, where $|\tau| \ll T$, the variance is given by

$$\text{Var}[\hat{\mu}_x] \approx \frac{1}{T} \int_{-\infty}^{\infty} C_{xx}(\tau) \, d\tau \tag{8.24}$$

Hence when the integral is finite valued, $\text{Var}[\hat{\mu}_x]$ approaches zero as T approaches infinity, proving that $\hat{\mu}_x$ is a *consistent* estimate of μ_x.

Consider the important special case where $\{x(t)\}$ is *bandwidth limited white noise* with mean value $\mu_x \neq 0$ and variance σ_x^2. Its autospectral density function can be described by

$$G_{xx}(f) = \begin{cases} \dfrac{\sigma_x^2}{B} + \mu_x^2 \delta(f) & 0 \le f \le B \\ 0 & f > B \end{cases} \tag{8.25}$$

where B is the bandwidth. The associated autocovariance function is given by

$$C_{xx}(\tau) = \int_0^{\infty} G_{xx}(f) \cos 2\pi f \tau \, d\tau - \mu_x^2$$

$$= \sigma_x^2 \left(\frac{\sin 2\pi B\tau}{2\pi B\tau} \right) \tag{8.26}$$

Note that $C_{xx}(0) = \sigma_x^2$ and $C_{xx}(\tau) = 0$ for $\tau = (n/2B)$, where n is an integer. Thus points $(1/2B)$ apart are uncorrelated. They will be statistically independent if $\{x(t)\}$ is also Gaussian. For this case, Equation (8.24) yields the approximate result when $BT \ge 5$ that

$$\text{Var}[\hat{\mu}_x] \approx \frac{\sigma_x^2}{2BT} \tag{8.27}$$

When $\mu_x \neq 0$, the normalized rms error is

$$\varepsilon[\hat{\mu}_x] \approx \frac{1}{\sqrt{2BT}} \left(\frac{\sigma_x}{\mu_x} \right) \tag{8.28}$$

From Equation (8.3), note that

$$E[\hat{\mu}_x^2] = \text{Var}[\hat{\mu}_x] + \mu_x^2 \tag{8.29}$$

Hence, using Equation (8.27), for bandwidth limited white noise,

$$E[\hat{\mu}_x^2] \approx \frac{\sigma_x^2}{2BT} + \mu_x^2 \tag{8.30}$$

For Gaussian data, when $\mu_x \neq 0$, the fourth-order moment from Equation (3.82) becomes

$$E\left[\hat{\mu}_x^4\right] = 3\left\{E\left[\hat{\mu}_x^2\right]\right\}^2 - 2\mu_x^4 \qquad (8.31)$$

Then, neglecting terms of order $(1/BT)^2$,

$$\text{Var}\left[\hat{\mu}_x^2\right] = E\left[\hat{\mu}_x^4\right] - \left\{E\left[\hat{\mu}_x^2\right]\right\}^2 \approx \frac{2\mu_x^2\sigma_x^2}{BT} \qquad (8.32)$$

Thus the normalized mean square random error is given by

$$\varepsilon_r^2\left[\hat{\mu}_x^2\right] = \frac{\text{Var}\left[\hat{\mu}_x^2\right]}{\mu_x^4} \approx \frac{2}{BT}\left(\frac{\sigma_x}{\mu_x}\right)^2$$

Now, comparing with Equation (8.28)

$$\varepsilon_r^2\left[\hat{\mu}_x^2\right] \approx 4\varepsilon^2\left[\hat{\mu}_x\right]$$

and

$$\varepsilon_r\left[\hat{\mu}_x^2\right] \approx 2\varepsilon[\hat{\mu}_x] \approx \frac{\sqrt{2}}{\sqrt{BT}}\left(\frac{\sigma_x}{\mu_x}\right) \qquad (8.33)$$

This agrees with the general relation of Equation (8.10).

***Example 8.2.* Random Error in Mean Value Estimate.** Consider a bandwidth limited white noise signal $x(t)$ with a bandwidth of $B = 100$ Hz, a mean value of $\mu_x = 0$, and a standard deviation of $\sigma_x = 2$. If the mean value of $x(t)$ is to be estimated by averaging over a record of length $T = 2$ sec, determine an interval that will include the mean value estimate with a probability of approximately 95%.

From Equation (8.27), the random error of the estimate $\hat{\mu}_x$ will have a standard deviation of

$$\sigma[\hat{\mu}_x] = \frac{2}{\sqrt{2(100)(2)}} = 0.10$$

It then follows from Equation (8.12) that the probability is about 95% that an estimate $\hat{\mu}_x$ will fall within the interval

$$[-0.2 \leq \hat{\mu}_x \leq 0.2]$$

8.2.2   *Mean Square Value Estimates*

As in Section 8.2.1, let $x(t)$ be a single sample time history record from a stationary (ergodic) random process $\{x(t)\}$. The mean square value of $\{x(t)\}$ can be estimated by time averaging over a finite time interval T as follows:

$$\hat{\psi}_x^2 = \frac{1}{T}\int_0^T x^2(t)\,dt \tag{8.34}$$

The true mean square value is

$$\psi_x^2 = E\left[x^2(t)\right] \tag{8.35}$$

and is independent of t since $\{x(t)\}$ is stationary. The expected value of the estimate $\hat{\psi}_x^2$ is

$$E\left[\hat{\psi}_x^2\right] = \frac{1}{T}\int_0^T E\left[x^2(t)\right]dt = \frac{1}{T}\int_0^T \psi_x^2\,dt = \psi_x^2 \tag{8.36}$$

Hence $\hat{\psi}_x^2$ is an *unbiased* estimate of ψ_x^2, independent of T.

The mean square error here is given by the variance

$$\mathrm{Var}\left[\hat{\psi}_x^2\right] = E\left[(\hat{\psi}_x^2 - \psi_x^2)^2\right] = E\left[\hat{\psi}_x^4\right] - \psi_x^4$$

$$= \frac{1}{T^2}\int_0^T\int_0^T \left(E\left[x^2(\xi)x^2(\eta)\right] - \psi_x^4\right)d\eta\,d\xi \tag{8.37}$$

Assume now that $\{x(t)\}$ is a *Gaussian* random process with mean value $\mu_x \neq 0$. Then the expected value in Equation (8.37) takes the special form, derived from Equation (3.81), as follows:

$$E\left[x^2(\xi)x^2(\eta)\right] = 2\left(R_{xx}^2(\eta - \xi) - \mu_x^4\right) + \psi_x^4 \tag{8.38}$$

From the basic relationship in Equation (8.21),

$$R_x^2(\eta - \xi) - \mu_x^4 = C_{xx}^2(\eta - \xi) + 2\mu_x^2 C_{xx}(\eta - \xi) \tag{8.39}$$

Hence

$$\mathrm{Var}\left[\hat{\psi}_x^2\right] = \frac{2}{T^2}\int_0^T\int_0^T \left(R_{xx}^2(\eta - \xi) - \mu_x^4\right)d\eta\,d\xi$$

$$= \frac{2}{T}\int_{-T}^T\left(1 - \frac{|\tau|}{T}\right)\left(R_{xx}^2(\tau) - \mu_x^4\right)d\tau$$

$$= \frac{2}{T}\int_{-T}^T\left(1 - \frac{|\tau|}{T}\right)\left(C_{xx}^2(\tau) + 2\mu_x^2 C_{xx}(\tau)\right)d\tau \tag{8.40}$$

For large T, where $|\tau| \ll T$, the variance becomes

$$\mathrm{Var}\left[\hat{\psi}_x^2\right] \approx \frac{2}{T} \int_{-\infty}^{\infty} \left(C_{xx}^2(\tau) + 2\mu_x^2 C_{xx}(\tau)\right) d\tau \qquad (8.41)$$

Thus $\hat{\psi}_x^2$ is a *consistent* estimate of ψ_x^2 since $\mathrm{Var}[\hat{\psi}_x^2]$ will approach zero as T approaches infinity assuming $C_{xx}^2(\tau)$ and $C_{xx}(\tau)$ are absolutely integrable over $(-\infty, \infty)$, as stated in Equation (5.117).

Consider the special case of bandwidth limited Gaussian white noise as defined by Equation (8.25). From Equation (8.26),

$$C_{xx}(\tau) = \sigma_x^2 \left(\frac{\sin 2\pi B\tau}{2\pi B\tau}\right) \qquad (8.42)$$

For this case, Equation (8.41) shows that

$$\mathrm{Var}\left[\hat{\psi}_x^2\right] \approx \frac{\sigma_x^4}{BT} + \frac{2}{BT}\mu_x^2\sigma_x^2 \qquad (8.43)$$

This is the variance of mean square value estimates, where B is the *total* bandwidth of the data and T is the *total* record length of the data. In general, for $\mu_x \neq 0$, the normalized rms error is

$$\varepsilon\left[\hat{\psi}_x^2\right] = \frac{\mathrm{s.d.}\left[\hat{\psi}_x^2\right]}{\psi_x^2} \approx \frac{1}{\sqrt{BT}}\left(\frac{\sigma_x}{\psi_x}\right)^2 + \frac{\sqrt{2}}{\sqrt{BT}}\left(\frac{\mu_x\sigma_x}{\psi_x^2}\right) \qquad (8.44)$$

For cases where $\mu_x = 0$, the quantity $\psi_x^2 = \sigma_x^2$ and results simplify to

$$\mathrm{Var}\left[\hat{\psi}_x^2\right] \approx \frac{\psi_x^4}{BT} = \frac{R_{xx}^2(0)}{BT} \qquad (8.45)$$

with

$$\varepsilon\left[\hat{\psi}_x^2\right] \approx \frac{1}{\sqrt{BT}} \qquad (8.46)$$

Corresponding results for rms value estimates $\hat{\psi}_x$ instead of $\hat{\psi}_x^2$ when $\mu_x = 0$ are found from Equation (8.10) to be

$$\varepsilon\left[\hat{\psi}_x\right] \approx \frac{1}{2\sqrt{BT}} \qquad (8.47)$$

Plots of Equations (8.46) and (8.47) versus the BT product are presented in Figure 8.1.

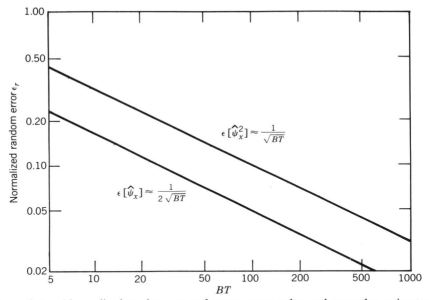

Figure 8.1 Normalized random error of mean square value and rms value estimates.

8.2.3 *Variance Estimates*

Variance estimates can be obtained from

$$\hat{\sigma}_x^2 = \hat{\psi}_x^2 - \hat{\mu}_x^2 \tag{8.48}$$

Now

$$\text{Var}[\hat{\sigma}_x^2] = E[\hat{\sigma}_x^4] - \left(E[\hat{\sigma}_x^2]\right)^2 \tag{8.49}$$

where

$$E[\hat{\sigma}_x^2] = E[\hat{\psi}_x^2] - E[\hat{\mu}_x^2] \tag{8.50}$$

$$E[\hat{\sigma}_x^4] = E[\hat{\psi}_x^4 - 2\hat{\psi}_x^2\hat{\mu}_x^2 + \hat{\mu}_x^4]$$

$$= E[\hat{\psi}_x^4] - 2E[\hat{\psi}_x^2\hat{\mu}_x^2] + E[\hat{\mu}_x^4] \tag{8.51}$$

Hence

$$\text{Var}[\hat{\sigma}_x^2] = \text{Var}[\hat{\psi}_x^2] + \text{Var}[\hat{\mu}_x^2] - 2\left(E[\hat{\psi}_x^2\hat{\mu}_x^2] - E[\hat{\psi}_x^2]E[\hat{\mu}_x^2]\right) \tag{8.52}$$

Unlike mean value and mean square value estimates, variance estimates based on Equation (8.48) will be *biased* estimates. Specifically, for bandwidth limited Gaussian white noise, substitution of Equations (8.36) and (8.30) into Equation (8.50) shows that

$$E\left[\hat{\sigma}_x^2\right] \approx \psi_x^2 - \left(\frac{\sigma_x^2}{2BT} + \mu_x^2\right)$$

$$\approx \sigma_x^2\left(1 - \frac{1}{2BT}\right) \tag{8.53}$$

Hence the bias error

$$b\left[\hat{\sigma}_x^2\right] = E\left[\hat{\sigma}_x^2\right] - \sigma_x^2 \approx \frac{-\sigma_x^2}{2BT} \tag{8.54}$$

This result agrees with Equation (4.11), letting $N = 2BT$.

Consider again the important special case of bandwidth limited Gaussian white noise where the variance terms for $\hat{\psi}_x^2$ and $\hat{\mu}_x^2$ are known from Equations (8.43) and (8.32), respectively. For Gaussian data, when $\mu_x \neq 0$, the fourth-order moment from Equation (3.82) becomes

$$E\left[\hat{\psi}_x^2\hat{\mu}_x^2\right] \approx E\left[\hat{\psi}_x^2\right]E\left[\hat{\mu}_x^2\right] + 2\left(E\left[\hat{\mu}_x^2\right]\right)^2 - 2\mu_x^4 \tag{8.55}$$

using the approximation $E[\hat{\psi}_x\hat{\mu}_x] \approx E[\hat{\mu}_x^2]$. Equation (8.30) gives

$$\left(E\left[\hat{\mu}_x^2\right]\right)^2 \approx \left(\frac{\sigma_x^2}{2BT} + \mu_x^2\right)^2 \approx \frac{\mu_x^2\sigma_x^2}{BT} + \mu_x^4 \tag{8.56}$$

neglecting terms of order $(1/BT)^2$. Hence

$$E\left[\hat{\psi}_x^2\hat{\mu}_x^2\right] - E\left[\hat{\psi}_x^2\right]E\left[\hat{\mu}_x^2\right] \approx \frac{2\mu_x^2\sigma_x^2}{BT} \tag{8.57}$$

Now, substitutions of Equations (8.32), (8.43), and (8.57) into Equation (8.52) yield

$$\mathrm{Var}\left[\hat{\sigma}_x^2\right] \approx \frac{\sigma_x^4}{BT} \tag{8.58}$$

which is independent of μ_x.

The normalized random error is

$$\varepsilon_r[\hat{\sigma}_x^2] = \frac{\text{s.d.}[\hat{\sigma}_x^2]}{\sigma_x^2} \approx \frac{1}{\sqrt{BT}} \tag{8.59}$$

This result applies even if $\mu_x \neq 0$, whereas the result of Equation (8.46) for mean square value estimates should be used only when $\mu_x = 0$. In either case, the data are assumed to behave like Gaussian bandwidth limited white noise. From Equation (8.10), one now has corresponding results for the standard deviation estimate $\hat{\sigma}_x$ instead of the variance estimate $\hat{\sigma}_x^2$, namely,

$$\varepsilon_r[\hat{\sigma}_x] \approx \frac{1}{2\sqrt{BT}} \tag{8.60}$$

Note that Figure 8.1 can also represent Equations (8.59) and (8.60) as the normalized random errors for variance estimates and for standard deviation estimates with arbitrary μ_x.

8.3 PROBABILITY DENSITY FUNCTION ESTIMATES

Consider a probability density measurement of a single sample time-history record $x(t)$ from a stationary (ergodic) random process $\{x(t)\}$. The probability that $x(t)$ assumes values between $x - (W/2)$ and $x + (W/2)$ during a time interval T may be estimated by

$$\hat{P}[x, W] = \widehat{\text{Prob}}\left[\left(x - \frac{W}{2}\right) \leq x(t) \leq \left(x + \frac{W}{2}\right)\right]$$

$$= \frac{1}{T} \sum_i \Delta t_i = \frac{T_x}{T} \tag{8.61}$$

where Δt_i is the time spent by $x(t)$ in this range during the ith entry into the range, and $T_x = \sum \Delta t_i$. The ratio T_x/T is the total fractional portion of the time spent by $x(t)$ in the range $[x - (W/2), x + (W/2)]$. It should be noted that T_x will usually be a function of the value x. The estimated probability $\hat{P}[x, W]$ will approach the true probability $P[x, W]$ as T approaches infinity. Moreover, this estimated probability is an unbiased estimate of the true probability. Hence

$$P[x, W] = E[\hat{P}[x, W]] = \lim_{T \to \infty} \hat{P}[x, W] = \lim_{T \to \infty} \frac{T_x}{T} \tag{8.62}$$

The probability density function $p(x)$ is defined by

$$p(x) = \lim_{W \to 0} \frac{P[x, W]}{W} = \lim_{\substack{T \to \infty \\ W \to 0}} \frac{\hat{P}[x, W]}{W} = \lim_{\substack{T \to \infty \\ W \to 0}} \hat{p}(x) \qquad (8.63)$$

where

$$\hat{p}(x) = \frac{\hat{P}[x, W]}{W} = \frac{T_x}{TW} \qquad (8.64)$$

is a sample estimate of $p(x)$. In terms of the probability density function $p(x)$, the probability of the time history $x(t)$ falling between any two values x_1 and x_2 is given by

$$\text{Prob}[x_1 \le x(t) \le x_2] = \int_{x_1}^{x_2} p(x) \, dx \qquad (8.65)$$

In particular

$$P[x, W] = \text{Prob}\left[x - \frac{W}{2} \le x(t) \le x + \frac{W}{2}\right] = \int_{x-(W/2)}^{x+(W/2)} p(\xi) \, d\xi \quad (8.66)$$

Then, from Equation (8.64)

$$E[\hat{p}(x)] = \frac{E[\hat{P}[x, W]]}{W} = \frac{P[x, W]}{W} = \frac{1}{W} \int_{x-(W/2)}^{x+(W/2)} p(\xi) \, d\xi \quad (8.67)$$

Thus for most $p(x)$,

$$E[\hat{p}(x)] \ne p(x) \qquad (8.68)$$

proving that $\hat{p}(x)$ is generally a *biased* estimate of $p(x)$.

The mean square error of the estimate $\hat{p}(x)$ is calculated from Equation (8.5) by

$$E\left[(\hat{p}(x) - p(x))^2\right] = \text{Var}[\hat{p}(x)] + b^2[\hat{p}(x)] \qquad (8.69)$$

where $\text{Var}[\hat{p}(x)]$ is the variance of the estimate as defined by

$$\text{Var}[\hat{p}(x)] = E\left[(\hat{p}(x) - E[\hat{p}(x)])^2\right] \qquad (8.70)$$

and $b[\hat{p}(x)]$ is the bias of the estimate as defined by

$$b[\hat{p}(x)] = E[\hat{p}(x)] - p(x) \tag{8.71}$$

8.3.1   *Bias of the Estimate*

An expression will now be derived for the bias term of Equation (8.71). In terms of the true probability density function, Equation (8.67) shows that

$$E[\hat{p}(x)] = \frac{1}{W} \int_{x-(W/2)}^{x+(W/2)} p(\xi) \, d\xi \tag{8.72}$$

By expanding $p(\xi)$ in a Taylor series about the point $\xi = x$, and retaining only the first three terms,

$$p(\xi) \approx p(x) + (\xi - x) p'(x) + \frac{(\xi - x)^2}{2} p''(x) \tag{8.73}$$

From the two relations

$$\int_{x-(W/2)}^{x+(W/2)} (\xi - x) \, d\xi = 0 \tag{8.74}$$

and

$$\int_{x-(W/2)}^{x+(W/2)} \frac{(\xi - x)^2}{2} \, d\xi = \frac{W^3}{24} \tag{8.75}$$

it follows that

$$E[\hat{p}(x)] \approx p(x) + \frac{W^2}{24} p''(x) \tag{8.76}$$

Thus a first-order approximation for the bias term is given by

$$b[\hat{p}(x)] \approx \frac{W^2}{24} p''(x) \tag{8.77}$$

where $p''(x)$ is the second derivative of $p(x)$ with respect to x.

***Example 8.3.* Bias in Probability Density Estimate of Gaussian Random Data.**   Probability density estimates are generally made using a window width of $W \leq 0.2\sigma_x$. However, consider the case where a crude probability density estimate is made using a window width of $W = \sigma_x$. Assume the data being analyzed are Gaussian random noise with a mean value of zero and a variance of unity. From Equation (8.67), the expected value of the estimate is given by

$$E[\hat{p}(x)] = \frac{1}{\sqrt{2\pi}} \int_{x-0.5}^{x+0.5} e^{-x^2/2} dx$$

where solutions of the integral are available from Table A.2 in Appendix A. For example, at the mean value ($x = 0$), $E[\hat{p}(0)] = 0.3830$. From Table A.1, however, $p(0) = 0.3989$. Hence, the actual bias error in the estimate of a Gaussian probability density function at its mean value is given by Equation (8.71) as

$$b[\hat{p}(0)] = 0.3830 - 0.3989 = -0.0159$$

where the minus sign means the estimate is less than the actual value.

Consider now the first-order approximation for the bias error given by Equation (8.77). For Gaussian data with zero mean value and unity variance,

$$p''(x) = \frac{-(1-x^2)}{\sqrt{2\pi}} e^{-x^2/2}$$

Hence at $x = 0$, the bias is approximated by

$$b[\hat{p}(0)] = \frac{-1}{24\sqrt{2\pi}} = -0.0166$$

which is within 5% of the actual bias error computed earlier.

8.3.2 Variance of the Estimate

To evaluate the variance of an estimate $\hat{p}(x)$, it is necessary to know the statistical properties of the time intervals Δt_i that constitute T_x. Unfortunately, such time statistics for a random process are very difficult to obtain. However, the general form of an appropriate variance expression for $\hat{p}(x)$ can be established by the following heuristic argument.

From Equation (8.64), the variance of $\hat{p}(x)$ is given by

$$\text{Var}[\hat{p}(x)] = \frac{1}{W^2} \text{Var}[\hat{P}(x, W)] \tag{8.78}$$

where $\hat{P}(x, W)$ is the estimate of a proportion $P(x, W)$. The variance for a proportion estimate based on N independent sample values is given by [Reference 8.3]

$$\text{Var}[\hat{P}(x, W)] = \frac{P(x, W)[1 - P(x, W)]}{N} \tag{8.79}$$

Substituting Equation (8.79) into Equation (8.78), and assuming $P(x, W) \approx Wp(x) \ll 1$, the variance of probability density estimates may be approximated by

$$\text{Var}[\hat{p}(x)] \approx \frac{p(x)}{NW} \qquad (8.80)$$

where N is still to be determined. Now from the time domain sampling theorem derived in Section 10.3.1, a sample record $x(t)$ of bandwidth B and length T can be completely reproduced with $N = 2BT$ discrete values. Of course, the N discrete values will not necessarily be statistically independent. Nevertheless, for any given stationary (ergodic) random process, each sample record will represent $n = N/c^2$ independent sample values (degrees of freedom), where c is a constant. Hence from Equation (8.80),

$$\text{Var}[\hat{p}(x)] \approx \frac{c^2 p(x)}{2BTW} \qquad (8.81)$$

The constant c is dependent on the autocorrelation function of the data and the sampling rate. For continuous bandwidth limited white noise, experimental studies indicate $c \approx 0.3$. If the bandwidth limited white noise is digitized to $N = 2BT$ discrete values, the experimental studies indicate $c = 1.0$, as would be expected from the results of Equation (8.26).

8.3.3   Normalized rms Error

The total mean square error of the probability density estimate $\hat{p}(x)$ is the sum of the variance defined in Equation (8.81) and the square of the bias defined in Equation (8.77). That is,

$$E\left[(\hat{p}(x) - p(x))^2\right] = \frac{c^2 p(x)}{2BTW} + \left[\frac{W^2 p''(x)}{24}\right]^2 \qquad (8.82)$$

Hence, the normalized mean square error is approximated by

$$\varepsilon^2[\hat{p}(x)] \approx \frac{c^2}{2BTWp(x)} + \frac{W^4}{576}\left[\frac{p''(x)}{p(x)}\right]^2 \qquad (8.83)$$

The square root gives the normalized rms error.

It is clear from Equation (8.83) that there are conflicting requirements on the window width W in probability density measurements. On the one hand, a large value of W is desirable to reduce the random error. On the other hand, a small value of W is needed to suppress the bias error. However, the total error will approach zero as $T \to \infty$ if W is restricted so that $W \to 0$ and $WT \to \infty$.

In practice, values of $W \leq 0.2\sigma_x$ will usually limit the normalized bias error to less than 1%. This is true because of the $p''(x)$ term in the bias portion of the error given by Equation (8.83). Probability density functions of common (approximately Gaussian) random data do not display abrupt or sharp peaks, which are indicative of a large second derivative.

8.3.4   *Joint Probability Density Estimates*

Joint probability density estimates for a pair of sample time-history records $x(t)$ and $y(t)$ from two stationary (ergodic) random processes $\{x(t)\}$ and $\{y(t)\}$ may be defined as follows. Analogous to Equation (8.61), let

$$\hat{P}\left[x, W_x; y, W_y\right] = \frac{T_{x,y}}{T} \tag{8.84}$$

estimate the joint probability that $x(t)$ is inside the interval W_x centered at x while simultaneously $y(t)$ is inside the interval W_y centered at y. This is measured by the ratio $T_{x,y}/T$, where $T_{x,y}$ represents the amount of time that these two events coincide in time T. Clearly, $T_{x,y}$ will usually be a function of both x and y. This estimated joint probability will approach the true probability $P[x, W_x; y, W_y]$ as T approaches infinity, namely,

$$P\left[x, W_x; y, W_y\right] = \lim_{T \to \infty} \hat{P}\left[x, W_x; y, W_y\right] = \lim_{T \to \infty} \frac{T_{x,y}}{T} \tag{8.85}$$

The joint probability density function $p(x, y)$ is now defined by

$$p(x, y) = \lim_{\substack{W_x \to 0 \\ W_y \to 0}} \frac{P\left[x, W_x; y, W_y\right]}{W_x W_y} = \lim_{\substack{T \to \infty \\ W_x \to 0 \\ W_y \to 0}} \frac{\hat{P}\left[x, W_x; y, W_y\right]}{W_x W_y} = \lim_{\substack{T \to \infty \\ W_x \to 0 \\ W_y \to 0}} \hat{p}(x, y)$$

$$\tag{8.86}$$

where

$$\hat{p}(x, y) = \frac{\hat{P}\left[x, W_x; y, W_y\right]}{W_x W_y} = \frac{T_{x,y}}{T W_x W_y} \tag{8.87}$$

Assume that W_x and W_y are sufficiently small that the bias errors are negligible. Then the mean square error associated with the estimate $\hat{p}(x, y)$ will be given by the variance of the estimate. As for first-order probability density estimates, this quantity is difficult to determine precisely by theoretical arguments alone. However, by using the same heuristic arguments that produced Equation (8.81), a general form for the variance can be approximated.

Specifically, for the special case where $x(t)$ and $y(t)$ are both bandwidth limited white noise with identical bandwidths B,

$$\text{Var}[\,\hat{p}(x, y)] \approx \frac{c^2 p(x, y)}{2\,BTW_x W_y} \tag{8.88}$$

where c is an unknown constant.

8.4   CORRELATION FUNCTION ESTIMATES

Consider now two sample time-history records $x(t)$ and $y(t)$ from two stationary (ergodic) random processes $\{x(t)\}$ and $\{y(t)\}$. The next statistical quantities of interest are the stationary autocorrelation functions $R_{xx}(\tau)$ and $R_{yy}(\tau)$ and the cross-correlation function $R_{xy}(\tau)$. To simplify the following derivation, the mean values μ_x and μ_y will be assumed to be zero. For continuous data, $x(t)$ and $y(t)$, which exist only over a time interval T, the sample cross-correlation estimate $\hat{R}_{xy}(\tau)$ can be defined by

$$\hat{R}_{xy}(\tau) = \begin{cases} \dfrac{1}{T - \tau} \displaystyle\int_0^{T-\tau} x(t)\,y(t + \tau)\,dt & 0 \le \tau < T \\[4mm] \dfrac{1}{T - |\tau|} \displaystyle\int_{|\tau|}^{T} x(t)\,y(t + \tau)\,dt & -T < \tau \le 0 \end{cases} \tag{8.89}$$

To avoid use of absolute value signs, τ will be considered positive henceforth since a similar proof applies for negative τ. The sample autocorrelation function estimates $\hat{R}_{xx}(\tau)$ and $\hat{R}_{yy}(\tau)$ are merely special cases when the two records coincide. That is,

$$\hat{R}_{xx}(\tau) = \frac{1}{T - \tau} \int_0^{T-\tau} x(t)\,x(t + \tau)\,dt \qquad 0 \le \tau < T$$

$$\hat{R}_{yy}(\tau) = \frac{1}{T - \tau} \int_0^{T-\tau} y(t)\,y(t + \tau)\,dt \qquad 0 \le \tau < T \tag{8.90}$$

Thus by analyzing the cross-correlation function estimate, one derives results that are also applicable to the autocorrelation function estimates.

If the data exist for time $T + \tau$ instead of only for time T, then an alternative definition for $\hat{R}_{xy}(\tau)$ is

$$\hat{R}_{xy}(\tau) = \frac{1}{T} \int_0^T x(t)\,y(t + \tau)\,dt \qquad 0 \le \tau < T \tag{8.91}$$

This formula has a fixed integration time T instead of a variable integration time as in Equation (8.89), and is the way the correlation functions have been defined previously. Note that for either Equation (8.89) or Equation (8.91), mean square estimates of $x(t)$ or $y(t)$ are merely special cases when $\tau = 0$. For simplicity in notation, Equation (8.91) will be used in the following development instead of Equation (8.89). The same final results are obtained for both definitions, assuming the data exists for time $T + \tau$.

The expected value of the estimate $\hat{R}_{xy}(\tau)$ is given by

$$E\left[\hat{R}_{xy}(\tau)\right] = \frac{1}{T}\int_0^T E\left[x(t)y(t+\tau)\right]\,dt$$

$$= \frac{1}{T}\int_0^T R_{xy}(\tau)\,dt = R_{xy}(\tau) \qquad (8.92)$$

Hence $\hat{R}_{xy}(\tau)$ is an *unbiased* estimate of $R_{xy}(\tau)$, independent of T. The mean square error is given by the variance

$$\mathrm{Var}\left[\hat{R}_{xy}(\tau)\right] = E\left[\left(\hat{R}_{xy}(\tau) - R_{xy}(\tau)\right)^2\right] = E\left[\hat{R}_{xy}^2(\tau)\right] - R_{xy}^2(\tau)$$

$$= \frac{1}{T^2}\int_0^T\int_0^T\left(E\left[x(u)y(u+\tau)x(v)y(v+\tau)\right] - R_{xy}^2(\tau)\right)\,dv\,du$$

$$(8.93)$$

At this point, in order to simplify the later mathematical analysis and to agree with many physical cases of interest, it will be assumed that the random processes $\{x(t)\}$ and $\{y(t)\}$ are jointly *Gaussian* for any set of fixed times. This restriction may be removed by substituting certain integrability conditions on the non-Gaussian parts of the random processes without altering in any essential way the results to be derived. When $\{x(t)\}$ and $\{y(t)\}$ are jointly Gaussian, it follows that $\{x(t)\}$ and $\{y(t)\}$ are separately Gaussian.

For Gaussian stationary random processes with zero mean values, the fourth-order statistical expression is obtained from Equation (5.130) as follows:

$$E\left[x(u)y(u+\tau)x(v)y(v+\tau)\right] = R_{xy}^2(\tau) + R_{xx}(v-u)R_{yy}(v-u)$$

$$+ R_{xy}(v-u+\tau)R_{yx}(v-u-\tau)$$

$$(8.94)$$

Hence the variance expression may be written as

$$\text{Var}\big[\hat{R}_{xy}(\tau)\big] = \frac{1}{T^2} \int_0^T \int_0^T \big(R_{xx}(v - u) R_{yy}(v - u)$$

$$+ R_{xy}(v - u + \tau) R_{yx}(v - u - \tau) \big) \, dv \, du$$

$$= \frac{1}{T} \int_{-T}^T \left(1 - \frac{|\xi|}{T} \right) \big(R_{xx}(\xi) R_{yy}(\xi)$$

$$+ R_{xy}(\xi + \tau) R_{yx}(\xi - \tau) \big) \, d\xi \qquad (8.95)$$

The second expression occurs from the first by letting $\xi = v - u$, $d\xi = dv$, and then reversing the order of integration between ξ and u. Now,

$$\lim_{T \to \infty} T \, \text{Var}\big[\hat{R}_{xy}(\tau)\big]$$

$$= \int_{-\infty}^{\infty} \big(R_{xx}(\xi) R_{yy}(\xi) + R_{xy}(\xi + \tau) R_{yx}(\xi - \tau) \big) \, d\xi < \infty \qquad (8.96)$$

assuming $R_{xx}(\xi) R_{yy}(\xi)$ and $R_{xy}(\xi) R_{yx}(\xi)$ are absolutely integrable over $(-\infty, \infty)$. This proves that $\hat{R}_{xy}(\tau)$ is a *consistent* estimate of $R_{xy}(\tau)$, and for large T, has a variance given by

$$\text{Var}\big[\hat{R}_{xy}(\tau)\big] \approx \frac{1}{T} \int_{-\infty}^{\infty} \big(R_{xx}(\xi) R_{yy}(\xi) + R_{xy}(\xi + \tau) R_{yx}(\xi - \tau) \big) \, d\xi$$

$$(8.97)$$

Several special cases of Equation (8.97) are worthy of note. For autocorrelation estimates, Equation (8.97) becomes

$$\text{Var}\big[\hat{R}_{xx}(\tau)\big] \approx \frac{1}{T} \int_{-\infty}^{\infty} \big(R_{xx}^2(\xi) + R_{xx}(\xi + \tau) R_{xx}(\xi - \tau) \big) \, d\xi$$

$$(8.98)$$

At the zero displacement point $\tau = 0$,

$$\text{Var}\big[\hat{R}_{xx}(0)\big] \approx \frac{2}{T} \int_{-\infty}^{\infty} R_{xx}^2(\xi) \, d\xi \qquad (8.99)$$

The assumption that $R_{xx}(\tau)$ approaches zero for large τ shows that

$$R_{xx}^2(\xi) \gg R_{xx}(\xi + \tau) R_{xx}(\xi - \tau) \qquad \text{for } \tau \gg 0$$

Hence for large τ,

$$\text{Var}\left[\hat{R}_{xx}(\tau)\right] \approx \frac{1}{T}\int_{-\infty}^{\infty} R_{xx}^2(\xi)\, d\xi \qquad (8.100)$$

which is one-half the value of Equation (8.99).

8.4.1 *Bandwidth Limited White Noise*

Consider the special case where $\{x(t)\}$ is bandwidth limited Gaussian white noise with a mean value $\mu_x = 0$ and a bandwidth B as defined in Equations (8.25) and (8.26). For sample records $x(t)$ of length T, from Equation (8.98), the variance for $\hat{R}_{xx}(\tau)$ is given conservatively by

$$\text{Var}\left[\hat{R}_{xx}(\tau)\right] \approx \frac{1}{2BT}\left[R_{xx}^2(0) + R_{xx}^2(\tau)\right] \qquad (8.101)$$

This reduces to Equation (8.45) at the point $\tau = 0$. Similarly, when both $x(t)$ and $y(t)$ are samples of length T from bandwidth limited Gaussian white noise with mean values $\mu_x = \mu_y = 0$ and identical bandwidths B, it follows from Equation (8.97) that

$$\text{Var}\left[\hat{R}_{xy}(\tau)\right] \approx \frac{1}{2BT}\left[R_{xx}(0)R_{xy}(0) + R_{xy}^2(\tau)\right] \qquad (8.102)$$

This result requires that T be sufficiently large that Equation (8.97) can replace Equation (8.95). Satisfactory conditions in practice are $T \geq 10|\tau|$ and $BT \geq 5$.
  For $\mu_x = \mu_y = 0$ and $R_{xy} \neq 0$, the normalized mean square error is given by

$$\varepsilon^2\left[\hat{R}_{xy}(\tau)\right] = \frac{\text{Var}\left[\hat{R}_{xy}(\tau)\right]}{R_{xy}^2(\tau)} \approx \frac{1}{2BT}\left[1 + \frac{R_{xx}(0)R_{yy}(0)}{R_{xy}^2(\tau)}\right] \qquad (8.103)$$

The square root gives the normalized rms error ε, which includes only a random error term since the bias error is zero if the record length is longer than $T + \tau$. Thus, for cross-correlation function estimates,

$$\varepsilon\left[\hat{R}_{xy}(\tau)\right] \approx \frac{1}{\sqrt{2BT}}\left[1 + \rho_{xy}^{-2}(\tau)\right]^{1/2} \qquad (8.104)$$

where

$$\rho_{xy}(\tau) = \frac{R_{xy}(\tau)}{\sqrt{R_{xx}(0)R_{yy}(0)}} \qquad (8.105)$$

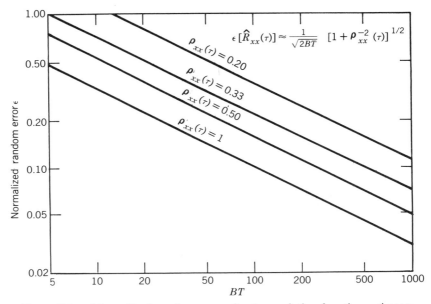

Figure 8.2   Normalized random error of autocorrelation function estimates.

is the cross-correlation coefficient function. Note that $R_{xy}(\tau) = C_{xy}(\tau)$ when $\mu_x = \mu_y = 0$. When $x(t) = y(t)$, at the special point $\tau = 0$, the quantity $\hat{R}_{xx}(0) = \hat{\psi}_x^2$ and $\rho_{xx}(0) = 1$ so that one obtains

$$\varepsilon[\hat{R}_{xx}(0)] = \varepsilon[\hat{\psi}_x^2] \approx \frac{1}{\sqrt{BT}} \qquad (8.106)$$

in agreement with Equation (8.46). In general, for autocorrelation function estimates,

$$\varepsilon[\hat{R}_{xx}(\tau)] \approx \frac{1}{\sqrt{2BT}}\left[1 + \rho_{xx}^{-2}(\tau)\right]^{1/2} \qquad (8.107)$$

where

$$\rho_{xx}(\tau) = \frac{R_{xx}(\tau)}{R_{xx}(0)} \qquad (8.108)$$

Equation (8.107) is plotted in Figure 8.2.

8.4.2   Noise-to-Signal Considerations

Some applications of Equation (8.104) are worthy of note. Suppose $x(t)$ and $y(t)$ are given by

$$x(t) = s(t) + m(t)$$
$$y(t) = s(t) + n(t) \qquad (8.109)$$

where $s(t)$, $m(t)$, and $n(t)$ are mutually uncorrelated. Then

$$R_{xy}(\tau) = R_{ss}(\tau) = R_{ss}(0)\rho_{ss}(\tau) = S\rho_{ss}(\tau)$$

$$R_{xx}(0) = R_{ss}(0) + R_{mm}(0) = S + M \qquad (8.110)$$

$$R_{yy}(0) = R_{ss}(0) + R_{nn}(0) = S + N$$

It follows from Equation (8.104) that the normalized mean square error is

$$\varepsilon^2\left[\hat{R}_{xy}(\tau)\right] \approx \frac{1}{2BT}\left[1 + \frac{(S+M)(S+N)}{S^2\rho_{ss}^2(\tau)}\right] \qquad (8.111)$$

At $\tau = 0$, where $\rho_{ss}(0) = 1$,

$$\varepsilon^2\left[\hat{R}_{xy}(0)\right] \approx \frac{1}{2BT}\left[2 + (M/S) + (N/S) + (M/S)(N/S)\right] \qquad (8.112)$$

These relations are useful for two-detector systems where $x(t)$ and $y(t)$ of Equation (8.109) measure the outputs of the two detectors containing a common signal $s(t)$ but uncorrelated noises $m(t)$ and $n(t)$.

CASE 1.

$M = 0$ where $(N/S) \gg 1$

$$\varepsilon^2\left[\hat{R}_{xy}(0)\right] \approx \frac{1}{2BT}\left(\frac{N}{S}\right) \qquad (8.113)$$

CASE 2.

$M = N$ where $(N/S) \gg 1$

$$\varepsilon^2\left[\hat{R}_{xy}(0)\right] \approx \frac{1}{2BT}\left(\frac{N}{S}\right)^2 \qquad (8.114)$$

These two cases can represent important physical applications. For example, in Case 1, $x(t)$ might be a noise-free reference signal being cross-correlated against a corrupted received signal $y(t)$. In Case 2, a corrupted received signal $y(t)$ is being autocorrelated against itself. For any given value of $(N/S) \gg 1$, a much larger value of BT is required in Case 2 than in Case 1 to achieve a desired mean square error.

Further discussion of these matters appears in Reference 8.1. Applications for these and other formulas in this chapter are given in Reference 8.2.

8.4.3   *Location Estimates of Peak Correlation Values*

Previous results in this section give random error formulas that indicate how well one can estimate the magnitudes of peak values of $R_{xy}(\tau)$ and $R_{xx}(\tau)$. There remains the difficult related problem of determining the precise location where these peak correlation values occur. For definiteness, assume that $R_{xx}(\tau)$ has the form associated with bandwidth limited white noise, namely,

$$R_{xx}(\tau) = R_{xx}(0)\left(\frac{\sin 2\pi B\tau}{2\pi B\tau}\right) \tag{8.115}$$

The maximum value of $R_{xx}(\tau)$ occurs at $\tau = 0$. Expansion of $\sin 2\pi B\tau$ near $\tau = 0$ yields

$$\sin 2\pi B\tau \approx (2\pi B\tau) - \frac{(2\pi B\tau)^3}{6} \tag{8.116}$$

Thus near $\tau = 0$, the estimate $\hat{R}_{xx}(\tau)$ is given by

$$\hat{R}_{xx}(\tau) \approx \left[1 - \frac{2(\pi B\tau)^2}{3}\right]R_{xx}(0) \tag{8.117}$$

where the mean value is

$$E\left[\hat{R}_{xx}(0)\right] = R_{xx}(0) \tag{8.118}$$

Hence $\hat{R}_{xx}(0)$ is an unbiased estimate of $R_{xx}(0)$. The variance in these estimates is given by

$$\mathrm{Var}\left[\hat{R}_{xx}(0)\right] = E\left[\left\{\hat{R}_{xx}(0) - R_{xx}(0)\right\}^2\right] \approx \tfrac{4}{9}(\pi B)^4 R_{xx}^2(0) E\left[\tau^4\right] \tag{8.119}$$

The normalized mean square error is then

$$\varepsilon^2\left[\hat{R}_{xx}(0)\right] = \frac{\mathrm{Var}\left[\hat{R}_{xx}(0)\right]}{R_{xx}^2(0)} \approx \tfrac{4}{9}(\pi B)^4 E\left[\tau^4\right] \tag{8.120}$$

Assume next that these values of τ are such that τ follows a Gaussian distribution with zero mean value and variance $\sigma_1^2(\tau)$, namely,

$$\mu_1(\tau) = E[\tau] = 0$$

$$\sigma_1^2(\tau) = E[\tau^2] \tag{8.121}$$

Then, the fourth-order moment in Equation (8.120) satisfies

$$E[\tau^4] = 3\sigma_1^4(\tau) \tag{8.122}$$

It follows that

$$\varepsilon^2[\hat{R}_{xx}(0)] \approx \tfrac{4}{3}(\pi B)^4 \sigma_1^4(\tau) \tag{8.123}$$

This proves that

$$\sigma_1(\tau) \approx \frac{0.93}{\pi B}\left\{\varepsilon[\hat{R}_{xx}(0)]\right\}^{1/2} \tag{8.124}$$

The 95% confidence interval for determining the location where the peak value occurs is now

$$[-2\sigma_1(\tau) \leq \tau \leq 2\sigma_1(\tau)] \tag{8.125}$$

***Example 8.4.* Time-Delay Estimate from Cross-Correlation Calculation.**
Assume two received time-history records, $x(t)$ and $y(t)$, contain a common signal $s(t)$ and uncorrelated noises, $m(t)$ and $n(t)$, such that

$$x(t) = s(t) + m(t)$$

$$y(t) = s(t + \tau_0) + n(t)$$

From the developments in Section 5.1.4, the maximum value in the cross-correlation function

$$R_{xy}(\tau) = R_{ss}(\tau - \tau_0)$$

will occur at $\tau = \tau_0$, which defines the time delay between the received signal $s(t)$ in $x(t)$ and $y(t)$. Now assume $s(t)$ represents bandlimited white noise with a bandwidth of $B = 100$ Hz, and the noise-to-signal ratios in $x(t)$ and $y(t)$ are given by $M/S = N/S = 10$. If the available record lengths for $x(t)$ and $y(t)$ are $T = 5$ sec, determine the accuracy of the time-delay estimate τ_0 based on the time of the maximum value of the correlation estimate $\hat{R}_{xy}(\tau)$.

From Equation (8.114), the normalized random error of the correlation estimate at its maximum value is approximated by

$$\varepsilon[\hat{R}_{xy}(\tau_0)] \simeq \varepsilon[\hat{R}_{ss}(0)] \simeq \left[\frac{10^2}{(2)(100)(5)}\right]^{1/2} = 0.32$$

Note that a more exact error is given by Equation (8.112) as $\varepsilon = 0.35$. Now the standard deviation of the estimate $\hat{\tau}_0$ is approximated by Equation (8.124) by

$$\sigma_1(\tau) \simeq \frac{0.93(0.35)^{1/2}}{\pi(100)} = 0.0017 \text{ sec}$$

Hence from Equation (8.125), an approximate 95% confidence interval for the time delay τ_0 in seconds is given by

$$[\hat{\tau}_0 - 0.0034 \leq \tau_0 \leq \hat{\tau}_0 + 0.0034]$$

8.5 AUTOSPECTRAL DENSITY FUNCTION ESTIMATES

A schematic diagram of a general filter device for estimating the autospectral density function associated with a sample time-history record $x(t)$ is displayed in Figure 8.3. The input sample record $x(t)$ is assumed to be averaged over a time interval T and to be drawn from a stationary (ergodic) random process with zero mean value. The tunable narrow-band filter is assumed to have a finite nonzero constant bandwidth B_e centered at a frequency f, which may be varied over the frequency range of interest. This *resolution bandwidth B_e* should not be confused with the full bandwidth occupied by the record $x(t)$. It turns out that in order to obtain a consistent estimate of $G_{xx}(f)$, one must introduce a filtering procedure that averages over a band of frequencies. The final estimate $\hat{G}_{xx}(f)$ describes the time average of $x^2(t)$ in terms of its frequency components lying inside the frequency band $f - (B_e/2)$ to $f + (B_e/2)$, divided by the bandwidth B_e. Note that because a nonzero mean value corresponds to a discrete spectral component at zero frequency, the zero mean value assumption in the following developments is critical only when B_e includes $f = 0$. For all other cases where B_e does not include $f = 0$, the results to follow apply to data with arbitrary nonzero mean values.

The mean square value of $x(t)$ within the bandwidth B_e centered at f is estimated by

$$\hat{\psi}_x^2(f, B_e) = \frac{1}{T} \int_0^T x^2(t, f, B_e) \, dt \qquad (8.126)$$

where $x(t, f, B_e)$ represents the narrow-band filter output and T is the

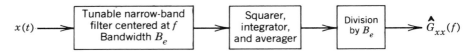

Figure 8.3 Constant bandwidth filter device for measuring autospectrum.

averaging time. It is shown in Section 8.2.2 that this estimated value will be an unbiased and consistent estimate of the true mean square value as T approaches infinity. Hence

$$\psi_x^2(f, B_e) = E\left[\hat{\psi}_x^2(f, B_e)\right] = \lim_{T \to \infty} \frac{1}{T} \int_0^T x^2(t, f, B_e)\, dt \quad (8.127)$$

where $\psi_x^2(f, B_e)$ is the mean square value of $x(t)$ associated with the filter bandwidth B_e centered at f.

The autospectral density function is defined by

$$G_{xx}(f) = \lim_{B_e \to 0} \frac{\psi_x^2(f, B_e)}{B_e} = \lim_{\substack{T \to \infty \\ B_e \to 0}} \frac{1}{B_e T} \int_0^T x^2(t, f, B_e)\, dt = \lim_{\substack{T \to \infty \\ B_e \to 0}} \hat{G}_{xx}(f)$$

$$(8.128)$$

where

$$\hat{G}_{xx}(f) = \frac{1}{B_e T} \int_0^T x^2(t, f, B_e)\, dt = \frac{\hat{\psi}_x^2(f, B_e)}{B_e} \quad (8.129)$$

is the sample estimate of $G_{xx}(f)$ determined by the procedure of Figure 8.3. In terms of the true autospectral density function $G_{xx}(f)$, the mean square value of $x(t)$ between any two frequency limits f_1 and f_2 is given by

$$\psi_x^2(f_1, f_2) = \int_{f_1}^{f_2} G_{xx}(f)\, df \quad (8.130)$$

In particular

$$\psi_x^2(f, B_e) = \int_{f-(B_e/2)}^{f+(B_e/2)} G_{xx}(\xi)\, d\xi \quad (8.131)$$

Equations (8.129) and (8.131) show that

$$E\left[\hat{G}_{xx}(f)\right] = \frac{\psi_x^2(f, B_e)}{B_e} = \frac{1}{B_e} \int_{f-(B_e/2)}^{f+(B_e/2)} G_{xx}(\xi)\, d\xi \quad (8.132)$$

Thus, for most $G_{xx}(f)$,

$$E\left[\hat{G}_{xx}(f)\right] \neq G_{xx}(f) \quad (8.133)$$

so that $\hat{G}_{xx}(f)$ is generally a *biased* estimate of $G_{xx}(f)$.

The mean square error of the estimate $\hat{G}_{xx}(f)$ is calculated from Equation (8.5) by

$$E\left[\left(\hat{G}_{xx}(f) - G_{xx}(f)\right)^2\right] = \text{Var}\left[\hat{G}_{xx}(f)\right] + b^2\left[\hat{G}_{xx}(f)\right] \quad (8.134)$$

where $\text{Var}[\hat{G}_{xx}(f)]$ is the variance of the estimate as defined by

$$\text{Var}[\hat{G}_{xx}(f)] = E\left[(\hat{G}_{xx}(f) - E[\hat{G}_{xx}(f)])^2\right] \qquad (8.135)$$

and $b[\hat{G}_{xx}(f)]$ is the bias of the estimate as defined by

$$b[\hat{G}_{xx}(f)] = E[\hat{G}_{xx}(f)] - G_{xx}(f) \qquad (8.136)$$

8.5.1   *Bias of the Estimate*

An expression for the bias term of Equation (8.136) may be derived by the procedure used to derive the bias term for probability density estimates in Section 8.3.1. Specifically, by expanding $G_{xx}(\xi)$ in a Taylor series about the point $\xi = f$ and retaining only the first three terms, it follows from Equation (8.132) that

$$E[\hat{G}_{xx}(f)] \approx G_{xx}(f) + \frac{B_e^2}{24}G_{xx}''(f) \qquad (8.137)$$

Thus the bias term is approximated by

$$b[\hat{G}_{xx}(f)] \approx \frac{B_e^2}{24}G_{xx}''(f) \qquad (8.138)$$

where $G_{xx}''(f)$ is the second derivative of $G_{xx}(f)$ with respect to f, and is related to $R_{xx}(\tau)$ by the expression

$$G_{xx}''(f) = -8\pi^2 \int_{-\infty}^{\infty} \tau^2 R_{xx}(\tau)e^{-j2\pi f\tau}\,d\tau \qquad (8.139)$$

It should be emphasized that Equation (8.138) is only a first-order approximation of the bias error, which is applicable for cases where $B_e^2 G_{xx}''(f) < G_{xx}(f)$. Since autospectra in practice often display sharp peaks reflecting large second derivatives, this may provide an inadequate measure of the bias error in some cases. Generally speaking, Equation (8.138) will exaggerate the degree of bias in estimates where $B_e^2 G_{xx}''(f)$ is large.

The bias error in Equation (8.138) is derived assuming the spectral estimation is accomplished using an ideal rectangular spectral window as defined in Equation (8.131) and illustrated in Figure 8.4. It will be seen in Section 11.5 that spectral estimation procedures in practice involve spectral windows that deviate widely from an ideal rectangular form. Nevertheless, Equation (8.138) constitutes a useful first-order approximation that correctly describes important qualitative results. In particular, the bias error increases as $G''(f)$ increases for a given B_e, or as B_e increases for a given $G''(f)$. Also, it is clear

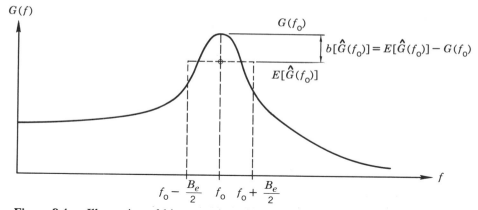

Figure 8.4 Illustration of bias error introduced by frequency smoothing of spectral density estimates.

from Figure 8.4 that the bias error is always in the direction of reduced dynamic range; that is, spectral density peaks are underestimated and spectral density valleys are overestimated.

Example 8.5. Illustration of Bias in Autospectrum Estimate. Assume white noise is applied to a single degree-of-freedom system defined in Section 2.4.1. From Example 6.3, the displacement response of the system will have an autospectral density function given by

$$G_{yy}(f) = \frac{G}{\left[1 - (f/f_n)^2\right]^2 + [2\zeta f/f_n]^2}$$

where G is the spectral density of the excitation in displacement units, f_n is the undamped natural frequency of the system, and ζ is the damping ratio of the system. As pointed out in Equation (2.25), the peak value of $G_{yy}(f)$ occurs at the resonance frequency $f_r = f_n\sqrt{1 - 2\zeta^2}$. For the case where $2\zeta^2 \ll 1$, the second derivative of $G_{yy}(f)$ at the frequency f_r yields

$$G''_{yy}(f_r) \approx \left(-8/B_r^2\right)G_{yy}(f_r)$$

where B_r is the half-power point bandwidth of the resonance peak given by Equation (2.27) as

$$B_r \approx 2\zeta f_r$$

Hence, the normalized bias error from Equation (8.138) becomes

$$\varepsilon_b\left[\hat{G}_{yy}(f_r)\right] = \frac{b\left[\hat{G}_{yy}(f_r)\right]}{G_{yy}(f_r)} \approx -\frac{1}{3}\left(\frac{B_e}{B_r}\right)^2$$

This result is often used to approximate the maximum bias error in autospectral density estimates of data representing the response of lightly damped mechanical and electrical systems. It is clear that the resolution bandwidth B_e should be less than the system bandwidth B_r to achieve small bias errors. A common practice is to choose $B_e \approx B_r/4$ so that a negligible bias error of $\varepsilon_b \approx -2\%$ is obtained.

8.5.2   Variance of the Estimate

The most direct way to arrive at a variance expression for autospectral density estimates is to apply the results derived in Section 8.2.2, as follows. From Equation (8.129), the estimate

$$B_e\hat{G}_{xx}(f) = \hat{\psi}_x^2(f, B_e) \tag{8.140}$$

is an unbiased estimate of the mean square value of $x(t)$ within the bandwidth B_e centered at f. The true value is given by $\psi_x^2(f, B_e) = B_eG_{xx}(f)$ when $G_{xx}(f)$ is constant over the bandwidth B_e. This will be approximately the case if B_e is sufficiently small. The result of Equation (8.45) applies to these estimates with $R_{xx}(0) = B_eG_{xx}(f)$. Hence

$$\text{Var}\left[B_e\hat{G}_{xx}(f)\right] \approx \frac{B_e^2G_{xx}^2(f)}{B_eT} \tag{8.141}$$

But, since B_e is a constant,

$$\text{Var}\left[B_e\hat{G}_{xx}(f)\right] = B_e^2\text{Var}\left[\hat{G}_{xx}(f)\right] \tag{8.142}$$

This gives for the variance of the estimate

$$\text{Var}\left[\hat{G}_{xx}(f)\right] \approx \frac{G_{xx}^2(f)}{B_eT} \tag{8.143}$$

Further discussion of the variance of autospectral density estimates is also presented in Section 8.5.4.

The result in Equation (8.143) is based on the assumption that the filtered data behave like bandwidth limited Gaussian white noise. This is an excellent assumption in practice when the filter resolution bandwidth B_e is small. The central limit theorem applies to indicate that the filtered data should be more

Gaussian than the input data, and the fact that B_e is small means that the output spectrum must be essentially constant. Hence, for small B_e, one can strongly state that the normalized random error will be

$$\varepsilon_r[\hat{G}_{xx}(f)] \approx \frac{1}{\sqrt{B_e T}} \qquad (8.144)$$

Note that this result is independent of frequency.

8.5.3 *Normalized rms Error*

The total mean square error of the autospectral density estimate $\hat{G}_{xx}(f)$ is the sum of the variance defined in Equation (8.143) and the square of the bias defined in Equation (8.138). That is,

$$E\left[(\hat{G}_{xx}(f) - G_{xx}(f))^2\right] \approx \frac{G_{xx}^2(f)}{B_e T} + \left[\frac{B_e^2 G_{xx}''(f)}{24}\right]^2 \qquad (8.145)$$

Hence, the normalized mean square error is approximated by

$$\varepsilon^2[\hat{G}_{xx}(f)] = \frac{E\left[(\hat{G}_{xx}(f) - G_{xx}(f))^2\right]}{G_{xx}^2(f)} \approx \frac{1}{B_e T} + \frac{B_e^4}{576}\left[\frac{G_{xx}''(f)}{G_{xx}(f)}\right]^2$$

$$(8.146)$$

The square root gives the normalized rms error.

Two important features of the error expression for autospectral density estimates should be noted. First, there are conflicting requirements on the resolution bandwidth B_e, namely, a small value of B_e is needed to suppress the bias portion of the error while a large value of B_e is desired to reduce the random portion of the error. This is similar to the situation discussed in Section 8.3.3 for the window width W in probability density measurements. The problem here, however, is more critical since autospectra in practice often display sharp peaks (large second derivatives), which aggravate the bias error problem. Second, the random portion of the error includes only the resolution bandwidth B_e, and not the total data bandwidth B. Hence the random portion of the error is a function primarily of analysis parameters rather than unknown data parameters. This greatly enhances the practical value of Equation (8.146) in experimental design and data analysis. The above issues are pursued further in Section 8.6.

8.5.4 *Estimates from Finite Fourier Transforms*

In Section 8.5.2, an expression is derived for the variance of autospectral density estimates obtained by filtering, squaring, and averaging operations, as would be performed by analog instruments. One can also arrive at a variance expression for spectral density estimates obtained by direct Fourier transform operations, as would be accomplished on a digital computer using the fast Fourier transform techniques discussed in Chapter 11. A brief summary of this second approach may add insight into the spectral density estimation error problem.

Consider the autospectral density function of a stationary (ergodic) Gaussian random process $\{x(t)\}$, as defined in Equation (5.67). Specifically, given a sample record $x(t)$ of unlimited length T, the autospectrum is

$$G_{xx}(f) = 2 \lim_{T \to \infty} \frac{1}{T} E\left[|X(f, T)|^2\right] \tag{8.147}$$

where $X(f, T)$ is the finite Fourier transform of $x(t)$, that is,

$$X(f, T) = \int_0^T x(t) e^{-j2\pi f t} \, dt \tag{8.148}$$

Now an estimate of $G_{xx}(f)$ can be obtained by simply omitting the limiting and expectation operations in Equation (8.147). This will yield the "raw" estimate

$$\tilde{G}_{xx}(f) = \frac{2}{T}|X(f, T)|^2 \tag{8.149}$$

with the narrowest possible resolution $\Delta f = (1/T)$.

To determine the variance of this estimate, observe that the finite Fourier transform $X(f, T)$ is defined by a series of components at frequencies $f = k/T$, $k = 1, 2, 3, \ldots$. Further observe that $X(f, T)$ is a complex number where the real and imaginary parts, $X_R(f, T)$ and $X_I(f, T)$, can be shown to be uncorrelated random variables with zero means and equal variances. Since a Fourier transformation is a linear operation, $X_R(f, T)$ and $X_I(f, T)$ will be Gaussian random variables if $x(t)$ is Gaussian. It follows that the quantity

$$|X(f, T)|^2 = X_R^2(f, T) + X_I^2(f, T) \tag{8.150}$$

is the sum of the squares of two independent Gaussian variables. Hence from the definition of Equation (4.16) as applied in Equation (4.37), each frequency component of the estimate $\tilde{G}_{xx}(f)$ will have a sampling distribution given by

$$\frac{\tilde{G}_{xx}(f)}{G_{xx}(f)} = \frac{\chi_2^2}{2} \tag{8.151}$$

where χ_2^2 is the chi-square variable with $n = 2$ degrees of freedom.

Note that the result in Equation (8.151) is independent of the record length T, that is, increasing the record length does not alter the distribution function defining the random error of the estimate. It only increases the number of spectral components in the estimate. If the record length is interpreted as a measure of the sample size for the estimate, this implies that *Equation* (8.149) *produces an inconsistent estimate of autospectral density functions*, as previously stated in Section 5.2.2. Furthermore, the random error of the estimate is substantial. Referring to Equations (4.19) and (4.20), the mean and variance of the chi-square variable are n and $2n$, respectively. Then the normalized standard error, which defines the random portion of the estimation error, is

$$\varepsilon_r[\tilde{G}_{xx}(f)] = \frac{\sigma[\tilde{G}_{xx}(f)]}{G_{xx}(f)} = \frac{\sqrt{2n}}{n} = \sqrt{\frac{2}{n}} \qquad (8.152)$$

For the case at hand, $n = 2$, so $\varepsilon_r = 1$, which means that the standard deviation of the estimate is as great as the quantity being estimated. This would be an unacceptable random error for most applications.

In practice, the random error of autospectra estimates produced by Equation (8.149) is reduced by computing an ensemble of estimates from n_d different (distinct, disjoint) subrecords, each of length T, and averaging the results to obtain a final "smooth" estimate given by

$$\hat{G}_{xx}(f) = \frac{2}{n_d T} \sum_{i=1}^{n_d} |X_i(f, T)|^2 \qquad (8.153)$$

Since each spectral calculation in Equation (8.153) adds two statistical degrees of freedom to the estimate, it follows that

$$\varepsilon_r[\hat{G}_{xx}(f)] = \sqrt{\frac{2}{2n_d}} = \frac{1}{\sqrt{n_d}} \qquad (8.154)$$

The minimum total record length required to compute the autospectrum estimate is clearly

$$T_r = n_d T \qquad (8.155)$$

and the resolution bandwidth of the analysis is approximated by

$$B_e \simeq \Delta f = 1/T \qquad (8.156)$$

Hence, $n_d = (T_r/T) \simeq B_e T_r$, so Equation (8.154) is equivalent to

$$\varepsilon_r[\hat{G}_{xx}(f)] \approx \frac{1}{\sqrt{B_e T_r}} \qquad (8.157)$$

in agreement with Equation (8.144) where T_r replaces T.

Referring back to Equation (8.151), the sampling distribution for an auto-spectral density estimate may now be written as

$$\frac{\hat{G}_{xx}(f)}{G_{xx}(f)} = \frac{\chi_n^2}{n} \qquad n = 2n_d \qquad (8.158)$$

From Equation 4.47, it follows that a $(1 - \alpha)$ confidence interval for $G_{xx}(f)$ based on an estimate $\hat{G}_{xx}(f)$ is given by

$$\left[\frac{n\hat{G}_{xx}(f)}{\chi_{n;\,\alpha/2}^2} \leq G_{xx}(f) \leq \frac{n\hat{G}_{xx}(f)}{\chi_{n;\,1-\alpha/2}^2} \right] \qquad n = 2n_d \qquad (8.159)$$

As before, if the normalized random error is relatively small, say $\varepsilon_r \leq 0.10$, then 95% confidence intervals can be approximated from Equations (8.14) and (8.154) by

$$\left[\left(1 - \frac{2}{\sqrt{n_d}} \right) \hat{G}_{xx}(f) \leq G_{xx}(f) \leq \left(1 + \frac{2}{\sqrt{n_d}} \right) \hat{G}_{xx}(f) \right] \qquad (8.160)$$

It should be mentioned that the random errors in spectral density calculations might be somewhat greater than indicated in the foregoing equations, depending on the exact details of the calculations, as discussed in Section 11.5.2.

8.6 RECORD LENGTH REQUIREMENTS

The error expressions derived in Sections 8.2–8.5 provide ways to assess the statistical accuracy of various parameter estimates after an experiment has been completed. It would be desirable if these error expressions could also be used to predict the accuracy of parameter estimates for future experimental data—in particular, to establish the record length required to obtain a predetermined degree of accuracy. The error formulas do relate the record length T to an error for each parameter estimate. These relationships, however, generally include other factors that are unknown prior to data collection. Hence, it is not feasible to use them directly as a basis for record length selections. Certain assumptions based upon a prior knowledge of the data properties and/or engineering judgment will also be required.

For example, the variance of a mean square value estimate for Gaussian data is related to the record length T in terms of the covariance function and mean value of the data, as detailed in Equation (8.41). If one were prepared to assume that all data of interest were bandwidth limited Gaussian white noise with a bandwidth of B and a mean value of zero, then the record length required for a specified normalized rms error ε would be given by Equation (8.46) as $T = (B\varepsilon^2)^{-1}$. Unfortunately, such an assumption is rarely justified in practice even as a first-order approximation.

There is one parameter estimate where reasonable assumptions often can be made which provide a usable relationship between record length and error, namely, an autospectral density estimate. From Equation (8.146), it is seen that the random portion of the normalized rms error of spectra estimates is a function only of the record length T and the resolution bandwidth B_e (assuming the spectral density is relatively smooth within B_e and the data is Gaussian). This implies that the record length needed to provide a specified normalized random error ε_r in such spectral density estimates can be predicted by $T = (B_e \varepsilon_r^2)^{-1}$, where B_e is a known parameter of the analysis procedure rather than an unknown parameter of the data. The remaining problem then is to select an appropriate value for B_e.

Again referring to Equation (8.146), the resolution bandwidth B_e is a primary factor in the bias portion of the spectral density estimation error. Also important in the determination of bias, however, is $G_{xx}''(f)$, which defines the "peakeness" of the spectrum within the bandwidth B_e. Hence a limiting value for $G_{xx}''(f)$ must be anticipated to relate the bias error ε_b to the bandwidth B_e. This is usually accomplished by an evaluation of the physics of the problem in question. For example, consider the special case of data that represent the response of a resonant physical system where the behavior of each resonance can be approximated by a simple second order system as described by Equation (2.19). If the spectrum of the system excitation is assumed to be approximately uniform over the bandwidth of any given resonance, then the response spectrum near a resonance can be anticipated to have the form $G_{xx}(f) = c|H(f)|^2$, where c is a constant and $H(f)$ is given in mechanical units by Equation (2.24). A bias error can now be calculated for spectral measurements at resonance frequencies in terms of a ratio of the resolution bandwidth B_e to the half power point bandwidth B_r of $H(f)$ as shown in Example 8.5.

The foregoing discussions of record length selections for autospectral densities function estimates are particularly important for two reasons. First, the spectrum is the single most important parameter of random data for many engineering applications. Second, spectral estimates are generally the most demanding of the various parameter estimates considered in this chapter from the viewpoint of required record length for a given error. This latter fact is easily verified by comparing the error expression for autospectra estimates to the error expressions for other parameter estimates. It is seen that the denominators of the error terms for other parameter estimates generally involve factors which are larger than the value B_e required for well resolved spectral estimates.

The various error analysis formulas listed in this chapter involve either a BT_r product or $B_e T_r$ product where T_r (in place of T) is the *total record length*, B is the *total bandwidth* occupied by the data (corresponding to bandwidth limited white noise with a specified variance), and B_e is the *resolution bandwidth* for spectral estimates. Instead of employing the $B_e T_r$ product, frequency domain estimates can be defined in terms of the number of averages $n_d = B_e T_r$ as used in the digital computational procedures of Chapter 11. Similarly, other

Table 8.1

Record Lengths and Averages for Basic Estimates

Estimate	Required Total Record Length	Required Number of Averages
$\hat{\mu}_x$	$T_r = \dfrac{1}{2B\varepsilon^2}\left(\dfrac{\sigma_x}{\mu_x}\right)^2$	$N = \dfrac{1}{\varepsilon^2}\left(\dfrac{\sigma_x}{\mu_x}\right)^2$
$\hat{\mu}_x^2$	$T_r = \dfrac{2}{B\varepsilon^2}\left(\dfrac{\sigma_x}{\mu_x}\right)^2$	$N = \dfrac{4}{\varepsilon^2}\left(\dfrac{\sigma_x}{\mu_x}\right)^2$
$\hat{\sigma}_x, \hat{\psi}_x$	$T_r = \dfrac{1}{4B\varepsilon^2}$	$N = \dfrac{1}{2\varepsilon^2}$
$\hat{\sigma}_x^2, \hat{\psi}_x^2$	$T_r = \dfrac{1}{B\varepsilon^2}$	$N = \dfrac{2}{\varepsilon^2}$
$\hat{p}(x)$	$T_r = \dfrac{1}{WBp(x)\varepsilon^2}$	$N = \dfrac{2}{Wp(x)\varepsilon^2}$
$\hat{R}_{xy}(\tau)$	$T_r = \dfrac{1}{2B\varepsilon^2}[1 + \rho_{xy}^{-2}(\tau)]$	$N = \dfrac{1}{\varepsilon^2}[1 + \rho_{xy}^{-2}(\tau)]$
$\hat{G}_{xx}(f)$	$T_r = \dfrac{1}{B_e\varepsilon^2}$	$n_d = \dfrac{1}{\varepsilon^2}$

formulas in this chapter can be expressed in terms of appropriate digital parameters using $N = 2BT_r$ for the number of independent estimates.

Table 8.1 lists the minimum required total record length T_r and the minimum required number of independent averages $N = 2BT_r$ or $n_d = B_eT_r$ to achieve a desired error ε in terms of other parameters in the data or in the analysis. It is assumed that bias errors are negligible and can be ignored so that the normalized random error ε_r is the same as the normalized rms error ε appearing in Table 8.1.

PROBLEMS

8.1 Which of the following parameter estimates usually involves a bias error as well as a random error?
(a) Mean value.
(b) Mean square value.
(c) Probability density function.

(d) Autocorrelation function.
(e) Autospectral density function.

8.2 The mean value of a bandwidth limited Gaussian white noise random signal $x(t)$ is estimated from a sample record of length T. Assume the standard deviation of $x(t)$ is $\sigma_x = 1$ and the standard deviation of the estimate is $\sigma[\hat{\mu}_x] = 0.1$. Determine the normalized random error ε associated with estimates of the following parameters from the same sample record.
(a) Mean square value.
(b) rms value.
(c) Standard deviation.

8.3 Assume $x(t)$ is a sample record of bandwidth limited Gaussian white noise where $\mu_x = 0.10$, $\sigma_x = 0.20$, $B = 200$ Hz, and $T = 2$ sec. Compute the normalized rms error for
(a) mean value estimates.
(b) mean square value estimates.

8.4 Assume the probability density function of a Gaussian random signal $x(t)$ is estimated using a window width of $W = 0.25\sigma_x$. Determine the normalized bias error of the estimate $\hat{p}(x)$ at $x = \mu_x + 2.5\sigma_x$,
(a) by exact calculations.
(b) using the first-order approximation of Equation (8.77).

8.5 Assume the probability distribution (not density) function of a Gaussian random process $\{x(t)\}$ is estimated by ensemble averaging procedures over $N = 100$ independent sample records. Determine the normalized random error of the estimate $\hat{P}(x)$ at $x = \mu$. Is there a bias error in this estimate?

8.6 Consider two bandwidth limited white noise records $x(t)$ and $y(t)$ of bandwidth B and length T where

$$x(t) = s(t) + n_1(t) \quad \text{and} \quad y(t) = s(t) + n_2(t)$$

Assume $s(t)$, $n_1(t)$, and $n_2(t)$ are mutually uncorrelated with zero mean values and mean square values of $\psi_s^2 = S$, $\psi_{n_1}^2 = N_1$, and $\psi_{n_2}^2 = N_2$. Determine
(a) the normalized rms error of the autocorrelation estimates $\hat{R}_{xx}(\tau)$ and $\hat{R}_{yy}(\tau)$ at $\tau = 0$.
(b) the normalized rms error of the cross-correlation estimate $\hat{R}_{xy}(\tau)$ at $\tau = 0$.
(c) the result in (b) if $N_1 \gg S$ and $N_2 \gg S$.
(d) the result in (b) if $N_1 = 0$ and $N_2 \gg S$.

8.7 Given the two bandwidth limited white noise signals defined in Problem 8.6, determine an approximate 95% probability interval for the time delay $\hat{\tau}$ corresponding to the maximum value of the estimated cross-correlation function $\hat{R}_{xy}(\tau)$.

8.8   Let $\lambda(f) = |G_{xx}(f)/G''_{xx}(f)|^{1/2}$ and define λ_m as the maximum value of $\lambda(f)$. Determine the requirements on B_e and T as a function of λ_m if the normalized bias error and the normalized random error of an autospectral density function estimate are each to be less than 5%.

8.9   Consider an autospectral density function estimate $\hat{G}_{xx}(f)$ computed from finite Fourier transform operations and the ensemble averaging procedure described in Section 8.5.4. Suppose a sample record $x(t)$ of total length $T_r = 60$ sec is divided into 12 independent contiguous segments. Determine the 95% confidence interval for the true value $G_{xx}(f)$ when $\hat{G}_{xx}(f) = 0.30$ units^2/Hz. What is the normalized random error of the estimate?

8.10 Assume a random process $\{x(t)\}$ has an autospectral density function given by

$$G_{xx}(f) = \frac{10}{\left[1 - (f/100)^2\right]^2 + [0.1f/100]^2}$$

If the autospectrum is to be estimated from sample data, determine the resolution bandwidth B_e and the total record length T_r required to limit the normalized bias error to $\varepsilon_b \leq 0.05$ and the normalized random error to $\varepsilon_r = 0.10$.

REFERENCES

8.1   Bendat, J. S., *Principles and Applications of Random Noise Theory*, Wiley, New York, 1958. Reprinted by Krieger, Melbourne, Florida, 1977.

8.2   Bendat, J. S., and Piersol, A. G., *Engineering Applications of Correlation and Spectral Analysis*, Wiley-Interscience, New York, 1980.

8.3   Dixon, W. J. and Massey, F. J., Jr., *Introduction to Statistical Analysis*, 3rd ed., McGraw-Hill, New York, 1969.

CHAPTER 9

STATISTICAL ERRORS IN ADVANCED ESTIMATES

This chapter continues the development from Chapter 8 on statistical errors in random data analysis. Emphasis is now on frequency domain properties of joint sample records from two different stationary (ergodic) random processes. The advanced parameter estimates discussed in this chapter include magnitude and phase estimates of cross-spectral density functions, followed by various quantities contained in single-input/output problems and multiple-input/output problems, as covered in Chapters 6 and 7. In particular, statistical error formulas are developed for

Frequency response function estimates (gain and phase)
Coherence function estimates
Coherent output spectrum estimates
Multiple coherence function estimates
Partial coherence function estimates

9.1 CROSS-SPECTRAL DENSITY FUNCTION ESTIMATES

Consider the cross-spectral density function between two stationary (ergodic) Gaussian random processes $\{x(t)\}$ and $\{y(t)\}$, as defined in Equation (5.66). Specifically, given a pair of sample records $x(t)$ and $y(t)$ of unlimited length T, the one-sided cross-spectrum is given by

$$G_{xy}(f) = \lim_{T \to \infty} \frac{2}{T} E[X^*(f, T)Y(f, T)] \tag{9.1}$$

where $X(f, T)$ and $Y(f, T)$ are the finite Fourier transforms of $x(t)$ and $y(t)$, respectively, that is

$$X(f) = X(f, T) = \int_0^T x(t) e^{-j2\pi ft} \, dt$$

$$Y(f) = Y(f, T) = \int_0^T y(t) e^{-j2\pi ft} \, dt \tag{9.2}$$

It follows that a "raw" estimate (no averages) of the cross-spectrum for a finite record length T is given by

$$\tilde{G}_{xy}(f) = \frac{2}{T}[X^*(f)Y(f)] \tag{9.3}$$

and will have a resolution bandwidth of

$$B_e \approx \Delta f = (1/T) \tag{9.4}$$

meaning that spectral components will be estimated *only* at the discrete frequencies

$$f_k = (k/T) \qquad k = 0, 1, 2, \ldots \tag{9.5}$$

As discussed in Section 8.5.1, the resolution $B_e \approx (1/T)$ in Equation (9.4) establishes the potential resolution bias error in an analysis. It will be assumed in this chapter, however, that T is sufficiently long to make the resolution bias error of the cross-spectrum estimates negligible.

As for the autospectrum estimates discussed in Section 8.5.4, the "raw" cross-spectrum estimate given by Equation (9.3) will have an unacceptably large random error for most applications. In practice, the random error is reduced by computing an essemble of estimates from n_d different (distinct, disjoint) subrecords, each of length T, and averaging the results to obtain a final "smooth" estimate given by

$$\hat{G}_{xy}(f) = \frac{2}{n_d T} \sum_{i=1}^{n_d} X_i^*(f) Y_i(f) \tag{9.6}$$

It follows that the minimum total record length required to compute the cross-spectrum estimate is $T_r = n_d T$. Special cases of Equation (9.6) produce the autospectra estimates $\hat{G}_{xx}(f)$ and $\hat{G}_{yy}(f)$ by letting $x(t) = y(t)$.

The quantities $X(f)$ and $Y(f)$ in Equation (9.2) can be broken down into real and imaginary parts as follows:

$$X(f) = X_R(f) - jX_I(f) \qquad Y(f) = Y_R(f) - jY_I(f) \tag{9.7}$$

where

$$X_R(f) = \int_0^T x(t)\cos 2\pi ft\, dt \qquad X_I(f) = \int_0^T x(t)\sin 2\pi ft\, dt \qquad (9.8)$$

$$Y_R(f) = \int_0^T y(t)\cos 2\pi ft\, dt \qquad Y_I(f) = \int_0^T y(t)\sin 2\pi ft\, dt \qquad (9.9)$$

If $x(t)$ and $y(t)$ are normally distributed with zero mean values, the quantities in Equations (9.8) and (9.9) will be normally distributed with zero mean values. From Equation (9.3), omitting the index i and the frequency f to simplify the notation, "raw" estimates are then given by

$$\tilde{G}_{xx} = \frac{2}{T}\left(X_R^2 + X_I^2\right) \qquad \tilde{G}_{yy} = \frac{2}{T}\left(Y_R^2 + Y_I^2\right) \qquad (9.10)$$

$$\tilde{G}_{xy} = \tilde{C}_{xy} - j\tilde{Q}_{xy} = |\tilde{G}_{xy}|e^{-j\theta_{xy}} \qquad (9.11)$$

where

$$\tilde{C}_{xy} = \frac{2}{T}(X_R Y_R + X_I Y_I) \qquad \tilde{Q}_{xy} = \frac{2}{T}(X_R Y_I - X_I Y_R) \qquad (9.12)$$

$$|\tilde{G}_{xy}|^2 = \tilde{C}_{xy}^2 + \tilde{Q}_{xy}^2 \qquad \tan\tilde{\theta}_{xy} = \tilde{Q}_{xy}/\tilde{C}_{xy} \qquad (9.13)$$

For values computed only at $f = f_k$ defined in Equation (9.5), one can verify from Equations (9.8) and (9.9) that

$$E[X_R X_I] = E[Y_R Y_I] = 0$$

$$E[X_R^2] = E[X_I^2] = (T/4)G_{xx}$$

$$E[Y_R^2] = E[Y_I^2] = (T/4)G_{yy} \qquad (9.14)$$

$$E[X_R Y_R] = E[X_I Y_I] = (T/4)C_{xy}$$

$$E[X_R Y_I] = -E[X_I Y_R] = (T/4)Q_{xy}$$

Hence $E[\tilde{C}_{xy}] = C_{xy}$ and $E[\tilde{Q}_{xy}] = Q_{xy}$ with

$$E[\tilde{G}_{xx}] = G_{xx} \qquad E[\tilde{G}_{yy}] = G_{yy} \qquad (9.15)$$

$$E[\tilde{G}_{xy}] = E[\tilde{C}_{xy}] - jE[\tilde{Q}_{xy}] = C_{xy} - jQ_{xy} = G_{xy} \qquad (9.16)$$

The Gaussian assumption for any four variables a_1, a_2, a_3, a_4 with zero mean values gives from Equation (3.72)

$$E[a_1 a_2 a_3 a_4] = E[a_1 a_2]E[a_3 a_4] + E[a_1 a_3]E[a_2 a_4] + E[a_1 a_4]E[a_2 a_3]$$

$$\tag{9.17}$$

Repeated application of this formula shows that

$$E[\tilde{G}_{xx}^2] = \frac{4}{T^2}E[(X_R^2 + X_I^2)^2]$$

$$= \frac{4}{T^2}E[X_R^4 + 2X_I^2 X_R^2 + X_I^4] = 2G_{xx}^2 \tag{9.18}$$

where the following relationships are used:

$$E[X_R^4] = 3(E[X_R^2])^2 = 3(T/4)^2 G_{xx}^2$$

$$E[X_I^2 X_R^2] = E[X_I^2]E[X_R^2] = (T/4)^2 G_{xx}^2 \tag{9.19}$$

$$E[X_I^4] = 3(E[X_I^2])^2 = 3(T/4)^2 G_{xx}^2$$

Similarly, one can verify that

$$E[\tilde{G}_{yy}^2] = 2G_{yy}^2 \tag{9.20}$$

$$E[\tilde{C}_{xy}^2] = \tfrac{1}{2}(G_{xx}G_{yy} + 3C_{xy}^2 - Q_{xy}^2) \tag{9.21}$$

$$E[\tilde{Q}_{xy}^2] = \tfrac{1}{2}(G_{xx}G_{yy} + 3Q_{xy}^2 - C_{xy}^2) \tag{9.22}$$

9.1.1 Variance Formulas

Basic variance error formulas will now be calculated. By definition, for any unbiased "raw" estimate $\tilde{A}$, where $E[\tilde{A}] = A$, its variance from Equation (8.3) is

$$\text{Var}[\tilde{A}] = E[\tilde{A}^2] - A^2 \tag{9.23}$$

Hence

$$\text{Var}[\tilde{G}_{xx}] = G_{xx}^2 \qquad \text{Var}[\tilde{G}_{yy}] = G_{yy}^2 \tag{9.24}$$

$$\text{Var}[\tilde{C}_{xy}] = \tfrac{1}{2}(G_{xx}G_{yy} + C_{xy}^2 - Q_{xy}^2) \tag{9.25}$$

$$\text{Var}[\tilde{Q}_{xy}] = \tfrac{1}{2}(G_{xx}G_{yy} + Q_{xy}^2 - C_{xy}^2) \tag{9.26}$$

Observe also from Equations (9.13), (9.21), and (9.22) that

$$E\left[|\tilde{G}_{xy}|^2\right] = E\left[\tilde{C}_{xy}^2\right] + E\left[\tilde{Q}_{xy}^2\right] = G_{xx}G_{yy} + |G_{xy}|^2 \qquad (9.27)$$

Hence

$$\text{Var}\left[|\tilde{G}_{xy}|\right] = G_{xx}G_{yy} = \frac{|G_{xy}|^2}{\gamma_{xy}^2} \qquad (9.28)$$

where γ_{xy}^2 is the ordinary coherence function defined by

$$\gamma_{xy}^2 = \frac{|G_{xy}|^2}{G_{xx}G_{yy}} \qquad (9.29)$$

In accordance with Equation (4.9), the corresponding variance errors for "smooth" estimates of all of the above "raw" estimates will be reduced by a factor of n_d when averages are taken over n_d statistically independent "raw" quantities. To be specific

$$\text{Var}\left[\hat{G}_{xx}\right] = G_{xx}^2/n_d \qquad \text{Var}\left[\hat{G}_{yy}\right] = G_{yy}^2/n_d \qquad (9.30)$$

$$\text{Var}\left[|\hat{G}_{xy}|\right] = |G_{xy}|^2/\gamma_{xy}^2 n_d \qquad (9.31)$$

From Equation (8.9), the normalized rms errors (which are here the same as the normalized random errors) become

$$\varepsilon\left[\hat{G}_{xx}\right] = \frac{1}{\sqrt{n_d}} \qquad \varepsilon\left[\hat{G}_{yy}\right] = \frac{1}{\sqrt{n_d}} \qquad (9.32)$$

$$\varepsilon\left[|\hat{G}_{xy}|\right] = \frac{1}{|\gamma_{xy}|\sqrt{n_d}} \qquad (9.33)$$

The quantity $|\gamma_{xy}|$ is the positive square root of γ_{xy}^2. Note that ε for the cross-spectrum magnitude estimate $|\hat{G}_{xy}|$ varies inversely with $|\gamma_{xy}|$ and approaches $(1/\sqrt{n_d})$ as γ_{xy}^2 approaches one. Reference 9.1 contains these results and some of the other formulas derived in this chapter.

A summary is given in Table 9.1 on the main normalized random error formulas for various spectral density estimates. The number of averages n_d represents n_d distinct (nonoverlapping) records, which are assumed to contain statistically different information from record to record. These records may occur by dividing a long stationary ergodic record into n_d parts, or they may

Table 9.1

Normalized Random Errors for Spectral Estimates

Estimate	Normalized Random Error, ε
$\hat{G}_{xx}(f), \hat{G}_{yy}(f)$	$\dfrac{1}{\sqrt{n_d}}$
$|\hat{G}_{xy}(f)|$	$\dfrac{1}{|\gamma_{xy}(f)|\sqrt{n_d}}$
$\hat{C}_{xy}(f)$	$\dfrac{\left[G_{xx}(f)G_{yy}(f) + C_{xy}^2(f) - Q_{xy}^2(f)\right]^{1/2}}{C_{xy}(f)\sqrt{2n_d}}$
$\hat{Q}_{xy}(f)$	$\dfrac{\left[G_{xx}(f)G_{yy}(f) + Q_{xy}^2(f) - C_{xy}^2(f)\right]^{1/2}}{Q_{xy}(f)\sqrt{2n_d}}$

occur by repeating an experiment n_d times under similar conditions. With the exception of autospectrum estimates, all error formulas are functions of frequency. Unknown true values of desired quantities are replaced by measured values when one applies these results to evaluate the random errors in actual measured data.

9.1.2 Covariance Formulas

A number of basic covariance error formulas will now be derived. By the definition in Equation (3.33), for two unbiased "raw" estimates $\tilde{A}$ and $\tilde{B}$, where $E[\tilde{A}] = A$ and $E[\tilde{B}] = B$, their covariance is

$$\text{Cov}(\tilde{A}, \tilde{B}) = E[(\Delta A)(\Delta B)] = E[(\tilde{A} - A)(\tilde{B} - B)]$$

$$= E[\tilde{A}\tilde{B}] - AB \tag{9.34}$$

The increments $\Delta A = (\tilde{A} - A)$ and $\Delta B = (\tilde{B} - B)$. The estimates $\tilde{A}$ and $\tilde{B}$ are said to be *uncorrelated* when $\text{Cov}(\tilde{A}, \tilde{B}) = 0$.

Application of Equation (9.17) shows that

$$E[\tilde{C}_{xy}\tilde{Q}_{xy}] = \frac{4}{T^2}E[(X_RY_R + X_IY_I)(X_RY_I - X_IY_R)] = 2C_{xy}Q_{xy} \tag{9.35}$$

Hence, by Equation (9.34), at every frequency f_k

$$\text{Cov}(\tilde{C}_{xy}, \tilde{Q}_{xy}) = C_{xy}Q_{xy} \tag{9.36}$$

Similarly, one can verify that

$$\text{Cov}(\tilde{G}_{xx}, \tilde{C}_{xy}) = G_{xx}C_{xy} \qquad \text{Cov}(\tilde{G}_{xx}, \tilde{Q}_{xy}) = G_{xx}Q_{xy}$$
$$\text{Cov}(\tilde{G}_{yy}, \tilde{C}_{xy}) = G_{yy}C_{xy} \qquad \text{Cov}(\tilde{G}_{yy}, \tilde{Q}_{xy}) = G_{yy}Q_{xy} \tag{9.37}$$

Also

$$E\left[\tilde{G}_{xx}\tilde{G}_{yy}\right] = \frac{4}{T^2}E\left[(X_R^2 + X_I^2)(Y_R^2 + Y_I^2)\right] = G_{xx}G_{yy} + |G_{xy}|^2 \tag{9.38}$$

Hence

$$\text{Cov}(\tilde{G}_{xx}, \tilde{G}_{yy}) = |G_{xy}|^2 = \gamma_{xy}^2 G_{xx}G_{yy} \tag{9.39}$$

The following covariance results will now be proved.

$$\text{Cov}(\tilde{G}_{xx}, \tilde{\theta}_{xy}) = 0 \qquad \text{Cov}(\tilde{G}_{yy}, \tilde{\theta}_{xy}) = 0 \tag{9.40}$$

$$\text{Cov}(|\tilde{G}_{xy}|, \tilde{\theta}_{xy}) = 0 \tag{9.41}$$

In words, $\tilde{\theta}_{xy}$ is uncorrelated with $\tilde{G}_{xx}$, $\tilde{G}_{yy}$ and $|\tilde{G}_{xy}|$ at every frequency f_k. To prove Equation (9.40), note that

$$\tan\theta_{xy} = \frac{Q_{xy}}{C_{xy}}$$

Then, differential increments of both sides yields

$$\sec^2\theta_{xy}\,\Delta\theta_{xy} \approx \frac{C_{xy}\,\Delta Q_{xy} - Q_{xy}\,\Delta C_{xy}}{C_{xy}^2}$$

where

$$\sec^2\theta_{xy} = \frac{|G_{xy}|^2}{C_{xy}^2}$$

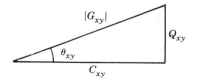

Thus

$$\Delta\theta_{xy} \approx \frac{C_{xy}\,\Delta Q_{xy} - Q_{xy}\,\Delta C_{xy}}{|G_{xy}|^2} \tag{9.42}$$

Now, one can set

$$\Delta\theta_{xy} = \tilde{\theta}_{xy} - \theta_{xy} \qquad \Delta Q_{xy} = \tilde{Q}_{xy} - Q_{xy} \qquad \Delta C_{xy} = \tilde{C}_{xy} - C_{xy}$$

Then, using Equation (9.37),

$$\mathrm{Cov}\big(\tilde{G}_{xx}, \tilde{\theta}_{xy}\big) = E\big[(\Delta G_{xx})(\Delta\theta_{xy})\big]$$

$$\approx \frac{C_{xy}}{|G_{xy}|^2}\mathrm{Cov}\big(\tilde{G}_{xx}, \tilde{Q}_{xy}\big) - \frac{Q_{xy}}{|G_{xy}|^2}\mathrm{Cov}\big(\tilde{G}_{xx}, \tilde{C}_{xy}\big) = 0$$

Similarly

$$\mathrm{Cov}\big(\tilde{G}_{yy}, \tilde{\theta}_{xy}\big) = 0$$

This proves Equation (9.40).

To prove Equation (9.41), note that

$$|G_{xy}|^2 = C_{xy}^2 + Q_{xy}^2$$

Taking differential increments of both sides yields

$$|G_{xy}||\Delta|G_{xy}| \approx C_{xy}\,\Delta C_{xy} + Q_{xy}\,\Delta Q_{xy}$$

Thus

$$\Delta|G_{xy}| \approx \frac{C_{xy}\,\Delta C_{xy} + Q_{xy}\,\Delta Q_{xy}}{|G_{xy}|} \tag{9.43}$$

Now, Equations (9.42) and (9.43) give

$$\mathrm{Cov}\big(|\tilde{G}_{xy}|, \tilde{\theta}_{xy}\big) = E\big[(\Delta|G_{xy}|)(\Delta\theta_{xy})\big]$$

$$\approx \frac{1}{|G_{xy}|^3}E\big[(C_{xy}\,\Delta C_{xy} + Q_{xy}\,\Delta Q_{xy})(C_{xy}\,\Delta Q_{xy} - Q_{xy}\,\Delta C_{xy})\big]$$

$$\approx \frac{1}{|G_{xy}|^3}\big\{(C_{xy}^2 - Q_{xy}^2)\mathrm{Cov}\big(\tilde{C}_{xy}, \tilde{Q}_{xy}\big)$$

$$+ C_{xy}Q_{xy}\big(\mathrm{Var}\big[\tilde{Q}_{xy}\big] - \mathrm{Var}\big[\tilde{C}_{xy}\big]\big)\big\} = 0$$

This proves Equation (9.41).

From Equation (9.43), one can also derive the following covariance results:

$$\text{Cov}\big(\tilde{G}_{xx}, |\tilde{G}_{xy}|\big) = E\big[(\Delta G_{xx})(\Delta|G_{xy}|)\big] \approx G_{xx}|G_{xy}|$$

$$\text{Cov}\big(\tilde{G}_{yy}, |\tilde{G}_{xy}|\big) = E\big[(\Delta G_{yy})(\Delta|G_{xy}|)\big] \approx G_{yy}|G_{xy}| \tag{9.44}$$

$$\text{Cov}\big(\tilde{C}_{xy}, |\tilde{G}_{xy}|\big) = E\big[(\Delta C_{xy})(\Delta|G_{xy}|)\big] \approx \frac{C_{xy}|G_{xy}|\big(1 + \gamma_{xy}^2\big)}{2\gamma_{xy}^2}$$

$$\text{Cov}\big(\tilde{Q}_{xy}, |\tilde{G}_{xy}|\big) = E\big[(\Delta Q_{xy})(\Delta|G_{xy}|)\big] \approx \frac{Q_{xy}|G_{xy}|\big(1 + \gamma_{xy}^2\big)}{2\gamma_{xy}^2} \tag{9.45}$$

Equation (9.43) offers a way to derive an approximate formula for $\text{Var}[|\tilde{G}_{xy}|]$ as follows:

$$\text{Var}\big[|\tilde{G}_{xy}|\big]$$

$$= E\big[(\Delta|G_{xy}|)^2\big] \approx \frac{1}{|G_{xy}|^2} E\big[(C_{xy}\,\Delta C_{xy} + Q_{xy}\,\Delta Q_{xy})^2\big]$$

$$\approx \frac{1}{|G_{xy}|^2}\Big\{ C_{xy}^2\,\text{Var}\big(\tilde{C}_{xy}\big) + 2C_{xy}Q_{xy}\text{Cov}\big(\tilde{C}_{xy}, \tilde{Q}_{xy}\big) + Q_{xy}^2\,\text{Var}\big(\tilde{Q}_{xy}\big) \Big\}$$

$$\approx \frac{1}{|G_{xy}|^2}\left\{ \frac{|G_{xy}|^2}{2} G_{xx}G_{yy} + \frac{|G_{xy}|^4}{2} \right\} \approx \frac{|G_{xy}|^2\big(1 + \gamma_{xy}^2\big)}{2\gamma_{xy}^2} \tag{9.46}$$

This approximate formula [Reference 9.2], based on differential increments, is inferior to the direct derivation producing the more exact formula of Equation (9.28), namely,

$$\text{Var}\big[|\tilde{G}_{xy}|\big] = \frac{|G_{xy}|^2}{\gamma_{xy}^2} \tag{9.47}$$

Note that the variance from Equation (9.47) will always be greater than the variance from Equation (9.46) since

$$\frac{1}{\gamma_{xy}^2} > \frac{1 + \gamma_{xy}^2}{2\gamma_{xy}^2} \qquad \text{for all } \gamma_{xy}^2 < 1$$

Consider the coherence function defined by

$$\gamma_{xy}^2 = \frac{|G_{xy}|^2}{G_{xx}G_{yy}} \tag{9.48}$$

Logarithms of both sides give

$$\log \gamma_{xy}^2 = 2\log|G_{xy}| - \log G_{xx} - \log G_{yy}$$

Then, differential increments of both sides show that

$$\frac{\Delta\gamma_{xy}^2}{\gamma_{xy}^2} \approx \frac{2\Delta|G_{xy}|}{|G_{xy}|} - \frac{\Delta G_{xx}}{G_{xx}} - \frac{\Delta G_{yy}}{G_{yy}}$$

where

$$\Delta\gamma_{xy}^2 = \tilde{\gamma}_{xy}^2 - \gamma_{xy}^2 \qquad \Delta|G_{xy}| = |\tilde{G}_{xy}| - G_{xy}$$

$$\Delta G_{xx} = \tilde{G}_{xx} - G_{xx} \qquad \Delta G_{yy} = \tilde{G}_{yy} - G_{yy}$$

Thus

$$\Delta\gamma_{xy}^2 \approx \gamma_{xy}^2 \left\{ \frac{2\Delta|G_{xy}|}{|G_{xy}|} - \frac{\Delta G_{xx}}{G_{xx}} - \frac{\Delta G_{yy}}{G_{yy}} \right\} \tag{9.49}$$

Now, using Equations (9.40) and (9.41), it follows directly from Equation (9.51) that

$$\text{Cov}\left(\tilde{\gamma}_{xy}^2, \tilde{\theta}_{xy}\right) = 0 \tag{9.50}$$

In words, the estimates $\tilde{\gamma}_{xy}^2$ and $\tilde{\theta}_{xy}$ are uncorrelated at every frequency f_k.

A summary is given in Table 9.2 on some of the main covariance formulas derived in this section.

9.1.3   Phase Angle Estimates

A formula for the variance of "raw" estimates $\tilde{\theta}_{xy}$ (expressed in radians) can be derived from Equation (9.42) as follows:

$$\text{Var}\left[\tilde{\theta}_{xy}\right] = E\left[(\Delta\theta_{xy})^2\right] \approx \frac{1}{|G_{xy}|^4} E\left[\left(C_{xy}\,\Delta Q_{xy} - Q_{xy}\,\Delta C_{xy}\right)^2\right]$$

$$= \frac{1}{|G_{xy}|^4}\left\{ C_{xy}^2\,\text{Var}\left(\tilde{Q}_{xy}\right) - 2C_{xy}Q_{xy}\text{Cov}\left(\tilde{C}_{xy}, \tilde{Q}_{xy}\right) + Q_{xy}^2\,\text{Var}\left(\tilde{C}_{xy}\right) \right\}$$

$$= \frac{1}{|G_{xy}|^4}\left\{ \frac{|G_{xy}|^2}{2}G_{xx}G_{yy} - \frac{|G_{xy}|^4}{2} \right\} = \frac{1 - \gamma_{xy}^2}{2\gamma_{xy}^2} \tag{9.51}$$

Table 9.2

Covariance Formulas for Various Estimates

$$\text{Cov}(\tilde{C}_{xy}, \tilde{Q}_{xy}) = C_{xy}Q_{xy}$$

$$\text{Cov}(\tilde{G}_{xx}, \tilde{C}_{xy}) = G_{xx}C_{xy}$$

$$\text{Cov}(\tilde{G}_{xx}, \tilde{Q}_{xy}) = G_{xx}Q_{xy}$$

$$\text{Cov}(\tilde{G}_{xx}, \tilde{G}_{yy}) = |G_{xy}|^2 = \gamma_{xy}^2 G_{xx}G_{yy}$$

$$\text{Cov}(\tilde{G}_{xx}, \tilde{\theta}_{xy}) = 0$$

$$\text{Cov}(|\tilde{G}_{xy}|, \tilde{\theta}_{xy}) = 0$$

$$\text{Cov}(\tilde{\gamma}_{xy}^2, \tilde{\theta}_{xy}) = 0$$

$$\text{Cov}(\tilde{G}_{xx}, |\tilde{G}_{xy}|) = G_{xx}|G_{xy}|$$

$$\text{Cov}(\tilde{C}_{xy}, |\tilde{G}_{xy}|) = \frac{C_{xy}|G_{xy}|\left(1 + \gamma_{xy}^2\right)}{2\gamma_{xy}^2}$$

$$\text{Cov}(\tilde{Q}_{xy}, |\tilde{G}_{xy}|) = \frac{Q_{xy}|G_{xy}|\left(1 + \gamma_{xy}^2\right)}{2\gamma_{xy}^2}$$

For "smooth" estimates of $\hat{\theta}_{xy}$ from n_d averages, one obtains

$$\text{s.d.}\left[\hat{\theta}_{xy}\right] = \frac{\text{s.d.}\left[\tilde{\theta}_{xy}\right]}{\sqrt{n_d}} \approx \frac{\left(1 - \gamma_{xy}^2\right)^{1/2}}{|\gamma_{xy}|\sqrt{2n_d}} \tag{9.52}$$

This gives the standard deviation of the phase estimate $\hat{\theta}_{xy}$. Here, one should not use the normalized random error, since θ_{xy} may be zero. The result in Equation (9.52) for the cross-spectrum phase angle, like the result in Equation (9.33) for the cross-spectrum magnitude estimate, varies with frequency. Independent of n_d for any $n_d > 1$, note that s.d.$[\hat{\theta}_{xy}]$ approaches zero as γ_{xy}^2 approaches one.

Example 9.1. **Illustration of Random Errors in Cross-Spectral Density Estimate.** Consider a pair of sample records $x(t)$ and $y(t)$, which represent Gaussian and stationary random processes with zero mean values. Assume the coherence function between the two random processes at a frequency of interest is $\gamma_{xy}^2 = 0.25$. If the cross-spectral density function $\hat{G}_{xy}(f)$ is estimated using $n_d = 100$ averages, determine the normalized random error in the magnitude estimate and the standard deviation of the phase estimate at the frequency of interest.

From Equation (9.33), the normalized random error in the magnitude of the cross-spectral density function estimate will be

$$\varepsilon\left[|\hat{G}_{xy}|\right] = \frac{1}{0.5\sqrt{100}} = 0.20$$

Note that this is twice as large as the normalized random errors in the autospectra estimates given by Equation (9.32). The standard deviation of the phase estimate is given by Equation (9.52) as

$$s.d.\left[\hat{\theta}_{xy}\right] = \frac{(0.75)^{1/2}}{(0.5)\sqrt{200}} = 0.12 \text{ radians}$$

or about 7 degrees.

9.2   SINGLE-INPUT/OUTPUT MODEL ESTIMATES

Consider the single-input/single-output model of Figure 9.1 where

$X(f)$ = Fourier transform of measured input signal $x(t)$, assumed noise free

$Y(f)$ = Fourier transform of measured output signal $y(t) = v(t) + n(t)$

$V(f)$ = Fourier transform of computed output signal $v(t)$

$N(f)$ = Fourier transform of computed output noise $n(t)$

$H_{xy}(f)$ = frequency response function of optimum constant parameter linear system estimating $y(t)$ from $x(t)$

Assume that $x(t)$ and $y(t)$ are the only records available for analysis and that they are representative members of zero mean value Gaussian random processes. Data can be either stationary random data or transient random data. For definiteness here, stationary data will be assumed and spectral results will be expressed using one-sided spectra. Normalized error formulas are the same for one-sided or two-sided spectra.

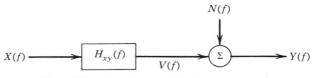

Figure 9.1   Single-input/output model.

From Section 6.1.4, the following equations apply to this model in Figure 9.1 to estimate various quantities of interest. The optimum frequency response function estimate is

$$\hat{H}_{xy}(f) = \frac{\hat{G}_{xy}(f)}{\hat{G}_{xx}(f)} \tag{9.53}$$

where $\hat{G}_{xx}(f)$ and $\hat{G}_{xy}(f)$ are "smooth" estimates of the input autospectral density function and the input/output cross-spectral density function, respectively. The associated ordinary coherence function estimate is

$$\hat{\gamma}_{xy}^2(f) = \frac{|\hat{G}_{xy}(f)|^2}{\hat{G}_{xx}(f)\hat{G}_{yy}(f)} \tag{9.54}$$

where $\hat{G}_{yy}(f)$ is a "smooth" estimate of the output autospectral density function. The coherent output spectrum estimate is

$$\hat{G}_{vv}(f) = |\hat{H}_{xy}(f)|^2\hat{G}_{xx}(f) = \hat{\gamma}_{xy}^2(f)\hat{G}_{yy}(f) \tag{9.55}$$

The noise output spectrum estimate is

$$\hat{G}_{nn}(f) = \left[1 - \hat{\gamma}_{xy}^2(f)\right]\hat{G}_{yy}(f) \tag{9.56}$$

It is also known from Equation (6.61) that

$$\hat{G}_{yy}(f) = \hat{G}_{vv}(f) + \hat{G}_{nn}(f) \tag{9.57}$$

since $v(t)$ and $n(t)$ are uncorrelated with $G_{vn}(f) = 0$ when $\hat{H}_{xy}(f)$ is computed by Equation (9.53).

In polar form, the frequency response function can be expressed as

$$\hat{H}_{xy}(f) = |\hat{H}_{xy}(f)|e^{-j\hat{\phi}_{xy}(f)} \tag{9.58}$$

where

$$|\hat{H}_{xy}(f)| = \frac{|\hat{G}_{xy}(f)|}{\hat{G}_{xx}(f)} = \text{system gain factor estimate} \tag{9.59}$$

$$\hat{\phi}_{xy}(f) = \tan^{-1}\left[\frac{\hat{Q}_{xy}(f)}{\hat{C}_{xy}(f)}\right] = \text{system phase factor estimate} \tag{9.60}$$

This quantity $\hat{\phi}_{xy}(f)$ is the same as the phase angle $\hat{\theta}_{xy}(f)$ in $\hat{G}_{xy}(f)$, and is also the phase that would be assigned to $\hat{\gamma}_{xy}(f)$ when $\hat{\gamma}_{xy}(f)$ is defined as the complex-valued function given by

$$\hat{\gamma}_{xy}(f) = |\hat{\gamma}_{xy}(f)|e^{-j\hat{\phi}_{xy}(f)} = +\sqrt{\hat{\gamma}_{xy}^2(f)}\,e^{-j\hat{\phi}_{xy}(f)} \qquad (9.61)$$

Note that all of the above "smooth" estimates can be computed from the original computed "smooth" estimates of $\hat{G}_{xx}(f)$, $\hat{G}_{yy}(f)$, and $\hat{G}_{xy}(f)$.

Random error formulas for all of these quantities will be derived in terms of the unknown true coherence function $\gamma_{xy}^2(f)$ and the number of independent averages n_d. To apply these results to evaluate measured data, one should use the computed coherence function estimate $\hat{\gamma}_{xy}^2(f)$ with an appropriate number of independent averages n_d. This will give practical results, particularly if the resulting random errors are less than 20%.

9.2.1 Bias in Frequency Response Functions Estimates

Estimation of the frequency response function and coherence function will generally involve bias errors from a number of sources as follows:

1. Bias inherent in the estimation procedure
2. Bias due to propagation time delays
3. Nonlinear and/or time-varying system parameters
4. Bias in autospectral and cross-spectral density estimates
5. Measurement noise at the input point (no bias problem from uncorrelated noise at the output point)
6. Other inputs that are correlated with the measured input (no bias problem from other uncorrelated inputs, which merely act as uncorrelated output noise with respect to the measured input)

The first noted source of bias error results from the fact that, in general,

$$E\left[\hat{H}_{xy}\right] = E\left[\frac{\hat{G}_{xy}}{\hat{G}_{xx}}\right] \neq \frac{E\left[\hat{G}_{xy}\right]}{E\left[\hat{G}_{xx}\right]}$$

$$E\left[\hat{\gamma}_{xy}^2\right] = E\left[\frac{|\hat{G}_{xy}|^2}{\hat{G}_{xx}\hat{G}_{yy}}\right] \neq \frac{E\left[|\hat{G}_{xy}|^2\right]}{E\left[\hat{G}_{xx}\right]E\left[\hat{G}_{yy}\right]} \qquad (9.62)$$

Hence $E[\hat{H}_{xy}] \neq H_{xy}$ and $E[\hat{\gamma}_{xy}^2] \neq \gamma_{xy}^2$. These bias errors are usually negligible compared to other possible errors that occur in practice, and can be ignored when combinations of n_d and $\hat{\gamma}_{xy}^2$ make associated normalized random errors small. As either $n_d \to \infty$ or $\gamma_{xy}^2 \to 1$, these inherent bias errors go to zero.

The second noted source of bias error occurs because it is required to measure $x(t)$ and $y(t)$ using a common time base and to correct for propagation time delays τ_1 that may occur between $x(t)$ and $y(t)$ if τ_1 is not negligible compared to the sample record lengths T. Assume

$$x(t) = x(t) \qquad\qquad 0 \le t \le T$$

$$y(t) = \begin{cases} \text{arbitrary} & 0 \le t < \tau_1 \\ x(t - \tau_1) & \tau_1 \le t \le T \end{cases} \qquad (9.63)$$

Then, to a first order of approximation, one can express the cross-correlation function estimate $\hat{R}_{xy}(\tau)$ by

$$\hat{R}_{xy}(\tau) \approx \left(1 - \frac{\tau_1}{T}\right) R_{xy}(\tau) \qquad (9.64)$$

showing that $\hat{R}_{xy}(\tau)$ is a *biased* estimate of $R_{xy}(\tau)$ since $E[\hat{R}_{xy}(\tau)] \ne R_{xy}(\tau)$. It follows that

$$\hat{G}_{xy}(f) \approx \left(1 - \frac{\tau_1}{T}\right) G_{xy}(f)$$

$$\hat{H}_{xy}(f) \approx \left(1 - \frac{\tau_1}{T}\right) H_{xy}(f) \qquad (9.65)$$

$$\hat{\gamma}_{xy}^2(f) \approx \left(1 - \frac{\tau_1}{T}\right)^2 \gamma_{xy}^2(f)$$

Thus $\hat{G}_{xy}(f)$, $\hat{H}_{xy}(f)$, and $\hat{\gamma}_{xy}^2(f)$ are also biased estimates. To remove these possible bias errors, the signal $y(t)$ should be shifted in time by the amount τ_1 so as to bring $x(t)$ and $y(t)$ into time coincidence. The time delay τ_1 can be estimated either from the physical geometry of the problem with a known velocity of propagation or from a separate measurement of $\hat{R}_{xy}(\tau)$ to determine where the first peak value occurs [Reference 9.3].

The third indicated source of bias results from violations of the basic assumptions that the system is a constant-parameter linear system. Even when the constant-parameter assumption is reasonably valid, the linearity assumption will often be violated if the operating range of interest is sufficiently wide. It should be noted, however, that the application of Equation (9.53) to nonlinear systems will produce the best linear approximation (in the least squares sense) for the frequency response function under the specified input and output conditions. This fact constitutes an important advantage to estimating frequency response functions from actual data rather than from laboratory or simulated data, which do not reproduce the actual input conditions.

The fourth source of bias occurs because of bias errors in spectral density estimates. As noted in Section 8.5.1, this error can be quite significant at frequencies where spectral peaks occur. These errors can be suppressed by obtaining properly resolved estimates of autospectra and cross-spectra, that is, by making B_e sufficiently narrow to accurately define peaks in the spectra. A quantitative formula is developed in Example 8.5 for spectra representing the response of a resonant system.

The fifth source of bias is measurement noise $m(t)$ at the input, where this noise occurs at the input point but does not actually pass through the system. Assuming the true input to be $u(t)$, the measured $\hat{G}_{xx} = \hat{G}_{uu} + \hat{G}_{mm}$. In place of Equations (9.53) and (9.54), one obtains

$$\hat{H}_{xy} = \frac{\hat{G}_{xy}}{\hat{G}_{uu} + \hat{G}_{mm}} \qquad \hat{\gamma}_{xy}^2 = \frac{|\hat{G}_{xy}|^2}{(\hat{G}_{uu} + \hat{G}_{mm})\hat{G}_{yy}} \qquad (9.66)$$

where the frequency f is omitted to simplify the notation. If all other bias errors are ignored, then $E[\hat{G}_{xy}] = G_{uy}$ and

$$E\left[\hat{H}_{xy}\right] = \frac{G_{uy}}{G_{uu} + G_{mm}} = H_{uy}\left[\frac{G_{uu}}{G_{uu} + G_{mm}}\right] \qquad (9.67)$$

$$E\left[\hat{\gamma}_{xy}^2\right] = \frac{|G_{uy}|^2}{(G_{uu} + G_{mm})G_{yy}} = \gamma_{uy}^2\left[\frac{G_{uu}}{G_{uu} + G_{mm}}\right] \qquad (9.68)$$

Hence, since $G_{mm} > 0$, these expected values will both be too low and the resulting bias errors will be negative. The normalized bias errors of Equation (8.9b) become here

$$\varepsilon_b\left[\hat{H}_{xy}\right] = \varepsilon_b\left[\hat{\gamma}_{xy}^2\right] = -\left[\frac{G_{mm}}{G_{uu} + G_{mm}}\right] \qquad (9.69)$$

For example, if $G_{mm} = 0.10G_{uu}$, then the resulting estimates would be biased downward by $(0.10/1.10) \approx 9\%$.

The sixth source of bias listed above is due to the contributions of other inputs that are correlated with the measured input. This error is readily illustrated for the simple case of one additional correlated input by the developments in Section 7.2. Specifically, when the problem is treated properly, the correct frequency response function for $H_1(f)$ is stated in Equation (7.23). On the other hand, when the problem is treated as a single-input/single-output case, the estimate $\hat{H}_1(f)$ is given by Equation (9.53). The expected value of this estimate does not yield the correct result, showing that it is biased. Also, the correct coherence function between x_1 and the output y when x_2 is present and correlated with x_1 is stated in Equation (7.34). When x_1 alone is assumed to be present, the estimate $\hat{\gamma}_{xy}^2(f)$ is given by Equation (9.54). Again, the expected value of this estimate does not yield the correct result.

The types of bias errors discussed above occur also in general multiple-input/output problems, where more complicated expressions are needed to evaluate the bias errors.

9.2.2 Coherent Output Spectrum Estimates

From Equation (9.57), assuming $\hat{G}_{vv}$ and $\hat{G}_{nn}$ are statistically independent, it follows that

$$\text{Var}\big[\hat{G}_{yy}\big] = \text{Var}\big[\hat{G}_{vv}\big] + \text{Var}\big[\hat{G}_{nn}\big] \qquad (9.70)$$

where the frequency f is omitted to simplify the notation. From Equation (9.30), the variances are

$$\text{Var}\big[\hat{G}_{yy}\big] = \frac{G_{yy}^2}{n_d} \qquad \text{Var}\big[\hat{G}_{nn}\big] = \frac{G_{nn}^2}{n_d} \qquad (9.71)$$

Hence, by substituting $G_{nn} = (1 - \gamma_{xy}^2)G_{yy}$ and $G_{vv} = \gamma_{xy}^2 G_{yy}$,

$$\text{Var}\big[\hat{G}_{vv}\big] = \frac{G_{yy}^2 - G_{nn}^2}{n_d} = \frac{\big(2 - \gamma_{xy}^2\big)G_{vv}^2}{\gamma_{xy}^2 n_d} \qquad (9.72)$$

Finally, the normalized random error becomes

$$\varepsilon\big[\hat{\gamma}_{xy}^2\hat{G}_{yy}\big] = \varepsilon\big[\hat{G}_{vv}\big] = \frac{\text{s.d.}[\hat{G}_{vv}]}{G_{vv}} = \frac{\big[2 - \gamma_{xy}^2\big]^{1/2}}{|\gamma_{xy}|\sqrt{n_d}} \qquad (9.73)$$

This derivation does not use any differential increments. Note that $\varepsilon[\hat{G}_{vv}] = \varepsilon[\hat{G}_{yy}] = (1/\sqrt{n_d})$ as $\gamma_{xy}^2 \to 1$. Note also that

$$\varepsilon\big[\hat{G}_{vv}\big] > \varepsilon\big[\hat{G}_{yy}\big] \qquad \text{for all } \gamma_{xy}^2 < 1 \qquad (9.74)$$

Figure 9.2 plots Equation (9.73) as a function of γ_{xy}^2 for the special case when $n_d = 100$. Figure 9.3 plots Equation (9.73) for arbitrary values of n_d and γ_{xy}^2. Table 9.3 contains appropriate values that must be satisfied between γ_{xy}^2 and n_d to achieve $\varepsilon[\hat{G}_{vv}] = 0.10$ using Equation (9.73).

Example 9.2. Illustration of Bias and Random Errors in Coherent Output Spectrum Estimate.   Consider the case where a number of independent acoustic noise sources in air produce an acoustic pressure signal $y(t)$ at a receiver location of interest. A coherent output spectrum is estimated based on an input measurement $x(t)$ representing a source located about 30 ft from the receiver. Assume the analysis is performed using $n_d = 400$ averages and

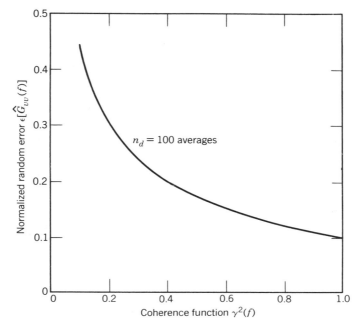

Figure 9.2 Normalized random error of coherent output spectral estimates when $n_d = 100$.

subrecord lengths of $T = 0.1$ sec ($B_e \approx 10$ Hz). Further assume that $x(t)$ and $y(t)$ are measured and analyzed on a common time base; that is, no precomputational delay is used. If the true coherence and the output spectral density at a frequency of interest are $\gamma_{xy}^2 = 0.4$ and $G_{yy} = 10$, determine the primary bias and random errors in the coherent output spectrum estimate.

First considering bias errors, the time required for acoustic noise to propagate about 30 ft in air ($c \approx 1120$ ft/sec) is approximately $\tau_1 = 0.03$ sec. Hence from Equation (9.65), the coherence estimate will be biased to yield on the average

$$\hat{\gamma}_{xy}^2 = \left(1 - \frac{0.03}{0.1}\right)^2 (0.4) = 0.2$$

rather than the correct value of $\gamma_{xy}^2 = 0.4$, and the coherent output spectrum estimate will be on the average

$$\hat{G}_{vv} = (0.2)10 = 2$$

rather than the correct value of $G_{vv} = 4$. The time delay bias error in this case causes an underestimate of the coherent output spectrum by 50% on the average.

Now considering the random error, it is the biased value of the coherence estimate that will control the random error. Hence, from Equation (9.73), the

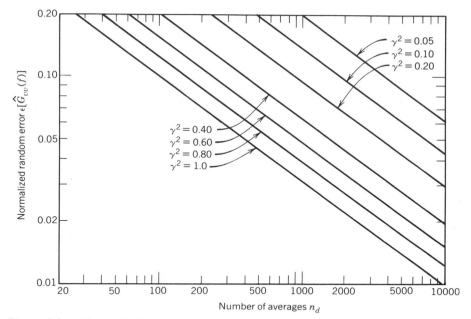

Figure 9.3   Normalized random error of coherent output spectral estimates versus number of averages.

normalized random error of the coherent output spectrum estimate at the frequency of interest is

$$\varepsilon\left[\hat{G}_{vv}\right] = \frac{[2 - 0.2]^{1/2}}{(0.45)(20)} = 0.15$$

9.2.3   *Coherence Function Estimates*

An approximate result for the bias error in coherence function estimates is derived in Reference 9.4. It is stated there that

$$b\left[\hat{\gamma}_{xy}^2\right] \approx \frac{1}{n_d}\left(1 - \hat{\gamma}_{xy}^2\right)^2 \tag{9.75}$$

Table 9.3

Conditions for $\varepsilon[\hat{G}_{vv}] = 0.10$

γ_{xy}^2	0.30	0.40	0.50	0.60	0.70	0.80	0.90	1.00
n_d	567	400	300	234	186	156	123	100

For example, if $n_d \geq 100$, as will often occur in practice, then for all $\hat{\gamma}_{xy}^2$,

$$b\left[\hat{\gamma}_{xy}^2\right] \leq 0.01$$

This bias error approaches zero as $n_d \to \infty$ or $\hat{\gamma}_{xy}^2 \to 1$.

To obtain a formula for the random error, start with the equation

$$G_{vv} = \gamma_{xy}^2 G_{yy} \qquad (9.76)$$

Then, differential increments of both sides yields

$$\Delta G_{vv} \approx \gamma_{xy}^2 \, \Delta G_{yy} + G_{yy} \, \Delta \gamma_{xy}^2 \qquad (9.77)$$

where, using "smooth" estimates here,

$$\Delta G_{vv} = \hat{G}_{vv} - G_{vv} \qquad \Delta G_{yy} = \hat{G}_{yy} - G_{yy} \qquad \Delta \gamma_{xy}^2 = \hat{\gamma}_{xy}^2 - \gamma_{xy}^2 \quad (9.78)$$

By definition, for unbiased estimates,

$$\mathrm{Var}\left[\hat{G}_{vv}\right] = E\left[\Delta G_{vv} \, \Delta G_{vv}\right] \qquad \mathrm{Var}\left[\hat{G}_{yy}\right] = E\left[\Delta G_{yy} \, \Delta G_{yy}\right]$$

$$\mathrm{Var}\left[\hat{\gamma}_{xy}^2\right] = E\left[\Delta \gamma_{xy}^2 \, \Delta \gamma_{xy}^2\right] \qquad (9.79)$$

Also

$$\mathrm{Cov}\left(\hat{G}_{vv}, \hat{G}_{yy}\right) = E\left[\Delta G_{vv} \, \Delta G_{yy}\right] = \mathrm{Var}\left[\hat{G}_{vv}\right] \qquad (9.80)$$

since

$$\Delta G_{yy} = \Delta G_{vv} + \Delta G_{nn} \qquad \text{and} \qquad E\left[\Delta G_{vv} \, \Delta G_{nn}\right] = 0$$

Solving Equation (9.77) for $G_{yy} \, \Delta \gamma_{xy}^2$ gives

$$G_{yy} \, \Delta \gamma_{xy}^2 \approx \Delta G_{vv} - \gamma_{xy}^2 \, \Delta G_{yy}$$

Squaring both sides and taking expected values yields

$$G_{yy}^2 \mathrm{Var}\left[\hat{\gamma}_{xy}^2\right] \approx \mathrm{Var}\left[\hat{G}_{vv}\right] - 2\gamma_{xy}^2 \mathrm{Cov}\left(\hat{G}_{vv}, \hat{G}_{yy}\right) + \gamma_{xy}^4 \mathrm{Var}\left[\hat{G}_{yy}\right]$$

Substitutions from Equations (9.71), (9.72), and (9.80) then show

$$\mathrm{Var}\left[\hat{\gamma}_{xy}^2\right] \approx \frac{2\gamma_{xy}^2 \left(1 - \gamma_{xy}^2\right)^2}{n_d} \qquad (9.81)$$

Finally

$$\varepsilon\left[\hat{\gamma}_{xy}^2\right] = \frac{\text{s.d.}\left[\hat{\gamma}_{xy}^2\right]}{\gamma_{xy}^2} \approx \frac{\sqrt{2}\left(1 - \gamma_{xy}^2\right)}{|\gamma_{xy}|\sqrt{n_d}} \qquad (9.82)$$

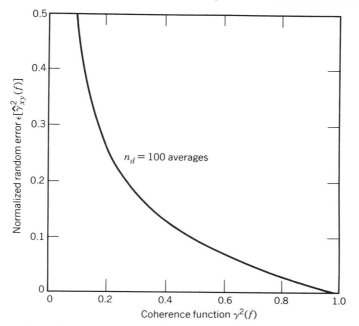

Figure 9.4   Normalized random error of coherence function estimates when $n_d = 100$.

For small ε, Equation (9.82) agrees with results in Reference 9.4. Note that $\varepsilon[\hat{\gamma}_{xy}^2]$ approaches zero as either $n_d \to \infty$ or $\gamma_{xy}^2 \to 1$. For any $n_d > 1$, the coherence function estimates can be more accurate than the autospectra and cross-spectra estimates used in their computation provided γ_{xy}^2 is close to unity. The restriction that $n_d > 1$ is of course needed since $\hat{\gamma}_{xy}^2 = 1$ at all f when $n_d = 1$ and gives a meaningless estimate for $\hat{\gamma}_{xy}^2$.

Figure 9.4 plots Equation (9.82) as a function of γ_{xy}^2 for the special case when $n_d = 100$. Figure 9.5 plots Equation (9.82) for arbitrary values of n_d and γ_{xy}^2. Table 9.4 contains appropriate values that must be satisfied between γ_{xy}^2 and n_d to achieve $\varepsilon[\hat{\gamma}_{xy}^2] = 0.10$ using Equation (9.82).

Example 9.3.   Illustration of Confidence Interval for Coherence Function Estimate.   Suppose the coherence function between two random signals $x(t)$ and $y(t)$ is estimated using $n_d = 100$ averages. Assume a value of $\hat{\gamma}_{xy}^2 = 0.5$ is estimated at a frequency of interest. Determine approximate 95% confidence intervals for the true value of γ_{xy}^2.

From Equation (9.82), the normalized random error of the coherence function estimate is approximated by using the estimate $\hat{\gamma}_{xy}^2$ in place of the unknown γ_{xy}^2. For the problem at hand, this yields

$$\varepsilon\left[\hat{\gamma}_{xy}^2\right] = \frac{\sqrt{2}\,(1 - 0.5)}{(0.71)\sqrt{100}} = 0.1$$

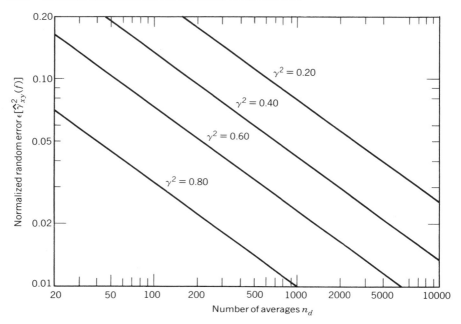

Figure 9.5 Normalized random error of coherence function estimates versus number of averages.

Hence, from Equation (8.14), an approximate 95% confidence interval for the true value of coherence at this frequency is

$$\left[0.4 \le \gamma_{xy}^2 \le 0.6\right]$$

9.2.4 Gain Factor Estimates

To establish the random error in gain factor estimates, since random error formulas are known for G_{xx} and G_{vv}, one can start with the equation

$$G_{vv} = |H_{xy}|^2 G_{xx} \tag{9.83}$$

Table 9.4

Conditions for $\varepsilon[\hat{\gamma}_{xy}^2] = 0.10$

γ_{xy}^2	0.30	0.40	0.50	0.60	0.70	0.80	0.90
n_d	327	180	100	54	26	10	3

Taking differential increments of both sides yields

$$\Delta G_{vv} \approx |H_{xy}|^2 \Delta G_{xx} + G_{xx} \Delta |H_{xy}|^2 \qquad (9.84)$$

The quantities ΔG_{xx} and ΔG_{vv} are as per Equation (9.78), while

$$\Delta |H_{xy}|^2 = |\hat{H}_{xy}|^2 - |H_{xy}|^2$$

and, assuming unbiased estimates,

$$\mathrm{Var}\left[|\hat{H}_{xy}|^2\right] = E\left[\Delta |H_{xy}|^2 \, \Delta |H_{xy}|^2\right]$$

Also, since $|G_{xv}|^2 = G_{xx}G_{vv}$, from Equation (9.39),

$$\mathrm{Cov}(\hat{G}_{xx}, \hat{G}_{vv}) = E[\Delta G_{xx} \, \Delta G_{vv}] = \frac{G_{xx}G_{vv}}{n_d} \qquad (9.85)$$

Solving Equation (9.84) for $G_{xx} \Delta |H_{xy}|^2$ gives

$$G_{xx} \Delta |H_{xy}|^2 \approx \Delta G_{vv} - |H_{xy}|^2 \Delta G_{xx}$$

Squaring both sides and taking expected values yields

$$G_{xx}^2 \mathrm{Var}\left[|\hat{H}_{xy}|^2\right] \approx \mathrm{Var}[\hat{G}_{vv}] - 2|H_{xy}|^2 \mathrm{Cov}(\hat{G}_{xx}, \hat{G}_{vv}) + |H_{xy}|^4 \mathrm{Var}[\hat{G}_{xx}]$$

Substitutions then show

$$\mathrm{Var}\left[|\hat{H}_{xy}|^2\right] \approx \frac{2\left(1 - \gamma_{xy}^2\right)|H_{xy}|^4}{\gamma_{xy}^2 n_d} \qquad (9.86)$$

For any estimate $\hat{A}$, as a first order of approximation,

$$\Delta A^2 \approx 2A \, \Delta A$$

where

$$\Delta A = \hat{A} - A \qquad \Delta A^2 = \hat{A}^2 - A^2$$

$$E[\Delta A] = 0 \qquad E[\Delta A^2] = 0$$

Now

$$(\Delta A^2)^2 \approx 4A^2(\Delta A)^2$$

$$E[\Delta A^2 \, \Delta A^2] \approx 4A^2 E[\Delta A \, \Delta A]$$

This is the same as

$$\mathrm{Var}[\hat{A}^2] \approx 4A^2 \mathrm{Var}[\hat{A}] \tag{9.87}$$

Dividing through by A^4 gives

$$\varepsilon^2[\hat{A}^2] = \frac{\mathrm{Var}[\hat{A}^2]}{A^4} \approx \frac{4\,\mathrm{Var}[\hat{A}]}{A^2} = 4\varepsilon^2[\hat{A}]$$

Hence

$$\varepsilon[\hat{A}^2] \approx 2\varepsilon[\hat{A}] \tag{9.88}$$

This useful result can compare random errors from mean square and rms estimates, as shown earlier in Equation (8.10).

Applying Equation (9.87) to Equation (9.86) gives the formula

$$\mathrm{Var}\big[|\hat{H}_{xy}|\big] \approx \frac{\left(1 - \gamma_{xy}^2\right)|H_{xy}|^2}{2\gamma_{xy}^2 n_d} \tag{9.89}$$

Finally

$$\varepsilon\big[|\hat{H}_{xy}|\big] = \frac{\mathrm{s.d.}\big[|\hat{H}_{xy}|\big]}{|H_{xy}|} \approx \frac{\left(1 - \gamma_{xy}^2\right)^{1/2}}{|\gamma_{xy}|\sqrt{2n_d}} \tag{9.90}$$

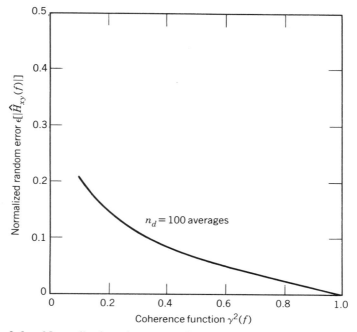

Figure 9.6 Normalized random error of gain factor estimates when $n_d = 100$.

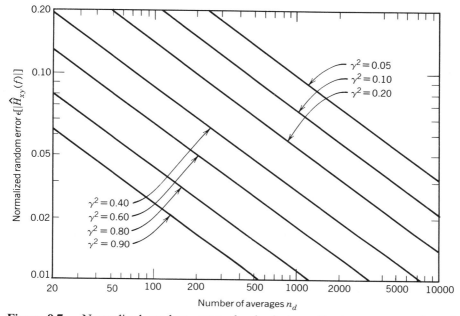

Figure 9.7 Normalized random error of gain factor estimates versus number of averages.

Note that this result is the same as Equation (9.52). It shows that $\varepsilon[|\hat{H}_{xy}|]$ approaches zero as γ_{xy}^2 approaches one, independent of the size of n_d, and also approaches zero as n_d becomes large, independent of the value of γ_{xy}^2. These results agree with the nature of much more complicated formulas in Reference 9.5, where error results were obtained using the F distribution.

Figure 9.6 plots Equation (9.90) as a function of γ_{xy}^2 for the special case when $n_d = 100$. Figure 9.7 plots Equation (9.90) for arbitrary values of n_d and γ_{xy}^2. Table 9.5 gives examples of values that must be satisfied between γ_{xy}^2 and n_d to achieve $\varepsilon[|\hat{H}_{xy}|] = 0.10$ using Equation (9.90).

Table 9.5

Conditions for $\varepsilon[|\hat{H}_{xy}|] = 0.10$

γ_{xy}^2	0.30	0.40	0.50	0.60	0.70	0.80	0.90
n_d	117	75	50	34	22	13	6

9.2.5 *Phase Factor Estimates*

Phase factor estimates $\hat{\phi}_{xy}(f)$ from Equation (9.60) are the same as phase angle estimates $\hat{\theta}_{xy}(f)$ whose standard deviation is derived in Equation (9.52). Hence

$$\text{s.d.}\left[\hat{\phi}_{xy}\right] \approx \frac{\left(1 - \gamma_{xy}^2\right)^{1/2}}{|\gamma_{xy}|\sqrt{2n_d}} \tag{9.91}$$

Equation (9.90) now shows that

$$\text{s.d.}\left[\hat{\phi}_{xy}\right] \approx \varepsilon\left[|\hat{H}_{xy}|\right] \tag{9.92}$$

In words, the standard deviation for the phase factor estimate $\hat{\phi}_{xy}$, measured in radians, is approximately the same as the normalized random error for the gain factor estimate $|\hat{H}_{xy}|$. This claim is reasonable whenever $\varepsilon[|\hat{H}_{xy}|]$ is small, say $\varepsilon \leq 0.20$, as demonstrated in Reference 9.6. Hence, one can state that Figure 9.6 plots the standard deviation s.d.$[\hat{\phi}_{xy}]$ in radians as a function of γ_{xy}^2 for the special case when $n_d = 100$. Also, Table 9.5 gives examples of values that must be satisfied between γ_{xy}^2 and n_d to achieve s.d.$[\hat{\phi}_{xy}] \approx 0.10$ radian.

***Example 9.4.* Illustration of Random Errors in Frequency Response Function Estimate.** Suppose the frequency response function between two random signals $x(t)$ and $y(t)$ is estimated using $n_d = 50$ averages. Assume the coherence function at one frequency of interest is $\gamma_{xy}^2(f_1) = 0.10$ and at a second frequency of interest is $\gamma_{xy}^2(f_2) = 0.90$. Determine the random errors in the frequency response function gain and phase estimates at the two frequencies of interest.

From Equation (9.90), the normalized random error in the gain factor estimates at frequencies f_1 and f_2 will be

$$\varepsilon\left[|\hat{H}_{xy}(f_1)|\right] = \frac{(1 - 0.10)^{1/2}}{(0.32)(10)} = 0.30$$

$$\varepsilon\left[|\hat{H}_{xy}(f_2)|\right] = \frac{(1 - 0.90)^{1/2}}{(0.95)(10)} = 0.033$$

From Equation (9.91), the results given above also constitute approximations for the standard deviation (not normalized) in the phase factor estimate as follows.

$$\text{s.d.}\left[\hat{\phi}_{xy}(f_1)\right] = 0.30 \text{ radians} \qquad \text{s.d.}\left[\hat{\phi}_{xy}(f_2)\right] = 0.033 \text{ radians}$$

$$= 17 \text{ degrees} \qquad\qquad\qquad = 1.9 \text{ degrees}$$

Table 9.6

Summary of Single-Input/Output Random Error Formulas

Function Being Estimated	Normalized Random Error, ε
$\hat{\gamma}_{xy}^2(f)$	$\dfrac{\sqrt{2}\left[1 - \gamma_{xy}^2(f)\right]}{\|\gamma_{xy}(f)\|\sqrt{n_d}}$
$\hat{G}_{vv}(f) = \hat{\gamma}_{xy}^2(f)\hat{G}_{yy}(f)$	$\dfrac{\left[2 - \gamma_{xy}^2(f)\right]^{1/2}}{\|\gamma_{xy}(f)\|\sqrt{n_d}}$
$\|\hat{H}_{xy}(f)\|$	$\dfrac{\left[1 - \gamma_{xy}^2(f)\right]^{1/2}}{\|\gamma_{xy}(f)\|\sqrt{2n_d}}$
$\hat{\phi}_{xy}(f)$	$\text{s.d.}[\hat{\phi}_{xy}(f)] \approx \dfrac{\left[1 - \gamma_{xy}^2(f)\right]^{1/2}}{\|\gamma_{xy}(f)\|\sqrt{2n_d}}$

Hence, the estimates made with a coherence of $\gamma_{xy}^2(f_2) = 0.90$ are almost ten times as accurate as those made with a coherence of $\gamma_{xy}^2(f_1) = 0.10$. This concludes the example.

A summary is given in Table 9.6 on the main normalized random error formulas for single-input/output model estimates. Some engineering measurements are shown in References 9.6 and 9.7. A general theory for resolution bias errors is presented in Reference 9.8.

9.3   MULTIPLE-INPUT/OUTPUT MODEL ESTIMATES

Consider the more general multiple-input/output model of Figure 9.8. All records should be measured simultaneously using a common time base. The first series of steps should be to replace this given model by the conditioned model of Figure 9.9 as defined in Chapter 7. Procedures for doing this are described by the iterative computational algorithms in Section 7.3. The only averaging required is in the computation of "smooth" estimates $\hat{G}_{ij}(f)$ of autospectra and cross-spectra from the original given data. All other quantities are then computed algebraically as follows (for simplicity in notation, the dependence on f will be omitted).

Compute first the initial set of conditioned estimates $\hat{G}_{ij\cdot 1}$ by the formula

$$\hat{G}_{ij\cdot 1} = \hat{G}_{ij} - \hat{L}_{1j}\hat{G}_{i1} \qquad (9.93)$$

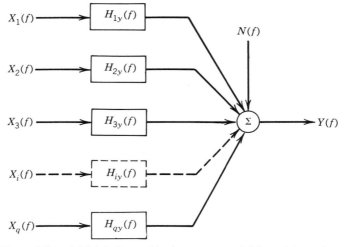

Figure 9.8 Multiple-input/single-output model for arbitrary inputs.

where

$$\hat{L}_{1j} = \frac{\hat{G}_{1j}}{\hat{G}_{11}} \qquad (9.94)$$

This is the only step that uses the original $\hat{G}_{ij}$. From these results, compute next the second set of conditioned estimates $\hat{G}_{ij \cdot 2!}$ by the same procedure exactly as in Equations (9.93) and (9.94), namely,

$$\hat{G}_{ij \cdot 2!} = \hat{G}_{ij \cdot 1} - \hat{L}_{2j}\hat{G}_{i2 \cdot 1} \qquad (9.95)$$

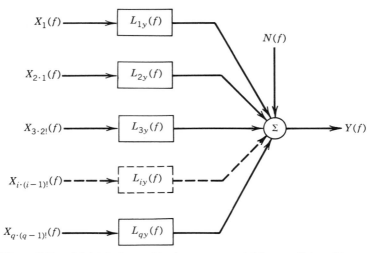

Figure 9.9 Multiple-input/single-output model for conditioned inputs.

where

$$\hat{L}_{2j} = \frac{\hat{G}_{2j \cdot 1}}{\hat{G}_{22 \cdot 1}} \qquad (9.96)$$

Compute next the third set of conditioned estimates $\hat{G}_{ij \cdot 3!}$ using the just obtained $\hat{G}_{ij \cdot 2!}$ by a similar procedure, and so on to as many terms as required. Further algebraic operations on these results yield estimates for all of the partial coherence functions, the multiple coherence function, and related quantities.

The computation of the multiple coherence function estimate $\hat{\gamma}_{y:x}^2$ can be carried out as follows:

$$\hat{\gamma}_{y:x}^2 = \frac{\hat{G}_{vv}}{\hat{G}_{yy}} = \frac{\hat{G}_{yy} - \hat{G}_{nn}}{\hat{G}_{yy}} = 1 - \frac{\hat{G}_{nn}}{\hat{G}_{yy}} \qquad (9.97)$$

Here, x represents all of the inputs $x_1, x_2, \ldots, x_q$. The output noise spectrum

$$\hat{G}_{nn} = \hat{G}_{yy \cdot q!} = \hat{G}_{yy}\left(1 - \hat{\gamma}_{1y}^2\right)\left(1 - \hat{\gamma}_{2y \cdot 1}^2\right) \cdots \left(1 - \hat{\gamma}_{qy \cdot (q-1)!}^2\right) \qquad (9.98)$$

using previously computed ordinary and partial coherence functions. Thus

$$\hat{\gamma}_{y:x}^2 = 1 - \left[\left(1 - \hat{\gamma}_{1y}^2\right)\left(1 - \hat{\gamma}_{2y \cdot 1}^2\right) \cdots \left(1 - \hat{\gamma}_{qy \cdot (q-1)!}^2\right)\right] \qquad (9.99)$$

The multiple coherent output spectrum estimate is given by

$$\hat{G}_{vv} = \hat{\gamma}_{y:x}^2 \hat{G}_{yy} = |\hat{L}_{1y}|^2 \hat{G}_{11} + |\hat{L}_{2y}|^2 \hat{G}_{22 \cdot 1} + \cdots + |\hat{L}_{qy}|^2 \hat{G}_{qq \cdot (q-1)!} \qquad (9.100)$$

from the system output terms in Figure 9.9. Note that this last formula can also be solved for $\hat{\gamma}_{y:x}^2$ to provide another way to estimate $\hat{\gamma}_{y:x}^2$ by computing

$$\hat{\gamma}_{y:x}^2 = \frac{\hat{G}_{vv}}{\hat{G}_{yy}} \qquad (9.101)$$

where $\hat{G}_{vv}$ is given by the expansion in Equation (9.100).

9.3.1 Multiple Coherence Function Estimates

The multiple coherence function estimate $\hat{\gamma}_{y:x}^2$ applies to the models in both Figure 9.8 and 9.9. If n_d averages are used to compute smooth estimates of autospectra and cross-spectra from the original given data, then analogous to Equation (9.82),

$$\varepsilon\left[\hat{\gamma}_{y:x}^2\right] \approx \frac{\sqrt{2}\left[1 - \gamma_{y:x}^2\right]}{|\gamma_{y:x}|\sqrt{n_d + 1 - q}} \qquad (9.102)$$

This is the same form as Equation (9.82) except that the ordinary coherence function γ_{xy}^2 is replaced by $\gamma_{y:x}^2$ and n_d is replaced by $(n_d + 1 - q)$.

9.3.2   *Multiple Coherent Output Spectrum Estimates*

Let $\hat{G}_{vv} = \hat{\gamma}^2_{y:x}\hat{G}_{yy}$ be the multiple coherent output spectrum estimate calculated from the product of $\hat{\gamma}^2_{y:x}$ and $\hat{G}_{yy}$ or by Equation (9.100). Similar to Equation (9.73),

$$\varepsilon\left[\hat{G}_{vv}\right] \approx \frac{\left[2 - \gamma^2_{y:x}\right]^{1/2}}{|\gamma_{y:x}|\sqrt{n_d + 1 - q}} \qquad (9.103)$$

As before, this is the same form as earlier except that $\gamma^2_{y:x}$ replaces γ^2_{xy} and $(n_d + 1 - q)$ replaces n_d.

9.3.3   *Single Conditioned-Input/Output Models*

Figure 9.9 can be broken down into a set of simple single conditioned-input/output models whose spectral nature is illustrated in Figure 9.10. The

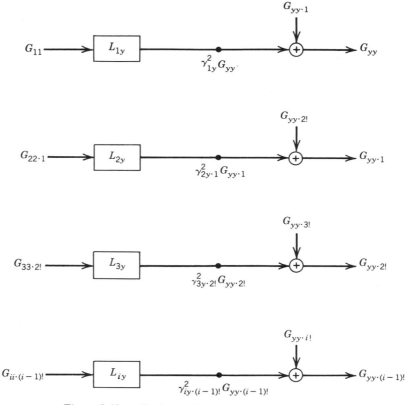

Figure 9.10   Single conditioned-input/output models.

very first system in Figure 9.10 is the simple single-input/output problem treated in Section 9.2. The succeeding systems are direct extensions, where it is obvious which terms should be related to terms in the first system. Computation of "smooth" conditioned terms shown in these models involves subtraction of terms, which lead to reducing the number of averages n_d by one for each successive step. Thus when $\hat{G}_{yy}$ is estimated using n_d averages, $\hat{G}_{yy \cdot 1}$ will be associated with $n_d - 1$ averages, $\hat{G}_{yy \cdot 2!}$ will be associated with $n_d - 2$ averages and, in general, $\hat{G}_{yy \cdot (i-1)!}$ will be associated with $n_d + 1 - i$ averages. Random error formulas will now be listed which are appropriate for these simple conditioned input/output models involving a fewer number of averages.

9.3.4   *Partial Coherence Function Estimates*

Similar to Equation (9.82), the following results are obtained:

$$\varepsilon\left[\hat{\gamma}_{1y}^2\right] \approx \frac{\sqrt{2}\left[1 - \gamma_{1y}^2\right]}{|\gamma_{1y}|\sqrt{n_d}}$$

$$\varepsilon\left[\hat{\gamma}_{2y \cdot 1}^2\right] \approx \frac{\sqrt{2}\left[1 - \gamma_{2y \cdot 1}^2\right]}{|\gamma_{2y \cdot 1}|\sqrt{n_d - 1}} \tag{9.104}$$

$$\varepsilon\left[\hat{\gamma}_{3y \cdot 2!}^2\right] \approx \frac{\sqrt{2}\left[1 - \gamma_{3y \cdot 2!}^2\right]}{|\gamma_{3y \cdot 2!}|\sqrt{n_d - 2}}$$

In general, for $i = 1, 2, \ldots, q$,

$$\varepsilon\left[\hat{\gamma}_{iy \cdot (i-1)!}^2\right] \approx \frac{\sqrt{2}\left[1 - \gamma_{iy \cdot (i-1)!}^2\right]}{|\gamma_{iy \cdot (i-1)!}|\sqrt{n_d + 1 - i}} \tag{9.105}$$

9.3.5   *Partial Coherent Output Spectrum Estimates*

Similar random error formulas to Equation (9.73) follow for partial coherent output spectrum estimates representing the system output terms shown in Figure 9.10. Specifically,

$$\varepsilon\left[|\hat{L}_{1y}|^2\hat{G}_{11}\right] = \varepsilon\left(\hat{\gamma}_{1y}^2\hat{G}_{yy}\right) \approx \frac{\left[2 - \gamma_{1y}^2\right]^{1/2}}{|\gamma_{1y}|\sqrt{n_d}}$$

$$\varepsilon\left[|\hat{L}_{2y}|^2\hat{G}_{22 \cdot 1}\right] = \varepsilon\left(\hat{\gamma}_{2y \cdot 1}^2\hat{G}_{yy \cdot 1}\right) \approx \frac{\left[2 - \gamma_{2y \cdot 1}^2\right]^{1/2}}{|\gamma_{2y \cdot 1}|\sqrt{n_d - 1}} \tag{9.106}$$

$$\varepsilon\left[|\hat{L}_{3y}|^2\hat{G}_{33 \cdot 2!}\right] = \varepsilon\left(\hat{\gamma}_{3y \cdot 2!}^2\hat{G}_{yy \cdot 2!}\right) \approx \frac{\left[2 - \gamma_{3y \cdot 2!}^2\right]^{1/2}}{|\gamma_{3y \cdot 2!}|\sqrt{n_d - 2}}$$

and so on.

9.3.6 *Gain Factor Estimates for Conditioned Models*

Analogous to Equation (9.90), the following results are obtained:

$$\varepsilon\left[|\hat{L}_{1y}|\right] \approx \frac{\left[1 - \gamma_{1y}^2\right]^{1/2}}{|\gamma_{1y}|\sqrt{2n_d}}$$

$$\varepsilon\left[|\hat{L}_{2y}|\right] \approx \frac{\left[1 - \gamma_{2y\cdot1}^2\right]^{1/2}}{|\gamma_{2y\cdot1}|\sqrt{2(n_d - 1)}} \qquad (9.107)$$

$$\varepsilon\left[|\hat{L}_{3y}|\right] \approx \frac{\left[1 - \gamma_{3y\cdot2!}^2\right]^{1/2}}{|\gamma_{3y\cdot2!}|\sqrt{2(n_d - 2)}}$$

and so on.

9.3.7 *Phase Factor Estimates for Conditioned Models*

Similar to Equation (9.92), the following formulas give approximate results for phase factor uncertainties, in radians, associated with the $\{L_{iy}\}$ systems in Figure 9.10:

$$\text{s.d.}\left[\hat{\phi}_{1y}\right] \approx \varepsilon\left[|\hat{L}_{1y}|\right]$$

$$\text{s.d.}\left[\hat{\phi}_{2y}\right] \approx \varepsilon\left[|\hat{L}_{2y}|\right] \qquad (9.108)$$

In general, for $i = 1, 2, \ldots, q$,

$$\text{s.d.}\left[\hat{\phi}_{iy}\right] \approx \varepsilon\left[|\hat{L}_{iy}|\right] \qquad (9.109)$$

Note that errors will be small whenever $\varepsilon[|\hat{L}_{iy}|]$ is small and that phase estimates are not normalized.

PROBLEMS

9.1 Consider two random signals $x(t)$ and $y(t)$ representing stationary random processes. Assume the autospectra of the two signals, as well as the cross-spectrum between the signals, is to be estimated using $n_d = 100$ averages. If the coherence function between $x(t)$ and $y(t)$ is $\gamma_{xy}^2 = 0.50$ at a frequency of interest, determine the normalized random error for
   (a) the autospectra estimates $\hat{G}_{xx}$ and $\hat{G}_{yy}$.
   (b) the cross-spectrum magnitude estimate $|\hat{G}_{xy}|$.

9.2 For two random signals $x(t)$ and $y(t)$, suppose that at the frequency of interest the autospectra values are $G_{xx} = G_{yy} = 1$ and the coincident and quadrature spectral values are $C_{xy} = Q_{xy} = 0.5$. Assuming $n_d = 100$ averages, determine the normalized random error for estimates of the coincident and quadrature spectral density values $\hat{C}_{xy}$ and $\hat{Q}_{xy}$.

9.3 For the two random signals in Problem 9.2, determine an approximate 95% probability interval for the estimated value of the phase factor $\hat{\theta}_{xy}$ at the frequency of interest.

9.4 For the two random signals in Problem 9.2, determine at the frequency of interest approximate 95% probability intervals for the estimated values of
(a) the coherence function $\hat{\gamma}_{xy}^2$.
(b) the coherent output spectrum $\hat{G}_{vv} = \hat{\gamma}_{xy}^2 \hat{G}_{yy}$.

9.5 Consider two bandwidth limited white noise random signals defined by

$$x(t) = u(t) + m(t)$$

$$y(t) = 10u(t - \tau_1) + n(t)$$

where $m(t)$ and $n(t)$ are statistically independent noise signals. Assume the spectral density values of the input signal and noise at a frequency of interest are equal, that is, $G_{uu} = G_{mm}$, and the delay value $\tau_1 = 0.1$ sec. If the frequency response function from $x(t)$ to $y(t)$ is estimated using a large number of averages and individual record lengths of $T = 0.5$ sec, determine the primary sources of bias errors and the magnitude of the bias errors.

9.6 For the two bandwidth limited white noise signals in Problem 9.5, assume $G_{mm} = 0$ and $G_{nn} = 100G_{uu}$ at a frequency of interest. If the output spectral density at the frequency of interest is $G_{yy} = 1$, determine
(a) the true value of the coherent output spectrum $G_{vv} = \gamma_{xy}^2 G_{yy}$.
(b) the bias error in the estimate $\hat{G}_{vv}$.
(c) the random error in the estimate $\hat{G}_{vv}$ assuming $n_d = 400$ averages.

9.7 Consider a single-input/single-output system where the coherence function between the input $x(t)$ and the output $y(t)$ is $\gamma_{xy}^2 = 0.75$ at a frequency of interest. Suppose the autospectra of the input and output are $G_{xx} = 2.1$ and $G_{yy} = 0.8$, respectively. What should be the minimum number of averages n_d in a frequency response function estimate to ensure that the estimated gain factor $|\hat{H}_{xy}|$ is within $\pm 10\%$ of the true gain factor $|H_{xy}|$ with a probability of 95%?

9.8 For the single-input/single-output system in Problem 9.7, what value of coherence is needed to provide a gain factor estimate and autospectra estimates with the same normalized random error?

9.9   Consider a two-input/one-output system as in Figure 7.2 where the inputs are $x_1(t)$ and $x_2(t)$ and the output is $y(t)$. Assume estimates are computed using $n_d = 100$ averages to obtain the following quantities at a frequency of interest: $\hat{G}_{11} = 20$, $\hat{G}_{22} = 25$, $\hat{G}_{yy} = 40$, $\hat{G}_{12} = 15$, and $\hat{G}_{1y} = \hat{G}_{2y} = 16 - j12$. Determine 95% confidence intervals for the gain and phase factors of the frequency response function H_{2y} between $x_2(t)$ and $y(t)$.

9.10   Using the data in Problem 9.9, determine the expected value and normalized random error for the multiple coherence function estimate $\hat{\gamma}_{y:x}^2$ between the two inputs and the output.

REFERENCES

9.1   Bendat, J. S., "Statistical Errors in Measurement of Coherence Functions and Input/Output Quantities," *Journal of Sound and Vibration*, Vol. 59, p. 405, 1978.

9.2   Jenkins, G. M. and Watts, D. G., *Spectral Analysis and Its Applications*, Holden-Day, San Francisco, 1968.

9.3   Seybert, A. F., and Hamilton, J. F., "Time Delay Bias Errors in Estimating Frequency Response and Coherence Functions," *Journal of Sound and Vibration*, Vol. 60, p. 1, 1978.

9.4   Carter, G. C., Knapp, C. H., and Nuttall, A. H., "Estimation of the Magnitude-Squared Coherence via Overlapped Fast Fourier Transform Processing," *IEEE Transactions on Audio and Electroacoustics*, Vol. AU-21, p. 337, August 1973.

9.5   Goodman, N. R., "Measurement of Matrix Frequency Response Functions and Multiple Coherence Functions," AFFDL TR 65-56, Air Force Flight Dynamics Laboratory, Wright-Patterson AFB, Ohio, February 1965.

9.6   Bendat, J. S. and Piersol, A. G., *Engineering Applications of Correlation and Spectral Analysis*, Wiley-Interscience, New York, 1980.

9.7   Herlufsen, H., "Dual Channel FFT Analysis (Parts I and II)," *Bruel & Kjaer Technical Review*, Nos. 1 and 2, 1984.

9.8   Schmidt, H., "Resolution Bias Errors in Spectral Density, Frequency Response and Coherence Function Measurements," *Journal of Sound and Vibration*, Vol. 101, p. 347, 1985.

CHAPTER 10

DATA ACQUISITION AND PROCESSING

Appropriate techniques for the acquisition and processing of random data are heavily dependent on the physical phenomenon represented by the data and the desired engineering goals of the processing. In broad terms, however, the required operations may be divided into five primary categories as follows:

Data collection
Data recording (including transmission)
Data preparation
Data qualification
Data analysis

Each of these categories involves a number of sequential steps as schematically illustrated in Figure 10.1. The purpose of this chapter is to summarize basic considerations associated with each of these key steps. The emphasis throughout is on potential sources of error over and above the statistical errors inherent in sampling considerations, as developed in Chapters 8 and 9. The digital computations required for data analysis are presented in Chapter 11.

10.1   DATA COLLECTION

The primary element in data collection is the instrumentation transducer. In general terms, a transducer is any device that translates power from one form to another. In an engineering context, this usually means the translation of some measure of a physical phenomenon of interest into an analog signal with a calibrated relationship between the input and output quantities. Referring to Figure 10.1(a), this translation may involve up to three basic operations: (a) a mechanical conversion of the physical quantity of interest into an intermediate

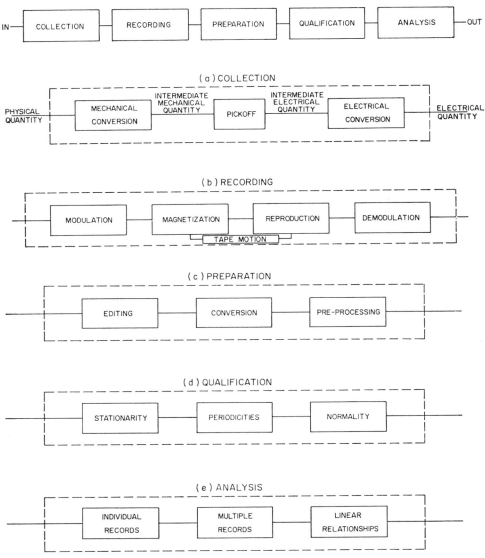

Figure 10.1 Key steps in data acquisition and processing.

mechanical quantity, (b) a pickoff step that converts the intermediate mechanical quantity into an intermediate electrical quantity, and (c) an electrical conversion into a final electrical quantity, usually voltage. Some transducers may combine any two or all three of the operations, depending on the physical quantity being measured and the specific nature of the transducer. For example, a thermocouple, which is widely used as a temperature transducer, converts a difference in temperature directly to a difference in voltage without intermediate steps. On the other hand, a resistance thermometer, another commonly used temperature transducer, first converts a change in temperature to a change in resistance, and then converts the resistance change to an electrical voltage change. For this case, there is a two-step operation that bypasses the conversion to an intermediate mechanical quantity. Seismic type motion transducers, which are often used to measure displacements, velocities, and accelerations, involve all three steps indicated in Figure 10.1(a). The input motion is converted to a relative displacement, which in turn is converted through a pickoff device to an intermediate electrical quantity such as change in resistance. Finally, the change in resistance is converted to an electrical voltage proportional to the input.

Ideally, the operations listed above would be accomplished without distortion or modification of the time history of the physical quantity being measured. In other words, if the input time history is $x(t)$ and the output time history is $y(t)$, the perfect transducer would provide an analog output $y(t) = cx(t)$, where c is a simple calibration constant. Unfortunately, this ideal situation is difficult to achieve in practice. Gain and phase modifications as well as distortion producing nonlinearities are often inherent in the transducer operations. This fact makes the transducer a potential source of error in any data acquisition and processing program.

***Example 10.1.* Open Loop Seismic Accelerometer.**   Consider an idealized open loop seismic accelerometer, as schematically illustrated in Figure 10.2(a). The mechanical conversion in this transducer is achieved through a spring-dashpot supported seismic mass, where an input acceleration at the foundation is converted to a relative displacement of the mass. The pickoff device is a potentiometer. The final electrical conversion is achieved by measuring the voltage across the potentiometer. Using the procedures outlined in Section 2.4.1, it is readily shown that the frequency response function of this system between the input acceleration and the output voltage is given by

$$H(f) = \frac{c}{4\pi^2 f_n^2 \left[1 - (f/f_n)^2 + j2\zeta f/f_n\right]} \tag{10.1}$$

where c is a calibration constant and ζ and f_n are the damping ratio and undamped natural frequency, respectively, as defined in Equation (2.22). The gain and phase factors of this system are plotted in Figures 10.2(b) and (c). Note that these gain and phase factors are identical in form to those for a simple mechanical system with force excitation, as plotted in Figure 2.3.

Consider first the gain factor given in Figure 10.2(b). It is seen that the accelerometer provides a relatively uniform gain for frequencies well below the undamped natural frequency of the transducer. For frequencies approaching the undamped natural frequency, the gain factor tends to peak and/or diminish, depending on the damping ratio. The widest frequency range with uniform gain is provided by a damping ratio of approximately $\zeta = 0.7$. The

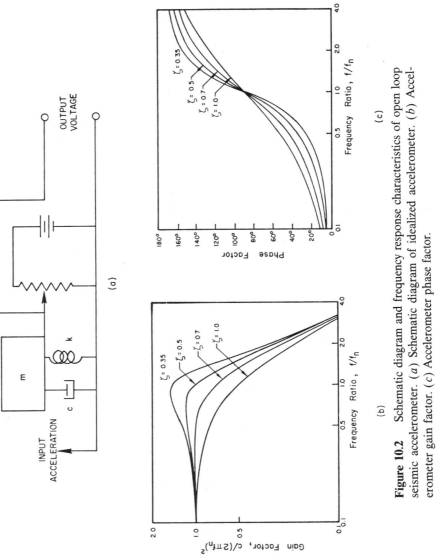

Figure 10.2 Schematic diagram and frequency response characteristics of open loop seismic accelerometer. (*a*) Schematic diagram of idealized accelerometer. (*b*) Accelerometer gain factor. (*c*) Accelerometer phase factor.

characteristics illustrated in Figure 10.2(*b*) limit the upper frequency range of seismic accelerometers to something less than the undamped natural frequency, depending on the amount of error (deviation from a constant gain factor) that can be accepted. The errors indicated for various conditions in Figure 10.2(*b*) might be either corrected by calibration or simply tolerated. Either way, it is essential that they be acknowledged to avoid unrecognized and perhaps serious magnitude errors in the final processed data.

Now consider the phase factor given in Figure 10.2(*c*). The accelerometer introduces a phase shift in the data that varies from zero degrees at zero frequency to 180 degrees at very high frequencies. The exact details of the phase shift are a function of the damping ratio. The important point here is that this phase shift corresponds to a time delay at any given frequency and, hence, can lead to phase distortion in the resulting data. Such distortion may cause errors in the analysis of single channel data of the complex periodic and transient form as well as multiple channel data of any type where joint properties are of interest. The single channel errors can be minimized by selecting a transducer with a phase factor of $\phi(f) \approx 0$. For the seismic accelerometer being considered, this would require a natural frequency that is very high relative to the highest frequency of interest in the data. The cross-channel errors can be controlled by using transducers with matched phase factors for the acquisition of multiple channel data. This concludes Example 10.1.

In summary, considerable care is warranted in the selection and use of data acquisition transducers. When such transducers are procured from commercial sources, supporting literature is usually provided that specifies the limitations on their use. These specifications are generally accurate, but it must be remembered that commercial manufacturers are not inclined to be pessimistic in the claims they make for their products. It is unwise to use a commercially procured transducer under conditions that exceed the limits of the manufacturers specifications unless its applicability to such use has been substantiated by appropriate studies. See References 10.1 and 10.2 for more detailed discussions of data collection and transducer systems.

10.2 DATA RECORDING

For some applications, it is possible to perform all desired data processing directly on the transducer signals in real time. For most applications, however, this is not practical and some form of storage (and perhaps remote transmission) of the transducer signals will be required. The most desirable and convenient type of data storage system is the magnetic tape recorder. Although other types of recorders could be used, the magnetic tape recorder has the advantages of being able to store large quantities of data and to reproduce them in electrical form. The most desirable way to transmit data signals is through electrical lines. There are obvious situations where this is not feasible; for example, retrieving data from a spacecraft in earth orbit. For such cases, radio transmission (telemetry) of the transducer signals is usually required.

Both magnetic tape recording and radio telemetry are major subjects within themselves and generally beyond the scope of this book. Basic material on these subjects is available from References 10.3, 10.4, and 10.5. There are two phases of the subjects, however, which should be discussed since they directly relate to fundamental problems in data processing. As noted in Figure 10.1(*b*), these two phases are the magnetization–reproduction operation in magnetic tape recording and the modulation–demodulation operation in both magnetic tape recording and telemetry.

10.2.1   *Magnetization – Reproduction Procedures*

The configuration for recording and reproducing information on magnetic tape for most commercially available recorders is as indicated in Figure 10.3(*a*). A moving plastic film with a magnetic coating slides over a record head. When an electrical current is passed through the record head, it generates a magnetic flux that magnetizes the particles on the tape. As the tape moves by the reproduce head, a signal is generated proportional to the rate of change of flux.

This type of record–reproduce system poses a number of troublesome problems, including the basic nonlinearity of the magnetization process, tape dropouts, and other side effects introduced by design compromises in the magnetic tape. From the viewpoint of basic recorder capabilities, however, the two most pertinent problems are as follows. First, the reproduce head responds to rate of change of flux on the magnetic tape and, hence, extracts a differentiated version of the input signal rather than a faithful reproduction of it. Second, the reproduce head averages the change of flux across the head gap. The first noted problem tends to limit the lower frequency response capabilities of the recorder as illustrated in Figure 10.3(*b*), while the second problem tends to limit the high frequency capabilities of the recorder as illustrated in Figure 10.3(*c*). Note that the high frequency response of the recorder can be improved by reducing the gap width, but this also reduces the sensitivity of the reproduce head. The gap depth may be reduced to bring the sensitivity back up, but this reduces head life. The sensitivity may also be increased by adding windings to the reproduce head, but this also attenuates the high frequency response.

Commercial recorders counteract the two problems discussed above by using equalizer networks to produce a resultant output with a uniform frequency response function over the widest possible frequency range. In quantitative terms, commercially available machines with a tape speed of 120 in./sec will provide a frequency response within ± 3 dB over a frequency range from about 400 to 2,000,000 Hz. The resulting signal to noise ratio, however, is only about 28 dB. Such record–reproduce capabilities are unsatisfactory for most instrumentation purposes for two reasons; first, the recorder is not capable of dc response and, second, the recorder has an inadequate signal to noise ratio. On

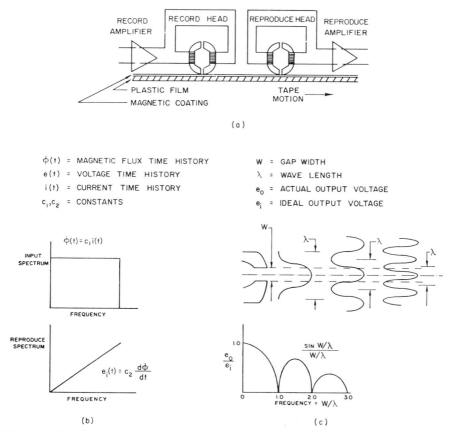

Figure 10.3 Basic principles and properties of magnetic tape recording. (*a*) Operating principles. (*b*) Recording frequency response. (*c*) Reproduce gap attenuation.

the other hand, the recorder does provide a generous frequency bandwidth. The problem then is to exchange bandwidth for the desirable features of dc response and improved signal to noise ratio. This is achieved by the modulation–demodulation procedure.

10.2.2 Modulation – Demodulation Procedures

The two most common modulation–demodulation procedures used for scientific data transmission and/or recording are frequency modulation and pulse code modulation.

Frequency modulation (FM) is the most widely used modulation procedure for analog telemetry and tape recording. The basic approach in FM is to make the frequency of a carrier signal analogous to the amplitude of the input data

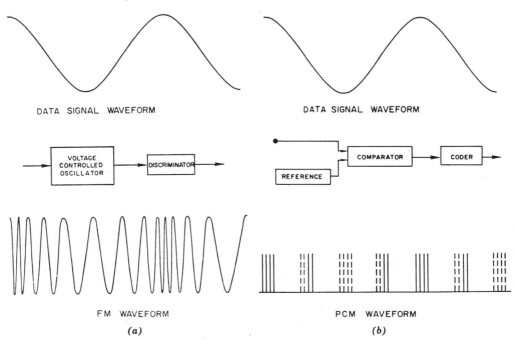

DATA SIGNAL WAVEFORM   DATA SIGNAL WAVEFORM

FM WAVEFORM   PCM WAVEFORM

(*a*)   (*b*)

Figure 10.4   Basic principles of data modulation procedures. (*a*) Frequency modulation (FM). (*b*) Pulse code modulation (PCM).

signal, as indicated in Figure 10.4(*a*). With zero input, the modulated signal is simply the carrier frequency itself. With a full scale dc (static) input, the modulated signal is a pure tone at the upper deviation frequency. With a sinusoidal input, the modulated signal theoretically looks like a sine wave that varies in frequency. For the case of tape recording, the modulated signal actually looks more like a square wave with varying width because FM signals are usually recorded at saturation. An important parameter in FM is the ratio of the deviation of the carrier frequency (Δf) to the modulating or data frequency (f_d). This ratio is called the *modulation index* (m_f), and is closely related to both the bandwidth capabilities and the output signal to noise ratio provided by FM transmission and recording. These relationships are summarized in Table 10.1, along with the basic advantages and disadvantages of FM.

Referring to Table 10.1, it is noted that FM is particularly sensitive to tape speed variations (time base errors) since these variations directly translate into amplitude errors. Of particular concern are dynamic time base errors. The most commonly discussed single-channel dynamic time base error is flutter, which may be defined as a variation in tape velocity from the nominal. The magnitude of flutter in high-quality multiple-channel instrumentation recorders is on the order of 0.25%. In general, the spectrum of the flutter noise is

Table 10.1

Basic Characteristics of Modulation Procedures

Characteristic	Frequency Modulation	Pulse Code Modulation
Bandwidth requirements	$B_r = 2f_d(1 + m_f)$	$B_r = ncf_d$
Signal-to-noise ratio	$S_o/N_o \simeq 6m_f^2 S_i/N_i$	$S_o/N_o \simeq 2^{2n}$
Advantages	dc response, good S/N improvement	dc response, very good S/N improvement, multiplexing convenience
Disadvantages	tape speed sensitivity, reduced bandwidth	very reduced bandwidth

B_r = required recorder (transmission) bandwidth in Hz
c = number of multiplexed channels
f_d = maximum frequency of data signal in Hz
m_f = modulation index in FM
n = number of pulses per code group in PCM
S_i/N_i = direct record signal-to-noise ratio
S_o/N_o = output signal-to-noise ratio

relatively flat out to some upper cutoff frequency, usually about 1000 Hz for the higher tape speeds. The most disturbing interchannel dynamic time base error is dynamic skew, which may be defined as a variation in the angle at which the tape passes over the head. This phenomenon clearly will affect interchannel analyses such as cross-correlation and cross-spectra. Dynamic skew problems can be minimized by recording data intended for joint analysis on adjacent record heads when feasible.

In pulse code modulation (PCM), the input analog signal is immediately converted to a digital code, which is then recorded as shown in Figure 10.4(b). Since the digital code consists of a sequence of pulses, there is no difficulty in retrieving the code off magnetic tape for input signals with frequencies down to dc. There is an upper frequency limit on the recorded data, which is determined by the sampling rate of the input analog-to-digital converter (ADC). This upper frequency limit is detailed in Section 10.3.2. The bandwidth and signal-to-noise requirements, as well as the basic advantages and disadvantages of PCM transmission and recording, are summarized in Table 10.1.

It is seen in Table 10.1 that the signal-to-noise ratio in PCM is a function only of the number of bits used in the digital code; it is completely independent of the direct record signal-to-noise ratio. This constitutes a major advantage over analog modulation procedures, such as FM. Also, multiplexing several channels of data onto a single transmission or recorder channel is straightforward with PCM since the coded data from several signals can easily be mixed along the time base. For FM, multiplexing must be accomplished by having the different data signals share the available frequency bandwidth of the telemetry or tape recorder channel. See Reference 10.5 for further details.

***Example 10.2.* Standard FM Recording Parameters.**   Most FM transmission and recording of scientific data is accomplished using standards established by the Inter-Range Instrumentation Group (IRIG), as detailed in Reference 10.6. These standards fix the modulation index at $m_f = 2.16$ and the percentage deviation of the carrier signal frequency for full scale deviation at $\pm 40\%$. One of the many IRIG standard carrier signal center frequencies is 108 kHz. Determine the maximum data frequency f_d that can be recorded with an FM tape recorder using the 108 kHz carrier signal. Also determine the resulting signal-to-noise ratio of the recording.

For a full scale deviation of $\pm 40\%$, it follows that the maximum deviation of the carrier signal frequency is $\Delta f = 43.2$ kHz. Then by definition, the maximum data frequency is $f_d = \Delta f / m_f = 20$ kHz. From Table 10.1, the signal-to-noise ratio of the recorded data will be $S_o/N_o \approx 28 S_i/N_i$. Assuming a typical direct record signal-to-noise ratio of $S_i/N_i \approx 630$ (28 dB), the output signal-to-noise ratio should be about $S_o/N_o \approx 17640$ (42.5 dB). In practice, FM tape recorders might provide signal-to-noise ratios of up to 6 dB better than the above figure depending on the details of the FM circuits.

10.3   DATA PREPARATION

The next key phase in data acquisition and processing is the preparation of the raw data for detailed analysis. The raw data usually consist of analog voltage signals directly from the transducers or from magnetic tape recordings of the transducer signals. A number of operations are needed at this point to make the transducer signals suitable for detailed analysis, as outlined in Figure 10.1.

The first data preparation operation indicated in Figure 10.1 is data editing. This refers to those preanalysis operations that are designed to detect and eliminate spurious and/or degraded data signals that might have resulted from acquisition and recording problems such as excessive noise, signal dropouts, and loss of signal because of transducer malfunctions. Editing is usually accomplished through visual inspection of the data time-history signals by a talented analyst before the data are converted to a digital format.

The next and most important step in data preparation is conversion of the analog data signals to a digital format (digitization). The digitization process consists of two separate and distinct operations: (a) sampling and (b) quantization. Sampling is the process of defining the instantaneous points in time at which the data are to be observed, while quantization is the conversion of data values at the sampling points into numerical form. Before discussing practical sampling considerations, it will be helpful to review sampling theorems for random data. These subjects are followed by discussions of analog-to-digital converters and preprocessing procedures.

10.3.1   *Sampling Theorems for Random Records*

Suppose a sample random time history record $x(t)$ from a random process $\{x_k(t)\}$ exists only for the time interval from 0 to T seconds, and is zero at all other times. Its Fourier transform is

$$X(f) = \int_0^T x(t) e^{-j2\pi ft} \, dt \tag{10.2}$$

Assume that $x(t)$ is continually repeated to obtain a periodic time function with a period of T seconds. The fundamental frequency increment is $f = 1/T$. By a Fourier series expansion

$$x(t) = \sum_{-\infty}^{\infty} A_n e^{j2\pi nt/T} \tag{10.3}$$

where

$$A_n = \frac{1}{T} \int_0^T x(t) e^{-j2\pi nt/T} \, dt \tag{10.4}$$

From Equation (10.2),

$$X\left(\frac{n}{T}\right) = \int_0^T x(t) e^{-j2\pi nt/T} \, dt = TA_n \tag{10.5}$$

Thus $X(n/T)$ determines A_n and, therefore, $x(t)$ at all t. In turn, this determines $X(f)$ for all f. This result is the *sampling theorem in the frequency domain*. The fundamental frequency increment $1/T$ is called a *Nyquist co-interval*.

Suppose that a Fourier transform $X(f)$ of some sample random time-history record $x(t)$ exists only over a frequency interval from $-B$ to B Hz, and is zero at all other frequencies. The actual realizable frequency band ranges from 0 to B Hz. The inverse Fourier transform yields

$$x(t) = \int_{-B}^{B} X(f) e^{j2\pi ft} \, df \tag{10.6}$$

Assume that $X(f)$ is continually repeated in frequency to obtain a periodic frequency function with a period of $2B$ Hz. The fundamental time increment is $t = 1/2B$. Now

$$X(f) = \sum_{-\infty}^{\infty} C_n e^{-j\pi n f/B} \tag{10.7}$$

where

$$C_n = \frac{1}{2B} \int_{-B}^{B} X(f) e^{j\pi n f/B} \, df \tag{10.8}$$

From Equation (10.6),

$$x\left(\frac{n}{2B}\right) = \int_{-B}^{B} X(f) e^{j\pi n f/B} \, df = 2BC_n \tag{10.9}$$

Thus $x(n/2B)$ determines C_n and, hence, $X(f)$ at all f. In turn, this determines $x(t)$ for all t. This result is the *sampling theorem in the time domain*. The fundamental time increment $1/2B$ is called a *Nyquist interval*.

Assume that the sample record $x(t)$ exists only for the time interval from 0 to T seconds, and suppose also that its Fourier transform $X(f)$ exists only in a frequency interval from $-B$ to B Hz. This dual assumption is not theoretically possible because of an *uncertainty principle* [Reference 10.7]. In practice, however, it may be closely approximated with finite time intervals and band-pass filters. Assuming $x(t)$ and $X(f)$ are so restricted in their time and frequency properties, only a finite number of discrete samples of $x(t)$ or $X(f)$ are required to describe $x(t)$ completely for all t. By sampling $X(f)$ at Nyquist co-interval points $1/T$ apart on the frequency scale from $-B$ to B, the number of discrete samples required to describe $x(t)$ is

$$N = \frac{2B}{1/T} = 2BT \tag{10.10}$$

By sampling $x(t)$ at Nyquist interval points $1/2B$ apart on the time scale from 0 to T, it follows that

$$N = \frac{T}{1/2B} = 2BT \tag{10.11}$$

Thus the same number of discrete samples are required when sampling the Nyquist co-intervals on the frequency scale, or when sampling the Nyquist intervals on the time scale.

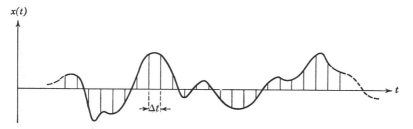

x(t)

Figure 10.5   Sampling of a continuous record.

10.3.2   *Sampling Procedures and Aliasing Errors*

Sampling of analog signals for digital data analysis is usually performed at equally spaced time intervals, as shown in Figure 10.5. The problem then is to determine an appropriate sampling interval Δt. From Equation (10.11), the minimum number of discrete sample values required to describe a data record of length T and bandwidth B is $N = 2BT$. It follows that the maximum sampling interval for equally spaced samples is $\Delta t = 1/(2B)$. On the one hand, sampling at points that are more closely spaced than $1/(2B)$ will yield correlated and highly redundant sample values and, thus, unnecessarily increase later computations. On the other hand, sampling at points that are further apart than $1/(2B)$ will lead to confusion between the low- and high-frequency components in the original data. This latter problem is called *aliasing*. It constitutes a potential source of error that does not arise in direct analog data processing, but is inherent in all digital processing that is preceded by an analog-to-digital conversion. Those who have viewed a motion picture of the classic western vintage have undoubtedly observed the apparent reversal in the direction of rotation of the stage coach wheels as the coach slows down or speeds up. That observation is a simple illustration of an aliasing error caused by the analog-to-digital conversion operation performed by a motion picture camera.

To be more explicit on this matter of aliasing, consider a continuous record that is sampled such that the time interval between sample values is Δt seconds, as shown in Figure 10.5. The sampling rate is then $1/(\Delta t)$ sps. However, at least two samples per cycle are required to define a frequency component in the original data. This follows directly from the sampling theorems in Section 10.3.1. Hence, the highest frequency that can be defined by a sampling rate of $1/\Delta t$ sps is $1/(2\Delta t)$ Hz. Frequencies in the original data above $1/(2\Delta t)$ Hz will appear below $1/(2\Delta t)$ Hz and be confused with the data in this lower-frequency range, as illustrated in Figure 10.6. This band-limiting frequency

$$f_c = \frac{1}{2\Delta t} \tag{10.12}$$

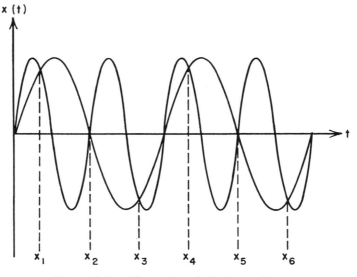

Figure 10.6   Illustration of aliasing problem.

is called the *Nyquist frequency* or *folding frequency*. Actually, the folding of data frequencies into the frequency range from 0 to f_c occurs in an accordion-pleated fashion, as indicated in Figure 10.7. Specifically, for any frequency f in the range $0 \leq f \leq f_c$, the higher frequencies which are aliased with f are defined by

$$(2f_c \pm f), (4f_c \pm f), \ldots, (2nf_c \pm f), \ldots \tag{10.13}$$

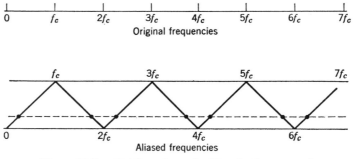

Figure 10.7   Folding about the Nyquist frequency f_c.

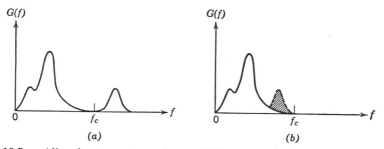

Figure 10.8   Aliased autospectrum due to folding. (a) True spectrum. (b) Aliased spectrum.

To prove this fact, observe that for $t = 1/2f_c$

$$\cos 2\pi ft = \cos 2\pi (2nf_c \pm f)\frac{1}{2f_c} = \cos\frac{\pi f}{f_c} \qquad (10.14)$$

Thus all data at frequencies $2nf_c \pm f$ have the same cosine function as data at frequency f when sampled at points $1/2f_c$ apart. For example, if $f_c = 100$ Hz, then data at 30 Hz would be aliased with data at the frequencies 170 Hz, 230 Hz, 370 Hz, 430 Hz, and so forth. Similarly, the power at these higher confounding frequencies is aliased with the power in the lower frequencies. This occurs because for $t = 1/2f_c$, the power quantities $\sin^2(2\pi ft)$ and $\cos^2(2\pi ft)$ do not distinguish between a frequency f and frequencies $2nf_c \pm f$. Hence when the Nyquist frequency f_c is as shown in Figure 10.8, a true autospectral density function as pictured in Figure 10.8(a) would be folded into the aliased autospectral density function as illustrated in Figure 10.8(b).

The only practical way to avoid aliasing errors in digital data analysis is to remove that information in the original analog data that might exist at frequencies above the Nyquist frequency f_c prior to the analog-to-digital conversion. This is done by restricting the frequency range of the original data with an analog low pass filter prior to the analog-to-digital conversion. Such low pass filters on analog-to-digital conversion equipment are referred to as "anti-aliasing filters." Since no low pass filter has an infinitely sharp rolloff, it is customary to set the anti-aliasing filter cutoff frequency at about 70–80% of f_c to assure that any data at frequencies above f_c are strongly suppressed.

10.3.3   *Quantization Errors*

In analog-to-digital conversion, since the magnitude of each data sample must be expressed by some fixed number of digits, only a fixed set of levels are available for approximating the infinite number of levels in the analog data.

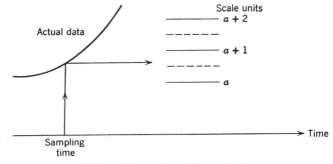

Figure 10.9   Illustration of quantization error.

No matter how fine the scale, a choice between two consecutive values will be required, as illustrated in Figure 10.9. If the quantization is done properly, the true level will be approximated by the quantizing level closest to it. The accuracy of the approximating process is a function of the number of available levels. Since most present day digitizers produce binary outputs compatible with computers, the number of levels may be described by the number of binary digits (bits) produced. Typical analog to digital conversion systems produce 6 to 16 bits, which corresponds to a range of 64 to 65,536 levels. For ideal conversion, the quantizing error will have a uniform probability distribution with a standard deviation of approximately $0.29\Delta x$, where Δx is the quantizing increment. This fact is easily shown as follows:

Let $p(x)$ be the quantization error probability density function defined by

$$p(x) = \begin{cases} 1 & -0.50 \le x \le 0.50 \\ 0 & \text{otherwise} \end{cases}$$

The mean value of the error is clearly zero since $p(x)$ is symmetric about $x = 0$, and the variance of the error is

$$\sigma_x^2 = \int_{-\infty}^{\infty} (x - \mu_x)^2 p(x) \, dx = \int_{-0.5}^{0.5} x^2 \, dx = \frac{1}{12}$$

Hence the standard deviation is

$$\sigma_x = \sqrt{\tfrac{1}{12}} \approx 0.29 \text{ scale unit} \tag{10.15}$$

This is the rms value of the quantization error, which may be considered an rms noise on desired signals. For example, suppose the full range of a signal is quantized at 256 scale units (8 bits). The peak signal to rms noise ratio in this case would be $(256\Delta x)/(0.29\Delta x) \approx 1000$, or about 60 dB.

In practice, the quantizing error is usually unimportant relative to other sources of error in the data acquisition and processing procedures. However, care must be exercised to assure that the range of the continuous data is set to occupy as much of the available quantizing range as possible. Otherwise, the resolution will be poor and the quantizing error could become significant.

10.3.4 *Analog - to - Digital Converters*

Commercial analog-to-digital converters (ADC) are usually either binary or ASCII (American Standard Code for Information Interchange) systems. A binary system maps a continuous signal onto a set of 2 digits (0 or 1), while an ASCII system maps the signal onto a set of 10 digits (0–9). Binary systems are simpler to build, but usually require machine-language programming to process data. Conversely, ASCII systems are more complicated to build and relatively inefficient, but have the advantage that data may be read directly by a computer program written in a common language such as FORTRAN. Also, ASCII systems can drive terminal or register displays directly.

Besides the sampling and quantization errors already discussed, other ADC errors of importance include

1. *Aperture error*—arising from the fact that the data sample is taken over a finite period of time rather than instantaneously
2. *Jitter*—arising from the fact that the time interval between samples can vary slightly in some random manner
3. *Nonlinearities*—arising from many sources such as misalignment of parts, bit dropouts, quantization spacing, and zero discontinuity

Commercially available ADC equipment can provide sampling rates of up to 10^8 sps for 8-bit words and 10^7 sps for 12-bit words. Of course, practical considerations such as the desired bit resolution and digital recording devices may reduce the actual sampling rate that can be employed for digital analysis. Furthermore, the ADC sampling rate must be divided by the number of multiplexed channels being simultaneously digitized to obtain the actual sampling rate for each individual channel of data.

One final point concerning ADC equipment should be mentioned. When more than one channel of data are being simultaneously digitized, the time interval between the sampling of each channel introduces an interchannel time base error in the digital data that can be significant at the higher frequencies. This problem may be avoided by the use of sample-hold circuits in the ADC equipment. These circuits simultaneously sample all input data channels and then hold the sampled signals until the ADC multiplexing switch passes the various channels.

Example 10.3. **Illustration of Digitization.** The random vibration of a structure is measured with a transducer that produces an analog voltage signal proportional to acceleration. The vibration record is to be converted to a digital format for analysis over the frequency range from 0 to 2000 Hz with a signal-to-noise ratio of at least 80 dB. Determine the sampling rate and number of bits per data point required in the analog-to-digital conversion.

First, to obtain a definition of the data up to 2000 Hz without aliasing, the data should be low pass filtered with a cutoff frequency of 2000 Hz. Because the anti-aliasing filters do not have an infinite rolloff, the Nyquist frequency should be set somewhat above 2000 Hz, say at 2500 Hz. Then from Equation (10.12), the required sampling interval is $\Delta t = 1/2f_c = 0.2$ msec, which gives a sampling rate of 5000 sps.

To obtain a peak signal-to-rms-noise ratio of 80 dB (10^4 in amplitude) it follows from Equation (10.15) that $2^n/0.289 = 10^4$, where n is the number of bits used for each data point. Taking logarithms to the base 10 of this equation, $0.301n = 3.46$. Hence, $n = 11.5$, so the required number of bits for the conversion is 12.

10.4   DATA QUALIFICATION

The correct procedures for analyzing random data, as well as interpreting the analyzed results, are strongly influenced by certain basic characteristics, which may or may not be exhibited by the data. The three most important of these basic characteristics are the stationarity of the data, the presence of periodicities in the data, and the normality of the data. Stationarity is of concern because the analysis procedures required for nonstationary data are generally more complicated than those which are appropriate for stationary data. Periodicities in the data should at least be identified to avoid erroneous interpretations of later results. The validity of an assumption that the data (excluding periodicities) have a Gaussian probability density function should be investigated since the normality assumption is vital to many analytical applications for random data. Referring back to Figure 10.1, qualification of sampled data in terms of these basic characteristics is indicated as a separate operation to be performed prior to detailed data analysis. In practice, however, it is often accomplished as an integral part of the data analysis phase. Practical considerations and procedures for such qualification will now be discussed.

10.4.1   *Test for Stationarity*

Perhaps the simplest way to evaluate the stationarity of sampled random data is to consider the physics of the phenomenon producing the data. If the basic physical factors that generate the phenomenon are time invariant, then stationarity of the resulting data generally can be accepted without further study. For example, consider the random data representing pressure fluctuations in the turbulent boundary layer generated by the flight of a high-speed aircraft. If the aircraft is flying at constant altitude and airspeed with a fixed configuration, it would be reasonable to assume that the resulting pressure data are stationary. On the other hand, if the aircraft is rapidly changing altitude, airspeed, and/or configuration, then nonstationarities in the resulting pressure data would be anticipated.

In practice, data are often collected under circumstances that do not permit an assumption of stationarity based on simple physical considerations. In such cases, the stationarity of the data must be evaluated by studies of available sample time history records. This evaluation might range from a visual inspection of the time histories by a talented analyst to detailed statistical tests of appropriate data parameters. In any case, there are certain important assumptions that must be made if the stationarity of data is to be ascertained from individual sample records. First, it must be assumed that any given sample record will properly reflect the nonstationary character of the random process in question. This is a reasonable assumption for those nonstationary random processes which involve deterministic trends, as discussed in Chapter 12. Second, it must be assumed that any given sample record is very long compared to the lowest-frequency component in the data, excluding a nonstationary mean. In other words, the sample record must be long enough to permit nonstationary trends to be differentiated from the random fluctuations of the time history.

Beyond these basic assumptions, it is convenient (but not necessary) to assume further that any nonstationarity of interest will be revealed by time trends in the mean square value of the data. Of course, one can readily contrive a nonstationary random process with a stationary mean square value; for example, a process where each sample function is a constant amplitude oscillation with continuously increasing frequency and random initial phase. Nevertheless, such cases are unusual in practice because it is highly unlikely for nonstationary data to have a time-varying autocorrelation function at any time displacement τ without the value at $\tau = 0$ varying. Since $R(0) = \psi^2$, the mean square value will usually reveal a time-varying autocorrelation. A similar argument applies to higher-order properties.

With these assumptions in mind, the stationarity of random data can be tested by investigating a single record $x(t)$ as follows.

1. Divide the sample record into N equal time intervals where the data in each interval may be considered independent
2. Compute a mean square value (or mean value and variance separately) for each interval and align these sample values in time sequence, as follows:

$$\overline{x_1^2}, \overline{x_2^2}, \overline{x_3^2}, \dots, \overline{x_N^2}$$

3. Test the sequence of mean square values for the presence of underlying trends or variations other than those due to expected sampling variations

The final test of the sample values for nonstationary trends may be accomplished in many ways. If the sampling distribution of the sample values is known, various statistical tests discussed in Chapter 4 could be applied. However, as noted in Section 8.2.2, the sampling distribution of mean square values requires a detailed knowledge of the frequency composition of the data. Such knowledge is generally not available at the time one wishes to establish whether or not data are stationary. Hence a nonparametric approach that does not require a knowledge of the sampling distributions of data parameters is more desirable. Two such nonparametric tests that are applicable to this problem are outlined in Section 4.7, namely, the run test and the reverse arrangements test. Of the two, the reverse arrangements test is the more powerful for detecting monotonic trends in a sequence of observations. The reverse arrangements test may be directly applied as a test for stationarity as follows.

Let it be hypothesized that the sequence of sample mean square values $(\overline{x_1^2}, \overline{x_2^2}, \overline{x_3^2}, \ldots, \overline{x_N^2})$ represents independent sample measurements of a stationary random variable with a mean square value of ψ_x^2. If this hypothesis is true, the variations in the sequence of sample values will be random and display no trends. Hence the number of reverse arrangements will be as expected for a sequence of independent random observations of the random variable, as given by Equation (4.56). If the number of reverse arrangements is significantly different from this number, the hypothesis of stationarity would be rejected. Otherwise, the hypothesis would be accepted. Note that the above testing procedure does not require a knowledge of either the frequency bandwidth of the data or the averaging time used to compute sample mean square values. Nor is it limited to a sequence of mean square values. It will work equally well on mean values, rms values, standard deviations, mean absolute values, or any other parameter estimate. Furthermore, it is not necessary for the data under investigation to be free of periodicities. Valid conclusions are obtained, even when periodicities are present, as long as the fundamental period is short compared to the averaging time used to compute sample values.

Example 10.4. **Test for Stationarity.**   To illustrate the application of the reverse arrangements test as a test for stationarity, consider the sequence of 20 mean square value measurements plotted in Figure 10.10. The measurements were made on the output of an analog random noise generator as the gain level was slowly increased by about 20% during the measurement sequence. From Section 4.7.2, the number of reverse arrangements in the sequence are found to be as follows:

$A_1 = 7$	$A_6 = 4$	$A_{11} = 7$	$A_{16} = 0$
$A_2 = 2$	$A_7 = 0$	$A_{12} = 6$	$A_{17} = 2$
$A_3 = 6$	$A_8 = 8$	$A_{13} = 3$	$A_{18} = 0$
$A_4 = 2$	$A_9 = 0$	$A_{14} = 0$	$A_{19} = 1$
$A_5 = 0$	$A_{10} = 2$	$A_{15} = 2$	

The total number of reverse arrangements is then $A = 52$.

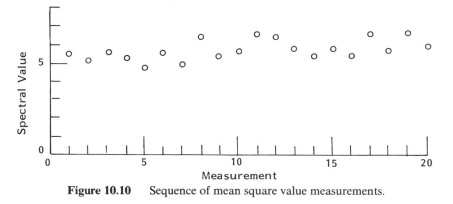

Figure 10.10   Sequence of mean square value measurements.

Now let it be hypothesized that the data are stationary. From Table A.7, this hypothesis would be accepted at the $\alpha = 0.01$ level of significance if the sequence of $N = 20$ measurements produced more than $A_{20;\,0.01} = 59$ reverse arrangements. Since the sequence actually produced only $A = 52$ reverse arrangements, the hypothesis of stationarity is rejected at the 1% level of significance, meaning the data are correctly identified as being nonstationary. This concludes Example 10.4.

An assumption of stationarity can often be supported (or rejected) by a simple nonparametric test of sample mean square values (or related sample parameters) computed from the available data. However, if one is not prepared to accept time invariance of mean square values as sufficient evidence of time invariance of autocorrelation functions, then tests can still be performed by further segmenting of the data in the frequency domain. Specifically, the data can be segmented into several contiguous frequency ranges by bandpass filtering, and the sample mean square values in each frequency interval can then be individually tested for time invariance. Since spectral density functions and correlation functions are Fourier transform pairs, time invariance of one directly implies time invariance of the other.

10.4.2   *Test for Periodicities*

Periodic and/or almost periodic components in otherwise random data will theoretically appear as delta functions in the autospectrum of the data. In practice, they will appear as sharp peaks in the autospectrum, which might be confused with narrow-band random contributions. Hence it is desirable to identify the presence of periodic components so they will not be misinterpreted as narrow-band random components with finite autospectral densities. If periodic components in the data are intense, their presence is usually obvious. However, less intense periodic components in random data may not be so obvious. The most effective procedures for detecting periodic components are those associated with the various analysis procedures that would be employed for random data analysis anyway. Hence in practical terms, a test for periodicities usually evolves from analysis procedures that would be performed assuming the data are random. Specifically, the presence of periodic components in otherwise random data may often be detected by visual inspection of an autospectral density function, an amplitude probability density function, and/or an autocorrelation function measured from stationary data. The autospectrum is the most commonly used analysis parameter for this application.

To be more specific, a highly resolved spectral density density estimate will reveal periodic components as sharp peaks, even when the periodicities are of relatively small intensity. A sharp peak in the autospectrum of sample data, however, may also represent narrow-band random data. These two cases can usually be distinguished from one another by repeating the autospectral density measurement with a narrower resolution filter bandwidth. If the measured spectral peak represents a sine wave, the indicated bandwidth of the peak will always be equal to the bandwidth of the analyzer filter, no matter how narrow the filter. Furthermore, the indicated spectral density will always increase in direct proportion to the reduction in filter bandwidth. This method method of detection will clearly not work unless the resolution bandwidth of the analysis is smaller than the bandwidth of possible narrow-band random data.

Example 10.5. **Spectrum of Sine Wave in Noise**.   To illustrate how an autospectrum can reveal the presence of a periodic component in otherwise random data, refer to Figure 10.11. In this example, the output of a thermal noise generator is mixed with a sinusoidal signal. The sinusoidal signal has an rms amplitude equal to one-twentieth that of the random signal. Figure 10.11(*a*), which was made using a relatively wide resolution bandwidth, gives little or no indication of the presence of the sinusoid. Figure 10.11(*b*), which was made using one-fifth of the previous resolution bandwidth, indicates a possible sinusoid quite clearly. Figure 10.11(*c*), which was made using a resolution bandwidth reduced by another factor of 5, gives a strong indication.

10.4.3   *Test for Normality*

Perhaps the most obvious way to test samples of stationary random data for normality is to measure the probability density function of the data and compare it to the theoretical normal distribution. If the sample record is sufficiently long to permit a measurement with small error compared to the deviations from normality, the lack of normality will be obvious. If the sampling distribution of the probability density estimate is known, various statistical tests for normality can be performed even when the random error is large. However, as for stationarity testing discussed in Section 10.4.1, a knowledge of the sampling distribution of probability density measurements requires frequency information for the data which may be difficult to obtain in

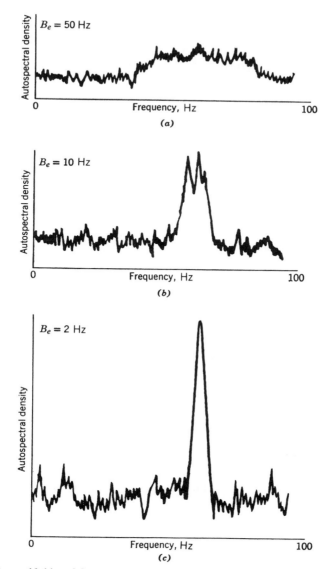

Figure 10.11   Measured autospectra of a sine wave in random noise.

practical cases. Hence a nonparametric test is desirable. One of the most convenient nonparametric tests for normality is the chi-square goodness-of-fit test outlined in Section 4.6. The details of applying the chi-square goodness-of-fit test with a numerical illustration are presented in Example 4.3.

10.5   DATA ANALYSIS

The procedures for analyzing the properties of random data may be divided logically into two categories: the procedure for analyzing individual sample records and the procedure for analyzing a collection of sample records given the properties of the individual records. Applicable data analysis procedures for these two categories are now outlined.

10.5.1   *Procedure for Analyzing Individual Records*

An overall procedure for analyzing the pertinent statistical properties of individual sample time history records is presented in Figure 10.12. Note that many of the suggested steps in the procedure might be omitted for some applications while additional steps would be required for other applications. Further note that the data qualification steps discussed in Section 10.4 are incorporated into the procedure to help clarify how these two parts of the overall data processing problem interact. Each block in Figure 10.12 will now be discussed.

MEAN AND MEAN SQUARE VALUE ANALYSIS.   The first step indicated by Block A is a mean and mean square value (or variance) measurement. This step is almost universally performed for one or both of two valid reasons. First, since the mean and mean square values are the basic measures of central tendency and dispersion, their calculation is generally required for even the most rudimentary applications. Second, the calculation of short time-averaged mean and mean square value estimates provides a basis for evaluating the stationarity of the data, as indicated in Figure 10.12 and illustrated in Section 10.4.1. The computation of mean and mean square values is discussed in Section 11.1. The statistical accuracy of mean and mean square value estimates is developed in Section 8.2.

AUTOCORRELATION ANALYSIS.   The autocorrelation function of stationary data is the inverse Fourier transform of the autospectral density function and, thus, produces no new information over the autospectrum. Since the autospectrum can be computed more efficiently and is generally easier to interpret for most applications, autocorrelation functions are often omitted. There may be special situations, however, where an autocorrelation estimate is desired. In such cases the autocorrelation function is usually computed from a spectral density estimate, as detailed in Section 11.4.2. The statistical accuracy of autocorrelation estimates is discussed in Section 8.4.

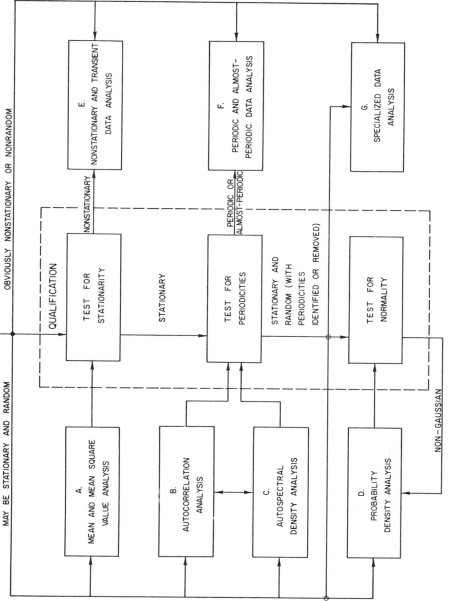

Figure 10.12 General procedure for analyzing individual sample records.

349

AUTOSPECTRAL DENSITY ANALYSIS.   Perhaps the most important single descriptive characteristic of stationary random data is the autospectral density function, which defines the frequency composition of the data. For constant-parameter linear physical systems, the output autospectrum is equal to the input autospectrum multiplied by the square of the gain factor of the system. Thus autospectra measurements can yield information concerning the dynamic characteristics of the system. The total area under an autospectrum is equal to the mean square value ψ^2. To be more general, the mean square value of the data in any frequency range of concern is determined by the area under the autospectrum bounded by the limits of that frequency range. Obviously, the measurement of autospectra data, as indicated by Block C, will be valuable for many analysis objectives.

The physical significance of the autospectral density function for input/output random data problems is shown clearly in Chapters 6 and 7. Other engineering applications are discussed in Reference 10.8. The computation of autospectra is detailed in Section 11.5, and the statistical accuracy of the computed estimates is developed in Section 8.5.

PROBABILITY DENSITY ANALYSIS.   The last fundamental analysis included in the procedure is probability density analysis, as indicated by Block D. Probability density analysis is often omitted from a data analysis procedure because of the tendency to assume that all random phenomena are normally distributed. In some cases, however, random data may deviate substantially from the Gaussian form, particularly when the data in question are the result of a nonlinear operation. The computation of probability density functions is presented in Section 11.3 and the statistical accuracy of the computed estimates is discussed in Section 8.3.

NONSTATIONARY AND TRANSIENT DATA ANALYSIS.   All of the analysis techniques discussed thus far apply only to sample records of stationary data. If the data are determined to be nonstationary during the qualification phase of the processing, then special analysis techniques will be required as indicated by Block E. The analysis of nonstationary and transient data is discussed in Chapter 12. Note that certain classes of nonstationary data can sometimes be analyzed using the same equipment or computer programs employed for stationary data analysis. However, the results of such analyses must be interpreted with caution as illustrated in Sections 12.5 and 12.6.

PERIODIC AND ALMOST-PERIODIC DATA ANALYSIS.   If sinusoids due to periodic or almost-periodic contributions are detected in the data during the qualification phase, then special attention is warranted. Specifically, one of two approaches should be followed. First, the sinusoidal components might be isolated from the random portion of the data by filtering operations and analyzed separately, as indicated by Block F. Second, the sinusoidal components might be analyzed along with the random portion of the data, and then simply accounted for in the results. For example, if an autospectrum is computed for data that include sinusoidal components, a delta function symbol might be superimposed on each spectral peak at the frequency of an identified sinusoid, and labeled with the mean square value of the sinusoid. The mean square value can be estimated from the spectral plot by multiplying the maximum indicated spectral density of the peak by the resolution bandwidth used for the analysis. If this is not done, the physical significance of such spectral peaks might be misinterpreted, as discussed in Section 10.4.2.

SPECIALIZED DATA ANALYSIS. Various other analyses of individual time-history records are often required, depending on the specific goals of the data processing. For example, studies of fatigue damage in mechanical systems usually involve the calculation of peak probability density functions of strain data, as discussed in Reference 10.9. An investigation of zero crossings or arbitrary level crossings might be warranted for certain communication noise problems, as detailed in Section 5.4.3 and Reference 10.7. The computation of Hilbert transforms may be desired for special problems discussed in Chapter 13. Such specialized analyses, as indicated by Block G, must be established in the context of the engineering problem of concern.

10.5.2   *Procedures for Analyzing a Set of Records*

The preceding section presented methods for analyzing each individual sample record from an experiment. A procedure for analyzing further pertinent statistical properties of a collection of sample records is presented in Figure 10.13. As for the analysis of individual sample records outlined in Figure 10.12, many of the suggested steps in Figure 10.13 might be omitted for some applications while additional steps would be required for others. Furthermore, the suggested steps assume the individual records are stationary. Each block in Figure 10.13 will now be discussed.

ANALYSIS OF INDIVIDUAL RECORDS. This first step in the procedure is to analyze the pertinent statistical properties of the individual sample records, as outlined in Figure 10.12. Hence the applicable portions of Figure 10.12 constitute Block A in Figure 10.13.

TEST FOR CORRELATION. The next step indicated by Block B is to determine whether or not the individual sample records are correlated. In many cases, this decision involves little more than a cursory evaluation of pertinent physical considerations. For example, if the collection of sample records represents measurements of a physical phenomenon over widely separated time intervals, then usually the individual records can be accepted as uncorrelated without further study. On the other hand, if the collection represents simultaneous measurements of the excitation and response of a physical system, then correlation would be anticipated. For those cases where a lack of correlation is not obvious from basic considerations, a test for correlation among the sample records should be performed using cross correlation functions or coherence functions, as indicated in Figure 10.13.

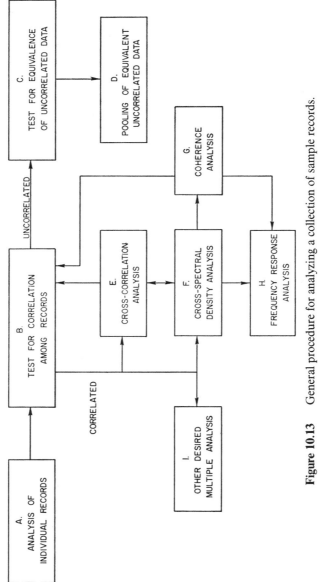

Figure 10.13 General procedure for analyzing a collection of sample records.

TEST FOR EQUIVALENCE OF UNCORRELATED DATA.   If sample records are found to be uncorrelated in Block B, then these records should be tested for equivalent statistical properties as indicated by Block C. This is an important but often overlooked step in the analysis of random data. Far too often the analyzed results for a large number of sample records are presented as individual plots when in fact the results differ only by amounts that fall within the acceptable limits of random error. The formal presentation of such redundant data is usually of no value, and can actually be detrimental in several ways. First, large quantities of analyzed data will sometimes tend to overwhelm the user and unnecessarily complicate the interpretation of the results. Second, the unsophisticated user might interpret the statistical scatter in individual results as physically meaningful differences. Third, more accurate results could be presented for the equivalent data if they were pooled prior to plotting, as will be discussed next. Note that for most applications, an equivalence of autospectra is a sufficient criterion for equivalence of sampled data. A procedure for testing the equivalence of autospectra is presented in Section 10.5.3.

POOLING OF EQUIVALENT UNCORRELATED DATA.   The analyzed results for individual sample records that are found to represent equivalent data should be pooled, as indicated by Block D. This is done by computing appropriately weighted averages of the results for the individual records. For example, assume two autospectral density function estimates were computed from two uncorrelated sample records that now are found to represent equivalent data. If $\hat{G}_1(f)$ and $\hat{G}_2(f)$ were the original autospectra estimates computed with n_{d1} and n_{d2} averages, respectively, a new pooled estimate of the autospectrum is given by

$$\hat{G}_p(f) = \frac{n_{d1}\hat{G}_1(f) + n_{d2}\hat{G}_2(f)}{n_{d1} + n_{d2}} \tag{10.16}$$

where $\hat{G}_p(f)$ is essentially computed with $n_{dp} = n_{d1} + n_{d2}$ averages. Equation (10.16) may be generalized for q estimates from uncorrelated but equivalent samples as follows.

$$\hat{G}_p(f) = \frac{\sum_{i=1}^{q} n_{di}\hat{G}_i(f)}{\sum_{i=1}^{q} n_{di}} \tag{10.17}$$

where $\hat{G}_p(f)$ is now computed with an equivalent number of averages given by

$$n_d = \sum_{i=1}^{q} n_{di} \tag{10.18}$$

Noting from Section 8.5.4 that the random error in an autospectral density estimate is approximated by $\varepsilon = 1/\sqrt{n_d}$, it follows from Equation (10.18) that the pooling operation produces an autospectrum estimate with a reduced random error. However, it also should be noted that the pooling operation generally will not suppress the systematic error (bias) in the autospectra estimates, as defined and discussed in Section 8.5.1. This fact often leads data analysts to reprocess sample records with equivalent statistical properties in a manner designed to reduce bias errors. For the case of autospectra estimates, the reprocessing might consist of a recomputation of autospectral density estimates from the original sample records using a greatly reduced resolution bandwidth to suppress the bias errors at the expense of increased random errors. The random errors in the individual estimates are then suppressed by the pooling operation.

CROSS-CORRELATION ANALYSIS. As for the case of autocorrelation and autospectral density functions, the cross-correlation and cross-spectral density functions are Fourier transform pairs. Hence the measurement of a cross-correlation function will technically not yield any new information over the cross-spectrum. However, it sometimes presents the desired information in a more convenient format. An example is the measurement of time delays between two measurement points. Such measurements are the basis for a wide range of applications summarized in Reference 10.8. Therefore, cross-correlation analysis is included in the procedures as a separate step indicated by Block E. Note that a cross-correlation estimate can be used as a test for correlation between two records.

Cross-correlation functions are usually estimated by computing the inverse Fourier transform of a cross-spectral density estimate, as detailed in Section 11.6.2. The statistical accuracy of cross-correlation estimates is discussed in Section 8.4.

CROSS-SPECTRAL DENSITY ANALYSIS. The most important joint measurement for a collection of correlated sample records is the cross-spectral density analysis indicated by Block F. Cross-spectral density functions provide information concerning the linear relationships that might exist among the collection of sample records. When interpreted into a physical context, such information often leads directly to problem solutions as developed in Chapters 6 and 7, and illustrated in Reference 10.8.

The computation of cross-spectral density functions is detailed in Section 11.6.3. The statistical accuracy of cross-spectra estimates is developed in Section 9.1.

COHERENCE FUNCTION ANALYSIS. Block G indicates the calculation of coherence functions based on autospectral and cross-spectral density estimates. Coherence functions of various types (ordinary, multiple, and partial) are valuable in several ways. First, they can be used to test for correlation among the collection of sample records. Second, they constitute a vital parameter in assessing the accuracy of frequency response function estimates. Third, they can provide a direct solution for certain types of problems.

The computation of coherence functions is detailed in Sections 11.6.6 and 11.7. Illustrations of their applications to engineering problems are presented in Reference 10.8. The statistical accuracy of coherence function estimates is developed in Sections 9.2.3 and 9.3.

FREQUENCY RESPONSE FUNCTION ANALYSIS.   The ultimate goal in the analysis of a collection of sample records is often to establish linear relationships among the data represented by the various records. The existence of such linear relationships can be detected from cross-correlation, cross-spectral density, or coherence function estimates. However, a meaningful description of the linear relationships is best provided by computing the frequency response functions of the relationships, as indicated by Block H.

The computation of frequency response functions is developed in Sections 11.6.4 and 11.7. The statistical accuracy of frequency response function estimates is discussed in Sections 9.2.4, 9.2.5 and 9.3.

OTHER DESIRED MULTIPLE ANALYSIS.   Block I indicates other joint analyses of a collection of sample records needed to satisfy special data processing goals. Included might be advanced spectral calculations such as generalized spectra used in nonstationary data analysis discussed in Chapter 12, or Hilbert transform techniques discussed in Chapter 13.

10.5.3 *Test for Equivalence of Autospectra*

The previous section discussed the need to establish the possible equivalence of autospectra estimates computed from two or more statistically independent sample records. A useful technique for this application will now be described.

An estimate $\hat{G}(f)$ of an autospectral density function $G(f)$ will have a sampling distribution that is approximately normal if the number of averages n_d is large, say $n_d \geq 30$. It is shown in Section 8.5 that the mean value (assuming no bias) and variance of the estimate are given by

$$E[\hat{G}(f)] \simeq G(f) \qquad (10.19)$$

$$\text{Var}[\hat{G}(f)] \simeq \frac{1}{n_d}G^2(f) \qquad (10.20)$$

Hence a $(1 - \alpha)$ confidence interval for $G(f)$ based on a measurement $\hat{G}(f)$ may be approximated by

$$\left[\hat{G}(f)\left(1 - z_{\alpha/2}\sqrt{\frac{1}{n_d}}\right) \leq G(f) \leq \hat{G}(f)\left(1 + z_{\alpha/2}\sqrt{\frac{1}{n_d}}\right)\right]$$

$$(10.21)$$

where $z_{\alpha/2}$ is the $100\alpha/2$ percentage point of the standardized normal distribution. To arrive at Equation (10.21), it is assumed that $z_{\alpha/2}\sqrt{1/n_d} \ll 1$, so that

$$\left(1 \pm z_{\alpha/2}\sqrt{\frac{1}{n_d}}\right)^{-1} \simeq \left(1 \mp z_{\alpha/2}\sqrt{\frac{1}{n_d}}\right) \tag{10.22}$$

A logarithmic transformation of the estimate $\hat{G}(f)$ to $\log \hat{G}(f)$ has the effect of producing a distribution that is closer to normal than the original distribution. The sample mean value and variance of $\log \hat{G}(f)$ become

$$E\left[\log \hat{G}(f)\right] \simeq \log G(f) \tag{10.23}$$

$$\mathrm{Var}\left[\log \hat{G}(f)\right] \simeq \frac{1}{n_d} \tag{10.24}$$

Thus the variance here is independent of frequency. Now, a $(1 - \alpha)$ confidence interval for $\log G(f)$ may be approximated by

$$\left[\log \hat{G}(f) - z_{\alpha/2}\sqrt{\frac{1}{n_d}}\right] \le \log G(f) \le \left[\log \hat{G}(f) + z_{\alpha/2}\sqrt{\frac{1}{n_d}}\right]$$

$$\tag{10.25}$$

This result can be derived directly from Equation (10.21) to provide a heuristic explanation for Equations (10.23) and (10.24). This derivation uses the assumption that $z_{\alpha/2}\sqrt{1/n_d} \ll 1$, so that

$$\log\left(1 \pm z_{\alpha/2}\sqrt{\frac{1}{n_d}}\right) \simeq \pm z_{\alpha/2}\sqrt{\frac{1}{n_d}} \tag{10.26}$$

Consider now two different autospectral density function estimates $\hat{G}_1(f)$ and $\hat{G}_2(f)$ obtained under different conditions; for example, from two different sample records or from two different parts of the same sample record. The problem is to decide whether or not these two autospectra are statistically equivalent over some frequency interval (f_a, f_b) of bandwidth $B = f_b - f_a$.

Assume each of the two autospectral density function estimates is based on a resolution bandwidth Δf, where N_f bandwidths are needed to cover the frequency range of interest. That is,

$$N_f = \frac{B}{\Delta f} \tag{10.27}$$

Further assume the number of averages for each estimate are n_{d1} and n_{d2}, respectively, meaning that the averaging time (record length) for each estimate may be different even though the resolution bandwidth is the same. From Equations (10.23) and (10.24), the sampling distributions of the logarithm of the estimates in the ith bandwidth are approximated by

$$\log \hat{G}_1(f_i) = y\left[\log G_1(f_i), \frac{1}{n_{d1}}\right]$$

$$\log \hat{G}_2(f_i) = y\left[\log G_2(f_i), \frac{1}{n_{d2}}\right] \tag{10.28}$$

where $y[\mu, \sigma^2]$ is a normally distributed random variable with a mean of μ and a variance of σ^2. Now, if the two sample records in question have the same autospectral density function $G(f) = G_1(f) = G_2(f)$, it follows from Equations (10.28) that

$$\log \frac{\hat{G}_1(f_i)}{\hat{G}_2(f_i)} = y\left[0, \frac{1}{n_{d1}} + \frac{1}{n_{d2}}\right] \tag{10.29}$$

Hence, from Section 4.6, the statistic

$$X^2 = \left[\frac{1}{n_{d1}} + \frac{1}{n_{d2}}\right]^{-1} \sum_{i=1}^{n}\left[\log \frac{\hat{G}_1(f_i)}{\hat{G}_2(f_i)}\right]^2 \tag{10.30}$$

has a chi-square distribution with N_f degrees of freedom. That is,

$$X^2 \simeq \chi_n^2 \qquad n = N_f \tag{10.31}$$

The result in Equation (10.30) provides a basis for testing the hypothesis that $G_1(f) = G_2(f)$. The region of acceptance for the hypothesis test is

$$X^2 \le \chi_{n;\,\alpha}^2 \qquad n = N_f \tag{10.32}$$

where α is the level of significance for the test, as detailed in Section 4.6.

10.5.4 Computational Considerations

Prior to about 1950, essentially all time series analysis of random data was accomplished using special-purpose analog instruments. With the growth of digital computers after that time, a movement toward digital data analysis procedures evolved. The first people to go to digital data analysis were generally those concerned with relatively low-frequency phenomena; for example, certain types of economic and biomedical data, oceanographic data, and atmospheric turbulence data. Those concerned with higher-frequency phenomena (in the audio frequency range and above) tended to continue using analog data analysis procedures because of the high sampling rates required and the relatively high cost of filtering operations on a digital computer. In 1965, the situation was changed radically by a single technical development, namely, the introduction of algorithms for the fast computation of Fourier series coefficients on a digital computer. These algorithms, commonly referred to as fast Fourier transform (FFT) procedures, permitted the filtering operations required for spectral density calculations to be accomplished with a dramatically increased speed and reduced cost.

Today, essentially all analysis of random data at frequencies up to at least 100 kHz is being accomplished using digital data analysis procedures. For data below 25 kHz in particular, a wide range of special-purpose digital instruments have become commercially available which perform various types of computations based on an FFT algorithm with remarkable speed. In light of these considerations, detailed procedures for analog data analysis covered in the first edition of this book have been deleted from the current edition. The modern digital procedures for the analysis of random data are presented in Chapter 11.

PROBLEMS

10.1. Suppose acceleration data are measured using a seismic accelerometer as illustrated in Figure 7.2, where $\zeta = 0.7$ and $f_n = 100$ Hz. Assume the spectral density function of the data is calculated over the frequency range from zero to 100 Hz. Determine the magnitude of the bias error in the resulting autospectrum at 100 Hz that can be attributed to the nonuniform gain factor of the accelerometer.

10.2. Assume the autocorrelation function of the acceleration data in Problem 1 is also calculated. Describe the nature of the error, if any, in the resulting autocorrelogram that can be attributed to the nonuniform phase factor of the accelerometer.

10.3. Suppose 10 random data signals are to be recorded over a frequency range from zero to 1000 Hz using pulse code modulation (PCM) with 12 pulses per code group. If the 10 signals are to be multiplexed onto a single recorder channel,
    (a) how much recorder bandwidth is required?
    (b) what is the resulting signal to noise ratio of the recorded data?

10.4. If random data signals must be recorded on a magnetic tape machine that is known to exhibit excessive flutter, which type of modulation would be preferable, FM or PCM?

10.5. Assume the autospectral density functions of sample time-history records are to be calculated by digital procedures. If it is known that the data exist only in the frequency range below 500 Hz,

(a)   what is the minimum sampling rate which can be used in digitizing the original time history?

(b)   is this minimum sampling rate also required for the calculation of autocorrelation functions?

(c)   is this minimum sampling rate also required for the calculation of probability density functions?

10.6.   In an analog-to-digital conversion operation on time-history data, how many quantization levels are required if the resulting digital data are to have a potential dynamic range (peak-signal-to-rms-noise ratio) of 100 : 1 (40 dB).

10.7.   Suppose an otherwise random data signal includes a sine wave with an rms value of 1 volt. Assume the autospectral density function of the signal is measured using a resolution filter bandwidth of $B_e = 10$ Hz. If the spectral density of the random portion of the signal in the vicinity of the sine wave frequency is $G = 0.1$ volts^2/Hz, what will be the indicated spectral density (on the average) at the frequency of the sine wave?

10.8.   For the combined random–sinusoidal signal in Problem 10.7, what resolution bandwidth would be required to produce an indicated spectral density at the frequency of the sine wave which would be 10 times greater (on the average) than the spectral density of the random portion of the data.

10.9.   Referring to the general data analysis procedure for single records presented in Figure 10.12, explain why a test for stationarity is suggested prior to a test for periodicities and/or normality.

10.10.   Referring to the general data analysis procedure for multiple records presented in Figure 10.13, explain why a coherence function estimate should always accompany a frequency response function calculation.

REFERENCES

10.1   Doeblin, E. O., *Measurement Systems: Application and Design*, McGraw-Hill, New York, 1966.

10.2   Beauchamp, K. G., and Yuen, C. K., *Data Acquisition for Signal Analysis*, Allan & Unwin, London, 1980.

10.3   Davies, G. L., *Magnetic Tape Instrumentation*, McGraw-Hill, New York, 1961.

10.4   Gregg, W. D., *Analog and Digital Communication*, Wiley, New York, 1977.

10.5   Fitzgerald, J., and Eason, T. S., *Fundamentals of Data Communication*, Wiley, New York, 1978.

10.6 Telemetry Standards IRIG Document 106-66. Secretariat Range Commanders Council, White Sands Missile Range, New Mexico, March 1966.

10.7 Bendat, J. S., *Principles and Applications of Random Noise Theory*, Wiley, New York, 1958. Reprinted by Krieger, Melbourne, Florida, 1977.

10.8 Bendat, J. S., and Piersol, A. G., *Engineering Applications of Correlation and Spectral Analysis*, Wiley-Interscience, New York, 1980.

10.9 Crandall, S. H., and Mark, W. D., *Random Vibration in Mechanical Systems*, Academic, Orlando, Florida, 1963.

CHAPTER 11

DIGITAL DATA ANALYSIS

This chapter details the basic digital operations required to estimate various time-series functions. The calculations assume the data to be processed are discrete time-series values representing sample records from stationary (ergodic) random processes. Techniques for nonstationary data analysis are considered in Chapter 12 and for Hilbert transforms in Chapter 13.

11.1 DATA PREPARATION

In some cases, the physical phenomenon of interest occurs directly in digital form, for example, neutron emission data or many forms of economic data. In most cases, however, the data originate in analog form and must be converted to a digital time series with proper concern for the quantization and aliasing problems detailed in Section 10.3. In either case, it is assumed that the digital data consist of N data values with an equally spaced sampling interval of Δt seconds, as previously illustrated in Figure 10.5. Specifically, let

$$\{u_n\} \qquad n = 1, 2, \ldots, N \tag{11.1}$$

be the data values associated with the times

$$t_n = t_0 + n\,\Delta t \qquad n = 1, 2, \ldots, N \tag{11.2}$$

where t_0 is arbitrary and does not enter into later formulas so long as the data represent a stationary random process. It follows that

$$u_n = u(t_0 + n\,\Delta t) \qquad n = 1, 2, \ldots, N \tag{11.3}$$

The total record length represented by the sample data is clearly $T = N\Delta t$. From the developments in Section 10.3, the Nyquist frequency associated with the sample data is given by

$$f_c = \frac{1}{2\,\Delta t} \qquad (11.4)$$

Various preliminary operations might be applied to the data as noted previously in Section 10.3. Special operations of common interest include (a) data standardization, (b) trend removal, and (c) filtering.

11.1.1 Data Standardization

The mean value of the sample data $\{u_n\}$, $n = 1, 2, \ldots, N$, is given by

$$\bar{u} = \frac{1}{N} \sum_{n=1}^{N} u_n \qquad (11.5)$$

For stationary ergodic data, the quantity $\bar{u}$ is an unbiased estimate of the mean value μ as demonstrated in Section 4.1. It is convenient for later calculations to transform the sample values $\{u_n\}$ to a new set of values $\{x_n\}$ that have a zero sample mean by computing

$$x_n = x(t_0 + n\,\Delta t) = u_n - \bar{u} \qquad n = 1, 2, \ldots, N \qquad (11.6)$$

All subsequent formulas will be stated in terms of the transformed data values $\{x_n\}$, where $\bar{x} = 0$.

The standard deviation of the transformed sample data $\{x_n\}$ is given by

$$s = \left[\frac{1}{N-1} \sum_{n=1}^{N} x_n^2 \right]^{1/2} \qquad (11.7)$$

The quantities s and s^2 are unbiased estimates of the standard deviation σ_x and the variance σ_x^2, respectively, as proved in Section 4.1. If later computer calculations are to be performed using fixed-point as opposed to floating-point arithmetic, it will be desirable to further standardize the data by transforming the values $\{x_n\}$ to a new set of values $\{z_n\}$ by computing

$$z_n = \frac{x_n}{s} \qquad n = 1, 2, \ldots, N \qquad (11.8)$$

11.1.2 Trend Removal

Situations sometimes occur where the sample data include spurious trends or low-frequency components with a wavelength longer than the record length $T_r = N\Delta t$. Common sources of spurious trends are data collection instrumentation drift and signal integration operations. If such trends are not removed from the data, large distortions can occur in the later processing of probability density, correlation, and spectral quantities. Caution is advised here, however, in that trend removal should be performed only if trends are physically expected or clearly apparent in the data.

The most common technique for trend removal is to fit a low-order polynomial to the data using the least squares procedures detailed in Section 4.8.2. Specifically, let the original data values $\{u_n\}$ be fit with a polynomial of degree K defined by

$$\tilde{u}_n = \sum_{k=0}^{K} b_k (n \, \Delta t)^k \qquad n = 1, 2, \ldots, N \tag{11.9}$$

A "least squares" fit is obtained by minimizing the squared discrepancies between the data values and the polynomial given by

$$Q = \sum_{n=1}^{N} (u_n - \tilde{u}_n)^2 = \sum_{n=1}^{N} \left[u_n - \sum_{k=0}^{K} b_k (n \, \Delta t)^k \right]^2 \tag{11.10}$$

Taking the partial derivatives of Q with respect to b_k and setting them equal to zero yields $K + 1$ equations of the form

$$\sum_{k=0}^{K} b_k \sum_{n=1}^{N} (n \, \Delta t)^{k+m} = \sum_{n=1}^{N} u_n (n \, \Delta t)^m \qquad m = 0, 1, 2, \ldots, K \tag{11.11}$$

which can be solved for the desired regression coefficients $\{b_k\}$. For example, when $K = 0$, Equation (11.11) becomes

$$b_0 \sum_{n=1}^{N} (n \, \Delta t)^0 = \sum_{n=1}^{N} u_n (n \, \Delta t)^0 \tag{11.12}$$

giving the result

$$b_0 = \frac{1}{N} \sum_{i=1}^{N} u_n = \bar{u} \tag{11.13}$$

For $K = 1$, Equation (11.11) becomes

$$b_0 \sum_{n=1}^{N} (n \, \Delta t)^m + b_1 \sum_{n=1}^{N} (n \, \Delta t)^{1+m} = \sum_{n=1}^{N} u_n (n \, \Delta t)^m \qquad m = 0, 1 \tag{11.14}$$

Noting the identities

$$\sum_{n=1}^{N} n = \frac{N(N + 1)}{2} \qquad \sum_{n=1}^{N} n^2 = \frac{N(N + 1)(2N + 1)}{6} \qquad (11.15)$$

Equation (11.14) yields the results

$$b_0 = \frac{2(2N + 1)\sum_{n=1}^{N} u_n - 6\sum_{n=1}^{N} n u_n}{N(N - 1)} \qquad (11.16a)$$

$$b_1 = \frac{12\sum_{n=1}^{N} n u_n - 6(N + 1)\sum_{n=1}^{N} u_n}{\Delta t\, N(N - 1)(N + 1)} \qquad (11.16b)$$

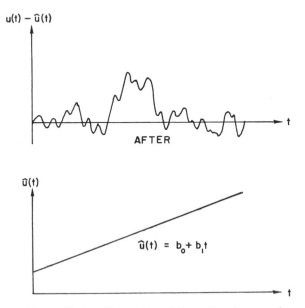

Figure 11.1 Illustration of linear trend removal.

Equation (11.16) defines a linear regression line with an intercept of b_0 and a slope of b_1, which should then be subtracted from the original data values $\{u_n\}$. Note that Equation (11.16) is equivalent to Equation (4.70), where $u_n = y_i$, $n\,\Delta t = x_i$, $b_0 = A$, and $b_1 = B$. An example of linear trend removal is illustrated in Figure 11.1. More complex trends can be removed by higher-order polynomial fits, but trend removal using an order of greater than $K = 3$ is generally not recommended.

11.1.3 *Digital Filtering*

The filtering of data prior to more detailed analyses may be desired for various reasons, including either the isolation or elimination of periodic components, as an integral step in "zoom" transform operations to be discussed in Section 11.5.4, or as an anti-aliasing step prior to data decimation. The last-mentioned application is of particular importance since decimation of sample data is often carried out to decrease the amount of data for later analysis. By definition, a d th-order decimation of sample data consists of keeping every d th data value and discarding all other data values. Hence, data that were originally sampled with a sampling interval of Δt are reduced to data with a new sampling interval of $d\,\Delta t$. It follows that the new Nyquist frequency becomes $f'_c = 1/(2d\,\Delta t)$ and all information above this frequency will be folded (aliased) back into the frequency interval from 0 to $1/(2d\,\Delta t)$. To avoid aliasing, the original sample data should be filtered prior to decimation to remove information in the frequency range above $1/(2d\,\Delta t)$ by using a low pass digital filter. Note that this does not negate the requirement to remove all information in the original data above $f_c = 1/(2\,\Delta t)$ by low pass analog filtering.

Digital filtering can be performed in either the time domain or the frequency domain. Frequency domain filtering corresponds to multiplying the Fourier transform of the data record $x(t)$ by the frequency response function $H(f)$ of the desired filter and then taking the inverse transform. Specifically, given a transformed data record $x(t)$, the filtered record is given by

$$y(t) = \text{IFT}[H(f)X(f)] \tag{11.17}$$

where IFT denotes the inverse Fourier transform and $X(f)$ is the Fourier transform of $x(t)$. This type of filtering has certain advantages, the primary ones being that it is simple to understand and no analytic expression is required for the filter frequency response function. However, the implementation tends to be both complex and time-consuming in the normal case where the time series is too long to be conveniently retained within the available computer memory. The problem usually requires individual filtering operations on contiguous segments of the original record.

Time domain filters can be divided into two types:

a. Nonrecursive, or finite impulse response (FIR) filters
b. Recursive, or infinite impulse response (IIR) filters

Nonrecursive (FIR) filters take the form

$$y_i = \sum_{k=0}^{M} h_k x_{i-k} \tag{11.18}$$

This is the digital equivalent of the convolution equation given in Equation (6.1), namely,

$$y(t) = \int_0^\infty h(\tau) x(t - \tau) \, d\tau \tag{11.19}$$

where $h(\tau)$ is the unit impulse response (weighting function) of the desired filter. In a like manner, $\{h_k\}$ defines the unit impulse response of the digital filter. Classical smoothing, interpolation, extrapolation, differentiation, and integration techniques are all examples of FIR filters.

A recursive (IIR) digital filter is that type of filter where the output time series is generated using not only a finite sum of input terms but also using previous outputs as input terms (a procedure engineers call feedback). A simple type of IIR filter is given by

$$y_n = cx_n + \sum_{k=1}^{M} h_k y_{n-k} \tag{11.20}$$

which uses M previous outputs and only one input. More general recursive filters involve M outputs and larger numbers of inputs. Equation (11.20) is illustrated in Figure 11.2. The triangles represent multiplication by the values shown within the triangles, the rectangles represent a delay of Δt from one point to another, and the circles represent summing operations.

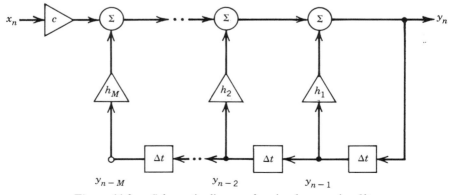

Figure 11.2   Schematic diagram for simple recursive filter.

The Fourier transform of Equation (11.20) yields the result

$$Y(f) = cX(f) + Y(f) \sum_{k=1}^{M} h_k e^{-j2\pi fk \, \Delta t} \tag{11.21}$$

where the summation involves a polynomial in powers of the exponential $\exp(-j2\pi f \Delta t)$. (Replacing this exponential by the letter z leads to a procedure for analyzing digital filters in terms of what is called z-transform theory.) From Equation (11.21), the frequency response function of the entire system is given by

$$H(f) = \frac{Y(f)}{X(f)} = \frac{c}{1 - \sum_{k=1}^{M} h_k e^{-j2\pi fk \, \Delta t}} \tag{11.22}$$

Studies of the properties of $H(f)$ are thus reduced to determining the location and nature of the poles in the denominator of this last result.

As an example of this procedure, consider the IIR filter defined by

$$y_n = (1 - a)x_n + ay_{n-1} \tag{11.23}$$

where $a = \exp(-\Delta t/RC)$. This can behave like a low pass filter as defined previously in Example 6.1. To verify this property, note that from Equation (11.22)

$$H(f) = \frac{1 - a}{1 - ae^{-j2\pi f \Delta t}} \tag{11.24}$$

The square of the filter gain factor is then given by

$$|H(f)|^2 = \frac{(1 - a)^2}{(1 + a^2) - 2a \cos 2\pi f \Delta t} \tag{11.25}$$

Observe that if $RC \gg \Delta t$, then $a = \exp(-\Delta t/RC) \approx 1 - (\Delta t/RC)$ and $(1 - a) \approx \Delta t/RC$. Also, if $2\pi f \Delta t \ll 1$, then $e^{-j2\pi f \Delta t}$ may be approximated by $(1 - j2\pi f \Delta t)$. For this situation

$$H(f) \approx \frac{1}{1 + j2\pi fRC}$$

and

$$|H(f)|^2 \approx \frac{1}{1 + (2\pi fRC)^2} \tag{11.26}$$

which are the usual low-pass RC filter results.

Recursive digital filters that give good approximations to Butterworth filters have been synthesized with the aid of Equation (11.22) by finding a set of numerical weights $\{h_k\}$ and a coefficient c such that the resulting $|H(f)|^2$ has the form

$$|H(f)|^2 = \frac{1}{1 + (\sin \pi f \Delta t / \sin \pi f_0 \Delta t)^{2M}} \qquad 0 \le f \le \frac{1}{2\Delta t} \qquad (11.27)$$

Note that $|H(f)|^2 = 1$ for $f = 0$ and $|H(f)|^2 = 0.5$ for $f = f_0$. At $f = 1/(2\Delta t)$, the Nyquist frequency, the quantity $H(f)$ approaches zero for large M. Thus, over the frequency range 0 to $1/(2\Delta t)$, the filter described by Equation (11.27) acts like a low-pass Butterworth filter of the form

$$|H(f)|^2 = \frac{1}{1 + (f/f_0)^K} \qquad (11.28)$$

where f_0 is the half-power point and K determines the rolloff rate. Fuller mathematical details for the synthesis of digital filters appear in References 11.1–11.4.

Example 11.1. Illustration of Recursive Digital Filter.   Assume digital data have been collected with a sampling rate of 1000 sps. It is now desired to low-pass filter the data using a simple RC type filter with a half-power point cutoff frequency of $f_0 = 10$ Hz. Determine an appropriate recursive digital filter to accomplish this task.

From Equation (11.26), the half-power point cutoff frequency f_0 occurs where $|H(f_0)|^2 = \frac{1}{2}|H(f)|^2_{\max} = \frac{1}{2}|H(0)|^2 = 0.5$. Since $RC \gg \Delta t = 10^{-3}$ for this case,

$$|H(10)|^2 \simeq \frac{1}{1 + (20\pi RC)^2} = 0.5$$

It follows that $RC = 0.016$ and $a = \exp(-0.063) = 0.94$. This result is checked by substituting $a = 0.94$ into Equation (11.25), which yields $|H(10)|^2 = 0.5$. Hence, the desired low-pass filter is given from Equation (11.23) as

$$y_n = 0.06x_n + 0.94y_{n-1}$$

11.2   FOURIER SERIES AND FAST FOURIER TRANSFORMS

Fourier series and Fourier transforms of data differ in their theoretical properties but not, for most practical purposes, in their digital computational details. This is because only a finite-range Fourier transform can actually be computed with digitized data, and this finite range can always be considered as the period of an associated Fourier series. From the discussions in Section 5.2.2, one of the main reasons for the importance of fast Fourier transforms is that they can be used to provide estimates of desired spectral density and correlation functions. Before explaining the basis for methods to compute fast Fourier transforms, it is instructive to review a standard Fourier series procedure.

11.2.1 *Standard Fourier Series Procedure*

If a stationary sample record $x(t)$ is periodic with a period of T_p and a fundamental frequency $f_1 = 1/T_p$, then $x(t)$ can be represented by the Fourier series

$$x(t) = \frac{a_0}{2} + \sum_{q=1}^{\infty} \left(a_q \cos 2\pi q f_1 t + b_q \sin 2\pi q f_1 t \right) \qquad (11.29)$$

where

$$a_q = \frac{2}{T} \int_0^T x(t) \cos 2\pi q f_1 t \, dt \qquad q = 0, 1, 2, \ldots$$

$$b_q = \frac{2}{T} \int_0^T x(t) \sin 2\pi q f_1 t \, dt \qquad q = 1, 2, 3, \ldots$$

Assume a sample record $x(t)$ is of finite length $T_r = T_p$, the fundamental period of the data. Further assume that the record is sampled at an even number of N equally spaced points a distance Δt apart, where Δt has been selected to produce a sufficiently high cutoff frequency $f_c = 1/2\,\Delta t$. Consider the initial point of the record to be zero and denote the transformed data values, as before, by

$$x_n = x(n\,\Delta t) \qquad n = 1, 2, \ldots, N \qquad (11.30)$$

Proceed now to calculate the finite version of a Fourier series that will pass through these N data values. For any point t in the interval $(0, T_p)$, the result is

$$x(t) = A_0 + \sum_{q=1}^{N/2} A_q \cos\left(\frac{2\pi q t}{T_p} \right) + \sum_{q=1}^{(N/2)-1} B_q \sin\left(\frac{2\pi q t}{T_p} \right) \qquad (11.31)$$

At the particular points $t = n\,\Delta t$, $n = 1, 2, \ldots, N$, where $T_p = N\Delta t$,

$$x_n = x(n\,\Delta t) = A_0 + \sum_{q=1}^{N/2} A_q \cos\left(\frac{2\pi q n}{N} \right) + \sum_{q=1}^{(N/2)-1} B_q \sin\left(\frac{2\pi q n}{N} \right) \qquad (11.32)$$

The coefficients A_q and B_q are given by

$$A_0 = \frac{1}{N} \sum_{n=1}^{N} x_n = \bar{x} = 0$$

$$A_q = \frac{2}{N} \sum_{n=1}^{N} x_n \cos \frac{2\pi qn}{N} \qquad q = 1, 2, \ldots, \frac{N}{2} - 1$$

$$A_{N/2} = \frac{1}{N} \sum_{n=1}^{N} x_n \cos n\pi$$

$$B_q = \frac{2}{N} \sum_{n=1}^{N} x_n \sin \frac{2\pi qn}{N} \qquad q = 1, 2, \ldots, \frac{N}{2} - 1$$

(11.33)

A digital computer program to obtain A_q and B_q involves the following steps:

1. Evaluate $\theta = 2\pi qn/N$ for fixed q and n
2. Compute $\cos\theta$, $\sin\theta$
3. Compute $x_n \cos\theta$, $x_n \sin\theta$
4. Accumulate both sums for $n = 1, 2, \ldots, N$
5. Increment q and repeat operations

This procedure requires a total of approximately N^2 *real* multiply–add operations.

For large N, these standard digital computation methods for determining the coefficients A_q and B_q can be time- and cost-consuming since both time and cost are functions of N^2. To greatly reduce these standard computational times, alternative methods have been proposed and developed, known as *fast Fourier transform* (FFT) procedures. These methods will now be discussed in some detail because of their importance in digital processing of random data.

11.2.2 Fast Fourier Transforms

An infinite-range Fourier transform of a real-valued or a complex-valued record $x(t)$ is defined by the complex-valued quantity

$$X(f) = \int_{-\infty}^{\infty} x(t)\, e^{-j2\pi ft}\, dt \tag{11.34}$$

Theoretically, as noted previously, this transform $X(f)$ will not exist for an $x(t)$ that is a representative member of a stationary random process when the infinite limits are used. However, by restricting the limits to a finite time interval of $x(t)$, say in the range $(0, T)$, then the finite-range Fourier transform will exist, as defined by

$$X(f, T) = \int_0^T x(t)\, e^{-j2\pi ft}\, dt \tag{11.35}$$

Assume now that this $x(t)$ is sampled at N equally spaced points a distance Δt apart, where Δt has been selected to produce a sufficiently high cutoff frequency. As before, the sampling times are $t_n = n\,\Delta t$. However, it is convenient here to start with $n = 0$. Hence in place of Equation (11.30), let

$$x_n = x(n\,\Delta t) \qquad n = 0, 1, 2, \ldots, N - 1$$

Then, for arbitrary f, the discrete version of Equation (11.35) is

$$X(f, T) = \Delta t \sum_{n=0}^{N-1} x_n \exp[-j2\pi fn\Delta t] \qquad (11.36)$$

The usual selection of discrete frequency values for the computation of $X(f, T)$ is

$$f_k = \frac{k}{T} = \frac{k}{N\Delta t} \qquad k = 0, 1, 2, \ldots, N - 1 \qquad (11.37)$$

At these frequencies, the transformed values give the Fourier components defined by

$$X_k = \frac{X(f_k)}{\Delta t} = \sum_{n=0}^{N-1} x_n \exp\left[-j\frac{2\pi kn}{N}\right] \qquad k = 0, 1, 2, \ldots, N - 1 \qquad (11.38)$$

where Δt has been included with $X(f_k)$ to have a scale factor of unity before the summation. Note that results are unique only out to $k = N/2$ since the Nyquist frequency occurs at this point. Fast Fourier transform methods are designed to compute these quantities, X_k, and can also be used to compute the coefficients A_q and B_q in Equation (11.33).

To simplify the notation further, let

$$W(u) = \exp\left[-j\frac{2\pi u}{N}\right] \qquad (11.39)$$

Observe that $W(N) = 1$ and for all u and v,

$$W(u + v) = W(u)W(v)$$

Also, let

$$X(k) = X_k \qquad \text{and} \qquad x(n) = x_n$$

Then Equation (11.38) becomes

$$X(k) = \sum_{n=0}^{N-1} x(n)W(kn) \qquad k = 0, 1, 2, \ldots, N - 1 \qquad (11.40)$$

Equations (11.38) and (11.40) should be studied so as to be easily recognized as the Fourier transform of $x(n)$ when $x(n)$ is expressed by a series of N terms. Such equations require a total of approximately N^2 *complex* multiply–add operations (where 1 complex multiply–add = 4 real multiply–adds) to compute all of the $X(k)$ terms involved.

BASIS FOR FFT PROCEDURES. The fast Fourier transform procedures are now based on decomposing N into its composite (nonunity) factors, and carrying out Fourier transforms over the smaller number of terms in each of the composite factors. In particular, if N is the product of p factors such that

$$N = \prod_{i=1}^{p} r_i = r_1 r_2 \cdots r_p \tag{11.41}$$

where the r's are all integers greater than unity, then as will be proved below, the $X(k)$ in Equation (11.40) can be found by computing in an iterative fashion the sum of p terms,

(N/r_1) Fourier transforms requiring $4r_1^2$ real operations each,
(N/r_2) Fourier transforms requiring $4r_2^2$ real operations each,

$$\vdots \tag{11.42}$$

(N/r_p) Fourier transforms requiring $4r_p^2$ real operations each.

Hence the total number of real operations becomes

$$4\left(Nr_1 + Nr_2 + Nr_3 + \cdots + Nr_p \right) = 4N \sum_{i=1}^{p} r_i \tag{11.43}$$

The resulting speed ratio of these FFT procedures to the standard method is then

$$\text{speed ratio} = \frac{N^2}{4N\sum_{i=1}^{p} r_i} = \frac{N}{4\sum_{i=1}^{p} r_i} \tag{11.44}$$

***Example 11.2.* FFT Speed Ratio for Powers of Two.** If $N = 2^p$, then $\sum_{i=1}^{p} r_i = 2p = 2\log_2 N$. In this case, the speed ratio by Equation (11.44) appears to be

$$\text{speed ratio} = \frac{N^2}{8Np} = \frac{N}{8p}$$

However, a doubling of the speed can be achieved in practice by noting that the values for $W(kn)$, when N is a power of 2, all turn out to be $+1$ or -1, so that multiplications are replaced by additions and subtractions. This yields a higher speed ratio of the order

$$\text{speed ratio} = \frac{N}{4p} \tag{11.45}$$

For example, if $N = 2^{13} = 8192$, Equation (11.45) gives a speed ratio of $(8192/52) \approx 158$. This result is deemed to be a conservative estimate since a further speed improvement of at most two can be obtained by dividing a single record into two parts and computing as indicated later in Section 11.2.4.

DERIVATION OF GENERAL PROCEDURE. To derive the result stated in Equation (11.42), express the indices k and n in Equation (11.40) in the following way:

$$k = \sum_{\nu=0}^{p-1} k_\nu \prod_{i=0}^{\nu} r_i \qquad \text{where } k_\nu = 0, 1, 2, \ldots, r_{\nu+1} - 1$$
$$r_0 = 1$$

$$(11.46)$$

$$n = \sum_{\nu=0}^{p-1} n_\nu \prod_{i=0}^{\nu} r_{m+1-i} \qquad \text{where } n_\nu = 0, 1, 2, \ldots, r_{p-\nu} - 1$$
$$r_{p+1} = 1$$

Note the indices k and n are replaced by new indices k_ν and n_ν as defined above. Equation (11.46) has the following interpretation:

$$k = k_0 + k_1 r_1 + k_2 r_1 r_2 + \cdots + k_{p-1}(r_1 r_2 \cdots r_{p-1})$$
$$n = n_0 + n_1 r_p + n_2 r_p r_{p-1} + \cdots + n_{p-1}(r_p r_{p-1} \cdots r_2)$$

$$(11.47)$$

where

$$k_0 = 0, 1, 2, \ldots, r_1 - 1 \qquad n_0 = 0, 1, 2, \ldots, r_p - 1$$

$$k_1 = 0, 1, 2, \ldots, r_2 - 1 \qquad n_1 = 0, 1, 2, \ldots, r_{p-1} - 1$$

$$\vdots \qquad \vdots \qquad\qquad \vdots \qquad \vdots$$

$$k_{p-1} = 0, 1, 2, \ldots, r_p - 1 \qquad n_{p-1} = 0, 1, 2, \ldots, r_1 - 1$$

By fixing values of k_ν and n_ν in turn, it is a straightforward exercise to verify that k and n will each vary from 0 to $N - 1$, where N is the product of all the

r's as per Equation (11.41). Equation (11.40) can now be rewritten as

$$X(k) = X(k_0, k_1, \ldots, k_{p-1})$$

$$= \sum_{n_0=0}^{r_p-1} \sum_{n_1=0}^{r_{p-1}-1} \cdots \sum_{n_{p-2}=0}^{r_2-1} \sum_{n_{p-1}=0}^{r_1-1} x(n_0, n_1, \ldots, n_{p-2}, n_{p-1}) W(kn) \tag{11.48}$$

where

$$W(kn) = W\Big(k\big[n_0 + n_1 r_p + \cdots + n_{p-\nu}(r_p r_{p-1} \cdots r_{\nu+1}) + \cdots$$

$$+ n_{p-1}(r_p r_{p-1} \cdots r_2)\big]\Big) \tag{11.49}$$

with k given by Equation (11.46).

An alternative way to write k is as follows.

$$k = (k_0 + k_1 r_1 + \cdots + k_{\nu-1} r_1 r_2 \cdots r_{\nu-1})$$

$$+ (r_1 r_2 \cdots r_\nu)(k_\nu + k_{\nu+1} r_{\nu+1} + \cdots + k_{p-1} r_{\nu+1} r_{\nu+2} \cdots r_{p-1})$$

Hence, a representative term in Equation (11.49) is such that

$$kn_{p-\nu}(r_p r_{p-1} \cdots r_{\nu+1}) = (k_0 + k_1 r_1 + \cdots + k_{\nu-1} r_1 r_2 \cdots r_{\nu-1})$$

$$\times n_{p-\nu}(r_p r_{p-1} \cdots r_{\nu+1})$$

$$+ Nn_{p-\nu}(k_\nu + k_{\nu+1} r_{\nu+1} + \cdots$$

$$+ k_{p-1} r_{\nu+1} r_{\nu+2} \cdots r_{p-1}) \tag{11.50}$$

Then, since W for any integer power of N is equal to unity, it follows for $\nu = 1, 2, \ldots, p$, that

$$W(kn_{p-\nu} r_p r_{p-1} \cdots r_{\nu+1}) = W\Big[(k_0 + k_1 r_1 + \cdots + k_{\nu-1} r_1 r_2 \cdots r_{\nu-1})$$

$$\times n_{p-\nu} r_p r_{p-1} \cdots r_{\nu+1}\Big]$$

$$\tag{11.51}$$

Equation (11.51) will now be interpreted. For $\nu = 1$, observe that

$$W\left(kn_{p-1}r_p r_{p-1} \cdots r_2\right) = W\left(k_0 n_{p-1}r_p r_{p-1} \cdots r_2\right)$$

$$= W\left(\frac{k_0 n_{p-1} N}{r_1}\right) = \exp\left[-j\frac{2\pi k_0 n_{p-1}}{r_1}\right] \quad (11.52)$$

This is the exponential required in a Fourier transform of $x(n_{p-1})$ as expressed by a series of r_1 terms, instead of N terms as shown in Equations (11.36) and (11.40). Note also that the indices k_0 and n_{p-1} each vary over the values $0, 1, 2, \ldots, r_1 - 1$, thus requiring a total of r_1^2 multiply–add operations to compute each of the associated $X(k_0)$ which might be involved here. For $\nu = 2, 3, \ldots, p$, Equation (11.51) with the aid of $W(u + v) = W(u)W(v)$ becomes the product of two factors, namely,

$$W\left[\left(k_0 + k_1 r_1 + \cdots + k_{\nu-2}r_1 r_2 \cdots r_{\nu-2}\right)n_{p-\nu}r_p r_{p-1} \cdots r_{\nu+1}\right]$$

multiplied by $W\left(k_{\nu-1}r_1 r_2 \cdots r_{\nu-1}n_{p-\nu}r_p r_{p-1} \cdots r_{\nu+1}\right)$ $\quad (11.53)$

where only the second factor contains $k_{\nu-1}$. This second factor is the quantity

$$W\left(\frac{k_{\nu-1}n_{p-\nu}N}{r_\nu}\right) = \exp\left[-j\frac{2\pi k_{\nu-1}n_{p-\nu}}{r_\nu}\right] \quad (11.54)$$

which is the exponent required in a Fourier transform of $x(n_{p-\nu})$ as expressed by a series of r_ν terms. Furthermore, the indices $k_{\nu-1}$ and $n_{p-\nu}$ each vary over the values $0, 1, 2, \ldots, r_\nu - 1$. Hence, a total of r_ν^2 multiply–add operations are needed to compute each of the associated $X(k_{\nu-1})$ which might be involved here.

THE FFT ALGORITHM. From the development in Equations (11.50) through (11.52), Equation (11.49) becomes

$$W(kn) = \prod_{\nu=1}^{p} T(k_0, k_1, \ldots, k_{\nu-2})W\left(\frac{k_{\nu-1}n_{p-\nu}N}{r_\nu}\right) \quad (11.55)$$

where

$$T(k_0, k_1, \ldots, k_{\nu-2}) = 1 \quad \text{for } \nu = 1$$

$$= W\left[\left(k_0 + k_1 r_1 + \cdots + k_{\nu-2}r_1 r_2 \cdots r_{\nu-2}\right)\right.$$

$$\left. \times \left(n_{p-\nu}r_p r_{p-1} \cdots r_{\nu+1}\right)\right]$$

$$\text{for } \nu = 2, 3, \ldots, p \quad (11.56)$$

Quantities such as the T terms in Equation (11.56) are often called *twiddle factors*. The result of Equation (11.55) should now be substituted into Equation (11.48). Regrouping of terms then yields

$$X(k_0, k_1, \ldots, k_{p-1}) = \sum_{n_0=0}^{r_p-1} T(k_0, k_1, \ldots, k_{p-2}) W\left(\frac{k_{p-1}n_0 N}{r_p}\right)$$

$$\text{times } \sum_{n_1=0}^{r_{p-1}-1} T(k_0, k_1, \ldots, k_{p-3}) W\left(\frac{k_{p-2}n_1 N}{r_{p-1}}\right)$$

times . . .

$$\text{times } \sum_{n_{p-2}=0}^{r_2-1} T(k_0) W\left(\frac{k_1 n_{p-2} N}{r_2}\right)$$

$$\text{times } \sum_{n_{p-1}=0}^{r_1-1} x(n_0, n_1, \ldots, n_{p-2}, n_{p-1}) W\left(\frac{k_0 n_{p-1} N}{r_1}\right) \qquad (11.57)$$

Thus, the desired Fourier transform can be computed in p successive iterative steps as shown by Equation (11.57). The principal remaining question of concern here is the understanding of these steps.

Consider the last innermost sum of terms in Equation (11.57). Let

$$A_1(k_0, n_0, n_1, \ldots, n_{p-2}) = \sum_{n_{p-1}=0}^{r_1-1} x(n_0, n_1, \ldots, n_{p-2}, n_{p-1}) W\left(\frac{k_0 n_{p-1} N}{r_1}\right)$$

$$(11.58)$$

Then, holding $n_0, n_1, \ldots, n_{p-2}$ fixed for each of their possible values, Equation (11.58) gives a total of (N/r_1) Fourier transforms of $x(n_{p-1})$ requiring r_1^2 operations each. At the next innermost sum of terms in Equation (11.57), let

$$A_2(k_0, k_1, n_0, n_1, \ldots, n_{p-3})$$

$$= \sum_{n_{p-2}=0}^{r_2-1} A_1(k_0, n_0, n_1, \ldots, n_{p-2}) T(k_0) W\left(\frac{k_1 n_{p-2} N}{r_2}\right) \quad (11.59)$$

Here, holding $k_0, n_0, n_1, \ldots, n_{p-3}$ fixed for each of their possible values, Equation (11.59) gives a total of (N/r_2) Fourier transforms of $x(n_{p-2})$ requiring r_2^2 operations each. Continuing in this way, at the νth step, where

$\nu = 2, 3, \ldots, p - 1$, let

$$A_\nu\left(k_0, k_1, \ldots, k_{\nu-1}, n_0, n_1, \ldots, n_{p-\nu-1}\right)$$

$$= \sum_{n_{p-\nu}=0}^{r_\nu - 1} A_{\nu-1}\left(k_0, k_1, \ldots, k_{\nu-2}, n_0, n_1, \ldots, n_{p-\nu}\right)$$

$$\times T(k_0, k_1, \ldots, k_{\nu-2}) W\left(\frac{k_{\nu-1} n_{p-\nu} N}{r_\nu}\right) \qquad (11.60)$$

Here, holding $k_0, k_1, \ldots, k_{\nu-2}, n_0, n_1, \ldots, n_{p-\nu-1}$ fixed for each of their possible values gives a total of (N/r_ν) Fourier transforms of $x(n_{p-\nu})$ requiring r_ν^2 operations each. At the last step, Equation (11.57) yields

$$X(k_0, k_1, \ldots, k_{p-1}) = A_p(k_0, k_1, \ldots, k_{p-1})$$

$$= \sum_{n_0=0}^{r_p - 1} A_{p-1}(k_0, k_1, \ldots, k_{p-2}, n_0)$$

$$\times T(k_0, k_1, \ldots, k_{p-2}) W\left(\frac{k_{p-1} n_0 N}{r_p}\right) \quad (11.61)$$

Holding $k_0, k_1, \ldots, k_{p-2}$ fixed for each of their possible values produces a total of (N/r_p) Fourier transforms of $x(n_0)$ requiring r_p^2 operations each. The sequence of steps in Equations (11.58) to (11.61) proves the result stated in Equation (11.42), making allowance for complex to real operations.

The formula derived in Equation (11.60) is a general *fast Fourier transform algorithm*, and is the basis for many Fourier transform computational procedures in use today. See Reference 11.5 for further details.

11.2.3 *Cooley–Tukey Procedure*

The Cooley–Tukey procedure, first introduced in 1965 in Reference 11.6, is a special case of the general algorithm of Equation (11.60) that is appropriate for binary digital computers. In particular, it applies to those situations where the number of data samples N is a power of 2, namely,

$$N = 2^p \qquad (11.62)$$

If necessary, zeros are added to the data sequence to satisfy this requirement. Here, the iterative procedures of Equation (11.57) become the sum of p terms, where every term involves $(N/2)$ Fourier transforms requiring 4 operations each. This gives a total of $2Np$ complex multiply–add operations.

It is desirable to write down the special equations that apply to this case because of their widespread importance. This will be done by substituting into previous results. Equation (11.47) becomes

$$k = k_0 + 2k_1 + 2^2 k_2 + \cdots + 2^{p-1} k_{p-1}$$

$$n = n_0 + 2n_1 + 2^2 n_2 + \cdots + 2^{p-1} n_{p-1} \tag{11.63}$$

where each k and n take on the values 0 and 1 only. Equation (11.54) becomes for all $v = 1, 2, \ldots, p$,

$$W\left(\frac{k_{v-1} n_{p-v} N}{r_v}\right) = \exp(-j\pi k_{v-1} n_{p-v}) \tag{11.64}$$

which takes on the values 1 and -1 only. Equation (11.56) becomes

$$T(k_0, k_1, \ldots, k_{v-2}) = 1 \qquad \text{for } v = 1$$

$$= W\left[\left(k_0 + 2k_1 + \cdots + 2^{v-2} k_{v-2}\right) 2^{p-v} n_{p-v}\right]$$

$$\text{for } v = 2, 3, \ldots, p \tag{11.65}$$

The Fourier transform iteration of Equation (11.57) can be expressed in this special case by the formula

$$X(k_0, k_1, \ldots, k_{p-1})$$

$$= \sum_{n_0=0}^{1} \sum_{n_1=0}^{1} \cdots \sum_{n_{p-2}=0}^{1} \sum_{n_{p-1}=}^{} x(n_0, n_1, \ldots, n_{p-2}, n_{p-1}) W(kn) \tag{11.66}$$

where

$$W(kn) = \prod_{v=1}^{p} T(k_0, k_1, \ldots, k_{v-2}) \exp(-j\pi k_{v-1} n_{p-v})$$

The first step in this iteration is now from Equation (11.58)

$$A_1(k_0, n_0, n_1, \ldots, n_{p-2}) = \sum_{n_{p-1}=0}^{1} x(n_0, n_1, \ldots, n_{p-2}, n_{p-1}) \exp(-j\pi k_0 n_{p-1})$$

$$\tag{11.67}$$

The νth step, for $\nu = 2, 3, \ldots, p - 1$, becomes from Equation (11.60)

$$A_\nu(k_0, k_1, \ldots, k_{\nu-1}, n_0, n_1, \ldots, n_{p-\nu-1})$$

$$= \sum_{n_{p-\nu}=0}^{1} A_{\nu-1}(k_0, k_1, \ldots, k_{\nu-2}, n_0, n_1, \ldots, n_{p-\nu})$$

$$\times T(k_0, k_1, \ldots, k_{\nu-2})\exp(-j\pi k_{\nu-1} n_{p-\nu}) \qquad (11.68)$$

This result is called the *Cooley–Tukey fast Fourier transform algorithm*. The last step in the iteration, Equation (11.61), is

$$X(k_0, k_1, \ldots, k_{p-1}) = A_p(k_0, k_1, \ldots, k_{p-1})$$

$$= \sum_{n_0=0}^{1} A_{p-1}(k_0, k_1, \ldots, k_{p-2}, n_0)$$

$$\times T(k_0, k_1, \ldots, k_{p-2})\exp(-j\pi k_{p-1} n_0) \quad (11.69)$$

and completes the procedure for this special case. A fuller discussion of these matters appears in References 11.2–11.6.

11.2.4 *Procedures for Real-Valued Records*

Fourier transforms of two real-valued records may be computed simultaneously by inserting one record $x(n)$ as the real part and one record $y(n)$ as the imaginary part of a complex record $z(n)$. In equation form, let

$$z(n) = x(n) + jy(n) \qquad n = 0, 1, \ldots, N - 1 \qquad (11.70)$$

The Fourier transform of $z(n)$ by Equation (11.38) is

$$Z(k) = \sum_{n=0}^{N-1} [x(n) + jy(n)]\exp\left[-j\frac{2\pi kn}{N}\right] \qquad k = 0, 1, 2, \ldots, N - 1$$

$$(11.71)$$

This can be computed by the fast Fourier transform procedures described. It is usually assumed in Equations (11.70) and (11.71) that N data points in $x(n)$ and $y(n)$ are transformed into N frequency points that are spaced $1/T$ apart. For these situations, the Nyquist frequency occurs when $k = N/2$ so that for N even, unique results occur only for $k = 0, 1, 2, \ldots, (N/2) - 1$. To obtain

$X(k)$ and $Y(k)$, observe that

$$\exp\left[j\frac{2\pi n(N-k)}{N}\right] = \exp\left[-j\frac{2\pi nk}{N}\right]$$

since $\exp[j2\pi n] = 1$ for any n. Hence, if $Z^*(k)$ is the complex conjugate of $Z(k)$, then

$$Z^*(N-k) = \sum_{n=0}^{N-1}[x(n) - jy(n)]\exp\left[-j\frac{2\pi nk}{N}\right]$$

It follows that

$$Z(k) + Z^*(N-k) = 2\sum_{n=0}^{N-1}x(n)\exp\left[-j\frac{2\pi nk}{N}\right] = 2X(k)$$

$$Z(k) - Z^*(N-k) = 2j\sum_{n=0}^{N-1}y(n)\exp\left[-j\frac{2\pi nk}{N}\right] = 2jY(k)$$

Thus, the two real-valued records $x(n)$ and $y(n)$ have Fourier transforms $X(k)$ and $Y(k)$ given by

$$X(k) = \frac{Z(k) + Z^*(N-k)}{2}$$
$$\qquad\qquad\qquad k = 0, 1, \ldots, N-1 \qquad (11.72)$$
$$Y(k) = \frac{Z(k) - Z^*(N-k)}{2j}$$

The same principle is used to calculate a double-length transform where a single, real-valued record $v(n)$, $n = 0, 1, \ldots, 2N - 1$, is divided into two records, one consisting of the values $v(n)$ where n is even and the other where n is odd. Specifically, let

$$x(n) = v(2n)$$
$$\qquad\qquad n = 0, 1, 2, \ldots, N-1 \qquad (11.73)$$
$$y(n) = v(2n + 1)$$

Now compute the Fourier transforms of $z(n) = x(n) + jy(n)$, $n = 0, 1, \ldots,$ $N - 1$, using Equation (11.71), and then the individual transforms, $X(k)$ and $Y(k)$, using Equation (11.72). These transforms can now be combined to obtain the desired Fourier transform of $v(n)$ by

$$V(k) = X(k) + Y(k)\exp\left[-j\frac{\pi}{N}\right]$$
$$V(N + k) = X(k) - Y(k)\exp\left[-j\frac{\pi}{N}\right] \qquad k = 0, 1, \ldots, N - 1 \quad (11.74)$$

Note that the operation calculates the Fourier transform of $2N$ real data values using a single N-point complex transform.

11.2.5 *Further Related Formulas*

Referring to Equation (11.34), the inverse Fourier transform of $X(f)$ is

$$x(t) = \int_{-\infty}^{\infty} X(f) \, e^{j2\pi ft} \, df \qquad (11.75)$$

This leads to the discrete inverse Fourier transform formula

$$x_n = \frac{1}{N} \sum_{k=0}^{N-1} X_k \exp\left[j\frac{2\pi kn}{N}\right] \qquad n = 0, 1, 2, \ldots, N - 1 \quad (11.76)$$

where the Fourier components X_k of Equation (11.38) are computed by the FFT procedures. The constant $(1/N)$ in Equation (11.76) is a scale factor only and is not otherwise important. This inverse Fourier transform can be computed by the same FFT procedures previously described by merely interchanging k and n and replacing x_n by X_k and $W(kn)$ by $W(-kn)$.

As a final topic here, it should be noted that the FFT computation always operates on a function with a nonnegative independent variable and produces a transformed function also with a nonnegative independent variable. Specifically, an input function of time is always defined as $x(t)$ over $0 \le t \le T$, not $-T/2 \le t \le T/2$. Similarly, the transformed function $X(f)$ is defined over $0 \le f \le 2f_c$, not $-f_c \le f \le f_c$. In digital terms, $x_n = x(n\,\Delta t)$, $n = 0, 1, \ldots,$ $N - 1$, and $X(k) = X(k/N\Delta t)/\Delta t$, $k = 0, 1, \ldots, N - 1$, where $k = N/2$ corresponds to the Nyquist frequency f_c defined in Equation (10.12). However, from Equation (11.72),

$$X(k) = X^*(N - k) \qquad k = 0, 1, \ldots, N - 1 \qquad (11.77)$$

where the values at frequencies above $N/2$ may be interpreted as the negative frequency values if desired to obtain a two-sided spectral function, as illustrated in Figure 11.3.

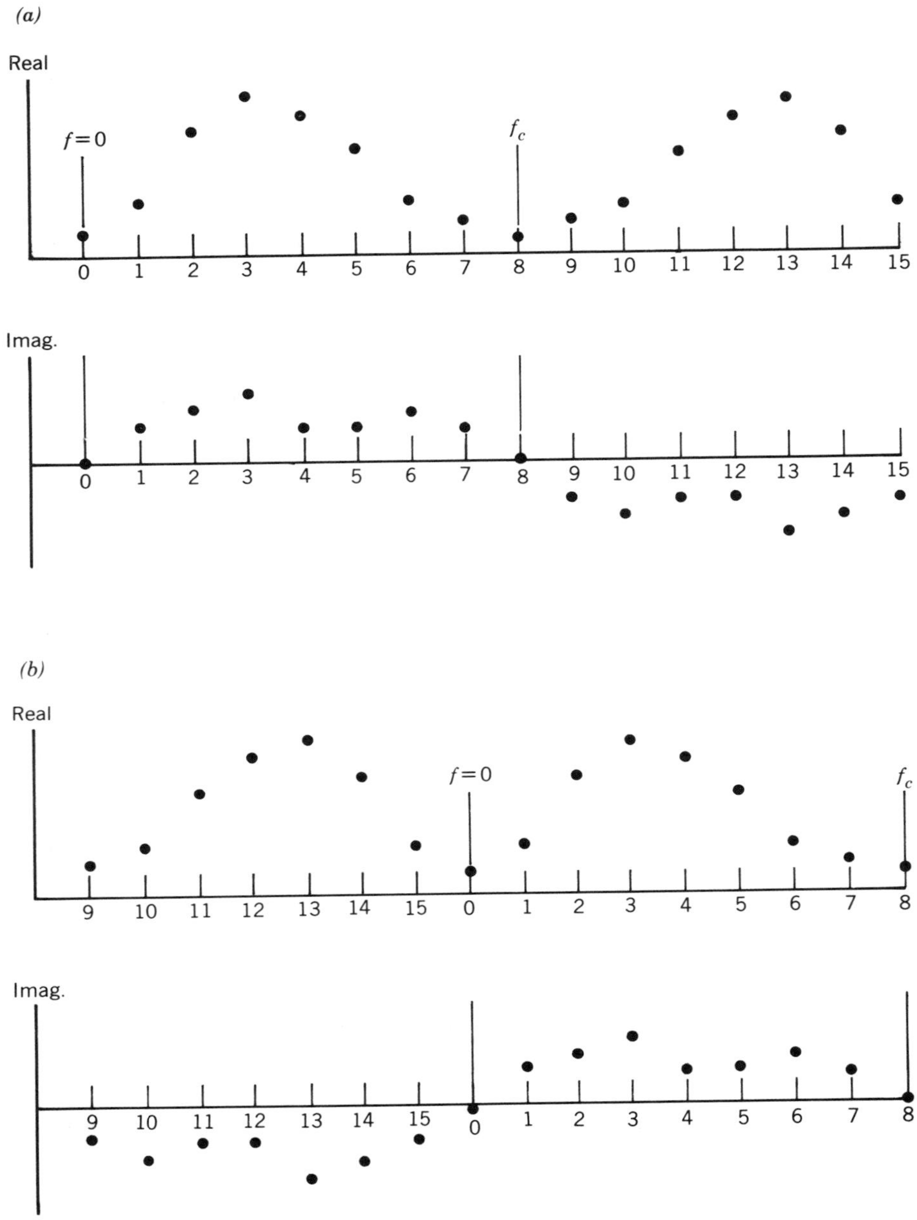

Figure 11.3 Illustration of frequency components for $N = 16$ point FFT. (*a*) Components as produced by FFT. (*b*) Transposed components defining two-sided spectrum.

11.2.6 *Winograd Fourier Transform*

New techniques for calculating discrete Fourier transforms are continually being devised. Most are modifications of the standard FFT optimized for a specific use or mechanization. One new procedure, introduced by Winograd in Reference 11.7, deserves mention because it involves a totally new approach. Essentially, the process maps the one-dimensional time sequence into a multi-dimensional array with each array dimension corresponding to one of the prime factors of N, the one-dimensional sequence length. The multidimensional array is then Fourier transformed and remapped back to the one-dimensional frequency sequence. The mapping is performed according to number-theoretic concepts using the so-called Chinese remainder theorem.

The theory and implementation of the Winograd Fourier transform is significantly more complex than that for the FFT described in Section 11.2.2. Furthermore, the mapping process expands the amount of memory needed to hold the data so that it is not possible to perform the algorithm in place. Despite these disadvantages, the Winograd algorithm is being used more and more because (a) it requires significantly less computational time and (b) it is easily generalized to radices other than two and, especially, to mixed radices. See Reference 11.7 for details.

11.3 PROBABILITY DENSITY FUNCTIONS

Consider N data values $\{x_n\}$, $n = 1, 2, \ldots, N$, from a transformed record $x(t)$ that is stationary with $\bar{x} = 0$. It follows from Equations (3.4) and (8.64) that the probability density function of $x(t)$ can be estimated by

$$\hat{p}(x) = \frac{N_x}{NW} \tag{11.78}$$

where W is a narrow interval centered at x and N_x is the number of data values that fall within the range $x \pm W/2$. Hence an estimate $\hat{p}(x)$ is obtained digitally by dividing the full range of x into an appropriate number of equal width class intervals, tabulating the number of data values in each class interval, and dividing by the product of the class interval width W and the sample size N. Note that the estimate $\hat{p}(x)$ is not unique, since it clearly is dependent on the number of class intervals and their width selected for the analysis.

A formal statement of this procedure will now be given. Let K denote the number of class intervals selected to cover the entire range of the data values from a to b. Then the width of each interval is given by

$$W = \frac{b - a}{K} \tag{11.79}$$

and the end point of the ith interval is defined by

$$d_i = a + iW \qquad i = 0, 1, 2, \ldots, K \qquad (11.80)$$

Note that $d_0 = a$ and $d_K = b$. Now define a sequence of $K + 2$ numbers $\{N_i\}$, $i = 0, 1, 2, \ldots, K + 1$, by the conditions

$$N_0 = [\text{number of } x \text{ such that } x \le d_0]$$

$$N_1 = [\text{number of } x \text{ such that } d_0 < x \le d_1]$$

$$\vdots$$

$$N_i = [\text{number of } x \text{ such that } d_{i-1} < x \le d_i]$$

$$\vdots$$

$$N_K = [\text{number of } x \text{ such that } d_{K-1} < x \le d_K]$$

$$N_{K+1} = [\text{number of } x \text{ such that } x > d_K] \qquad (11.81)$$

This procedure will sort out the N data values of x so that the number sequence $\{N_i\}$ satisfies

$$N = \sum_{i=0}^{K+1} N_i \qquad (11.82)$$

One method of doing this sorting on a digital computer is to examine each value x_n, $n = 1, 2, \ldots, N$, in turn as follows.

1. If $x_n \le a$, add the integer 1 to N_0.
2. If $a < x_n \le b$, compute $I = (x_n - a)/W$. Then, select i as the largest integer less than or equal to I, and add the integer 1 to N_i.
3. If $x_n > b$, add the integer 1 to N_{K+1}.

Four output forms of the sequence $\{N_i\}$ can be used. The first output is the *histogram*, which is simply the sequence $\{N_i\}$ without changes. The second output is the sample percentage of data in each class interval defined for $i = 0, 1, 2, \ldots, K + 1$, by

$$\hat{P}_i = \text{Prob}[d_{i-1} < x \le d_i] = \frac{N_i}{N} \qquad (11.83)$$

The third output is the sequence of sample probability density estimates$\{\hat{p}_i\}$

defined at the midpoints of the K class intervals in $[a, b]$ by

$$\hat{p}_i = \frac{\hat{P}_i}{W} = \left(\frac{N_i}{N}\right)\left(\frac{K}{b-a}\right) \qquad i = 1, 2, \ldots, K \qquad (11.84)$$

The fourth output is the sequence of sample probability distribution estimates $\{\hat{P}(i)\}$ defined at the class interval end points where $i = 0, 1, 2, \ldots, K + 1$, by

$$\hat{P}(i) = \mathrm{Prob}[-\infty < x \le d_i] = \sum_{j=0}^{i} \hat{P}_j = W \sum_{j=0}^{i} \hat{p}_j \qquad (11.85)$$

11.4 AUTOCORRELATION FUNCTIONS

There are two ways to compute autocorrelation estimates. The first is the direct method, involving the computation of average products among the sample data values. The second way is the indirect approach of first computing an auto-spectral density estimate using FFT procedures, and then computing the inverse transform of the autospectrum. The direct method is the easier to program and represents the more logical approach from the viewpoint of basic definitions. The second approach takes advantage of the dramatic computational efficiency of FFT algorithms and hence is much less expensive to execute.

11.4.1 *Autocorrelation Estimates via Direct Computations*

Consider N data values $\{x_n\}$, $n = 1, 2, \ldots, N$, sampled at equally spaced time intervals Δt from a transformed record $x(t) = x(n\,\Delta t)$ that is stationary with $\bar{x} = 0$. From the basic definition in Equation (8.90), the autocorrelation function of $x(t)$ will be estimated from the sample values at the time delay $r\,\Delta t$ by

$$\hat{R}_{xx}(r\,\Delta t) = \frac{1}{N-r} \sum_{n=1}^{N-r} x_n x_{n+r} \qquad r = 0, 1, 2, \ldots, m \qquad (11.86)$$

where r is called the *lag number* and m is the maximum lag number ($m < N$). Note that the number of possible products at each lag number r in Equation (11.86) is only $N - r$. Hence, the division by $N - r$ is needed to obtain an unbiased estimate of the autocorrelation function. The number of real multiply–add operations required to compute the autocorrelation estimate is approximately Nm assuming $m \ll N$.

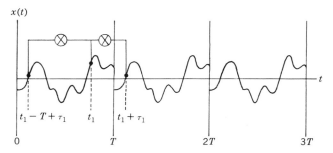

Figure 11.4 Illustration of circular effect in correlation analysis via FFT calculations.

11.4.2 *Autocorrelation Estimates via FFT Computations*

The indirect method of computing autocorrelation estimates is based on the Weiner–Khinchine relations defined in Equation (5.28). Specifically, the auto-correlation function is computed by taking the inverse Fourier transform of the autospectrum estimate. However, due to the underlying periodic assumption of the finite Fourier transform, the autocorrelation function computed by this procedure is "circular" in character. This occurs because the FFT algorithm essentially treats a record of length $T = N\Delta t$ as one period of an ongoing periodic function. Hence the resulting correlation function appears as if it were calculated from a periodic function, as illustrated in Figure 11.4. For a time t_1 and delay τ_1 such that $T - t_1 < \tau_1$, the product $x(t_1)x(t_1 + \tau_1) = x(t_1)x(t_1 - T + \tau_1)$. It follows that the resulting correlation function at any delay τ will be a composite of terms involving $R_{xx}(\tau)$ and $R_{xx}(T - \tau)$.

To evaluate this case, let the finite Fourier transform of a record $x(t)$, $0 \le t \le T$, and its complex conjugate be given by

$$X^*(f) = \int_0^T x(\alpha)\, e^{j2\pi f\alpha}\, d\alpha \qquad X(f) = \int_0^T x(\beta)\, e^{-j2\pi f\beta}\, d\beta \quad (11.87)$$

It follows that

$$|X(f)|^2 = \int_0^T \int_0^T x(\alpha)x(\beta)[e^{-j2\pi f(\beta - \alpha)}]\, d\beta\, d\alpha \qquad (11.88)$$

With the transformation of variables $\tau = \beta - \alpha$, $d\tau = d\beta$, and $\beta = \alpha + \tau$, Equation (11.88) becomes

$$|X(f)|^2 = \int_0^T \int_{-\alpha}^{T-\alpha} x(\alpha)x(\alpha + \tau)\, e^{-j2\pi f\tau}\, d\tau\, d\alpha \qquad (11.89)$$

From Equation (5.67), the two-sided autospectral density function of $x(t)$ is estimated by

$$\hat{S}_{xx}(f) = \frac{1}{T}E\left[|X(f)|^2\right] = \frac{1}{T}\int_0^T\int_{-\alpha}^{T-\alpha}\hat{R}_{xx}(\tau)\,e^{-j2\pi f\tau}\,d\tau\,d\alpha \quad (11.90)$$

By reversing the order of integration as shown in the sketch below, Equation

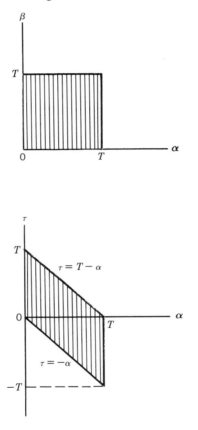

(11.90) can be written as

$$\hat{S}_{xx}(f) = \frac{1}{T}\int_{-T}^0\int_{-\tau}^T\hat{R}_{xx}(\tau)\,e^{-j2\pi f\tau}\,d\alpha\,d\tau$$

$$+ \frac{1}{T}\int_0^T\int_0^{T-\tau}\hat{R}_{xx}(\tau)\,e^{-j2\pi f\tau}\,d\alpha\,d\tau$$

$$= \frac{1}{T}\int_{-T}^0(T+\tau)\hat{R}_{xx}(\tau)\,e^{-j2\pi f\tau}\,d\tau$$

$$+ \frac{1}{T}\int_0^T(T-\tau)\hat{R}_{xx}(\tau)\,e^{-j2\pi f\tau}\,d\tau \quad (11.91)$$

In the first term of Equation (11.91), let $u = -\tau$ and $du = -d\tau$ to obtain

$$\int_{-T}^{0} (T + \tau)\hat{R}_{xx}(\tau)e^{-j2\pi f\tau}\,d\tau = \int_{0}^{T}(T - u)\hat{R}_{xx}(u)\,e^{j2\pi fu}\,du$$

where $\hat{R}_{xx}(-u)$ is replaced by $\hat{R}_{xx}(u)$. Next, change variables again by letting $\tau = T - u$, $d\tau = -du$, to give

$$\int_{0}^{T}(T - u)\hat{R}_{xx}(u)\,e^{j2\pi fu}\,du = \int_{0}^{T}\hat{R}_{xx}(T - \tau)\,e^{-j2\pi f\tau}\,d\tau$$

Here use is made of the fact that $e^{j2\pi fT} = 1$ for any $f = f_k = k\,\Delta f = (k/T)$ where k is an integer. Thus Equation (11.91) is the same as

$$\hat{S}_{xx}(f) = \frac{1}{T}\int_{0}^{T}\tau\hat{R}_{xx}(T - \tau)\,e^{-j2\pi f\tau}\,d\tau + \frac{1}{T}\int_{0}^{T}(T - \tau)\hat{R}_{xx}(\tau)\,e^{-j2\pi f\tau}\,d\tau$$

$$= \int_{0}^{T}\hat{R}_{xx}^{c}(\tau)\,e^{-j2\pi f\tau}\,d\tau \tag{11.92}$$

where

$$\hat{R}_{xx}^{c}(\tau) = \frac{(T - \tau)}{T}\hat{R}_{xx}(\tau) + \frac{\tau}{T}\hat{R}_{xx}(T - \tau) \tag{11.93}$$

It follows that the inverse Fourier transform of $\hat{S}_{xx}(f)$ in Equation (11.92) will yield $\hat{R}_{xx}^{c}(\tau)$ as defined in Equation (11.93). For a digital sample $\{x_n\}$, $n = 0, 1, \ldots, N - 1$, the *circular correlation function* in Equation (11.93) becomes

$$\hat{R}_{xx}^{c}(r\Delta t) = \frac{N - r}{N}\hat{R}_{xx}(r\Delta t) + \frac{r}{N}\hat{R}_{xx}[(N - r)\Delta t] \tag{11.94}$$

The two parts of Equation (11.94) are illustrated in Figure 11.5.

In practice, for correlation functions that decay rapidly, the circular effect is not of great concern for maximum lag values of, say, $m < 0.1N$. In any case, the problem can be avoided by adding zeros to the original data. The effect of adding zeros to the data is to spread the two portions of the circular correlation function. In particular, if N zeros are added to the original N data values, then the two portions will separate completely giving

$$\hat{R}_{xx}^{s}(r\Delta t) = \begin{cases} \dfrac{(N - r)}{N}\hat{R}_{xx}(r\Delta t) & r = 0, 1, \ldots, N - 1 \\[3mm] \dfrac{(r - N)}{N}\hat{R}_{xx}[(2N - r)\Delta t] & r = N, N + 1, \ldots, 2N - 1 \end{cases}$$

$$\tag{11.95}$$

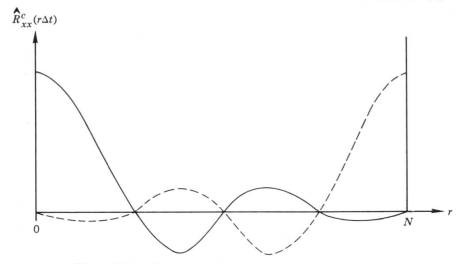

$\hat{R}^c_{xx}(r\Delta t)$

Figure 11.5 Illustration of circular correlation function.

The two parts of Equation (11.94) are shown in Figure 11.6. Note that the first half of the estimate where $0 \le r \le N - 1$ represents the autocorrelation function values for positive lags ($0 \le r \le m$) while the second half of the estimate where $N \le r \le 2N - 1$ constitutes the autocorrelation function values for negative lags ($-m \le r \le 0$). However, since autocorrelation functions are always even functions of r, the second half of the estimate can be discarded, so the final unbiased autocorrelation estimate is computed from

$$\hat{R}_{xx}(r\Delta t) = \frac{N}{N - r}\hat{R}^s_{xx}(r\Delta t) \qquad r = 0, 1, \ldots, N - 1 \qquad (11.96)$$

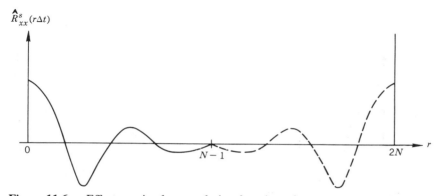

$\hat{R}^s_{xx}(r\Delta t)$

Figure 11.6 Effect on circular correlation function when N zeros are added.

The correlation estimate given by Equation (11.96) is statistically equivalent to the directly computed correlation estimate defined in Equation (11.86). However, depending on the maximum lag value m, the indirect FFT approach might require substantially less calculation. Specifically, the indirect method requires first the computation of a spectral density estimate, which involves FFT calculations over n_d independent records, each with N data values augmented by N zeros for a total of $2N$ values for each FFT. This is followed by an inverse FFT over the $2N$ data points of the averaged spectral density estimate to give a total of $(n_d + 1)$ FFT calculations, each requiring $(4Np)$ real operations as detailed in Section 11.2.2. For an equivalent total record length of $n_d N$ data points, the direct method requires approximately $mn_d N$ real operations. Hence for a similar maximum lag value, $m = N$, the speed ratio is

$$\text{Speed ratio} = \frac{n_d N^2}{(n_d + 1)4Np} \simeq \frac{N}{4p} \tag{11.97}$$

For example, if $N = 1024 = 2^{10}$, the speed ratio is $1024/40 \simeq 26$. In practice, since only real number sequences are involved in autocorrelation and autospectra estimates, an additional speed increase of almost two can be achieved by using the FFT procedures detailed in Section 11.2.4.

In summary, the following steps are recommended to compute the autocorrelation function via FFT procedures. The available sample size for the record $x(n \Delta t)$ is assumed to be Nn_d where $N = 2^p$.

1. Determine the maximum lag number m of interest and divide the available data record into n_d blocks, each consisting of $N \geq m$ data values.
2. Augment each block of N data values, $\{x_n\}$, $n = 1, 2, \ldots, N$, with N zeros to obtain a new sequence of $2N$ values.
3. Compute the $2N$-point FFT giving $X(f_k)$, $k = 0, 1, \ldots, 2N - 1$, using Equation (11.36) with $2N$ replacing N.
4. Compute the two-sided autospectral density estimate $\hat{S}_{xx}(f_k)$ for $k = 0, 1, \ldots, 2N - 1$ using Equation (11.101), to follow.
5. Compute the inverse FFT of $\hat{S}_{xx}(f_k)$ to obtain $\hat{R}^s_{xx}(r\Delta t)$ for $r = 0, 1, \ldots, 2N - 1$, using Equation (11.76) with N replaced by $2N$, $X_k = \hat{S}_{xx}(f_k)$, and $x_n = \hat{R}^s_{xx}(r\Delta t)$.
6. Discard the last half of $\hat{R}^s_{xx}(r\Delta t)$ to obtain results for $r = 0, 1, \ldots, N - 1$.
7. Multiply $\hat{R}^s_{xx}(r\Delta t)$, $r = 1, 2, \ldots, N - 1$, by the scale factor $N/(N - r)$ to obtain the desired $\hat{R}_{xx}(r\Delta t)$.

It should be noted that the inverse transform in step 5 requires that all values of the spectral estimate be used; that is, $\hat{S}_{xx}(f_k)$, $k = 0, 1, \ldots, 2N - 1$, even though the Nyquist frequency occurs at $k = N$. This matter is discussed further in Section 11.5.1. It should also be noted from Section 8.4 that the variance of an autocorrelation estimate is inversely proportional to $2B_eT_r = Nn_d$. Hence an acceptably accurate autocorrelation estimate can often be obtained from a single FFT spectrum ($n_d = 1$) if N is sufficiently large. Finally, the calculation of the autospectra estimate in step 4 should be accomplished without tapering operations of the type discussed in Section 11.5.2.

11.5 AUTOSPECTRAL DENSITY FUNCTIONS

In earlier years, autospectral density functions were usually estimated by procedures based on the definition in Equation (5.34), that is, an autocorrelation function was first calculated using Equation (11.86) and then Fourier transformed over appropriate lag values to obtain a spectral estimate. This approach, commonly called the *Blackman–Tukey procedure*, requires many more computer operations than the direct FFT calculation procedure based on the definition in Equation (5.67) and, hence, is no longer recommended. The basic calculations for the direct FFT approach are straightforward, but there are various "grooming" operations that are often added to the calculations to improve the quality of the resulting estimates. These matters will now be discussed.

11.5.1 *Basic Autospectra Estimation Procedures*

Consider a transformed data record $x(t)$ of total length T_r that is stationary with $\bar{x} = 0$. Let the record be divided into n_d contiguous segments, each of length T, as shown in Figure 11.7. It follows that each segment of $x(t)$ is $x_i(t)$, $(i - 1)T \le t \le iT$, $i = 1, 2, \ldots, n_d$. Using Equation (8.153) and dividing by 2, an estimate $\hat{S}_{xx}(f)$ of the two-sided autospectral density function $\hat{S}_{xx}(f)$ for an arbitrary f is given by

$$\hat{S}_{xx}(f) = \frac{1}{n_d T} \sum_{i=1}^{n_d} |X_i(f, T)|^2 \tag{11.98}$$

where

$$X_i(f, T) = \int_0^T x_i(t) e^{-j2\pi ft} dt$$

The averaging operation over the n_d records in Equation (11.98) approximates the expected value operation in Equation (5.67).

In digital terms, let each record segment $x_i(t)$ be represented by N data values $\{x_{in}\}$, $n = 0, 1, \ldots, N - 1$, $i = 1, 2, \ldots, n_d$. The finite Fourier trans-

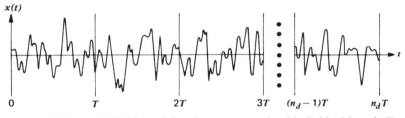

Figure 11.7 Subdivision of data into n_d records of individual length T.

form in Equation (11.98) will produce values at the discrete frequencies

$$f_k = \frac{k}{T} = \frac{k}{N\Delta t} \qquad k = 0, 1, \ldots, N - 1 \qquad (11.99)$$

The Fourier components for each segment are then given from Equation (11.38) by

$$X_i(f_k) = \Delta t\, X_{ik} = \Delta t \sum_{n=0}^{N-1} x_{in} \exp\left[\frac{-j2\pi kn}{N}\right] \qquad (11.100)$$

The two-sided autospectral density function estimate of Equation (11.98) now becomes

$$\hat{S}_{xx}(f_k) = \frac{1}{n_d N\Delta t} \sum_{i=1}^{n_d} |X_i(f_k)|^2 \qquad k = 0, 1, \ldots, N - 1 \quad (11.101)$$

As discussed in Section 11.2.4, when FFT procedures are used, the Nyquist frequency f_c occurs where $k = N/2$. Hence, the first $(N/2) + 1$ spectral values at $k = 0, 1, \ldots, N/2$ define the autospectral density estimate in the frequency range from $0 \le f_k \le f_c$, while the last $(N/2) - 1$ spectral values at $k = (N/2) + 1, (N/2) + 2, \ldots, N - 1$, can be interpreted as the autospectral density estimate in the frequency range from $-f_c < f < 0$. Since autospectra functions are always real valued, it follows from Equation (11.77) that $\hat{S}_{xx}(f_k) = \hat{S}_{xx}(2f_c - f_k)$.

The one-sided autospectral density function is estimated directly from Equation (11.101) by

$$\hat{G}_{xx}(f_k) = \frac{2}{n_d N\Delta t} \sum_{i=1}^{n_d} |X_i(f_k)|^2 \qquad k = 0, 1, \ldots, \frac{N}{2} \quad (11.102)$$

The number of data values N used for each FFT in Equations (11.101) and (11.102) is often called the *block size* for the calculation and is the key parameter in determining the resolution of the analysis given by

$$\Delta f = 1/T = 1/(N\Delta t) \qquad (11.103)$$

On the other hand, the number of averages n_d determines the random error of the estimate as detailed in Section 8.5.4. Note that for a Cooley–Tukey FFT algorithm, it is convenient to select an analysis block size that is a power of 2, that is $N = 2^p$, as discussed in Section 11.2.3.

11.5.2 *Side-Lobe Leakage Suppression Procedures*

The finite Fourier transform of $x(t)$ defined in Equation (11.98) can be viewed as the Fourier transform of an unlimited time history record $v(t)$ multiplied by a rectangular time window $u(t)$ where

$$u(t) = \begin{cases} 1 & 0 \le t \le T \\ 0 & \text{otherwise} \end{cases} \qquad (11.104)$$

In other words, the sample time-history record $x(t)$ can be considered to be the product

$$x(t) = u(t)v(t) \qquad (11.105)$$

as illustrated in Figure 11.8. It follows that the Fourier transform of $x(t)$ is the *convolution* of the Fourier transforms of $u(t)$ and $v(t)$, namely,

$$X(f) = \int_{-\infty}^{\infty} U(\alpha)V(f-\alpha)\,d\alpha \qquad (11.106)$$

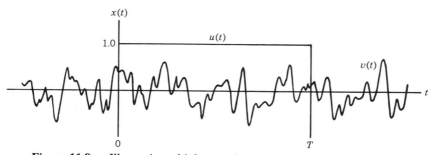

Figure 11.8 Illustration of inherent time window in spectral analysis.

For the case where $u(t)$ is the rectangular function defined by Equation (11.104), its Fourier transform is

$$U(f) = T\left(\frac{\sin \pi fT}{\pi fT}\right)e^{-j\pi fT} \tag{11.107}$$

A plot of $|U(f)|$ is shown in Figure 11.9. Note that the first side-lobe is about 13 dB down from the main lobe, and the side lobes falls off at a rate of 6 dB/octave thereafter. This function constitutes the basic "spectral window" of the analysis. The large side lobes of $|U(f)|$ allow leakage of power at frequencies well separated from the main lobe of the spectral window and may introduce significant distortions of the estimated spectra, particularly when the data are narrow-band in character. The leakage problem will not arise in the analysis of periodic data with period T_p as long as the record length T is an exact number of periods, that is, $T = kT_p$, $k = 1, 2, 3, \ldots$. In this case, the Fourier components at $f = kf_p = (k/T_p)$ cannot leak into the main lobe, because $U(f)$ in Equation (11.107) is always zero at these frequencies. However, if $T \neq kT_p$, then leakage will occur in the analysis of periodic data as well.

TIME HISTORY TAPERING.   To suppress the leakage problem, it is common in practice to introduce a time window that tapers the time-history data to eliminate the discontinuities at the beginning and end of the records to be analyzed. There are numerous such windows in current use, but one of the earliest and still commonly employed is a full cosine tapering window, called

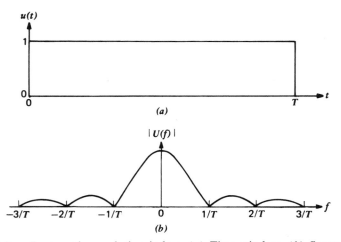

Figure 11.9   Rectangular analysis window. (*a*) Time window. (*b*) Spectral window.

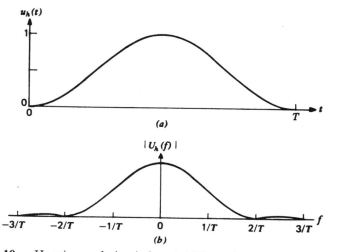

Figure 11.10   Hanning analysis window. (a) Time window. (b) Spectral window.

the *cosine squared* or *Hanning* window, which is given by

$$u_h(t) = \begin{cases} \frac{1}{2}\left(1 - \cos\dfrac{2\pi t}{T}\right) = 1 - \cos^2\left(\dfrac{\pi t}{T}\right) & 0 \le t \le T \\ 0 & \text{otherwise} \end{cases} \tag{11.108}$$

as shown in Figure 11.10a. The Fourier transform of Equation (11.108) is

$$U_h(f) = \tfrac{1}{2}U(f) - \tfrac{1}{4}U(f - f_1) - \tfrac{1}{4}U(f + f_1) \tag{11.109}$$

where $f_1 = 1/T$ and $U(f)$ is as defined in Equation (11.107). Note that

$$U(f - f_1) = -T\left[\frac{\sin \pi(f - f_1)T}{\pi(f - f_1)T}\right]e^{-j\pi fT}$$

$$U(f + f_1) = -T\left[\frac{\sin \pi(f + f_1)T}{\pi(f + f_1)T}\right]e^{-j\pi fT} \tag{11.110}$$

A plot of $|U_h(f)|$ is shown in Figure 11.10b. The first side lobe of the Hanning spectral window is some 32 dB below the mainlobe, and the side lobes fall off at 18 dB/octave thereafter.

Consider now any function $v(t)$ that is not periodic of period T and let

$$x(t) = u_h(t)v(t) \tag{11.111}$$

The Fourier transform of Equation (1.111) is

$$X(f) = \int_0^T x(t)e^{-j2\pi ft}\,dt = \int_{-\infty}^{\infty} U_h(\alpha)V(f-\alpha)\,dt \qquad (11.112)$$

At the discrete frequency values $f_k = (k/T)$ for $k = 0, 1, 2, \ldots, (N/2)$, one obtains

$$X(f_k) = \tfrac{1}{2}V(f_k) - \tfrac{1}{4}V(f_{k-1}) - \tfrac{1}{4}V(f_{k+1}) \qquad (11.113)$$

where

$$V(f_k) = \int_0^T v(t)e^{-j2\pi kt/T}\,dt \qquad (11.114)$$

To proceed further, assume that $v(t)$ behaves similar to bandwidth limited white noise over the frequency resolution bandwidth $\Delta f = (1/T)$. It then follows that for any two discrete frequencies f and g calculated at the points $k\,\Delta f = (k/T)$, expected value operations on $V^*(f)$ and $V(g)$ will give

$$E[V^*(f)V(g)] = \begin{cases} 0 & \text{for } f \neq g \\ 1 & \text{for } f = g \end{cases} \qquad (11.115)$$

Applying these properties to Equation (11.113) yields

$$E\left[|X(f_k)|^2\right] = \left(\tfrac{1}{2}\right)^2 + \left(\tfrac{1}{4}\right)^2 + \left(\tfrac{1}{4}\right)^2 = \tfrac{3}{8} \qquad (11.116)$$

for any $f_k = (k/T)$, $k = 0, 1, 2, \ldots, (N/2)$. This represents a loss factor due to using the Hanning window of Equation (11.108) to compute spectral density estimates by Fourier transform techniques. Hence one should multiply Equation (11.100) by the scale factor $\sqrt{\tfrac{8}{3}}$ to obtain the correct magnitudes in later spectral density estimates using Equations (11.101) and (11.102). Specifically, the autospectral density estimate with Hanning is computed by Equations (11.101) and (11.102) using

$$X_i(f_k) = \Delta t\sqrt{\tfrac{8}{3}} \sum_{n=0}^{N-1} x_{in}\left(1 - \cos^2\frac{\pi n}{N}\right)\exp\left[-\frac{j2\pi kn}{N}\right] \qquad (11.117)$$

where $f_k = k/(N\Delta t)$, $k = 0, 1, \ldots, (N/2)$. See References 11.8 and 11.9 for discussions of other tapering operations.

***Example 11.3.* Spectral Errors Due to Side-Lobe Leakage.**   To illustrate the effect of side-lobe leakage on spectral density estimates, consider the two autospectra estimates shown in Figure 11.11. Both spectra were computed from the same record of particle velocities measured in a water basin by a laser velocimeter during a long sequence of stationary wave activity scaled to ten times real time rates. The spectra were computed using $n_d = 400$ averages and a resolution of $\Delta f = 0.0488$ Hz ($T = 20.49$ sec) from digital data sampled at a rate of 50 sps ($f_c = 25$ Hz). The only difference between the two spectral density estimates is that the solid line was computed with cosine squared tapering (Hanning) of the time-history records and the dashed line was calculated with no tapering.

The results in Figure 11.11 clearly demonstrate the errors in autospectral density estimates that can be caused by spectral side-lobe leakage. Specifically, the spectral estimates at frequencies off a spectral peak are increased in value by leakage of power through the side lobes positioned at the frequency of a spectral peak. Note that the frequency resolution in this example ($\Delta f =$

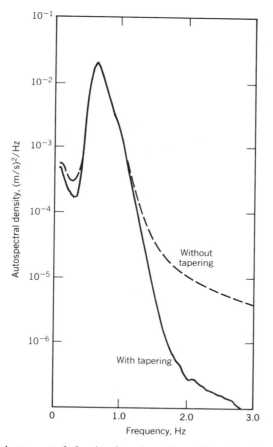

Figure 11.11 Autospectral density function of particle velocity in simulated ocean waves computed with and without tapering. These data resulted from studies funded by Shell Internationale Petroleum Maatschappij B. V., The Hague, Netherlands.

0.0488 Hz) complies with the requirement for negligible bias error given in Example 8.5. Thus, the leakage error can be a problem even when the usual resolution bias error is controlled. This concludes Example 11.3.

OVERLAPPED PROCESSING. From Figures 11.9 and 11.10, it is seen that the time-history tapering used to suppress side-lobe leakage also increases the width of the main lobe of the spectral window; that is, it reduces the basic resolving power of the analysis. For Hanning, the increase in the half-power bandwidth of the main lobe is about 60%. This is generally an acceptable penalty to pay for the suppression of leakage from frequencies outside the region of the main lobe. However, the time-history tapering operation also causes an increase in the variability of the resulting spectral estimates. This occurs because the tapering essentially discards relevant information near the beginning and end of each record. Assuming the data to be analyzed have an approximately uniform spectrum, a Hanning tapering of the time-history records will increase the variance of the resulting spectral estimates by a factor of about 2 [Reference 11.10]. That is, instead of a normalized random error of $\varepsilon = 1/\sqrt{n_d}$ as derived in Chapter 8, the error with Hanning will be approximately $\varepsilon = \sqrt{2/n_d}$.

To counteract the increase in variability caused by time-history tapering for side-lobe suppression, overlapped processing techniques are sometimes used. Specifically, instead of dividing a record $x(t)$ into n_d independent segments, $x_i(t), (i-1)T \leq t \leq iT, i = 1, 2, \ldots, n_d$, the record is divided into overlapped segments $x_i(t)$ covering the time intervals

$$[q(i-1)]T \leq t \leq [q(i-1)+1]T \qquad i = 1, 2, \ldots, (n_d/q) \qquad q < 1$$

$$(11.118)$$

A common selection in overlapped processing is $q = 0.5$, which produces 50% overlapping, as illustrated in Figure 11.12. This will retrieve about 90% of the stability lost due to the tapering operation but also will double the required number of FFT operations. See Reference 11.10 for details.

CORRELATION FUNCTION TAPERING. Another method for suppressing side-lobe leakage involves tapering the autocorrelation function of the data

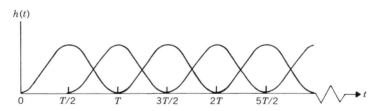

Figure 11.12 Sequence of tapered time windows for 50% overlapped processing.

rather than the original time history. This approach, sometimes called *lag weighting*, causes very little increase in the variance of the resulting spectral estimates, and hence eliminates the need for overlapped processing. It also requires fewer computations. The general procedure is as follows.

a. Compute the two-sided autospectral density function $\hat{S}_{xx}(f)$ from long, untapered time-history records using Equation (11.101).
b. Compute the autocorrelation function $\hat{R}_{xx}(\tau)$ by taking the inverse FFT of the autospectrum estimate as described in Section 11.4.2 (use of the circular correlation function is acceptable for this application).
c. Apply a taper $u(\tau)$ to the autocorrelation estimate such that $u(\tau) = 1$ at $\tau = 0$ and $u(\tau) = 0$ at $\tau = \tau_{max} \ll T$.
d. Based on Equation (5.34), recompute the smoothed, one-sided autospectral density function $\hat{G}_{xx}(f)$ by taking the FFT of $u(\tau)\hat{R}_{xx}(\tau)$ over the interval $0 \le \tau \le \tau_{max}$ and multiplying by 4.

Since the mean square value of the data is determined by the value of the autocorrelation function at $\tau = 0$, the tapering operation described above causes no change in the mean square value and hence no loss in the spectral values of the calculated spectrum. Also, if $\tau_{max} \ll T$, there is essentially no increase in the variance of the spectral estimate and hence no need for overlapped processing.

In computing spectra with lag weighting, one should calculate the original autospectrum from Equation (11.101) with a longer record length than needed to obtain the desired resolution in the final autospectrum. This will yield spectral estimates with smaller values of Δf and n_d than desired in the final estimates. However, since the final spectral estimates are given by the FFT of $u(\tau)R_{xx}(\tau)$, the resolution of the final estimates will be

$$\Delta f = \frac{1}{\tau_{max}} = \frac{1}{m\,\Delta t} \tag{11.119}$$

where $m < N$ is the maximum lag number. Furthermore, the normalized random error of the final estimates will be approximated by

$$\varepsilon = \sqrt{\frac{\tau_{max}}{n_d T}} = \sqrt{\frac{m}{n_d N}} \tag{11.120}$$

Hence, the resulting spectral estimates will have essentially the same variance as if they were computed using time-history tapering and overlapped processing procedures with the same resolution.

As for time-history tapering, a full cosine tapering window (Hanning) is often used for correlation function tapering. In this case,

$$u_h(\tau) = \begin{cases} \dfrac{1}{2}\left(1 + \cos\dfrac{\pi\tau}{\tau_{\max}}\right) = \dfrac{1}{2}\left(1 + \cos\dfrac{\pi r}{m}\right) & r = 0, 1, 2, \ldots, m \\ 0 & r > m \end{cases} \quad (11.121)$$

Hanning of the correlation function yields less spectral side-lobe suppression than Hanning of the original time-history data. Specifically, Hanning in the correlation domain corresponds to a spectral window where the first side lobe is 16 dB below the main lobe, and the side lobes fall off at 9 dB/octave thereafter. However, lag weighting functions that provide substantially greater spectral side-lobe suppression are available, as detailed in Reference 11.10.

11.5.3   Recommended Computational Steps

The following steps are recommended to compute autospectral density functions.

1. Divide the available data record for x_n, $n = 0, 1, \ldots, Nn_d$, into n_d blocks, each consisting of N data values.
2. If needed to suppress side-lobe leakage, taper the data values in each block $\{x_n\}$, $n = 0, 1, \ldots, N-1$, by the Hanning taper described in Equation (11.108), or some other appropriate tapering function (the correlation function tapering described in Section 11.5.2 may be used as an alternative).
3. Compute the N-point FFT for each block of data giving $X(f_k)$, $k = 0, 1, \ldots, (N-1)$. If necessary to reduce the variance increase caused by tapering, compute the FFTs for overlapped records as described in Equation (11.118).
4. Adjust the scale factor of $X(f_k)$ for the loss due to tapering (for Hanning tapering, multiply by $\sqrt{\frac{8}{3}}$)
5. Compute the autospectral density estimate from the n_d blocks of data using Equation (11.101) for a two-sided estimate or Equation (11.102) for a one-sided estimate.

11.5.4   Zoom Transform Procedures

A major problem in spectral analysis using FFT procedures is the computation of spectral values at relatively high frequencies with a very small resolution, that is, those cases where $f_c/\Delta f$ is very large. Of course, if there are no limits on the block size N that can be Fourier transformed, any desired resolution at higher frequencies can be obtained since

$$f_c/\Delta f = 1/(2\,\Delta t)\big/1/(N\Delta t) = N/2 \qquad (11.122)$$

However, there are practical limits on the block size N that can be Fourier transformed by a computer (even virtual memory machines) and the required computations using a Cooley–Tukey FFT algorithm increase in proportion to Np when $N = 2^p$. Hence it is desirable to use a computational technique that will permit the calculation of an autospectrum with a folding frequency of f_c and a resolution Δf using fewer data points than suggested in Equation (11.122). Such techniques are commonly referred to as *zoom transform* procedures.

There are several ways to achieve a zoom transform, but the most common approach is first to convert the original data into a collection of bandlimited records, and then to apply the principles of complex demodulation to move the lower frequency limit of each bandlimited record down to zero frequency. To be specific, let the original data record $x(t)$ be bandpass filtered to produce a new record

$$y(t) = \begin{cases} x(t) & f_0 - (B/2) \le f \le f_0 + (B/2) \\ 0 & \text{otherwise} \end{cases} \qquad (11.123)$$

Now multiply $y(t)$ by an exponential function to obtain the modulated signal

$$v(t) = y(t)e^{j2\pi f_1 t} \qquad (11.124)$$

where f_1 is the modulating frequency. The Fourier transform of $v(t)$ yields

$$V(f) = \int_0^T y(t)e^{j2\pi f_1 t}e^{-j2\pi f t}\,dt$$

$$= \int_0^T y(t)e^{-j2\pi(f-f_1)t}\,dt = Y(f - f_1) \qquad (11.125)$$

so the autospectral density function of $v(t)$ becomes

$$G_{vv}(f) = \lim_{T\to\infty} \frac{2}{T} E\big[|V(f)|^2\big] = G_{yy}(f - f_1) \qquad (11.126)$$

It follows from Equation (11.126) that if the modulating frequency f_1 is set at $f_1 = f_0 - B/2$, then the autospectrum of the original bandlimited data will be transposed in frequency down to the frequency range, $0 \le f \le B$, from the original frequency range, $f_0 - (B/2) \le f \le f_0 + (B/2)$, as indicated in Figure 11.13. The data can now be sampled at a minimum rate of $2B$ sps without aliasing, as opposed to a minimum rate of $2(f_0 + B/2)$ for the original data. Hence, a much finer resolution can be obtained for a given block size N,

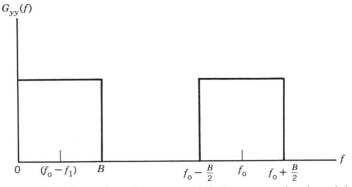

Figure 11.13   Illustration of frequency shift due to complex demodulation.

namely,

$$\text{original data:} \quad \Delta f \geq 2(f_0 + B/2)/N$$

$$\text{zoom data:} \quad \Delta f \geq 2B/N \tag{11.127}$$

Of course, the analysis must be repeated for individual zoom transforms of the data in all frequency ranges of interest.

In practice, the bandpass filtering and demodulation operations required to perform a zoom transform are sometimes accomplished using analog devices prior to the digitization of the data. This permits a relatively low-rate analog-to-digital converter (ADC) to be used to analyze data at frequencies well above the ADC rate. With modern high-speed ADC equipment (see Section 10.3.4), it is now common to sample the data at a rate appropriate for the original data $x(t)$, and then accomplish the zoom transform operations digitally. Specifically, the original record $x(n\,\Delta t)$, $n = 1, 2, \ldots, N$, is first bandpass filtered by digital filtering procedures outlined in Section 11.1.3 to obtain $y(n\,\Delta t)$, $n = 1, 2, \ldots, N$, in the frequency range $k_1/(N\Delta t) \leq f_k \leq k_2/(N\Delta t)$. The values of $y(n\,\Delta t)$ are next modulated to obtain

$$v(n\,\Delta t) = y(n\,\Delta t)\exp[-j2\pi nk_1/N] \qquad n = 1, 2, \ldots, N \tag{11.128}$$

The sample values $v(n\,\Delta t)$ can now be decimated by a ratio of

$$d = k_2/(k_2 - k_1) \tag{11.129}$$

to obtain $v(nd\,\Delta t)$, $n = 1, 2, \ldots, N$. Note that the required record length for each block is now $T = Nd\,\Delta t$ seconds. The Fourier transform of the decimated data then yields

$$Y\left(\frac{k}{Nd\,\Delta t}\right) = d\,\Delta t \sum_{n=0}^{N} v(nd\,\Delta t)\exp[-j2\pi kn/Nd] \qquad k = 0,1,\ldots,N/2$$

(11.130)

These operations are repeated over n_d independent records of $x(t)$, and the autospectrum is computed using Equation (11.102). The resulting spectrum will have a resolution of $\Delta f = 1/(Nd\,\Delta t)$ over the frequency range $k_1/(N\Delta t) \leq f_k \leq k_2/(N\Delta t)$, where $f_k = k/(Nd\,\Delta t)$, $k = dk_1$ to dk_2.

Example 11.4. **Parameter Selections for Zoom Transform.** Assume the autospectrum of stationary data $x(t)$ is to be computed with a normalized random error of $\varepsilon = 0.10$ and a resolution of $\Delta f = 1$ Hz over the frequency range 0 to 5120 Hz. Further assume the calculations are to be accomplished using a fixed block size of $N = 1024$ data points. Determine the total amount of data needed for the analysis.

To obtain a resolution of $\Delta f = 1$ Hz with $N = 1024$ data points, it follows from Equation (11.102) that the sampling interval must be

$$\Delta t = 1/(N\Delta f) = 0.000977$$

which provides a Nyquist frequency of

$$f_c = \frac{1}{2\,\Delta t} = 512 \text{ Hz}$$

Hence to analyze the data over the frequency range 0 to 5120 Hz, it will be necessary to zoom transform the data sequentially into 10 contiguous frequency bands of width $B = 512$ Hz. From Equation (8.154), to achieve an error of $\varepsilon = 0.10$, a total of $n_d = 100$ averages in each frequency band will be required. Hence, the total amount of data needed for the analysis in each frequency band is

$$T_r = n_d N\Delta t = 100 \text{ sec}$$

If the original time-history data are available in storage, then all 10 frequency bands can be analyzed with the 100 sec of data. If the analysis is being performed on line, however, then a total of 1000 sec of data would be needed to accomplish a sequential analysis in each of the 10 frequency bands.

11.5.5 *Other Spectral Analysis Procedures*

This chapter summarizes only conventional spectral analysis techniques that involve Fourier transforms that inherently produce spectral estimates with a minimum resolution of $\Delta f = 1/T$. Problems often arise in practice, where it is difficult or impossible to acquire a sufficiently long record to ensure the resolution needed for proper analysis. This is particularly true in the analysis of seismic, oceanographic, atmospheric turbulence, and some biomedical data. In recent years, a number of spectral estimation techniques based on modeling procedures have been developed in an effort to resolve this problem. Such techniques are often referred to as *parametric* spectral estimation procedures.

Essentially, these techniques develop a model for the process generating the data from the data itself. The model assumes that white noise is played through a linear system with a frequency response function $H(f)$ to produce the data. The frequency response function is of the form

$$H(f) = \frac{\sum\limits_{l=0}^{N} a_l z^l}{\sum\limits_{k=0}^{M} b_k z^k} \tag{11.131}$$

where $z = \exp(-j2\pi f \Delta t)$ is the z transform mentioned in Section 11.1.3. The model is then described by the difference equation

$$y_i = \sum_{l=0}^{N} a_l x_{i-l} + \sum_{k=1}^{M} b_k y_{i-k} \tag{11.132}$$

If the b_k are identically zero, the model is termed a *moving average* (MA). If all the a_l except a_0 are identically zero, then the model is called *autoregressive* (AR). If neither the a_l or b_k is zero, the model is referred to as ARMA. Determination of the type of model selected (AR, MA, or ARMA) must be made by the data analyst prior to attempting spectral estimation. This is usually done from some knowledge of the physical phenomenon being measured. If the wrong model is selected, the estimated spectra will be misleading and probably worse than those obtained by standard techniques.

After model selection, the model coefficients are derived by least squares techniques. The AR least squares technique is termed *maximum entropy spectral analysis* (MESA) because it was derived from the principle that the spectral estimate obtained must be the most random (have maximum entropy, in communication theory terms) of any autospectrum consistent with the measured data. An MA model gives rise to what is known as *maximum likelihood spectral analysis*, while no special name is given to the technique used in developing the ARMA model coefficients. The most difficult task in parametric spectral estimation is the order selection. This is a classic problem in least squares polynomial approximation. Tests have been devised to determine, in the statistical sense, an optimum model order. See References 11.11 and 11.12 for detailed discussions of these procedures.

11.6   JOINT RECORD FUNCTIONS

In the formulas to follow, it is assumed that two time-history records $u(t)$ and $v(t)$ are from stationary (ergodic) random processes and exist only for $t_0 \le t \le t_0 + T$, where t_0 is arbitrary and does not enter into later formulas because of the stationary assumption. Assume the sampling interval is Δt, which corresponds to a Nyquist frequency of $f_c = 1/(2\, \Delta t)$. Let the respective sample values of $u(t)$ and $v(t)$ be denoted

$$u_n = u(t_0 + n\, \Delta t) \qquad n = 1, 2, \ldots, N$$

$$v_n = v(t_0 + n\, \Delta t) \qquad T = N\Delta t$$
(11.133)

The first quantities to compute are the sample mean values given by

$$\bar{u} = \frac{1}{N} \sum_{n=1}^{N} u_n \qquad \bar{v} = \frac{1}{N} \sum_{n=1}^{N} v_n$$
(11.134)

The transformed data values can then be calculated by

$$x_n = u_n - \bar{u} \qquad y_n = v_n - \bar{v} \qquad n = 1, 2, \ldots, N$$
(11.135)

where $\bar{x} = 0$ and $\bar{y} = 0$. Various preliminary operations might also be performed, as summarized in Section 11.1.

11.6.1   *Joint Probability Density Functions*

It follows from Equations (3.25) and (8.87) that the joint probability density function of two stationary records $x(t)$ and $y(t)$ can be estimated from digitized data by

$$\hat{p}(x, y) = \frac{N_{x,y}}{N W_x W_y}$$
(11.136)

where W_x and W_y are narrow intervals centered on x and y, respectively, and $N_{x,y}$ is the number of pairs of data values that simultaneously fall within these intervals. Hence an estimate $\hat{p}(x, y)$ is obtained by dividing the full ranges of x and y into appropriate numbers of equal width class intervals forming two-dimensional rectangular cells, tabulating the number of data values in each cell, and dividing by the product of the cell area $W_x W_y$ and the sample size N. Computer procedures for sorting the data values into appropriate cells are similar to those outlined for probability density estimates in Section 11.3.

11.6.2 *Cross-Correlation Functions*

As for autocorrelation functions, there are two basic approaches to the estimation of cross-correlation functions, namely, the direct approach and the roundabout FFT approach. Procedures for both cases will now be discussed.

DIRECT PROCEDURES. Similar to the development presented in Section 11.4.1, unbiased estimates of the sample cross-correlation functions at lag numbers $r = 0, 1, 2, \ldots, m$ with $m < N$ are defined by

$$\hat{R}_{xy}(r\Delta t) = \frac{1}{N - r} \sum_{n=1}^{N-r} x_n y_{n+r} \tag{11.137}$$

$$\hat{R}_{yx}(r\Delta t) = \frac{1}{N - r} \sum_{n=1}^{N-r} y_n x_{n+r} \tag{11.138}$$

Note that the two cross-correlation functions $\hat{R}_{xy}(r\Delta t)$ and $\hat{R}_{yx}(r\Delta t)$ differ by the interchange of the x_n and y_n data values.

The sample cross-correlation function $\hat{R}_{xy}(r\Delta t)$ may be normalized to have values between plus and minus one through a division by $\sqrt{\hat{R}_{xx}(0)}\sqrt{\hat{R}_{yy}(0)}$. This defines a sample cross-correlation coefficient function

$$\hat{\rho}_{xy}(r\Delta t) = \frac{\hat{R}_{xy}(r\Delta t)}{\sqrt{\hat{R}_{xx}(0)}\sqrt{\hat{R}_{yy}(0)}} \qquad r = 0, 1, 2, \ldots, m \tag{11.139}$$

which theoretically should satisfy $-1 \le \hat{\rho}_{xy}(r\Delta t) \le 1$, as proved in Section 5.1.3. A similar formula exists for $\hat{\rho}_{yx}(r\Delta t)$.

VIA FAST FOURIER TRANSFORMS. Similar to the development outlined in Section 11.4.2, the cross-correlation function can also be computed via FFT procedures. The initial sample size for both $x(t)$ and $y(t)$ is assumed to be $N = 2^p$. For these computations, the cross-correlation function is obtained from the cross-spectral density function and involves two separate sets of FFTs, one for $x(t)$ and one for $y(t)$. These two sets of FFTs may be computed simultaneously by using the method in Section 11.2.4.

In summary, the following steps are recommended to compute the cross-correlation function via FFT procedures. The available sample size for the two records $x(n \Delta t)$ and $y(n \Delta t)$ is assumed to be $n_d N$, where $N = 2^p$.

1. Determine the maximum lag number m of interest and divide the available data records into n_d blocks, each consisting of $N \ge m$ data values.
2. Augment each block of N data values, $\{x_n\}$ and $\{y_n\}$, $n = 1, 2, \ldots, N$, with N zeros to obtain a new sequence of $2N$ data values.
3. Compute the $2N$-point FFT giving $Z(k)$ for $k = 0, 1, \ldots, 2N - 1$, using the FFT procedure of Equation (11.71).

4. Compute the $X(k)$ and $Y(k)$ values for $k = 0, 1, \ldots, 2N - 1$, using Equation (11.72).
5. Compute the two-sided cross-spectral density function estimate $\hat{S}_{xy}(f_k)$ for $k = 0, 1, \ldots, 2N - 1$, using the procedure in Section 11.6.3.
6. Compute the inverse FFT of $\hat{S}_{xy}(f_k)$ to obtain $\hat{R}^s_{xy}(r\Delta t)$ for $r = 0, 1, \ldots, 2N - 1$, using Equation (11.76).
7. Multiply $\hat{R}^s_{xy}(r\Delta t)$, $r = 0, 1, \ldots, (N - 1)$, by the scale factor $N/(N - r)$ to obtain the unbiased cross-correlation estimate $\hat{R}_{xy}(r\Delta t)$ for positive lag values.
8. Multiply $\hat{R}^s_{xy}(r\Delta t)$, $r = N + 1, N + 2, \ldots, 2N - 1$, by the scale factor $N/(r - N)$ to obtain the unbiased cross-correlation estimate $\hat{R}_{xy}(r\Delta t)$ for negative lag values.

The justifications for these various steps are similar to those discussed for autocorrelation analysis in Section 11.4.2.

11.6.3 Cross-Spectral Density Functions

Similar to the development for autospectral density functions in Section 11.5, the following steps are recommended to compute the cross-spectral density function.

1. Divide the available data records for x_n and y_n into n_d pairs of blocks, each consisting of N data values.
2. If needed to suppress side-lobe leakage, taper the data values in each pair of blocks, x_n and y_n, $n = 0, 1, \ldots, N - 1$, by the Hanning taper described in Section 11.5.2 or some other appropriate tapering function (the correlation function tapering described in Section 11.5.2 may be used as an alternative).
3. Store the tapered x_n values in the real part and the tapered y_n values in the imaginary part of $z_n = x_n + jy_n$, $n = 0, 1, \ldots, N - 1$.
4. Compute the N-point FFT for each block of data by Equation (11.71) giving $Z(k)$, $k = 0, 1, \ldots, N - 1$. If necessary to reduce the variance increase caused by tapering, compute the FFTs for overlapped records as described in Section 11.5.2.
5. Compute $X(k)$ and $Y(k)$, $k = 0, 1, \ldots, N - 1$, for each block of data using Equation (11.72).
6. Adjust the scale factor of $X(k)$ and $Y(k)$ for the loss due to tapering (for Hanning tapering, multiply by $\sqrt{\frac{8}{3}}$).
7. Compute the raw cross-spectral density estimate for each pair of blocks of data from $X^*(f_k) = \Delta t\, X^*(k)$ and $Y(f_k) = \Delta t\, Y(k)$ by

$$\tilde{S}_{xy}(f_k) = \frac{1}{N\Delta t} [X^*(f_k)Y(f_k)] \qquad k = 0, 1, \ldots, N - 1 \quad (11.140)$$

for a two-sided estimate, or

$$\tilde{G}_{xy}(f_k) = \frac{2}{N\Delta t}[X^*(f_k)Y(f_k)] \qquad k = 0, 1, \ldots, N/2 \quad (11.141)$$

for a one-sided estimate.
8. Average the raw cross-spectral density estimates from the n_d blocks of data to obtain the final smooth estimate of $\hat{S}_{xy}(f_k)$, $k = 0, 1, \ldots, N - 1$, or $\hat{G}_{xy}(f_k)$, $k = 0, 1, \ldots, N/2$. The smooth estimate

$$\hat{G}_{xy}(f_k) = \hat{C}_{xy}(f_k) - j\hat{Q}_{xy}(f_k) = |\hat{G}_{xy}(f_k)|e^{-j\theta_{xy}(f_k)}$$

11.6.4   *Frequency Response Functions*

For either single-input/single-output linear systems where extraneous noise is present only at the output or multiple-input/single-output linear systems where the inputs are uncorrelated, the recommended method for estimating the system frequency response function (including both gain and phase factors) is given by Equation (9.53), namely,

$$\hat{H}(f) = \frac{\hat{G}_{xy}(f)}{\hat{G}_{xx}(f)} = |\hat{H}(f)|e^{-j\phi(f)} \qquad (11.142)$$

It follows that

$$|\hat{H}(f)| = \frac{|\hat{G}_{xy}(f)|}{\hat{G}_{xx}(f)} \qquad \text{and} \qquad \hat{\phi}(f) = \hat{\theta}_{xy}(f) \qquad (11.143)$$

Hence, in terms of digital calculations at the discrete frequencies $f_k = k/(N\Delta t)$, $k = 0, 1, \ldots, N/2$, the gain factor and phase factor can be estimated by

$$|\hat{H}(f_k)| = \frac{[\hat{C}_{xy}^2(f_k) + \hat{Q}_{xy}^2(f_k)]^{1/2}}{\hat{G}_{xx}(f_k)} \qquad (11.144)$$

$$\hat{\phi}(f_k) = \tan^{-1}[\hat{Q}_{xy}(f_k)/\hat{C}_{xy}(f_k)] \qquad (11.145)$$

where $\hat{G}_{xx}(f_k)$ is the autospectrum estimate computed from x_n, $n = 0, 1, \ldots, N - 1$, as detailed in Section 11.5, and $\hat{C}_{xy}(f_k)$ and $\hat{Q}_{xy}(f_k)$ are the real and imaginary parts, respectively, of the cross-spectrum estimate $\hat{G}_{xy}(f_k)$ computed from x_n and y_n, $n = 0, 1, \ldots, N - 1$, as outlined in Section 11.6.3.

11.6.5 *Unit Impulse Response (Weighting) Functions*

The inverse Fourier transform of $\hat{H}(f)$ yields a circular biased estimate of the unit impulse response (weighting) function $\hat{h}(\tau)$ similar to the way the inverse Fourier transforms of $\hat{S}_{xx}(f)$ and $\hat{S}_{xy}(f)$ yield circular biased estimates of $\hat{R}_{xx}(\tau)$ and $\hat{R}_{xy}(\tau)$, respectively. If one desires to obtain $\hat{h}(\tau)$ from $\hat{H}(f)$, the two quantities $\hat{G}_{xx}(f)$ and $\hat{G}_{xy}(f)$ in Equation (11.142) should be replaced by two-sided quantities obtained by padding the original data with N zeros as described in Sections 11.4.2 and 11.6.2. The inverse Fourier transform of this new ratio will then yield $\hat{h}^s(r\Delta t)$ for values of $r = 0, 1, 2, \ldots, 2N - 1$, where the last half should be discarded. Multiplication by the scale factor $N/(N - r)$ gives the desired $\hat{h}(r\Delta t)$ for $r = 0, 1, 2, \ldots, N - 1$.

11.6.6 *Ordinary Coherence Functions*

From Equation (9.54), the ordinary coherence function $\gamma_{xy}^2(f)$ between two stationary records $x(t)$ and $y(t)$ is estimated by

$$\hat{\gamma}_{xy}^2(f_k) = \frac{|\hat{G}_{xy}(f)|^2}{\hat{G}_{xx}(f)\hat{G}_{yy}(f)} \tag{11.146}$$

where $\hat{G}_{xx}(f)$ and $\hat{G}_{yy}(f)$ are the estimated autospectral density functions of $x(t)$ and $y(t)$, respectively, and $\hat{G}_{xy}(f)$ is the estimated cross-spectral density function between $x(t)$ and $y(t)$. Hence, in terms of digital calculations at the discrete frequencies $f_k = k/(N\,\Delta t)$, $k = 0, 1, \ldots, N/2$, the ordinary coherence function is estimated by

$$\hat{\gamma}_{xy}^2(f_k) = \frac{|\hat{G}_{xy}(f_k)|^2}{\hat{G}_{xx}(f_k)\hat{G}_{yy}(f_k)} \tag{11.147}$$

where $\hat{G}_{xx}(f_k)$ and $\hat{G}_{yy}(f_k)$ are the autospectra estimates computed from x_n and y_n, $n = 0, 1, \ldots, N - 1$, as detailed in Section 11.5, and $\hat{G}_{xy}(f_k)$ is the cross-spectrum estimate computed from x_n and y_n, $n = 0, 1, \ldots, N - 1$, as outlined in Section 11.6.3.

11.7 MULTIPLE-INPUT/OUTPUT FUNCTIONS

Iterative algebraic procedures will now be listed to solve multiple-input/ single-output problems using the formulas derived in Section 7.3. These results are based on the algorithm in Section 7.3.4 that computes conditioned spectral density functions from the original spectral density functions. Multiple-input/multiple-output problems should be treated as combinations of these

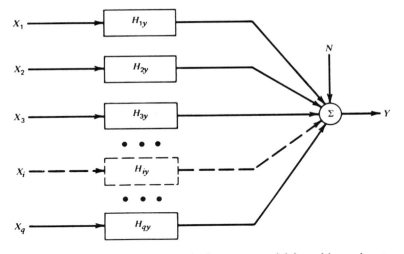

Figure 11.14   Multiple-input/single-output model for arbitrary inputs.

multiple-input/single-output problems by merely repeating the same procedures on each different desired output record.

Multiple-input/single-output models for arbitrary inputs and for an ordered
set of conditioned inputs are shown in Figures 11.14 and 11.15, respectively.
Wherever possible, the original collected arbitrary input records should be
ordered so as to agree with physical cause-and-effect conditions. Otherwise, as
noted in Section 7.2.4, a general rule is to order the input records based on

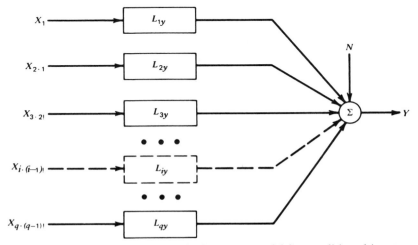

Figure 11.15   Multiple-input/single-output model for conditioned inputs.

their ordinary coherence functions between each input record and the output record. It is sufficient to do this ordering at selected frequencies of interest, usually corresponding to peaks in the output autospectrum.

11.7.1   *Fourier Transforms and Spectral Functions*

For every stationary random input record $x_i(t)$, $i = 1, 2, \ldots, q$, and for the stationary random output record $y(t) = x_{q+1}(t)$, divide their total record lengths T_r into n_d disjoint parts, each of length T, so that $T_r = n_d T$. Finite Fourier transforms should be computed for every subrecord of length T from all of the $(q + 1)$ input and output records at the discrete frequencies $f_k = k \, \Delta f$, $k = 1, 2, \ldots, (N/2)$, where $\Delta f = 1/(N \Delta t)$ is the frequency resolution of the analysis. Hence, one will obtain a grand total of $(q + 1)n_d$ different finite Fourier transforms, each of which is computed at $(N/2)$ different frequencies. This provides the basic information to compute estimates of autospectral and cross-spectral density functions based on n_d averages of similar quantities calculated at each of the $(N/2)$ different frequencies. One-sided spectral density functions will be denoted by

$$G_{ij}(f_k) = G_{x_i x_j}(f_k) \qquad i, j = 1, 2, \ldots, q, q + 1$$

$$G_{iy}(f_k) = G_{x_i y}(f_k) \qquad k = 1, 2, \ldots, (N/2)$$

(11.148)

The $G_{iy}(f_k)$ terms can be obtained by setting $j = q + 1 = y$. This gives the augmented $(q + 1) \times (q + 1)$ input/output measured spectral density matrix $\{G_{ij}\}$ that can be stored, as shown in Figure 11.16, at the successive frequencies f_k where $k = 1, 2, \ldots, (N/2)$. Note that this is a Hermitian matrix, where terms on the main diagonal are real-valued and terms off the main diagonal are complex conjugates of each other. This Hermitian property is true also for all of the conditioned spectral density matrices in the next section.

11.7.2   *Conditioned Spectral Density Functions*

The general algorithm derived in Section 7.3.4 to obtain conditioned spectral density functions by algebraic operations only is as follows. For any $j \geq i$ and any $r < j$, where $i = 1, 2, \ldots, q, q + 1$ and $r = 1, 2, \ldots, q$, and at any fixed frequency f_k,

$$G_{ij \cdot r!}(f_k) = G_{ij \cdot (r-1)!}(f_k) - L_{rj}(f_k) G_{ir \cdot (r-1)!}(f_k)$$

$$L_{rj}(f_k) = \frac{G_{rj \cdot (r-1)!}(f_k)}{G_{rr \cdot (r-1)!}(f_k)}$$

(11.149)

This algorithm yields results for the output $y(t)$ by letting $y(t) = x_{q+1}(t)$. To

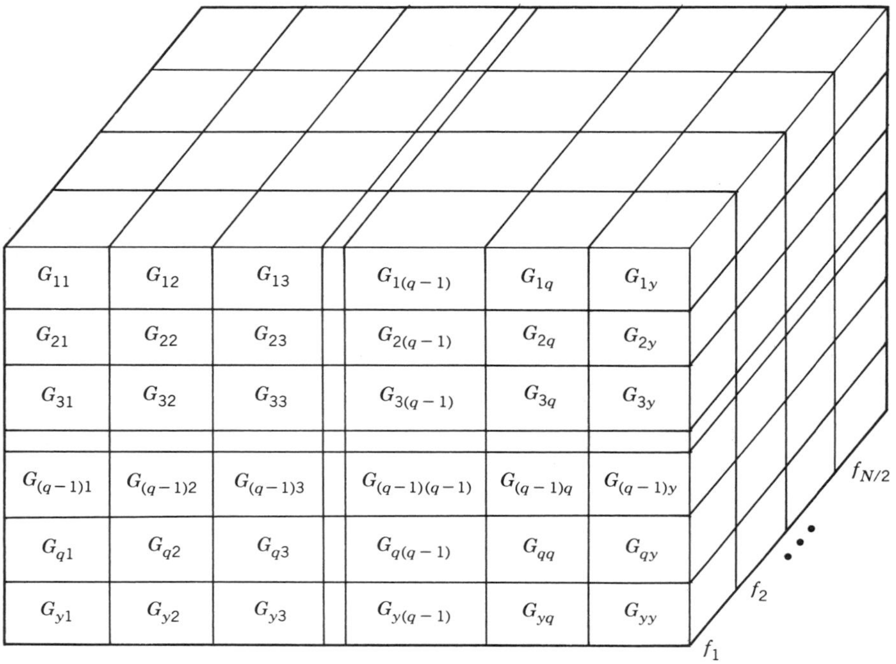

Figure 11.16   Augmented spectral matrix $\{G_{ij}\}$.

simplify the notation, the dependence on frequency will now be omitted.

Starting from $r = 1$, results are computed for successive terms from previous terms. For $r = 1$, the algorithm in Equation (11.149) gives at any frequency f_k the result

$$G_{ij\cdot 1} = G_{ij} - L_{1j}G_{i1}$$

$$L_{1j} = \frac{G_{1j}}{G_{11}} \qquad i, j = 2, 3, \ldots, q, q + 1 \qquad (11.150)$$

Note that the L_{1j} terms are known from the first row in Figure 11.16. Equation (11.150) gives the $\{G_{ij\cdot 1}\}$ terms from the original $\{G_{ij}\}$ terms. For $j = q + 1 = y$, one obtains

$$G_{iy\cdot 1} = G_{iy} - L_{1y}G_{i1}$$

$$L_{1y} = \frac{G_{1y}}{G_{11}} \qquad (11.151)$$

$G_{22 \cdot 1}$	$G_{23 \cdot 1}$	$G_{2(q-1) \cdot 1}$	$G_{2q \cdot 1}$	$G_{2y \cdot 1}$
$G_{32 \cdot 1}$	$G_{33 \cdot 1}$	$G_{3(q-1) \cdot 1}$	$G_{3q \cdot 1}$	$G_{3y \cdot 1}$
$G_{(q-1)2 \cdot 1}$	$G_{(q-1)3 \cdot 1}$	$G_{(q-1)(q-1) \cdot 1}$	$G_{(q-1)q \cdot 1}$	$G_{(q-1)y \cdot 1}$
$G_{q2 \cdot 1}$	$G_{q3 \cdot 1}$	$G_{q(q-1) \cdot 1}$	$G_{qq \cdot 1}$	$G_{qy \cdot 1}$
$G_{y2 \cdot 1}$	$G_{y3 \cdot 1}$	$G_{y(q-1) \cdot 1}$	$G_{yq \cdot 1}$	$G_{yy \cdot 1}$

Figure 11.17 Conditioned spectral matrix $\{G_{ij \cdot 1}\}$.

This yields the $q \times q$ conditioned spectral matrix $\{G_{ij \cdot 1}\}$ that can be stored at each frequency f_k as shown in Figure 11.17.

For $r = 2$, the algorithm gives at any frequency f_k the result

$$G_{ij \cdot 2!} = G_{ij \cdot 1} - L_{2j} G_{i2 \cdot 1}$$

$$L_{2j} = \frac{G_{2j \cdot 1}}{G_{22 \cdot 1}} \qquad i, j = 3, 4, \ldots, q, q + 1 \qquad (11.152)$$

Note that the L_{2j} terms are found from the first row in Figure 11.17. Equation (11.152) gives the $\{G_{ij \cdot 2!}\}$ terms from the $\{G_{ij \cdot 1}\}$ terms. For $j = q + 1 = y$, one obtains

$$G_{iy \cdot 2!} = G_{iy \cdot 1} - L_{2y} G_{i2 \cdot 1}$$

$$L_{2y} = \frac{G_{2y \cdot 1}}{G_{22 \cdot 1}} \qquad (11.153)$$

This yields the $(q - 1) \times (q - 1)$ conditioned spectral matrix $\{G_{ij \cdot 2!}\}$ that can be stored at each frequency f_k as shown in Figure 11.18.

$G_{33 \cdot 2!}$	$G_{3(q-1) \cdot 2!}$	$G_{3q \cdot 2!}$	$G_{3y \cdot 2!}$
$G_{(q-1)3 \cdot 2!}$	$G_{(q-1)(q-1) \cdot 2!}$	$G_{(q-1)q \cdot 2!}$	$G_{(q-1)y \cdot 2!}$
$G_{q3 \cdot 2!}$	$G_{q(q-1) \cdot 2!}$	$G_{qq \cdot 2!}$	$G_{qy \cdot 2!}$
$G_{y3 \cdot 2!}$	$G_{y(q-1) \cdot 2!}$	$G_{yq \cdot 2!}$	$G_{yy \cdot 2!}$

Figure 11.18 Conditioned spectral matrix $\{G_{ij \cdot 2!}\}$.

$G_{(q-1)(q-1)\cdot(q-2)!}$	$G_{(q-1)q\cdot(q-2)!}$	$G_{(q-1)y\cdot(q-2)!}$
$G_{q(q-1)\cdot(q-2)!}$	$G_{qq\cdot(q-2)!}$	$G_{qy\cdot(q-2)!}$
$G_{y(q-1)\cdot(q-2)!}$	$G_{yq\cdot(q-2)!}$	$G_{yy\cdot(q-2)!}$

Figure 11.19   Conditioned spectral matrix $\{G_{ij\cdot(q-2)!}\}$.

For $r = 3$, the algorithm gives at any frequency f_k the result

$$G_{ij\cdot 3!} = G_{ij\cdot 2!} - L_{3j}G_{i3\cdot 2!}$$

$$L_{3j} = \frac{G_{3j\cdot 2!}}{G_{33\cdot 2!}} \qquad i, j = 4, 5, \ldots, q, q+1 \quad (11.154)$$

Here the L_{3j} terms are known from the first row in Figure 11.18. Thus one obtains the $\{G_{ij\cdot 3!}\}$ terms from the $\{G_{ij\cdot 2!}\}$ terms, and so on.

This procedures continues until, after $(q - 2)$ steps, one obtains from Equation (11.149) the result

$$G_{ij\cdot(q-2)!} = G_{ij\cdot(q-3)!} - L_{(q-2)j}G_{i(q-2)\cdot(q-3)!}$$

$$L_{(q-2)j} = \frac{G_{(q-2)j\cdot(q-3)!}}{G_{(q-2)(q-2)\cdot(q-3)!}} \qquad i, j = q - 1, q, q+1$$

$$(11.155)$$

This gives the 3×3 conditioned spectral matrix $\{G_{ij\cdot(q-2)!}\}$ at each frequency f_k, as shown in Figure 11.19.

The $(q - 1)$th step from Equation (11.149) yields

$$G_{ij\cdot(q-1)!} = G_{ij\cdot(q-2)!} - L_{(q-1)j}G_{i(q-1)\cdot(q-2)!}$$

$$L_{(q-1)j} = \frac{G_{(q-1)j\cdot(q-2)!}}{G_{(q-1)(q-1)\cdot(q-2)!}} \qquad i, j = q, q+1 \quad (11.156)$$

Note that the $L_{(q-1)j}$ terms are found from the first row in Figure 11.19. Equation (11.156) gives the 2×2 conditioned spectral matrix $\{G_{ij\cdot(q-1)!}\}$ at each frequency f_k, as shown in Figure 11.20.

$G_{qq\cdot(q-1)!}$	$G_{qy\cdot(q-1)!}$
$G_{yq\cdot(q-1)!}$	$G_{yy\cdot(q-1)!}$

Figure 11.20   Conditioned spectral matrix $\{G_{ij\cdot(q-1)!}\}$.

$\boxed{G_{yy \cdot q!}}$ **Figure 11.21** Conditioned term $G_{yy \cdot q!}$

The final qth step from Equation (11.149) yields at each frequency f_k the single term in Figure 11.21, namely

$$G_{yy \cdot q!} = G_{yy \cdot (q-1)!} - L_{qy}G_{yq \cdot (q-1)!}$$

$$L_{qy} = \frac{G_{qy \cdot (q-1)!}}{G_{qq \cdot (q-1)!}} \tag{11.157}$$

Note that $G_{yy \cdot q!}$ is the output noise term G_{nn} in the q-input/single-output models of Figures 11.14 and 11.15. Thus, G_{nn} can be computed from the original measured data by following all of the steps outlined above, even though G_{nn} cannot be measured directly.

The information contained in Figures 11.16–11.21 represents the basis for identifying and interpreting various useful system properties and other relations from the measured multiple-input/output data. These matters are treated in succeeding parts of this section.

11.7.3 *Frequency Response and Coherence Functions*

Optimum frequency response functions $L_{ij} = L_{ij}(f_k)$ are computed at any frequency f_k by the formulas

$$L_{ij} = \frac{G_{ij \cdot (i-1)!}}{G_{ii \cdot (i-1)!}} \qquad i, j = 1, 2, \ldots, q, q+1 \tag{11.158}$$

When $j = q + 1 = y$, one obtains $L_{i(q+1)} = L_{iy}$ as shown in Figure 11.15, namely,

$$L_{iy} = \frac{G_{iy \cdot (i-1)!}}{G_{ii \cdot (i-1)!}} \qquad i = 1, 2, \ldots, q+1 \tag{11.159}$$

Special cases of Equation (11.158) show that

$$L_{1j} = \frac{G_{1j}}{G_{11}} \qquad j = 1, 2, \ldots, q+1 \tag{11.160}$$

$$L_{2j} = \frac{G_{2j \cdot 1}}{G_{22 \cdot 1}} \qquad j = 2, 3, \ldots, q+1 \tag{11.161}$$

$$L_{3j} = \frac{G_{3j \cdot 2!}}{G_{33 \cdot 2!}} \qquad j = 3, 4, \ldots, q+1 \tag{11.162}$$

and so on. Note that

$$L_{ii} = 1 \qquad \text{for all } i = 1, 2, \ldots, q + 1 \qquad (11.163)$$

Ordinary coherence functions $\gamma_{ij}^2 = \gamma_{ij}^2(f_k)$ are computed at any frequency f_k by the formulas

$$\gamma_{ij}^2 = \frac{|G_{ij}|^2}{G_{ii}G_{jj}} \qquad i, j = 1, 2, \ldots, q + 1 \qquad (11.164)$$

For cases where $i = 1$, one obtains

$$\gamma_{1j}^2 = \frac{|G_{1j}|^2}{G_{11}G_{jj}} \qquad j = 1, 2, \ldots, q + 1 \qquad (11.165)$$

When $j = q + 1 = y$, the equations given above become

$$\gamma_{iy}^2 = \frac{|G_{iy}|^2}{G_{ii}G_{yy}} \qquad i = 1, 2, \ldots, q + 1 \qquad (11.166)$$

In particular

$$\gamma_{1y}^2 = \frac{|G_{1y}|^2}{G_{11}G_{yy}} \qquad (11.167)$$

Note that

$$\gamma_{ii}^2 = 1 \qquad \text{for all } i \qquad (11.168)$$

Partial coherence functions $\gamma_{ij\cdot(i-1)!}^2 = \gamma_{ij\cdot(i-1)!}^2(f_k)$ are computed at any frequency f_k by the formulas

$$\gamma_{ij\cdot(i-1)!}^2 = \frac{|G_{ij\cdot(i-1)!}|^2}{G_{ii\cdot(i-1)!}G_{jj\cdot(i-1)!}} \qquad i = 1, 2, \ldots, q + 1 \qquad j \geq i \quad (11.169)$$

When $j = q + 1 = y$, one obtains

$$\gamma_{iy\cdot(i-1)!}^2 = \frac{|G_{iy\cdot(i-1)!}|^2}{G_{ii\cdot(i-1)!}G_{yy\cdot(i-1)!}} \qquad i = 1, 2, \ldots, q + 1 \qquad (11.170)$$

Note that $\gamma_{ii\cdot(i-1)!}^2 = 1$ for all i.

Multiple coherence functions $\gamma^2_{j:(j-1)!} = \gamma^2_{j:(j-1)!}(f_k)$ for $(j-1)$-input/ one-output models are computed at any frequency f_k by the formulas

$$\gamma^2_{j:(j-1)!} = 1 - \frac{G_{jj \cdot (j-1)!}}{G_{jj}} \qquad j = 1, 2, \ldots, q+1 \qquad (11.171)$$

When $j = q + 1 = y$, one obtains the multiple coherence function for the full q-input/one-output model given by

$$\gamma^2_{y:q!} = 1 - \frac{G_{yy \cdot q!}}{G_{yy}} \qquad (11.172)$$

11.7.4 *Coherent Output Spectral Functions*

Ordinary coherent output spectral functions $G_{j:1}$ state how much of the autospectrum of $x_j(t)$ is due to $x_1(t)$. These are computed at any frequency f_k by the formulas

$$G_{j:1} = \gamma^2_{1j}G_{jj} = |L_{1j}|^2 G_{11} \qquad j = 1, 2, \ldots, q+1 \qquad (11.173)$$

When $j = q + 1 = y$, Equation (11.173) becomes

$$G_{y:1} = \gamma^2_{1y}G_{yy} = |L_{1y}|^2 G_{11} \qquad (11.174)$$

Partial coherent output spectral functions $G_{j:2 \cdot 1}$ state how much of the autospectrum of $x_j(t)$ is due to $x_{2 \cdot 1}(t)$. These are computed at any frequency f_k by the formulas

$$G_{j:2 \cdot 1} = \gamma^2_{2j \cdot 1}G_{jj \cdot 1} = |L_{2j}|^2 G_{22 \cdot 1} \qquad j = 2, 3, \ldots, q+1 \qquad (11.175)$$

When $j = q + 1 = y$, Equation (11.175) reduces to

$$G_{y:2 \cdot 1} = \gamma^2_{2y \cdot 1}G_{yy \cdot 1} = |L_{2y}|^2 G_{22 \cdot 1} \qquad (11.176)$$

Partial coherent output spectral functions $G_{j:i \cdot (i-1)!}$ for $i = 2, 3, \ldots, q$ with $j \geq i$ state how much of the autospectrum of $x_j(t)$ is due to $x_i(t)$ when the linear effects of $x_1(t)$, $x_2(t)$, up to $x_{i-1}(t)$ are removed from $x_i(t)$. These are computed at any frequency f_k by the formulas

$$G_{j:i \cdot (i-1)!} = \gamma^2_{ij \cdot (i-1)!}G_{jj \cdot (i-1)!}$$

$$= |L_{ij}|^2 G_{ii \cdot (i-1)!} \qquad i = 2, 3, \ldots, q+1 \quad j \geq i \quad (11.177)$$

Note that Equations (11.173) and (11.175) are special cases of Equation (11.177) when $i = 1$ and $i = 2$, respectively.

When $j = q + 1 = y$, Equation (11.177) becomes

$$G_{y:i\cdot(i-1)!} = \gamma^2_{iy\cdot(i-1)!}G_{yy\cdot(i-1)!}$$

$$= |L_{iy}|^2 G_{ii\cdot(i-1)!} \qquad i = 2, 3, \ldots, q+1 \qquad (11.178)$$

For $i = q$, this shows that

$$G_{y:q\cdot(q-1)!} = \gamma^2_{qy\cdot(q-1)!}G_{yy\cdot(q-1)!} = |L_{qy}|^2 G_{qq\cdot(q-1)!} \qquad (11.179)$$

which defines how much of G_{yy} is due to the measured $x_q(t)$ when the linear effects of $x_1(t)$, $x_2(t)$, up to $x_{q-1}(t)$ are removed from $x_q(t)$.

Multiple coherent output spectral functions $G_{j:(j-1)!}$ state how much of the autospectrum of $x_j(t)$ is due to all of the preceding $x_1(t)$ to $x_{j-1}(t)$. These are computed at any frequency f_k by the formulas

$$G_{j:(j-1)!} = \gamma^2_{j:(j-1)!}G_{jj}$$

$$= G_{jj} - G_{jj\cdot(j-1)!} \qquad j = 2, 3, \ldots, q+1 \qquad (11.180)$$

When $j = q + 1 = y$, it follows that

$$G_{y:q!} = \gamma^2_{y:q!}G_{yy} = G_{yy} - G_{yy\cdot q!} \qquad (11.181)$$

This yields how much of G_{yy} is due to the q measured inputs $x_1(t)$, $x_2(t)$, up to $x_q(t)$.

11.7.5   *Decomposition of Measured Spectral Functions*

Any measured autospectral density function G_{jj} for $j = 1, 2, \ldots, q+1$ can be decomposed into physically meaningful quantities at any frequency f_k by the formulas

$$G_{jj} = \sum_{i=1}^{j} |L_{ij}|^2 G_{ii\cdot(i-1)!}$$

$$= \sum_{i=1}^{j-1} |L_{ij}|^2 G_{ii\cdot(i-1)!} + G_{jj\cdot(j-1)!} \qquad j = 1, 2, \ldots, q+1 \quad (11.182)$$

When $j = q + 1 = y$, Equation (11.182) becomes

$$G_{yy} = \sum_{i=1}^{q} |L_{iy}|^2 G_{ii\cdot(i-1)!} + G_{yy\cdot q!} \qquad (11.183)$$

The terms in G_{jj} and G_{yy} are the same as the partial coherent output spectral functions of Equations (11.177) and (11.178), respectively. For any value of j, Equation (11.182) decomposes the output spectrum as follows:

$$G_{jj} = G_{j:1} + G_{j:2\cdot1} + G_{j:3\cdot2!}$$

$$+ \cdots + G_{j:(j-1)\cdot(j-2)!} + G_{jj\cdot(j-1)!} \tag{11.184}$$

Similarly, Equation (11.183) is the same as

$$G_{yy} = G_{y:1} + G_{y:2\cdot1} + G_{y:3\cdot2!}$$

$$+ \cdots + G_{y:(q-1)\cdot(q-2)!} + G_{yy\cdot q!} \tag{11.185}$$

As desired for different applications, one can compute mean square values of the data contained within specified frequency ranges from f_a to f_b by integrating $G_{yy}(f)$ over those frequency ranges. Such mean square values will be more stable (i.e., have smaller random error) than individual frequency lines of $G_{yy}(f)$. For example, when $q = 3$, let

$$A_{yy} = \int_{f_a}^{f_b} G_{yy}(f)\, df \tag{11.186}$$

$$A_{y:1} = \int_{f_a}^{f_b} G_{y:1}(f)\, df \tag{11.187}$$

$$A_{y:2\cdot1} = \int_{f_a}^{f_b} G_{y:2\cdot1}(f)\, df \tag{11.188}$$

$$A_{y:3\cdot2!} = \int_{f_a}^{f_b} G_{y:3\cdot2!}(f)\, df \tag{11.189}$$

$$A_{yy\cdot3!} = \int_{f_a}^{f_b} G_{yy\cdot3!}(f)\, df \tag{11.190}$$

One now obtains

$$A_{yy} = A_{y:1} + A_{y:2\cdot1} + A_{y:3\cdot2!} + A_{yy\cdot3!} \tag{11.191}$$

In words, A_{yy} is the mean square value of $y(t)$ contained in the frequency range from f_a to f_b. The term $A_{y:1}$ is the part of A_{yy} due to $x_1(t)$; the term $A_{y:2\cdot1}$ is the part of A_{yy} due to $x_{2\cdot1}(t)$; the term $A_{y:3\cdot2!}$ is the part of A_{yy} due to $x_{3\cdot2!}(t)$; finally, the last term $A_{yy\cdot3!}$ is the part of A_{yy} that is not due to $x_1(t)$, $x_2(t)$, or $x_3(t)$.

***Example 11.5.* Three-Input / One-Output System.** Consider the special case of a three-input/one-output system. Start with the augmented spectral matrix corresponding to Figure 11.16 and then apply the algorithm of Equa-

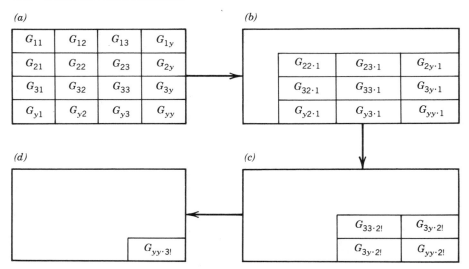

Figure 11.22 Conditioned results for three-input/one-output system.

tion (11.149) three times to obtain the conditioned results shown in Figure 11.22 at each frequency f_k of interest.

The $\{G_{ij \cdot 1}\}$ terms are computed by the formulas

$$G_{22 \cdot 1} = G_{22} - L_{12}G_{21} \qquad L_{12} = (G_{12}/G_{11})$$

$$G_{23 \cdot 1} = G_{23} - L_{13}G_{21} \qquad L_{13} = (G_{13}/G_{11})$$

$$G_{2y \cdot 1} = G_{2y} - L_{1y}G_{21} \qquad L_{1y} = (G_{1y}/G_{11})$$

$$G_{33 \cdot 1} = G_{33} - L_{13}G_{31}$$

$$G_{3y \cdot 1} = G_{3y} - L_{1y}G_{31}$$

$$G_{yy \cdot 1} = G_{yy} - L_{1y}G_{y1}$$

$$G_{32 \cdot 1} = G_{23 \cdot 1}^* \qquad G_{y2 \cdot 1} = G_{2y \cdot 1}^* \qquad G_{y3 \cdot 1} = G_{3y \cdot 1}^*$$

The $\{G_{ij \cdot 2!}\}$ terms are computed by the formulas

$$G_{33 \cdot 2!} = G_{33 \cdot 1} - L_{23}G_{32 \cdot 1} \qquad L_{23} = (G_{23 \cdot 1}/G_{22 \cdot 1})$$

$$G_{3y \cdot 2!} = G_{3y \cdot 1} - L_{2y}G_{32 \cdot 1} \qquad L_{2y} = (G_{2y \cdot 1}/G_{22 \cdot 1})$$

$$G_{yy \cdot 2!} = G_{yy \cdot 1} - L_{2y}G_{y2 \cdot 1}$$

$$G_{y3 \cdot 2!} = G_{3y \cdot 2!}^*$$

Finally, the $G_{yy\cdot3!}$ term is computed from the formula

$$G_{yy\cdot3!} = G_{yy\cdot2!} - L_{3y}G_{y3\cdot2!}$$

where

$$L_{3y} = \frac{G_{3y\cdot2!}}{G_{33\cdot2!}}$$

The following ordinary and partial coherence functions can now be computed at each frequency f_k of interest.

$$\gamma_{1y}^2 = \frac{|G_{1y}|^2}{G_{11}G_{yy}}$$

$$\gamma_{2y\cdot1}^2 = \frac{|G_{2y\cdot1}|^2}{G_{22\cdot1}G_{yy\cdot1}}$$

$$\gamma_{3y\cdot2!}^2 = \frac{|G_{3y\cdot2!}|^2}{G_{33\cdot2!}G_{yy\cdot2!}}$$

The multiple coherence function is given by

$$\gamma_{y:3!}^2 = 1 - \frac{G_{yy\cdot3!}}{G_{yy}}$$

Various coherent output spectral functions of interest are computed by the formulas

$$G_{y:1} = \gamma_{1y}^2 G_{yy} = \frac{|G_{1y}|^2}{G_{11}}$$

$$G_{y:2\cdot1} = \gamma_{2y\cdot1}^2 G_{yy\cdot1} = \frac{|G_{2y\cdot1}|^2}{G_{22\cdot1}}$$

$$G_{y:3\cdot2!} = \gamma_{3y\cdot2!}^2 G_{yy\cdot2!} = \frac{|G_{3y\cdot2!}|^2}{G_{33\cdot2!}}$$

The total $G_{yy} = G_{yy}(f_k)$ is decomposed here into the sum of four terms as follows.

$$G_{yy} = G_{y:1} + G_{y:2\cdot1} + G_{y:3\cdot2!} + G_{yy\cdot3!}$$

When $y = x_4$, the result for G_{yy} can be expressed as

$$G_{44} = G_{4:1} + G_{4:2\cdot1} + G_{4:3\cdot2!} + G_{44\cdot3!}$$

Earlier relations have also been computed involving a one-input/one-output system between x_1 as the input and x_2 as the output, where

$$G_{22} = G_{2:1} + G_{22\cdot1} = \gamma_{12}^2 G_{22} + G_{22\cdot1}$$

and a two-input/one-output system with x_1 and x_2 as inputs and x_3 as the output where

$$G_{33} = G_{3:1} + G_{3:2\cdot1} + G_{33\cdot2!}$$

$$= \gamma_{13}^2 G_{33} + \gamma_{23\cdot1}^2 G_{33\cdot1} + G_{33\cdot2!}$$

The sum of the computed terms on the right-hand sides of the last three equations should equal the original measured autospectral density functions on the left-hand sides of these equations to provide an overall check on the work.

PROBLEMS

11.1 A set of sample values $\{u_n\}$, $n = 1, 2, \ldots, N$, have a standard deviation of s_u. Determine the standard deviation of the transformed data values $\{x_n\} = \{u_n - \bar{u}\}$, $n = 1, 2, \ldots, N$.

11.2 Assume a set of sample values $\{u_n\}$, $n = 1, 2, \ldots, N$, are to be detrended using a polynomial of degree $K = 2$. Determine the coefficients (b_0, b_1, b_2) required for a least squares fit of the polynomial to the sample data.

11.3 Determine the equation for a first-order IIR (recursive) filter that will behave like a high pass RC filter.

11.4 Consider a sequence of $N = 16,384$ data points that are complex numbers. How many real operations would be required to Fourier transform the data values
    (a) using conventional calculations.
    (b) using Cooley–Tukey FFT procedures.

11.5 Assume an autocorrelation function is estimated from a sample record $x(t)$ of stationary random data where the autocorrelation function can be approximated by

$$\hat{R}_{xx}(\tau) = e^{-2|\tau|} \cos 18.85\tau$$

If the sample record is $T = 1$ sec long, what is the equation for the circular correlation function that would be obtained by computing the inverse Fourier transform of an autospectrum estimate without zero padding?

11.6 A sequence of $N_r = 8192$ sample values of a stationary random signal are available to compute an autocorrelation function estimate. It is desired to compute the autocorrelation estimate by the indirect FFT approach to a maximum lag value of $m = 256$. This could be accomplished by using a single block of $N = 8192$ data points versus $n_d = 32$ blocks of $N = 256$ data points. Which approach would require fewer calculations?

11.7 In computing an autospectral density function estimate by FFT procedures, suppose a record of total length $T_r = 4$ seconds is digitized at a sampling rate of 4096 sps. If a resolution of $\Delta f = 16$ Hz is desired, determine
(a) the approximate number of real operations required to perform the calculations.
(b) the normalized random error of the resulting estimate.

11.8 The Fourier coefficients of a record $x(t) = A \sin 100\pi t$ are calculated from a sample record of length $T = 1.015$ seconds. Assuming a very high sampling rate, determine the magnitude of the Fourier coefficient with the largest value that would be calculated using
(a) a rectangular time window.
(b) a Hanning time window.

11.9 Assume a cross-correlation function is to be estimated from two sample records, each digitized into a sequence of $N_r = 4096$ data points. Further assume the estimate is desired for lag values out to $m = 256$. Determine the speed ratio between the direct computational approach and the indirect FFT approach.

11.10 The *noise bandwidth* of a spectral window is defined as the bandwidth of an ideal rectangular bandpass filter that would produce the same output mean square value as the spectral window when analyzing white noise. Determine the noise bandwidth of the spectral window produced by a rectangular (untapered) time window of length T.

REFERENCES

11.1 Hamming, R. W., *Digital Filters*, 2nd ed., Prentice-Hall, Englewood Cliffs, New Jersey, 1983.

11.2 Rabiner, L. R., and Gold, B., *Theory and Applications of Digital Signal Processing*, Prentice-Hall, Englewood Cliffs, New Jersey, 1975.

11.3 Oppenheim, A. V., and Schafer, R. W., *Digital Signal Processing*, Prentice-Hall, Englewood

Cliffs, New Jersey, 1975.

11.4   Otnes, R. K., and Enockson, L., *Applied Time Series Analysis*, Vol. 1, Wiley, New York, 1978.

11.5   Brigham, E. O., *The Fast Fourier Transform*, Prentice-Hall, Englewood Cliffs, New Jersey, 1974.

11.6   Cooley, J. W., and Tukey, J. W., "An Algorithm for the Machine Calculation of Complex Fourier Series," *Mathematics of Computation*, Vol. 19, p 297, April 1965.

11.7   Winograd, S., "On Computing the Discrete Fourier Transform," *Mathematics of Computation*, Vol. 32, p. 175, January 1978.

11.8   Harris, F. J., "On the Use of Windows for Harmonic Analysis with the Discrete Fourier Transform," *Proceedings of the IEEE*, Vol. 66, p. 51, January 1978.

11.9   Nuttall, A. H., "Some Windows with Very Good Side Lobe Behavior," *IEEE Transactions on Acoustics, Speech, and Signal Processing*, Vol. ASSP-29, p. 84, February 1981.

11.10 Nuttall, A. H., and Carter, G. C., "Spectral Estimation Using Combined Time and Lag Weighting," *Proceedings of the IEEE*, Vol. 70, p. 1115, September 1982.

11.11 Childers, D. G., *Modern Spectral Analysis*, IEEE Press, New York, 1978.

11.12 Robinson, E. A., and Treitel, S., *Geophysical Signal Analysis*, Prentice-Hall, Englewood Cliffs, New Jersey, 1980.

CHAPTER 12

NONSTATIONARY DATA
ANALYSIS

The material presented in previous chapters has been restricted largely to the measurement and analysis of stationary random data, that is, data with statistical properties that are invariant with translations in time (or any other independent variable of the data). The theoretical ideas, error formulas, and processing techniques do not generally apply when the data are nonstationary. Special considerations are required in these cases. Such considerations are the subject of this chapter.

12.1 CLASSES OF NONSTATIONARY DATA

Much of the random data of interest in practice is nonstationary when viewed as a whole. Nevertheless, it is often possible to force the data to be at least piecewise stationary for measurement and analysis purposes. To repeat the example from Section 10.4.1, the pressure fluctuations in the turbulent boundary layer generated by a high-speed aircraft during a typical mission will generally be nonstationary since they depend on the airspeed and altitude, which vary during the mission. However, one can easily fly the aircraft under a specific set of fixed flight conditions so as to produce stationary boundary layer pressures for measurement purposes. The flight conditions can then be changed sequentially to other specific sets of fixed conditions, producing stationary data for measurement purposes until the entire mission environment has been represented in adequate detail by piecewise stationary segments. Such procedures for generating stationary data to represent a generally nonstationary phenomenon are commonly used and are strongly recommended to avoid the need for nonstationary data analysis procedures.

There are a number of situations where the approach to data collection and analysis described above is not feasible, and individual sample records of data

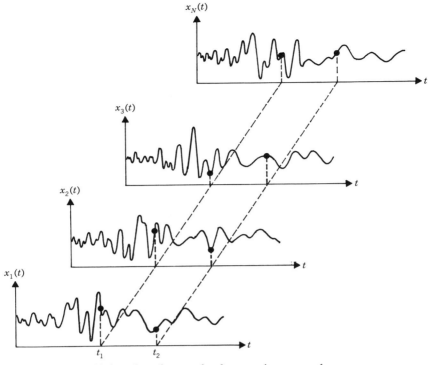

Figure 12.1   Sample records of nonstationary random process.

must be analyzed as nonstationary data. From a purely computational view-point, the most desirable situation is that in which an experiment producing the nonstationary data of interest can be repeated under statistically similar conditions. This allows an ensemble of sample records to be measured on a common time base, as illustrated in Figure 12.1. A more common situation, however, is that in which the nonstationary phenomenon of interest is unique and cannot be reproduced under statistically similar conditions. Examples include nonstationary ocean waves, atmospheric turbulence, and economic time-series data. The basic factors producing such data are too complex to allow the performance of repeated experiments under similar conditions. The analysis of data in these cases must be accomplished by calculations on single sample records.

An appropriate general methodology does not exist for analyzing the properties of all types of nonstationary random data from individual sample records. This is due partly to the fact that a nonstationary conclusion is a negative statement specifying only a lack of stationary properties, rather than a positive statement defining the precise nature of the nonstationarity. It follows that special techniques must be developed for nonstationary data that apply

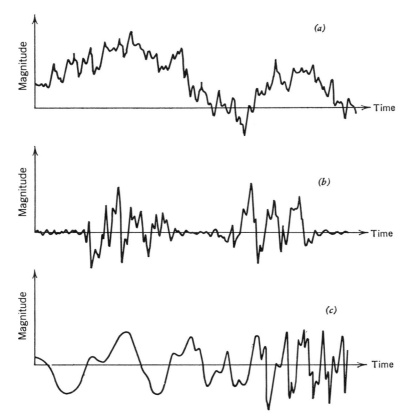

Figure 12.2 Examples of nonstationary data. (*a*) Time-varying mean value. (*b*) Time-varying mean square value. (*c*) Time-varying frequency structure.

only to limited classes of these data. The usual approach is to hypothesize a specific model for each class of nonstationary data of interest that consists of deterministic factors operating on an otherwise stationary random process. Three examples are shown in Figure 12.2. These nonstationary time-history records are constructed from

$$\text{(a)} \quad x(t) = a(t) + u(t)$$

$$\text{(b)} \quad x(t) = a(t)u(t) \tag{12.1}$$

$$\text{(c)} \quad x(t) = u(t^n)$$

where $u(t)$ is a sample record of a stationary random process $\{u(t)\}$ and $a(t)$ is a deterministic function that is repeated exactly on each record. Such elementary nonstationary models can be combined or extended to generate more complex models as required to fit various physical situations.

12.2   PROBABILITY STRUCTURE OF NONSTATIONARY DATA

For a nonstationary random process $\{x(t)\}$ as illustrated in Figure 12.1, statistical properties over the ensemble at any time t are not invariant with respect to translations in t. Hence at any value of $t = t_1$, the probability structure of the random variable $x(t_1)$ would be a function of t_1. To be precise, the *nonstationary probability density function* $p(x, t_1)$ of $\{x(t_1)\}$ is defined by

$$p(x, t_1) = \lim_{\Delta x \to 0} \frac{\text{Prob}\left[x < x(t_1) \le x + \Delta x\right]}{\Delta x} \tag{12.2}$$

and has the following basic properties for any t:

$$1 = \int_{-\infty}^{\infty} p(x, t)\, dx$$

$$\mu_x(t) = E[x(t)] = \int_{-\infty}^{\infty} xp(x, t)\, dx \tag{12.3}$$

$$\psi_x^2(t) = E[x^2(t)] = \int_{-\infty}^{\infty} x^2 p(x, t)\, dx$$

$$\sigma_x^2(t) = E\left[\{x(t) - \mu_x(t)\}^2\right] = \psi_x^2(t) - \mu_x^2(t)$$

These formulas also apply to stationary cases where $p(x, t) = p(x)$, independent of t. The *nonstationary probability distribution function* $P(x, t_1)$ is defined by

$$P(x, t_1) = \text{Prob}\left[-\infty < x(t_1) \le x\right] \tag{12.4}$$

Similar relationships exist for $P(x, t_1)$ as for the probability distribution function $P(x)$ discussed in Chapter 3.

If the nonstationary random process $\{x(t)\}$ is Gaussian at $t = t_1$, then $p(x, t_1)$ takes the special form

$$p(x, t_1) = \left[\sigma_x(t_1)\sqrt{2\pi}\right]^{-1} \exp\left\{\frac{-[x - \mu_x(t_1)]^2}{2\sigma_x^2(t_1)}\right\} \tag{12.5}$$

which is completely determined by the nonstationary mean and mean square values of $x(t)$ at $t = t_1$. This result indicates that the measurement of these two quantities may be quite significant in many nonstationary applications just as in previous stationary applications.

12.2.1 *Higher-Order Probability Functions*

For a pair of times t_1 and t_2, the *second-order nonstationary probability density function* of $x(t_1)$ and $x(t_2)$ is defined by

$$p(x_1, t_1; x_2, t_2)$$

$$= \lim_{\substack{\Delta x_1 \to 0 \\ \Delta x_2 \to 0}} \frac{\text{Prob}[x_1 < x(t_1) \le x_1 + \Delta x_1 \text{ and } x_2 < x(t_2) \le x_2 + \Delta x_2]}{(\Delta x_1)(\Delta x_2)}$$

$$(12.6)$$

and has the following basic properties for any t_1, t_2:

$$1 = \iint_{-\infty}^{\infty} p(x_1, t_1; x_2, t_2)\, dx_1\, dx_2$$

$$p(x_1, t_1) = \int_{-\infty}^{\infty} p(x_1, t_1; x_2, t_2)\, dx_2$$

$$(12.7)$$

$$p(x_2, t_2) = \int_{-\infty}^{\infty} p(x_1, t_1; x_2, t_2)\, dx_1$$

$$R_{xx}(t_1, t_2) = E[x(t_1)x(t_2)] = \iint_{-\infty}^{\infty} x_1 x_2 p(x_1, t_1; x_2, t_2)\, dx_1\, dx_2$$

For stationary cases, $p(x_1, t_1; x_2, t_2) = p(x_1, 0; x_2, t_2 - t_1)$. *Second order nonstationary probability distribution functions* may be defined analogous to Equation (12.4) by the quantity

$$P(x_1, t_1; x_2, t_2) = \text{Prob}[-\infty < x(t_1) \le x_1 \text{ and } -\infty < x(t_2) \le x_2] \quad (12.8)$$

Continuing in this way, higher-order nonstationary probability distribution and density functions may be defined that describe the nonstationary random process $\{x(t)\}$ in greater and greater detail. This procedure supplies a rigorous characterization of the nonstationary random process $\{x(t)\}$.

Consider next two different nonstationary random processes $\{x(t)\}$ and $\{y(t)\}$. For $x(t_1)$ and $y(t_2)$, the *joint (second-order) nonstationary probability density function* is defined by

$$p(x, t_1; y, t_2) = \lim_{\substack{\Delta x \to 0 \\ \Delta y \to 0}} \frac{\text{Prob}[x < x(t_1) \le x + \Delta x \text{ and } y < y(t_2) \le y + \Delta y]}{(\Delta x)(\Delta y)}$$

$$(12.9)$$

and has basic properties similar to Equation (12.7). In particular, the nonstationary cross-correlation function, which is discussed in Section 12.5, satisfies the relation

$$R_{xy}(t_1, t_2) = E[x(t_1)y(t_2)] = \iint_{-\infty}^{\infty} xyp(x, t_1; y, t_2)\, dx\, dy \quad (12.10)$$

For stationary cases, $p(x, t_1; y, t_2) = p(x, 0; y, t_2 - t_1)$.

The measurement of nonstationary probability density functions can be a formidable task. Even for the first-order density function defined in Equation (12.2), all possible combinations of x and t_1 must be considered. This will require the analysis of a large collection of sample records. If a Gaussian assumption can be made, Equation (12.5) reduces the problem of measuring $p(x, t_1)$ to measuring $\mu_x(t_1)$ and $\sigma_x^2(t_1)$, which is a much simpler undertaking. Nevertheless, ensemble averaging of a collection of sample records is still generally required, as discussed in Sections 12.3 and 12.4.

12.2.2   Time-Averaged Probability Functions

It often occurs in practice that only one or a very few sample records of data are available for a nonstationary random process of interest. There may be a strong temptation in such cases to analyze the data by time-averaging procedures as would be appropriate if the data were a sample record from a stationary (ergodic) random process. For some nonstationary data parameters, time-averaging analysis procedures can produce meaningful results in certain special cases, as will be discussed later. For the case of probability density functions, however, time-averaging procedures will generally produce severely distorted results. In particular, the probability density function computed by time-averaging data with a nonstationary mean square value will tend to exaggerate the probability density of low- and high-amplitude values at the expense of intermediate values, as demonstrated in the illustration to follow.

***Example 12.1.* Illustration of Time-Averaged Probability Density Function.** Assume a sample record of data consists of a normally distributed stationary time history with zero mean and variance σ_1^2 over the first half of the record, and a second normally distributed stationary time history with zero mean and variance $\sigma_2^2 > \sigma_1^2$ over the second half of the record. Then the time history $x(t)$ is given by

$$x(t) = \begin{cases} x_1(t) & 0 \le t \le T/2 \\ x_2(t) & T/2 \le t \le T \end{cases}$$

and the probability density function of $x(t)$ is given by

$$p(x, t) = \begin{cases} \dfrac{1}{\sigma_1\sqrt{2\pi}} e^{-x^2/2\sigma_1^2} & 0 \le t \le T/2 \\[2mm] \dfrac{1}{\sigma_2\sqrt{2\pi}} e^{-x^2/2\sigma_2^2} & T/2 \le t \le T \end{cases}$$

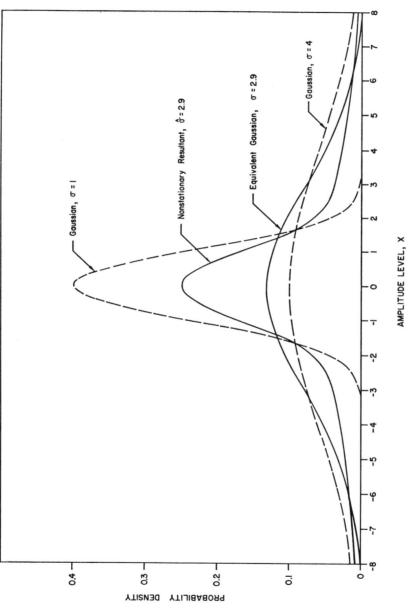

AMPLITUDE LEVEL, X

Gaussian, σ = 1

Nonstationary Resultant, σ̂ = 2.9

Equivalent Gaussian, σ = 2.9

Gaussian, σ = 4

PROBABILITY DENSITY

Figure 12.3   Illustration of probability density function of nonstationary data.

Now, if this lack of stationarity is ignored in the computation of a probability density function for $x(t)$, $0 \le t \le T$, the resulting density calculated at any level x will simply be the average of the densities for the two halves of the record at level x. That is,

$$\hat{p}(x) = \frac{1}{2\sqrt{2\pi}} \left[\frac{1}{\sigma_1} e^{-x^2/2\sigma_1^2} + \frac{1}{\sigma_2} e^{-x^2/2\sigma_2^2} \right]$$

For example, let $\sigma_1^2 = 1$ and $\sigma_2^2 = 16$. The nonstationary resultant probability density function that would be obtained for this case is computed and illustrated graphically in Figure 12.3. Observe that $\hat{\sigma} = 2.9$ in $\hat{p}(x)$ since $\hat{\sigma}^2 = \frac{1}{2}(\sigma_1^2 + \sigma_2^2) = 8.5$. The equivalent Gaussian probability density function for $\sigma = 2.9$ is shown also in Figure 12.3.

12.3   NONSTATIONARY MEAN VALUES

Consider the problem of estimating the time-varying mean value of nonstationary data. Given a collection of sample records $x_i(t)$, $0 \le t \le T$, $i = 1, 2, \ldots, N$, from a nonstationary process $\{x(t)\}$, the mean value at any time t is estimated by the ensemble average

$$\hat{\mu}_x(t) = \frac{1}{N} \sum_{i=1}^{N} x_i(t) \tag{12.11}$$

The estimate $\hat{\mu}_x(t)$ will differ over different choices of the N samples $\{x_i(t)\}$. Consequently, one must investigate for every t how closely an arbitrary estimate will approximate the true mean value. The expected value of $\hat{\mu}_x(t)$ is given by

$$E[\hat{\mu}_x(t)] = \frac{1}{N} \sum_{i=1}^{N} E[x_i(t)] = \mu_x(t) \tag{12.12}$$

where

$$\mu_x(t) = E[x_i(t)] \tag{12.13}$$

is the true mean value of the nonstationary process at time t. Hence $\hat{\mu}_x(t)$ is an *unbiased* estimate of $\mu_x(t)$ for all t, independent of N. The variance of the estimate $\hat{\mu}_x(t)$ is given by

$$\text{Var}[\hat{\mu}_x(t)] = E[\{\hat{\mu}_x(t) - \mu_x(t)\}^2] \tag{12.14}$$

Mean values of nonstationary random processes can be estimated using a special purpose instrument or a digital computer, as illustrated in Figure 12.4.

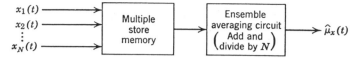

Figure 12.4   Procedure for nonstationary mean value measurement.

Two main steps are involved in the measurement. The first step is to obtain and store each record $x_i(t)$ as a function of t. This may be done continuously for all t in the range $0 \leq t \leq T$ or discretely by some digitizing procedure. After this has been done for N records, the next step is to perform an ensemble average by adding the records together and dividing by N. If each $x_i(t)$ is digitized in, say, M steps, then the total number of stored values would be MN.

12.3.1   *Independent Samples*

In most practical applications, the N sample functions used to compute $\hat{\mu}_x(t)$ will be statistically independent. Hence independence will be assumed here. Upon expanding Equation (12.14), as in the derivation of Equation (4.9), it is seen that the sample variance at time t is given by

$$\text{Var}[\hat{\mu}_x(t)] = \frac{\sigma_x^2(t)}{N} \tag{12.15}$$

where $\sigma_x^2(t)$ is the variance associated with the underlying nonstationary process $\{x(t)\}$. Thus the sample variance approaches zero as N approaches infinity, so that $\hat{\mu}_x(t)$ is a *consistent* estimate of $\mu_x(t)$ for all t.

Confidence intervals for the nonstationary mean value $\mu_x(t)$ can be constructed based on the estimate $\hat{\mu}_x(t)$ using the procedures detailed in Section 4.4. Specifically, the $(1 - \alpha)$ confidence interval at any time t is

$$\left[\hat{\mu}_x(t) - \frac{\hat{\sigma}_x(t)t_{n;\,\alpha/2}}{\sqrt{N}} \leq \mu_x(t) < \hat{\mu}_x(t) + \frac{\hat{\sigma}_x(t)t_{n;\,\alpha/2}}{\sqrt{N}} \right] \tag{12.16}$$

where $\hat{\sigma}_x(t)$ is an unbiased estimate of the standard deviation of $\{x(t)\}$ at time t given by Equation (4.12) as

$$\hat{\sigma}_x(t) = \left\{ \frac{1}{N-1} \sum_{i=1}^{N} [x_i(t) - \hat{\mu}_x(t)]^2 \right\}^{1/2} \tag{12.17}$$

and $t_{n;\,\alpha/2}$ is the $\alpha/2$ percentage point of Student's t variable with $n = N - 1$ degrees of freedom defined in Section 4.2.3. Note that Equation (12.16) applies even when $\{x(t)\}$ is not normally distributed, assuming the sample size is greater than, say, $N = 10$. This follows from the central limit theorem in Section 3.3.1, which applies to the nonstationary mean value calculation in Equation (12.11).

12.3.2 Correlated Samples

Consider a general situation where sample functions $x_i(t)$, $0 \leq t \leq T$, $i = 1, 2, \ldots, N$, from a nonstationary random process are correlated such that, for every t,

$$E[x_i(t)x_j(t)] = R_{xx}(k, t) \quad \text{where } k = j - i \quad (12.18)$$

The quantity $R_{xx}(k, t)$ is called a *nonstationary spatial cross-correlation function* at time t between all pairs of records $x_i(t)$ and $x_j(t)$ satisfying $k = j - i$. It follows from the definition of Equation (12.18) that, by interchanging i and j,

$$R_{xx}(-k, t) = R_{xx}(k, t) \quad (12.19)$$

When the sample functions $x_i(t)$ and $x_j(t)$ are independent, for $i \neq j$ corresponding to $k \neq 0$,

$$R_{xx}(k, t) = E[x_i(t)x_j(t)] = E[x_i(t)]E[x_j(t)] = \mu_x^2(t) \quad \text{for } k \neq 0$$
$$(12.20)$$

At $k = 0$, Equation (12.18) becomes

$$R_{xx}(0, t) = E[x_i^2(t)] = \sigma_x^2(t) + \mu_x^2(t) \quad (12.21)$$

These relations yield the independent sample case in the preceding section. For correlated samples, Equation (12.15) now takes the general form

$$\text{Var}[\hat{\mu}_x(t)] = \frac{\sigma_x^2(t)}{N} + \frac{1}{N^2} \sum_{\substack{i, j = 1 \\ i \neq j}}^{N} E\{[x_i(t) - \mu_x(t)][x_j(t) - \mu_x(t)]\}$$

$$= \frac{\sigma_x^2(t)}{N} + \frac{1}{N^2} \sum_{\substack{i, j = 1 \\ i \neq j}}^{N} [R_{xx}(j - i, t) - \mu_x^2(t)] \quad (12.22)$$

The next problem is to simplify the double sum appearing in Equation (12.22). The index $k = j - i$ takes on values $k = 1, 2, \ldots, N - 1$. Altogether, there are $N^2 - N$ terms. Since $R_{xx}(-k, t) = R_{xx}(k, t)$, the $N^2 - N$ terms in this double sum can be arranged so that there are two terms where $k = N - 1$

of the form $R_{xx}(N-1, t)$, four terms where $k = N-2$ of the form $R_{xx}(N-2, t), \ldots$, and $2(N-1)$ terms where $k = 1$ of the form $R_{xx}(1, t)$. Thus one derives the simplified expression

$$\sum_{\substack{i, j=1 \\ i \neq j}}^{N} R_{xx}(j - i, t) = 2 \sum_{k=1}^{N-1} (N - k) R_{xx}(k, t) \tag{12.23}$$

As a check, note that the sum

$$2 \sum_{k=1}^{N-1} (N - k) = N^2 - N \tag{12.24}$$

Substitution of Equation (12.23) into Equation (12.22) now yields

$$\text{Var}[\hat{\mu}_x(t)] = \frac{\sigma_x^2(t)}{N} + \frac{2}{N^2} \sum_{k=1}^{N-1} (N - k)[R_{xx}(k, t) - \mu_x^2(t)] \tag{12.25}$$

Equation (12.20) shows that Equation (12.25) reduces to Equation (12.15) when the records are independent, providing another check on the validity of Equation (12.25). The result of Equation (12.25) is an important extension of Equation (12.15) and should be used in place of Equation (12.15) for correlated samples.

A special situation of *complete dependence* between all samples is worthy of mention. For this case,

$$R_{xx}(k, t) = R_{xx}(0, t) = \sigma_x^2(t) + \mu_x^2(t) \qquad \text{for all } k \tag{12.26}$$

Equation (12.25) now becomes

$$\text{Var}[\hat{\mu}_x(t)] = \frac{\sigma_x^2(t)}{N} + \frac{1}{N^2}(N^2 - N)\sigma_x^2(t) = \sigma_x^2(t) \tag{12.27}$$

Thus no reduction in variance occurs when the samples are completely dependent.

For physical situations where a partial correlation may exist between the different samples, the following example may be helpful in giving quantitative results.

***Example 12.2.* Variance of Mean Value Estimate for Exponential Correlation Between Samples**.   An exponential form for the nonstationary cross-correlation function $R_{xx}(k, t)$ will now be assumed so as to obtain some quantitative results to characterize different degrees of correlation. To be specific, assume that

$$R_{xx}(k, t) = \mu_x^2(t) + \sigma_x^2(t)e^{-kc}$$

where k and c are positive constants. Determine the corresponding sample variance for nonstationary mean value estimates.

From Equation (12.25), the sample variance is given by

$$\text{Var}[\hat{\mu}_x(t)] = \frac{\sigma_x^2(t)}{N} + \frac{2\sigma_x^2(t)}{N^2} \sum_{k=1}^{N-1} (N-k)e^{-kc}$$

To evaluate the sum above, let

$$f(c) = \sum_{k=1}^{N-1} e^{-kc} = \frac{1 - e^{-(N-1)c}}{e^c - 1}$$

Then

$$f'(c) = -\sum_{k=1}^{N-1} ke^{-kc} = \frac{Ne^{-(N-2)c} - (N-1)e^{-(N-1)c} - e^c}{(e^c - 1)^2}$$

Now

$$F(c) = \sum_{k=1}^{N-1} (N-k)e^{-kc} = Nf(c) + f'(c) = \frac{(N-1)e^c - N + e^{-(N-1)c}}{(e^c - 1)^2}$$

Substitution into the variance expression gives

$$\text{Var}[\hat{\mu}_x(t)] = \frac{\sigma_x^2(t)}{N} + \frac{2\sigma_x^2(t)}{N^2}\left[\frac{(N-1)e^c - N + e^{-(N-1)c}}{(e^c - 1)^2}\right]$$

The result given above can be used to generate a set of curves for different values of $\sigma_x^2(t)$, N, and c. Experimental results would enable one to estimate the constant c for application of these curves.

12.3.3 *Analysis Procedures for Single Records*

As noted in Section 12.1, it often occurs in practice that only one sample record of data is available for a nonstationary process of interest. In such cases, nonstationary mean values are often estimated from a single sample record by one of several operations equivalent to low pass filtering. This technique can be employed profitably for certain classes of nonstationary data. To be more specific, consider a nonstationary random process of the sum form

$$\{x(t)\} = a(t) + \{u(t)\} \tag{12.28}$$

where $a(t)$ is a deterministic function and $\{u(t)\}$ is a random process with a stationary mean value of zero. It follows that the mean value of the process $\{x(t)\}$ at any time t is given by

$$E[\{x(t)\}] = E[a(t) + \{u(t)\}] = E[a(t)] + E[\{u(t)\}] = a(t) \quad (12.29)$$

If it is assumed that the variations of $a(t)$ are very slow compared to the lowest frequency in $\{u(t)\}$, then $a(t)$ can be separated from $\{u(t)\}$ by low-pass filtering operations on a single sample record $x(t)$. Such filtering operations may be physically accomplished in several ways, including the following:

a. Digital low-pass filtering with either recursive or nonrecursive filters as discussed in Section 11.1.3
b. Polynomial curve fitting (regression analysis) as introduced in Section 4.8 and discussed for trend removal applications in Section 11.1.2
c. Segmented mean value estimates (short time-averaging operations)

In any case, the resulting mean value estimates will involve a bias error that is a function of the low-pass filter cutoff frequency (the number of terms in the polynomial fit or the short averaging time) relative to the rate of variation of $a(t)$.

For example, consider the short time-averaged estimate of the mean value $\mu_x(t)$ given by

$$\hat{\mu}_x(t) = \int_{t-T/2}^{t+T/2} x(t)\, dt = \int_{t-T/2}^{t+T/2} [a(t) + u(t)]\, dt \quad (12.30)$$

where T is a short averaging time. It is readily shown that

$$E[\hat{\mu}_x(t)] = E\left[\int_{t-T/2}^{t+T/2} [a(t) + u(t)]\, dt\right] = \int_{t-T/2}^{t+T/2} \{E[a(t)] + E[u(t)]\}\, dt$$

$$= \int_{t-T/2}^{t+T/2} a(t)\, dt \neq a(t) \quad (12.31)$$

Hence this estimate $\hat{\mu}_x(t)$ will generally be *biased*. A development similar to that presented in Section 8.3.1 yields a first-order approximation for the bias error at any time t as

$$b[\hat{\mu}_x(t)] = \frac{T^2}{24} a''(t) \quad (12.32)$$

where $a''(t)$ is the second derivative of $a(t)$ with respect to t. It is clear from Equation (12.32) that the bias error diminishes as the averaging time T becomes small. However, the random error increases as T becomes small in a manner similar to that developed for stationary data in Section 8.2.1. Thus the selection of an appropriate averaging time T involves a compromise between random and bias errors. In most cases, this compromise is best arrived at by trial-and-error procedures.

12.4   NONSTATIONARY MEAN SQUARE VALUES

A similar analysis to the one given in Section 12.3 will now be carried out to determine how the nonstationary mean square values change with time. This can be estimated by using a special-purpose instrument or a computer that performs the following operation to calculate a sample mean square value from a sample of size N. Specifically, for N samples functions $x_i(t)$, $0 \leq t \leq T$, $i = 1, 2, 3, \ldots, N$, from a nonstationary process $\{x(t)\}$, fix t and compute the ensemble average estimate

$$\hat{\psi}_x^2(t) = \frac{1}{N} \sum_{i=1}^{N} x_i^2(t) \tag{12.33}$$

Independent of N, the quantity $\hat{\psi}_x^2(t)$ is an *unbiased* estimate of the true mean square value of the nonstationary process $\{x(t)\}$ at any time t since the expected value

$$E[\hat{\psi}_x^2(t)] = \frac{1}{N} \sum_{i=1}^{N} E[x_i^2(t)] = \psi_x^2(t) \tag{12.34}$$

The quantity

$$\psi_x^2(t) = E[x_i^2(t)] = \mu_x^2(t) + \sigma_x^2(t) \tag{12.35}$$

is the true mean square value of the nonstationary process at time t. Figure 12.4 indicates how to measure $\hat{\psi}_x^2(t)$ by merely replacing $x_i(t)$ by $x_i^2(t)$.

12.4.1   *Independent Samples*

It will now be assumed that the N sample functions $x_i(t)$ are independent, so that for all i and j,

$$E[x_i(t)x_j(t)] = E[x_i(t)]E[x_j(t)] = \mu_x^2(t) \tag{12.36}$$

The sample variance associated with the estimates $\hat{\psi}_x^2(t)$ are calculated as follows. By definition,

$$\text{Var}[\hat{\psi}_x^2(t)] = E[\{\hat{\psi}_x^2(t) - \psi_x^2(t)\}^2] = E[\{\hat{\psi}_x^2(t)\}^2] - \psi_x^4(t) \tag{12.37}$$

where $\psi_x^2(t)$ is given by Equation (12.35) and

$$E\left[\left\{\hat{\psi}_x^2(t)\right\}^2\right] = \frac{1}{N^2} \sum_{i,j=1}^{N} E\left[x_i^2(t)x_j^2(t)\right]$$

$$= \frac{1}{N^2}\left[\sum_{i=1}^{N} E\left[x_i^4(t)\right] + \sum_{\substack{i,j=1 \\ i\neq j}}^{N} E\left[x_i^2(t)x_j^2(t)\right]\right] \quad (12.38)$$

Thus the problem reduces to evaluation of the expected values appearing in Equation (12.38).

In order to obtain reasonable closed-form answers, it will be assumed now that the random process $\{x_i(t)\}$ at any time t follows a Gaussian distribution with mean value $\mu_x(t)$ and variance $\sigma_x^2(t)$. One can then derive

$$E\left[x_i^4(t)\right] = 3\psi_x^4(t) - 2\mu_x^4(t) \quad (12.39)$$

$$E\left[x_i^2(t)x_j^2(t)\right] = \psi_x^4(t) \quad \text{for } i \neq j \quad (12.40)$$

The derivation of Equations (12.39) and (12.40) is based on a nonstationary form of the fourth-order Gaussian relation of Equation (3.82), namely,

$$E\left[x_i(t)x_j(t)x_m(t)x_n(t)\right] = E\left[x_i(t)x_j(t)\right]E\left[x_m(t)x_n(t)\right]$$

$$+ E\left[x_i(t)x_m(t)\right]E\left[x_j(t)x_n(t)\right]$$

$$+ E\left[x_i(t)x_n(t)\right]E\left[x_j(t)x_m(t)\right] - 2\mu_x^4(t)$$

$$(12.41)$$

Substitution into Equations (12.37) and (12.38) yields the result

$$\text{Var}\left[\hat{\psi}_x^2(t)\right] = \frac{2}{N}\left[\psi_x^4(t) - \mu_x^4(t)\right] \quad (12.42)$$

Thus the sample variance approaches zero as N approaches infinity, so that $\hat{\psi}_x^2(t)$ is a *consistent* estimate of $\psi_x^2(t)$ for all t.

To arrive at confidence intervals for $\hat{\psi}_x^2(t)$, it is more convenient to work with the nonstationary variance estimate given from Equation (12.35) by

$$\hat{\sigma}_x^2(t) = \hat{\psi}_x^2 - \hat{\mu}_x^2 \quad (12.43)$$

Assuming $x_i(t)$, $i = 1, 2, \ldots, N$, is normally distributed, $\hat{\sigma}_x^2(t)$ will have a sampling distribution for each value of t given from Section 4.3.2 as

$$\hat{\sigma}_x^2(t) = \sigma_x^2(t)\chi_n^2/n \qquad n = N - 1 \tag{12.44}$$

where χ_n^2 is the chi-square variable with $n = N - 1$ degrees of freedom defined in Section 4.2.2. Hence, the $(1 - \alpha)$ confidence interval at any time t is

$$\left[\frac{n\hat{\sigma}_x^2(t)}{\chi_{n;\,\alpha/2}^2} \le \sigma_x^2(t) < \frac{n\hat{\sigma}_x^2(t)}{\chi_{n;\,1\text{-}\alpha/2}^2} \right] \qquad n = N - 1 \tag{12.45}$$

12.4.2 Correlated Samples

For situations involving correlated samples, it is assumed as in Section 12.3.2 that the sample records satisfy the relation

$$E\big[x_i(t)x_j(t)\big] = R_{xx}(k, t) \qquad \text{where } k = j - i \tag{12.46}$$

Equation (12.40) where $i \ne j$ is now replaced by

$$E\big[x_i^2(t)x_j^2(t)\big] = \psi_x^4(t) + 2\big[R_{xx}^2(k, t) - \mu_x^2(t)\big] \tag{12.47}$$

where $k = j - i \ne 0$. When $i = j$, Equation (12.46) becomes

$$E\big[x_i^2(t)\big] = R_{xx}(0, t) = \psi_x^2(t) \tag{12.48}$$

Proper steps for including $R_{xx}(k, t)$ in the analysis are developed in Section 12.3.2. A similar procedure here yields the result

$$\text{Var}\big[\hat{\psi}_x^2(t)\big] = \frac{2}{N}\big[\psi_x^4(t) - \mu_x^4(t)\big] + \frac{4}{N^2}\sum_{k=1}^{N-1}(N - k)\big[R_{xx}^2(k, t) - \mu_x^4(t)\big] \tag{12.49}$$

which is a useful generalization of Equation (12.42).

Example 12.3. **Variance of Mean Square Value Estimate for Exponential Correlation Between Samples.** In order to obtain some quantitative expressions corresponding to Equation (12.42) that will characterize different degrees of correlation, assume that $R_{xx}(k, t)$ has an exponential form such that

$$R_{xx}(k, t) = \mu_x^2(t) + \sigma_x^2(t)e^{-kc}$$

where k and c are positive constants. Determine the corresponding sample variance for mean square value estimates.

The sample variance is given by Equation (12.49). To carry out this evaluation, let $f(c)$ and $F(c)$ be defined as in Example 12.2. It then follows that

$$R_{xx}^2(k, t) - \mu_x^4(t) = \sigma_x^2(t)\left[2\mu_x^2(t)e^{-kc} + \sigma_x^2(t)e^{-2kc}\right]$$

Hence the second term in Equation (12.49) is

$$\frac{4\sigma_x^2(t)}{N^2} \sum_{k=1}^{N-1} (N - k)\left[2\mu_x^2(t)e^{-kc} + \sigma_x^2(t)e^{-2kc}\right]$$

$$= \frac{4\sigma_x^2(t)}{N^2}\left[2\mu_x^2(t)F(c) + \sigma_x^2(t)F(2c)\right]$$

The desired variance is now given by the above result and the first term in Equation (12.49).

12.4.3 *Analysis Procedures for Single Records*

For certain classes of data, the time-varying mean square value can be estimated from a single sample record by low pass filtering operations, as listed in Section 12.3.3. For this case, the applicable model is a nonstationary random process of the product form

$$\{x(t)\} = a(t)\{u(t)\} \tag{12.50}$$

where $a(t)$ is a deterministic function and $\{u(t)\}$ is a random process with a stationary mean and variance of zero and unity, respectively. Hence the mean square value of the process $\{u(t)\}$ at any time t is given by

$$\text{Var}\big[\{x(t)\}\big] = \text{Var}\big[a(t)\{u(t)\}\big] = a^2(t)\text{Var}\big[\{u(t)\}\big] = a^2(t) \tag{12.51}$$

As for mean value estimates, if the variations of $a(t)$ are very slow compared to the lowest frequency of $\{u(t)\}$, then $a^2(t)$ can be separated by low-pass filtering operations on a single sample record of $x(t)$. This may be physically accomplished by applying any of the techniques discussed in Section 12.3.3 to the squared data signal $x^2(t)$. The resulting mean square value estimates, of course, will generally be *biased*. For example, if a short time-averaging procedure is used

$$E\big[\hat{\psi}_x^2(t)\big] = E\left[\int_{t-T/2}^{t+T/2} a^2(t)u^2(t)\, dt\right] = \int_{t-T/2}^{t+T/2} E\big[a^2(t)\big]E\big[u^2(t)\big]\, dt$$

$$= \int_{t-T/2}^{t+T/2} a^2(t)\, dt \neq a^2(t) \tag{12.52}$$

Example 12.4. **Illustration of Short Time-Averaged Mean Square Value Estimate**.   Consider the structural vibration environment of a spacecraft during liftoff. One would anticipate that the vibration at any point on the spacecraft structure is a nonstationary random process since the excitation forces during liftoff (primarily acoustic noise generated by turbulent mixing of the rocket exhaust gases with the ambient air) are a function of rapidly changing parameters, such as vehicle speed and distance from ground reflections. The measurement of a time-varying mean square value by ensemble averaging procedures in accordance with Equation (12.33) is not economically feasible, since N spacecraft launches would be required to obtain the N sample records for ensemble averaging. Can short time-averaging procedures be applied to this problem to obtain meaningful results?

For any given spacecraft configuration, the structural vibration during liftoff will display a repeatable time trend since the parameters governing the excitation forces change in the same way from one launch to the next. Furthermore, this time trend is relatively slow (several seconds for a half cycle of variation) relative to the lowest frequency of significant energy in the excitation (several Hertz). Hence, as a first-order of approximation, the model of Equation (12.50) should apply to this problem.

To demonstrate its application, consider the results from actual test data presented in Figure 12.5. The data in this figure are short time-averaged rms values of the vibration measured at the same structural location on three

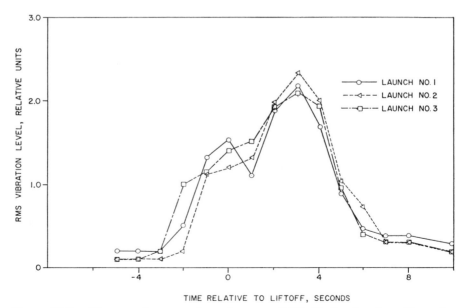

Figure 12.5   Time-varying rms values of spacecraft liftoff vibration data. These data were provided by the Jet Propulsion Laboratory, Pasadena, California.

spacecraft of similar configuration during liftoff. The averaging time for each computation is about 1 sec and the frequency range of the data is several hundred Hertz. The close agreement between the results for the three separate launches indicates that the sequence of short time-averaged mean square values provides a useful approximation for the time-varying mean square value $\psi^2(t)$, remembering, of course, that a bias error is present, as indicated in Equation (12.52).

12.5   CORRELATION STRUCTURE OF NONSTATIONARY DATA

Consider any pair of real-valued nonstationary random processes $\{x(t)\}$ and $\{y(t)\}$. The *mean values* at arbitrary fixed times t are defined by the expected values

$$\mu_x(t) = E[x(t)]$$

$$\mu_y(t) = E[y(t)]$$

(12.53)

Original data can always be transformed to have zero mean values by replacing $x(t)$ by $x(t) - \mu_x(t)$ and $y(t)$ by $y(t) - \mu_y(t)$. This will be assumed henceforth.

12.5.1   *Double Time Correlation Functions*

The *correlation functions* at any pair of fixed times t_1 and t_2 are defined by the expected values

$$R_{xx}(t_1, t_2) = E[x(t_1)x(t_2)]$$

$$R_{yy}(t_1, t_2) = E[y(t_1)y(t_2)]$$

(12.54)

$$R_{xy}(t_1, t_2) = E[x(t_1)y(t_2)]$$

(12.55)

The quantities $R_{xx}(t_1, t_2)$ and $R_{yy}(t_1, t_2)$ are called *nonstationary autocorrelation functions*, whereas $R_{xy}(t_1, t_2)$ is called a *nonstationary cross-correlation function*. For stationary random data, these results would not be functions of t_1 and t_2 but only of their difference $(t_2 - t_1)$, as developed in Section 5.1.1.

A proof similar to that used for stationary data in Section 5.1.3 shows that for any values of t_1 and t_2, an upper bound for the nonstationary cross-correlation function $R_{xy}(t_1, t_2)$ is given by the *cross-correlation inequality*

$$\left|R_{xy}(t_1, t_2)\right|^2 \leq R_{xx}(t_1, t_1)R_{yy}(t_2, t_2)$$

(12.56)

From the basic definitions, it is clear that

$$R_{xx}(t_2, t_1) = R_{xx}(t_1, t_2)$$
$$R_{yy}(t_2, t_1) = R_{yy}(t_1, t_2)$$
$$\quad (12.57)$$

$$R_{xy}(t_2, t_1) = R_{yx}(t_1, t_2) \quad (12.58)$$

Consider the problem of measuring $R_{xx}(t_1, t_2)$ using a set of N sample functions $x_i(t)$, $i = 1, 2, \ldots, N$, from the nonstationary random process. In place of Equation (12.54), one would compute the ensemble average estimate

$$\hat{R}_{xx}(t_1, t_2) = \frac{1}{N} \sum_{i=1}^{N} x_i(t_i) x_i(t_2) \quad (12.59)$$

A recommended procedure to perform this computation is as follows. Let $t_1 = t$, and let $t_2 = t - \tau$ where τ is a fixed time-delay value. This yields

$$\hat{R}_{xx}(t, t - \tau) = \frac{1}{N} \sum_{i=1}^{N} x_i(t) x_i(t - \tau) \quad (12.60)$$

which for stationary processes would be a function of τ only, but for nonstationary processes would be a function of both t and τ. For each fixed delay value τ and each record $x_i(t)$, calculate and store the product $x_i(t) x_i(t - \tau)$ for all t. Repeat for all N records, and then perform an ensemble average to yield the estimate of Equation (12.60). This whole operation must be repeated for every different τ of concern. Figure 12.6 illustrates this procedure for measuring nonstationary autocorrelation functions. A similar procedure may be followed for nonstationary cross-correlation function measurements.

12.5.2   *Alternative Double Time Correlation Functions*

A different double time correlation structure can be defined by making the following transformations. Let

$$t_1 = t - \frac{\tau}{2} \qquad t_2 = t + \frac{\tau}{2} \quad (12.61)$$

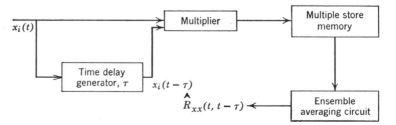

Figure 12.6   Procedure for nonstationary autocorrelation measurement.

Then

$$\tau = t_2 - t_1 \qquad t = \frac{t_1 + t_2}{2} \qquad (12.62)$$

Here τ is the time difference between t_1 and t_2, and t is the midtime between t_1 and t_2. Now

$$R_{xy}(t_1, t_2) = R_{xy}\left(t - \frac{\tau}{2}, t + \frac{\tau}{2}\right)$$

$$= E\left[x\left(t - \frac{\tau}{2}\right) y\left(t + \frac{\tau}{2}\right)\right] = \mathscr{R}_{xy}(\tau, t) \qquad (12.63)$$

Also

$$\mathscr{R}_{xx}(\tau, t) = E\left[x\left(t - \frac{\tau}{2}\right) x\left(t + \frac{\tau}{2}\right)\right]$$

$$\qquad (12.64)$$

$$\mathscr{R}_{yy}(\tau, t) = E\left[y\left(t - \frac{\tau}{2}\right) y\left(t + \frac{\tau}{2}\right)\right]$$

A script $\mathscr{R}$ is used in place of R to distinguish the (τ, t) plane from the (t_1, t_2) plane. Note that at the point $\tau = 0$, assuming $\mu_x(t) = \mu_y(t) = 0$,

$$\mathscr{R}_{xx}(0, t) = E\left[x^2(t)\right] = \sigma_x^2(t)$$

$$\qquad (12.65)$$

$$\mathscr{R}_{yy}(0, t) = E\left[y^2(t)\right] = \sigma_y^2(t)$$

$$\mathscr{R}_{xy}(0, t) = E\left[x(t) y(t)\right] = \sigma_{xy}(t) \qquad (12.66)$$

where $\sigma_x^2(t)$ and $\sigma_y^2(t)$ are the variances of $x(t)$ and $y(t)$ at time t and $\sigma_{xy}(t)$ is the covariance between $x(t)$ and $y(t)$ at time t. For any $\mathscr{R}(\tau, t)$, one has the relations

$$\mathscr{R}_{xx}(\tau, 0) = E\left[x\left(-\frac{\tau}{2}\right) x\left(\frac{\tau}{2}\right)\right] \qquad (12.67)$$

Also,

$$\mathscr{R}_{xx}(-\tau, t) = \mathscr{R}_{xx}(\tau, t)$$

$$\qquad (12.68)$$

$$\mathscr{R}_{yy}(-\tau, t) = \mathscr{R}_{yy}(\tau, t)$$

$$\mathscr{R}_{xy}(-\tau, t) = \mathscr{R}_{yx}(\tau, t) \qquad (12.69)$$

Equation (12.68) shows that $\mathscr{R}_{xx}(\tau, t)$ is an even function of τ.

In the (τ, t) plane, it sometimes may be possible to separate nonstationary correlation functions into stationary and nonstationary parts. Specifically, one might be able to write

$$\mathscr{R}(\tau, t) = A(t)R(\tau) = A\left(\frac{t_1 + t_2}{2}\right)R(t_2 - t_1) \qquad (12.70)$$

where $A(t)$ is a slowly varying nonnegative function of t at the midpoint (average) of the points t_1 and t_2, while $R(\tau)$ is a stationary correlation function at the time difference $(t_2 - t_1)$. When $\mathscr{R}(\tau, t)$ can be expressed in this way, the random process is said to be *locally stationary*. These situations are discussed in Section 12.6.4.

The *time-averaged cross-correlation function* $\overline{R}_{xy}(\tau)$ is defined from $\mathscr{R}_{xy}(\tau, t)$ by computing

$$\overline{R}_{xy}(\tau) = \lim_{T \to \infty} \frac{1}{T} \int_0^T \mathscr{R}_{xy}(\tau, t)\, dt \qquad (12.71)$$

The *time-averaged autocorrelation function* $\overline{R}_{xx}(\tau)$ is defined from $\mathscr{R}_{xx}(\tau, t)$ by

$$\overline{R}_{xx}(\tau) = \lim_{T \to \infty} \frac{1}{T} \int_0^T \mathscr{R}_{xx}(\tau, t)\, dt \qquad (12.72)$$

Since $\mathscr{R}_{xx}(-\tau, t) = \mathscr{R}_{xx}(\tau, t)$, it follows that

$$\overline{R}_{xx}(-\tau) = \overline{R}_{xx}(\tau) \qquad (12.73)$$

Thus $\overline{R}_{xx}(\tau)$ is a real-valued even function of τ, representing the usual autocorrelation function of stationary random data. From Equation (12.69), since $\mathscr{R}_{xy}(-\tau, t) = \mathscr{R}_{yx}(\tau, t)$, one obtains

$$\overline{R}_{xy}(-\tau) = \overline{R}_{yx}(\tau) \qquad (12.74)$$

so that $\overline{R}_{xy}(\tau)$ represents the usual cross-correlation function of stationary random data.

***Example 12.5.* Double Time Autocorrelation Function of a Periodic Signal.** Consider a periodic function defined by

$$x(t) = A \cos 2\pi f_0 t$$

where A and f_0 are constants. From Equation (12.54),

$$R_{xx}(t_1, t_2) = E[x(t_1)x(t_2)] = A^2 \cos 2\pi f_0 t_1 \cos 2\pi f_0 t_2$$

$$= \frac{A^2}{2}\left[\cos 2\pi f_0 (t_2 - t_1) + \cos 4\pi f_0\left(\frac{t_1 + t_2}{2}\right)\right]$$

It follows from Equation (12.62) that

$$\mathcal{R}_{xx}(\tau, t) = \frac{A^2}{2}\left[\cos 2\pi f_0 \tau + \cos 4\pi f_0 t\right]$$

where $\tau = (t_2 - t_1)$ and $t = (t_1 + t_2)/2$. In this example, $\mathcal{R}_{xx}(\tau, t)$ is the sum of a stationary component $(A^2/2)\cos 2\pi f_0 \tau$ and a nonstationary component $(A^2/2)\cos 4\pi f_0 t$.

Example 12.6. **Double Time Autocorrelation Function of Modulated Random Data.** Consider a modulated random process defined by

$$\{x(t)\} = [\cos 2\pi f_0 t]\{u(t)\}$$

where f_0 is a constant and $\{u(t)\}$ is a zero mean value stationary random process. From Equations (12.54) and (12.64),

$$R_{xx}(t_1, t_2) = E\left[x(t_1)x(t_2)\right] = \left[\cos 2\pi f_0 t_1 \cos 2\pi f_0 t_2\right]R_{uu}(t_2 - t_1)$$

$$\mathcal{R}_{xx}(\tau, t) = E\left[x\left(t - \frac{\tau}{2}\right)x\left(t + \frac{\tau}{2}\right)\right]$$

$$= \tfrac{1}{2}\left[\cos 2\pi f_0 \tau + \cos 4\pi f_0 t\right]R_{uu}(\tau)$$

For this example, the nonstationary component in $\mathcal{R}_{xx}(\tau, t)$ separates into a function of t and τ multiplied by a function of τ alone. Note that, for all t,

$$\mathcal{R}_{xx}(0, t) = E\left[x^2(t)\right] = \tfrac{1}{2}\left[1 + \cos 4\pi f_0 t\right]R_{uu}(0) \geq 0$$

In general, however, for $\tau \neq 0$, the quantity $\mathcal{R}_{xx}(\tau, t)$ may be positive or negative.

12.6 SPECTRAL STRUCTURE OF NONSTATIONARY DATA

Two distinct theoretical methods will be studied that can define the spectral structure of nonstationary data. Each of these methods has very special relationships and properties that can make one technique more suitable than the other for different applications. They are the following:

a. Double frequency (generalized) spectra
b. Frequency–time (instantaneous) spectra

Double frequency (generalized) spectra are discussed in Section 12.6.1 and 12.6.2. Frequency–time (instantaneous) spectra, also known as the Wigner distribution, are covered in some detail in Section 12.6.3. Physical cases of frequency–time spectra are discussed in Section 12.6.4 for the product model nonstationary data.

12.6.1 Double Frequency Spectral Functions

Assume that any $x(t)$ and any $y(t)$ from real-valued nonstationary random processes $\{x(t)\}$ and $\{y(t)\}$ have finite Fourier transforms given by

$$X(f, T) = \int_0^T x(t) e^{-j2\pi ft} dt$$

$$Y(f, T) = \int_0^T y(t) e^{-j2\pi ft} dt$$

(12.75)

where $x(t)$ and $y(t)$ are assumed to be zero outside of the range $(0, T)$. For simplicity in notation, the dependence on T will be omitted by letting

$$X(f) = X(f, T) \qquad Y(f) = Y(f, T) \qquad (12.76)$$

Also, the limits in Equation (12.75) and in following formulas will be omitted.

Spectral density functions at any pair of fixed frequencies f_1 and f_2 are defined by the expected values

$$S_{xx}(f_1, f_2) = E[X^*(f_1)X(f_2)]$$

$$S_{yy}(f_1, f_2) = E[Y^*(f_1)Y(f_2)]$$

(12.77)

$$S_{xy}(f_1, f_2) = E[X^*(f_1)Y(f_2)] \qquad (12.78)$$

where X^* and Y^* are complex conjugates of X and Y. The quantities $S_{xx}(f_1, f_2)$ and $S_{yy}(f_1, f_2)$ are called *double frequency* (*generalized*) *autospectral density functions*, whereas $S_{xy}(f_1, f_2)$ is called a *double frequency* (*generalized*) *cross-spectral density function*. Note that these functions are complex valued where f_1 and f_2 can take on any positive or negative values in the range $(-\infty, \infty)$.

For any values of f_1 and f_2, a proof similar to the one detailed for stationary data in Section 5.2.4 shows that an upper bound for this double frequency cross-spectral density function is given by the *cross-spectrum inequality*

$$|S_{xy}(f_1, f_2)|^2 \leq S_{xx}(f_1, f_2)S_{yy}(f_1, f_2) \qquad (12.79)$$

From the basic definitions, it is clear that

$$S_{xx}(f_2, f_1) = S_{xx}^*(f_1, f_2)$$
$$S_{yy}(f_2, f_1) = S_{yy}^*(f_1, f_2)$$

(12.80)

$$S_{xy}(f_2, f_1) = S_{yx}^*(f_1, f_2)$$

(12.81)

Equations (12.77) and (12.80) show that $S_{xx}(f, f)$ is a real-valued positive even function of f.

From Equation (12.75), one may write

$$X^*(f_1)Y(f_2) = \left[\int x(t_1) e^{j2\pi f_1 t_1} \, dt_1\right]\left[\int y(t_2) e^{-j2\pi f_2 t_2} \, dt_2\right]$$

(12.82)

Taking expected values of both sides of Equation (12.82) shows that

$$S_{xy}(f_1, f_2) = \int\int R_{xy}(t_1, t_2) e^{j2\pi(f_1 t_1 - f_2 t_2)} \, dt_1 \, dt_2$$

(12.83)

Hence $S_{xy}(f_1, f_2)$ does *not* equal the double Fourier transform of $R_{xy}(t_1, t_2)$, which is given by

$$\text{DFT}\big[R_{xy}(t_1, t_2)\big] = \int\int R_{xy}(t_1, t_2) e^{-j2\pi(f_1 t_1 + f_2 t_2)} \, dt_1 \, dt_2$$

(12.84)

Instead, it is the inverse Fourier transform of $R_{xy}(t_1, t_2)$ over t_1 followed by the direct Fourier transform over t_2. Equation (12.83) when $y(t) = x(t)$ shows how $S_{xx}(f_1, f_2)$ can be obtained from $R_{xx}(t_1, t_2)$.

The inverse single Fourier transform pairs to Equation (12.75) are

$$x(t) = \int X(f) e^{j2\pi ft} \, df$$

(12.85)

$$y(t) = \int Y(f) e^{j2\pi ft} \, df$$

where limits of integration may be from $-\infty$ to ∞. Since $x(t)$ is real valued, one can also write

$$x(t) = \int X^*(f) e^{-j2\pi ft} \, df$$

(12.86)

Now, from Equations (12.85) and (12.86), it follows that

$$x(t_1)y(t_2) = \left[\int X^*(f_1) e^{-j2\pi f_1 t_1} \, df_1\right]\left[\int Y(f_2) e^{j2\pi f_2 t_2} \, df_2\right]$$

(12.87)

Taking expected values of both sides of Equation (12.87) yields

$$R_{xy}(t_1, t_2) = \int\int S_{xy}(f_1, f_2) e^{-j2\pi(f_1 t_1 - f_2 t_2)} \, df_1 \, df_2 \qquad (12.88)$$

This is *not* the inverse double Fourier transform of $S_{xy}(f_1, f_2)$, which is given by

$$\text{IDFT}\big[S_{xy}(f_1, f_2)\big] = \int\int S_{xy}(f_1 f_2) e^{j2\pi(f_1 t_1 + f_2 t_2)} \, df_1 \, df_2 \qquad (12.89)$$

Instead, it is the direct Fourier transform of $S_{xy}(f_1, f_2)$ over f_1 followed by the inverse Fourier transform over f_2. Equation (12.88) when $y(t) = x(t)$ shows how $R_{xx}(t_1, t_2)$ can be obtained from $S_{xx}(f_1, f_2)$.

12.6.2   *Alternative Double Frequency Spectral Functions*

A different double frequency spectral structure can be defined by making the following transformations of variables. Let

$$f_1 = f - \frac{g}{2} \qquad\qquad f_2 = f + \frac{g}{2} \qquad (12.90)$$

Then

$$g = f_2 - f_1 \qquad\qquad f = \frac{f_1 + f_2}{2} \qquad (12.91)$$

Now the double frequency cross-spectrum can be written

$$S_{xy}(f_1, f_2) = S_{xy}\left(f - \frac{g}{2}, f + \frac{g}{2}\right)$$

$$= E\left[X^*\left(f - \frac{g}{2}\right)Y\left(f + \frac{g}{2}\right)\right] = \mathscr{S}_{xy}(f, g) \qquad (12.92)$$

where script $\mathscr{S}$ is used in place of S to distinguish the (f, g) plane from the (f_1, f_2) plane. For the double frequency autospectra,

$$\mathscr{S}_{xx}(f, g) = E\left[X^*\left(f - \frac{g}{2}\right)X\left(f + \frac{g}{2}\right)\right]$$

$$\qquad (12.93)$$

$$\mathscr{S}_{yy}(f, g) = E\left[Y^*\left(f - \frac{g}{2}\right)Y\left(f + \frac{g}{2}\right)\right]$$

Note that at the point $g = 0$, the functions

$$\mathscr{S}_{xx}(f,0) = E\left[|X(f)|^2\right]$$

$$\mathscr{S}_{yy}(f,0) = E\left[|Y(f)|^2\right]$$

(12.94)

$$\mathscr{S}_{xy}(f,0) = E\left[X^*(f)Y(f)\right]$$

(12.95)

represent the *energy autospectral density functions* of $x(t)$ and $y(t)$ at frequency f, and the *energy cross-spectral density function* between $x(t)$ and $y(t)$ at frequency f. Observe that

$$\mathscr{S}_{xx}(-f, g) = E\left[X^*\left(-f - \frac{g}{2}\right)X\left(-f + \frac{g}{2}\right)\right]$$

$$= E\left[X^*\left(f - \frac{g}{2}\right)X\left(f + \frac{g}{2}\right)\right] = \mathscr{S}_{xx}(f, g) \quad (12.96)$$

Hence $\mathscr{S}_{xx}(f, g)$ is an even function of f. Also

$$\mathscr{S}_{xx}^*(f, g) = E\left[X^*\left(f + \frac{g}{2}\right)X\left(f - \frac{g}{2}\right)\right] = \mathscr{S}_{xx}(f, -g) \quad (12.97)$$

shows that $\mathscr{S}_{xx}(f, -g)$ is the complex conjugate of $\mathscr{S}_{xx}(f, g)$. Similarly, it follows for arbitrary $x(t)$ and $y(t)$ that

$$\mathscr{S}_{xy}(-f, g) = \mathscr{S}_{yx}(f, g)$$

$$\mathscr{S}_{xy}^*(f, g) = \mathscr{S}_{yx}(f, -g)$$

(12.98)

Equation (12.98) gives Equations (12.96) and (12.97) when $x(t) = y(t)$.

Referring back to Equation (12.75) and letting $t_1 = t - (\tau/2)$ and $dt_1 = -(d\tau/2)$, it follows that

$$X^*\left(f - \frac{g}{2}\right) = \int x(t_1)e^{j2\pi(f-g/2)t_1}\,dt_1$$

$$= \int x\left(t - \frac{\tau}{2}\right)e^{j2\pi(f-g/2)(t-\tau/2)}\left(\frac{d\tau}{2}\right)$$

Similarly, from Equation (12.85), by letting $u = f + (g/2)$ and $du = (dg/2)$

$$y\left(t + \frac{\tau}{2}\right) = \int Y(u)e^{j2\pi(t+\tau/2)u}\,du$$

$$= \int Y\left(f + \frac{g}{2}\right)e^{j2\pi(f+g/2)(t+\tau/2)}\left(\frac{dg}{2}\right)$$

Now, since

$$\left(f + \frac{g}{2}\right)\left(t + \frac{\tau}{2}\right) = \left(f - \frac{g}{2}\right)\left(t - \frac{\tau}{2}\right) + f\tau + gt$$

it follows that

$$y\left(t + \frac{\tau}{2}\right)e^{-j2\pi f\tau} = \int Y\left(f + \frac{g}{2}\right)e^{j2\pi(f-g/2)(t-\tau/2)}e^{j2\pi gt}\left(\frac{dg}{2}\right)$$

Multiplication of both sides by $x(t - \tau/2)$ and integration over τ yields

$$\int x\left(t - \frac{\tau}{2}\right)y\left(t + \frac{\tau}{2}\right)e^{-j2\pi f\tau}\, d\tau$$

$$= \int\left[\int x\left(t - \frac{\tau}{2}\right)e^{j2\pi(f-g/2)(t-\tau/2)}\left(\frac{d\tau}{2}\right)\right]Y\left(f + \frac{g}{2}\right)e^{j2\pi gt}\, dg$$

$$= \int X^*\left(f - \frac{g}{2}\right)Y\left(f + \frac{g}{2}\right)e^{j2\pi gt}\, dg$$

Taking expected values of both sides proves that

$$\int \mathscr{R}_{xy}(\tau, t)e^{-j2\pi f\tau}\, d\tau = \int \mathscr{S}_{xy}(f, g)e^{j2\pi gt}\, dg \tag{12.99}$$

Special cases occur when $y(t) = x(t)$ to show how $\mathscr{R}_{xx}(\tau, t)$ relates to $\mathscr{S}_{xx}(f, g)$.

The left-hand side of Equation (12.99) is the Fourier transform of $\mathscr{R}_{xy}(\tau, t)$ with respect to τ, holding f and t fixed. The right-hand side of Equation (12.99) is the inverse Fourier transform of $\mathscr{S}_{xy}(f, g)$ with respect to g, holding f and t fixed. Either of these operations defines the *frequency-time* function.

$$\mathscr{W}_{xy}(f, t) = \int \mathscr{R}_{xy}(\tau, t)e^{-j2\pi f\tau}\, d\tau \tag{12.100}$$

which will be studied in the next section. This quantity $\mathscr{W}_{xy}(f, t)$ should never be confused with $\mathscr{S}_{sy}(f, g)$. It follows also from Equation (12.99) that

$$\mathscr{S}_{xy}(f, g) = \int\int \mathscr{R}_{xy}(\tau, t)e^{-j2\pi(f\tau + gt)}\, d\tau\, dt \tag{12.101}$$

In words, $\mathscr{S}_{xy}(f, g)$ is the double Fourier transform of $\mathscr{R}_{xy}(\tau, t)$ with respect to τ and t.

For the special case of stationary random data, the previous nonstationary relations simplify since the two-parameter correlation results of Equations

(12.55) and (12.63) become one-parameter results, namely,

$$R_{xy}(t_1, t_2) = R_{xx}(t_2 - t_1) \tag{12.102}$$

$$\mathcal{R}_{xy}(\tau, t) = R_{xy}(\tau) \tag{12.103}$$

with corresponding special results when $x(t) = y(t)$. From Equation (12.88), in general,

$$R_{xy}(t, t) = \iint S_{xy}(f_1, f_2) e^{-j2\pi t(f_1 - f_2)} \, df_1 \, df_2 \tag{12.104}$$

where the dependence on t appears in the exponent. For stationary data, one obtains

$$R_{xy}(t, t) = R_{xy}(0) = \iint S_{xy}(f_1, f_2) \, df_1 \, df_2$$

$$= \int S_{xy}(f_1) \, df_1 \tag{12.105}$$

where

$$S_{xy}(f_1) = \int S_{xy}(f_1, f_2) \, df_2 \tag{12.106}$$

Hence, for stationary data

$$S_{xy}(f_1, f_2) = S_{xy}(f_1) \, \delta_1(f_2 - f_1) \tag{12.107}$$

where $\delta_1(f)$ is a finite delta function defined by

$$\delta_1(f) = \begin{cases} T & (-1/2T) < f < (1/2T) \\ 0 & \text{otherwise} \end{cases} \tag{12.108}$$

It follows that $S_{xy}(f_1, f_2)$ exists only on the line $f_2 = f_1$ in the (f_1, f_2) plane, assuming frequencies f_1 and f_2 are spaced $(1/T)$ apart. Thus

$$E[X^*(f_1)Y(f_2)] = 0 \qquad \text{for } f_2 \neq f_1$$

$$E[X^*(f_1)Y(f_1)] = TS_{xy}(f_1) \qquad \text{for } f_2 = f_1 \tag{12.109}$$

which shows that

$$S_{xy}(f_1) = \frac{1}{T} E[X^*(f_1)Y(f_1)] \tag{12.110}$$

This development proves that, for consistency, one must define cross-spectral density functions for stationary random data by Equation (12.110), as done previously in Section (5.59).

Now considering the double frequency result of Equation (12.92), it is clear that

$$\mathscr{S}_{xy}(f, g) = S_{xy}(f)\,\delta_1(g) \tag{12.111}$$

with $\delta_1(g)$ satisfying Equation (12.108). This means that $\mathscr{S}_{xy}(f, g)$ exists only on the line $g = 0$ in the (f, g) plane, assuming frequencies of g are spaced $(1/T)$ apart. Thus, for stationary random data,

$$\mathscr{S}_{xy}(f, g) = E\left[X^*\left(f - \frac{g}{2}\right)Y\left(f + \frac{g}{2}\right)\right] = 0 \qquad \text{for } g \neq 0$$
$$\tag{12.112}$$
$$\mathscr{S}_{xy}(f, 0) = E\left[X^*(f)Y(f)\right] = TS_{xy}(f) \qquad \text{for } g = 0$$

Again

$$S_{xy}(f) = \frac{1}{T}E\left[X^*(f)Y(f)\right] \tag{12.113}$$

in agreement with Equation (12.110). The definition of autospectral density functions for stationary random data is merely a special case, namely,

$$S_{xx}(f) = \frac{1}{T}E\left[X^*(f)X(f)\right] = \frac{1}{T}E\left[|X(f)|^2\right] \tag{12.114}$$

Example 12.7. Double Frequency Autospectrum of a Periodic Signal. Consider a periodic function defined by

$$x(t) = A\cos 2\pi f_0 t$$

where A and f_0 are constants. The Fourier transform here yields

$$X(f) = \frac{A}{2}\left[\delta(f - f_0) + \delta(f + f_0)\right]$$

where $\delta(f)$ is the usual delta function. From Equation (12.77),

$$S_{xx}(f_1, f_2) = E\left[X^*(f_1)X(f_2)\right]$$
$$= \frac{A^2}{4}\left[\delta(f_1 - f_0) + \delta(f_1 + f_0)\right]\left[\delta(f_2 - f_0) + \delta(f_2 + f_0)\right]$$
$$= \frac{A^2}{2}\delta(f_2 - f_1) + \frac{A^2}{4}\left[\delta(f_2 - f_1 - 2f_0) + \delta(f_2 - f_1 + 2f_0)\right]$$

The function $S_{xx}(f_1, f_2)$ exists only along the three lines $f_2 = f_1$ and $f_2 = f_1 \pm 2f_0$ in the (f_1, f_2) plane. The associated

$$\mathscr{S}_{xx}(f, g) = E\left[X^*\left(f - \frac{g}{2}\right)X\left(f + \frac{g}{2}\right)\right]$$

$$= \frac{A^2}{4}\left[\delta\left(f - \frac{g}{2} - f_0\right) + \delta\left(f - \frac{g}{2} + f_0\right)\right]$$

$$\times \left[\delta\left(f + \frac{g}{2} - f_0\right) + \delta\left(f + \frac{g}{2} + f_0\right)\right]$$

$$= \frac{A^2}{4}\delta(g)\left[\delta(f - f_0) + \delta(f + f_0)\right]$$

$$+ \frac{A^2}{4}\delta(f)\left[\delta(g - 2f_0) + \delta(g + 2f_0)\right]$$

Note that $\mathscr{S}_{xx}(f, g)$ exists only along the three lines $g = 0$ and $g = \pm 2f_0$ in the (f, g) plane.

Example 12.8. **Double Frequency Autospectrum of Modulated Random Data.** Consider a modulated random process defined by

$$\{x(t)\} = [\cos 2\pi f_0 t]\{u(t)\}$$

where f_0 is a constant and $\{u(t)\}$ is a zero mean value stationary random process. From Example 12.6,

$$\mathscr{R}_{xx}(\tau, t) = \tfrac{1}{2}[\cos 2\pi f_0 \tau + \cos 4\pi f_0 t] R_{uu}(\tau)$$

Then from Equation (12.101)

$$\mathscr{S}_{xx}(f, g) = \int\int \mathscr{R}_{xx}(\tau, t) e^{-j2\pi(f\tau + gt)} \, d\tau \, dt$$

$$= \tfrac{1}{2}\delta(g)\int(\cos 2\pi f_0 \tau) R_{uu}(\tau) \, d\tau$$

$$+ \tfrac{1}{2}\int(\cos 4\pi f_0 t) e^{-j2\pi gt} \, dt \int R_{uu}(\tau) e^{-j2\pi f\tau} \, d\tau$$

$$= \tfrac{1}{4}\delta(g)[S_{uu}(f - f_0) + S_{uu}(f + f_0)]$$

$$+ \tfrac{1}{4}[\delta(g - 2f_0) + \delta(g + 2f_0)]S_{uu}(f)$$

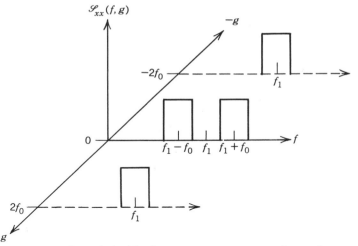

Figure 12.7 Illustration of double frequency autospectrum for cosine modulated narrow-band random noise.

This shows that the function $\mathscr{S}_{xx}(f, g)$ exists only along the three lines $g = 0$ and $g = \pm 2f_0$ in the (f, g) plane. The autospectrum $S_{uu}(f)$ is shifted by $\pm f_0$ along the line $g = 0$, and is unchanged along the lines $g = \pm 2f_0$. See Figure 12.7 for a plot of $\mathscr{S}_{xx}(f, g)$ for positive f when $S_{uu}(f)$ is narrow-band noise.

12.6.3 Frequency–Time Spectral Functions

Consider any pair of nonstationary random processes $\{x(t)\}$ and $\{y(t)\}$ with zero mean values. As defined in Sections 12.5.2 and 12.6.2, assume that one can compute the nonstationary correlation functions $\mathscr{R}_{xx}(\tau, t)$, $\mathscr{R}_{yy}(\tau, t)$, and $\mathscr{R}_{xy}(\tau, t)$, where

$$\mathscr{R}_{xy}(\tau, t) = E\left[x\left(t - \frac{\tau}{2}\right)y\left(t + \frac{\tau}{2}\right)\right] \qquad (12.115)$$

and the nonstationary spectral density functions $\mathscr{S}_{xx}(f, g)$, $\mathscr{S}_{yy}(f, g)$, and $\mathscr{S}_{xy}(f, g)$, where

$$\mathscr{S}_{xy}(f, g) = E\left[X^*\left(f - \frac{g}{2}\right)Y\left(f + \frac{g}{2}\right)\right] \qquad (12.116)$$

The Fourier transform of the nonstationary cross-correlation function $\mathscr{R}_{xy}(\tau, t)$ with respect to τ while holding t constant is given by

$$\mathscr{W}_{xy}(f, t) = \int \mathscr{R}_{xy}(\tau, t)e^{-j2\pi f\tau}\, d\tau \qquad (12.117)$$

This defines a *frequency–time spectral density function*, which is often called the *instantaneous (power) spectrum* and is the same as the Wigner distribution of Reference 12.1. The previous derivation of Equation (12.99) proves that

$$\mathcal{W}_{xy}(f, t) = \int \mathcal{S}_{xy}(f, g) e^{j2\pi g t} \, dg \tag{12.118}$$

Thus $\mathcal{W}_{xy}(f, t)$ is also the inverse Fourier transform of $\mathcal{S}_{xy}(f, g)$ with respect to g while holding f constant.

Corresponding to Equations (12.117) and (12.118), the inverse relations can be written

$$\mathcal{R}_{xy}(\tau, t) = \int \mathcal{W}_{xy}(f, t) e^{j2\pi f \tau} \, df \tag{12.119}$$

$$\mathcal{S}_{xy}(f, g) = \int \mathcal{W}_{xy}(f, t) e^{-j2\pi g t} \, dt \tag{12.120}$$

Substitution of Equation (12.117) into Equation (12.120) and Equation (12.118) into Equation (12.119) yields

$$\mathcal{S}_{xy}(f, g) = \int \int \mathcal{R}_{xy}(\tau, t) e^{-j2\pi(f\tau + gt)} \, d\tau \, dt \tag{12.121}$$

$$\mathcal{R}_{xy}(\tau, t) = \int \int \mathcal{S}_{xy}(f, g) e^{j2\pi(f\tau + gt)} \, df \, dg \tag{12.122}$$

Hence, $\mathcal{S}_{xy}(f, g)$ *is* the double Fourier transform of $\mathcal{R}_{xy}(\tau, t)$, and $\mathcal{R}_{xy}(\tau, t)$ *is* the inverse double Fourier transform of $\mathcal{S}_{xy}(f, g)$. These results should be compared with the previous relations of Equations (12.83) and (12.88).

Special results occur for nonstationary autocorrelation functions $\mathcal{R}_{xx}(\tau, t)$. For these situations, $\mathcal{W}_{xx}(f, t)$ is defined by

$$\mathcal{W}_{xx}(f, t) = \int \mathcal{R}_{xx}(\tau, t) e^{-j2\pi f \tau} \, d\tau$$

$$= \int \mathcal{R}_{xx}(\tau, t) \cos 2\pi f \tau \, d\tau \tag{12.123}$$

since $\mathcal{R}_{xx}(\tau, t)$ is an even function of τ. This result proves that

$$\mathcal{W}_{xx}^{*}(f, t) = \mathcal{W}_{xx}(f, t)$$

$$\mathcal{W}_{xx}(-f, t) = \mathcal{W}_{xx}(f, t) \tag{12.124}$$

in words, $\mathcal{W}_{xx}(f, t)$ is a real-valued even function of f. The inverse relation of

Equation (12.123) gives

$$\mathscr{R}_{xx}(\tau, t) = \int \mathscr{W}_{xx}(f, t) e^{j2\pi f\tau} df$$

$$= \int \mathscr{W}_{xx}(f, t) \cos 2\pi f\tau \, df \qquad (12.125)$$

since $\mathscr{W}_{xx}(f, t)$ is an even function of f.

From Equation (12.118), for nonstationary autospectral density functions $\mathscr{S}_{xx}(f, g)$, the associated relation with the aid of Equations (12.96) and (12.97) becomes

$$\mathscr{W}_{xx}(f, t) = \int_{-\infty}^{\infty} \mathscr{S}_{xx}(f, g) e^{j2\pi gt} dg$$

$$= \int_{0}^{\infty} [\mathscr{S}_{xx}(f, g) + \mathscr{S}_{xx}^{*}(f, g)] \cos 2\pi gt \, dg$$

$$+ j \int_{0}^{\infty} [\mathscr{S}_{xx}(f, g) - \mathscr{S}_{xx}^{*}(f, g)] \sin 2\pi gt \, dg \quad (12.126)$$

The inverse relation of Equation (12.126) gives

$$\mathscr{S}_{xx}(f, g) = \int \mathscr{W}_{xx}(f, t) e^{-j2\pi gt} dt$$

$$= \int \mathscr{W}_{xx}(f, t) \cos 2\pi gt \, dt - j \int \mathscr{W}_{xx}(f, t) \sin 2\pi gt \, dt \quad (12.127)$$

where t may vary over $(-\infty, \infty)$.

From Equation (12.124), it follows that

$$\mathscr{S}_{xx}(-f, g) = \mathscr{S}_{xx}(f, g)$$

$$\mathscr{S}_{xx}^{*}(f, g) = \mathscr{S}_{xx}(f, -g) \qquad (12.128)$$

in agreement with Equations (12.96) and (12.97).

It is worth noting that one can change nonstationary formulas from two-dimensional correlation time spaces to two-dimensional spectra frequency spaces by using the frequency–time spectra and appropriate changes of variables. The order can go in both directions, as shown in Table 12.1.

Table 12.1

Nonstationary Correlation and Spectral Functions

Function	Space
$R_{xy}(t_1, t_2)$	(time, time)
$\updownarrow$	$\updownarrow$
$\mathscr{R}_{xy}(\tau, t)$	(time, time)
$\updownarrow$	$\updownarrow$
$\mathscr{W}_{xy}(f, t)$	(frequency, time)
$\updownarrow$	$\updownarrow$
$\mathscr{S}_{xy}(f, g)$	(frequency, frequency)
$\updownarrow$	$\updownarrow$
$S_{xy}(f_1, f_2)$	(frequency, frequency)

Now consider special properties of the frequency–time spectrum $\mathscr{W}_{xx}(f, t)$. At the special point $\tau = 0$, it follows from Equations (12.65) and (12.125) that

$$\mathscr{R}_{xx}(0, t) = \int \mathscr{W}_{xx}(f, t)\, df = E[x^2(t)] \qquad (12.129)$$

Thus, *integration of $\mathscr{W}_{xx}(f, t)$ over all f gives the nonstationary mean square value (instantaneous signal power) of $\{x(t)\}$ at time t.* Also, at $g = 0$, from Equations (12.94) and (12.127),

$$\mathscr{S}_{xx}(f, 0) = \int \mathscr{W}_{xx}(f, t)\, dt = E[|X(f)|^2] \qquad (12.130)$$

Thus, *integration of $\mathscr{W}_{xx}(f, t)$ over all t gives the energy autospectral density function of $\{x(t)\}$ at frequency f.*

Let t vary over $(-\infty, \infty)$ and f vary over $(-\infty, \infty)$. Then the energy contained in $\{x(t)\}$ in the time interval from t_a to t_b is given by

$$\int_{t_a}^{t_b} E[x^2(t)]\, dt = \int_{t_a}^{t_b} \left[\int_{-\infty}^{\infty} \mathscr{W}_{xx}(f, t)\, df \right] dt \qquad (12.131)$$

On the other hand, the energy contained in $\{x(t)\}$ in the frequency interval from f_a to f_b is given by

$$\int_{f_a}^{f_b} E[|X(f)|^2]\, df = \int_{f_a}^{f_b} \left[\int_{-\infty}^{\infty} \mathscr{W}_{xx}(f, t)\, dt \right] df \qquad (12.132)$$

The total energy in $\{x(t)\}$ over the whole (f, t) plane is given by

$$\int_{-\infty}^{\infty} \int \mathscr{W}_{xx}(f, t) \, df \, dt = \int_{-\infty}^{\infty} \int \mathscr{W}_{xx}(f, t) \, dt \, df \qquad (12.133)$$

These relations show the physical importance of the frequency–time autospectrum $\mathscr{W}_{xx}(f, t)$ to describe the properties of $\{x(t)\}$.

Now consider the frequency–time cross-spectrum $\mathscr{W}_{xy}(f, t)$. From Equations (12.66) and (12.119),

$$\mathscr{R}_{xy}(0, t) = \int \mathscr{W}_{xy}(f, t) \, df = \sigma_{xy}(t) \qquad (12.134)$$

In words, *integration of* $\mathscr{W}_{xy}(f, t)$ *over all f gives the covariance between* $\{x(t)\}$ *and* $\{y(t)\}$ *at time t*. Also, from Equations (12.95) and (12.120),

$$\mathscr{S}_{xy}(f, 0) = \int \mathscr{W}_{xy}(f, t) \, dt = E[X^*(f)Y(f)] \qquad (12.135)$$

In words, *integration of* $\mathscr{W}_{xy}(f, t)$ *over all t gives the energy cross-spectral density function between* $\{x(t)\}$ *and* $\{y(t)\}$ *at frequency f*.

Returning to Equation (12.118), taking the complex conjugate of both sides yields, with the aid of Equation (12.98),

$$\mathscr{W}_{xy}^*(f, t) = \int \mathscr{S}_{xy}^*(f, g)e^{-j2\pi gt} \, dg$$

$$= \int \mathscr{S}_{yx}(f, -g)e^{-j2\pi gt} \, dg$$

$$= \int \mathscr{S}_{yx}(f, g)e^{j2\pi gt} \, dg = \mathscr{W}_{yx}(f, t) \qquad (12.136)$$

Also, from Equation (12.98) and (12.118), one obtains

$$\mathscr{W}_{xy}(-f, t) = \int \mathscr{S}_{xy}(-f, g)e^{j2\pi gt} \, dg$$

$$= \int \mathscr{S}_{yx}(f, g)e^{j2\pi gt} \, dg = \mathscr{W}_{yx}(f, t) \qquad (12.137)$$

When $x(t) = y(t)$, these equations become

$$\mathscr{W}_{xx}^*(f, t) = \mathscr{W}_{xx}(f, t)$$

$$\mathscr{W}_{xx}(-f, t) = \mathscr{W}_{xx}(f, t)$$

(12.138)

to give another proof of Equation (12.124) that $\mathscr{W}_{xx}(f, t)$ is a real-valued even function of f. Note that there is no restriction that $\mathscr{W}_{xx}(f, t)$ must be nonnegative. In fact, $\mathscr{W}_{xx}(f, t)$ *can* take on negative values, as shown in later examples.

The *time-averaged cross-spectral density function* $\bar{S}_{xy}(f)$ is defined from $\mathscr{W}_{xy}(f, t)$ by computing

$$\bar{S}_{xy}(f) = \lim_{T \to \infty} \frac{1}{T} \int_0^T \mathscr{W}_{xy}(f, t) \, dt$$

(12.139)

The *time-averaged autospectral density function* $\bar{S}_{xx}(f)$ is defined from $\mathscr{W}_{xx}(f, t)$ by

$$\bar{S}_{xx}(f) = \lim_{T \to \infty} \frac{1}{T} \int_0^T \mathscr{W}_{xx}(f, t) \, dt$$

(12.140)

Since $\mathscr{W}_{xy}(f, t) = FT[\mathscr{R}_{xy}(\tau, t)]$, assuming that various limiting operations may be interchanged,

$$\bar{S}_{xy}(f) = FT[\bar{R}_{xy}(\tau)]$$

$$\bar{S}_{xx}(f) = FT[\bar{R}_{xx}(\tau)]$$

(12.141)

where the time-averaged correlation functions are defined in Equations (12.71) and (12.72).

Equation (12.130) indicates that the quantity $\bar{S}_{xx}(f)$ will be nonnegative for all values of f, since, for large T,

$$\bar{S}_{xx}(f) = \frac{1}{T} E\left[|X(f)|^2\right] \geq 0$$

(12.142)

This is the usual definition for the autospectral density function of stationary random data. Similarly, from Equations (12.135) and (12.139), one obtains, for large T,

$$\bar{S}_{xy}(f) = \frac{1}{T} E[X^*(f)Y(f)]$$

(12.143)

This is the usual definition for cross-spectral density functions of stationary random data. One-sided time-averaged spectral density functions are given as before by

$$\overline{G}_{xy}(f) = \begin{cases} 2\overline{S}_{xy}(f) & f \geq 0 \\ 0 & f < 0 \end{cases} \tag{12.144}$$

$$\overline{G}_{xx}(f) = \begin{cases} 2\overline{S}_{xx}(f) & f \geq 0 \\ 0 & f < 0 \end{cases} \tag{12.145}$$

For the special case of stationary random data, the nonstationary correlation function becomes

$$\mathscr{R}_{xy}(\tau, t) = R_{xy}(\tau) \tag{12.146}$$

Hence

$$\mathscr{W}_{xy}(f, t) = \int \mathscr{R}_{xy}(\tau, t) e^{-j2\pi f\tau} \, d\tau = S_{xy}(f) \tag{12.147}$$

This shows that the frequency–time cross-spectrum $\mathscr{W}_{xy}(f, t)$ is now independent of t and is the usual stationary cross-spectral density function $S_{xy}(f)$. Similarly, for stationary random data, the nonstationary spectral density function

$$\mathscr{S}_{xy}(f, g) = S_{xy}(f) \, \delta(g) \tag{12.148}$$

It follows again that

$$\mathscr{W}_{xy}(f, t) = \int \mathscr{S}_{xy}(f, g) e^{j2\pi gt} \, dg = S_{xy}(f) \tag{12.149}$$

Thus $\mathscr{W}_{xy}(f, t)$ includes $S_{xy}(f)$ as a special case when data are stationary.

***Example 12.9.* Instantaneous Spectrum of a Periodic Signal.** Consider a periodic signal defined by

$$x(t) = A \cos 2\pi f_0 t$$

where A and f_0 are constants. From Example 12.5,

$$\mathscr{R}_{xx}(\tau, t) = E\left[x\left(t - \frac{\tau}{2} \right) x\left(t + \frac{\tau}{2} \right) \right]$$

$$= \frac{A^2}{2} [\cos 2\pi f_0 \tau + \cos 4\pi f_0 t]$$

The frequency–time spectrum is then given by Equation (12.117) as

$$\mathscr{W}_{xx}(f,t) = \int \mathscr{R}_{xx}(\tau,t) e^{-j2\pi f\tau} \, d\tau$$

$$= \frac{A^2}{4} \left[\delta(f - f_0) + \delta(f + f_0) + 2\,\delta(f)\cos 4\pi f_0 t \right]$$

Observe that

a. $\mathscr{W}_{xx}(f,t)$ has stationary components at $f = \pm f_0$ plus a nonstationary component at $f = 0$, which is periodic with frequency $2f_0$.

b. $\mathscr{W}_{xx}(f,t)$ can take on negative values.

The *instantaneous power* in this signal is given by Equation (12.65) as

$$\mathscr{R}_{xx}(0,t) = E\left[x^2(t)\right] = \frac{A^2}{2}\left[1 + \cos 4\pi f_0 t\right] = A^2\cos^2 2\pi f_0 t$$

This agrees with Equation (12.129), namely,

$$\int \mathscr{W}_{xx}(f,t)\,df = A^2\cos^2 2\pi f_0 t$$

Time-averaged results for this example from Equations (12.72) and (12.140) yield

$$\overline{R}_{xx}(\tau) = \lim_{T\to\infty} \frac{1}{T}\int_0^T \mathscr{R}_{xx}(\tau,t)\,dt = \frac{A^2}{2}\cos 2\pi f_0\tau$$

$$\overline{S}_{xx}(f) = \lim_{T\to\infty} \frac{1}{T}\int_0^T \mathscr{W}_{xx}(f,t)\,dt = \frac{A^2}{4}\left[\delta(f - f_0) + \delta(f + f_0)\right]$$

which agree with the results computed previously in Examples 5.1 and 5.5. Note that

$$\overline{S}_{xx}(f) = FT\left[\overline{R}_{xx}(\tau)\right]$$

and

$$\overline{R}_{xx}(0) = \frac{A^2}{2} = \int \overline{S}_{xx}(f)\,df$$

Example 12.10. **Instantaneous Spectrum of Modulated Random Data.** Consider a modulated random process defined by

$$\{x(t)\} = [\cos 2\pi f_0 t]\{u(t)\}$$

where f_0 is a constant and $\{u(t)\}$ is a zero mean value stationary random process. From Example 12.6,

$$\mathcal{R}_{xx}(\tau, t) = \tfrac{1}{2}[\cos 2\pi f_0 \tau + \cos 4\pi f_0 t] R_{uu}(\tau)$$

Then from Equation (12.123)

$$\mathcal{W}_{xx}(f, t) = \int \mathcal{R}_{xx}(\tau, t) e^{-j2\pi f \tau} \, d\tau$$

$$= \tfrac{1}{4}[S_{uu}(f - f_0) + S_{uu}(f + f_0)] + \tfrac{1}{2}(\cos 4\pi f_0 t) S_{uu}(f)$$

Observe that

a. The stationary component $S_{uu}(f)$ is shifted by $\pm f_0$.
b. The nonstationary component in $\mathcal{W}_{xx}(f, t)$ is periodic with frequency $2f_0$.
c. $\mathcal{W}_{xx}(f, t)$ can take on negative values.

The *instantaneous power* in these data is given by Equation (12.65) as

$$\mathcal{R}_{xx}(0, t) = E[x^2(t)] = R_{uu}(0)\cos^2 2\pi f_0 t$$

This agrees with integrating $\mathcal{W}_{xx}(f, t)$ over all f. The *energy autospectral density function* in these data is given by Equation (12.130) as

$$\mathcal{S}_{xx}(f, 0) = E\left[|X(f)|^2\right]$$

where

$$X(f) = \tfrac{1}{2}[U(f - f_0) + U(f + f_0)]$$

Hence, for finite T, where $E[|U(f)|^2] = TS_{uu}(f)$, one obtains

$$\mathcal{S}_{xx}(f, 0) = \frac{T}{4}[S_{uu}(f - f_0) + S_{uu}(f + f_0)]$$

This agrees with integrating $\mathcal{W}_{xx}(f, t)$ over t for finite T. A plot of $\mathcal{W}_{xx}(f, t)$ for positive f when $S_{uu}(f)$ is narrow-band noise is shown in Figure 12.8.
Time-averaged results for this example are

$$\bar{R}_{xx}(\tau) = \frac{1}{T} \int_0^T \mathcal{R}_{xx}(\tau, t) \, dt = \tfrac{1}{2}(\cos 2\pi f_0 \tau) R_{uu}(\tau)$$

$$\bar{S}_{xx}(f) = \frac{1}{T} \int_0^T \mathcal{W}_{xx}(f, t) \, dt = \tfrac{1}{4}[S_{uu}(f - f_0) + S_{uu}(f + f_0)]$$

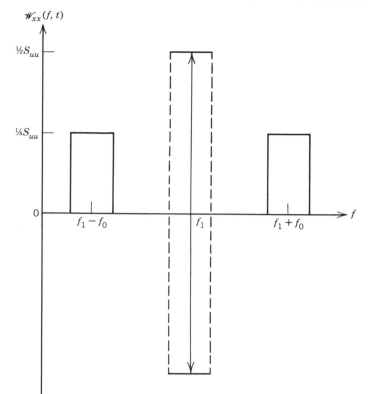

Figure 12.8 Illustration of frequency–time autospectrum for cosine modulated narrow-band random noise.

Note that

$$\overline{S}_{xx}(f) = FT\left[\overline{R}_{xx}(\tau)\right]$$

and

$$R_{xx}(0) = \tfrac{1}{2}R_{uu}(0) = \int \overline{S}_{xx}(f)\, df$$

12.6.4 *Product Model Nonstationary Data*

Consider a nonstationary random process $\{x(t)\}$ that represents the product of a deterministic signal $a(t)$ multiplied by a stationary random process $\{u(t)\}$, namely,

$$\{x(t)\} = a(t)\{u(t)\} \qquad (12.150)$$

This simple nonstationary process, commonly called the *product model*, provides an acceptable approximation to certain types of nonstationary data in practice, for example, atmospheric turbulence and boundary layer turbulence data [References 12.2 and 12.3].

It follows from Equation (12.64) that the nonstationary autocorrelation function of $\{x(t)\}$ in Equation (12.150) is given by

$$\mathscr{R}_{xx}(\tau, t) = E\left[x\left(t - \frac{\tau}{2}\right)x\left(t + \frac{\tau}{2}\right)\right] = R_{aa}(\tau, t)R_{uu}(\tau) \quad (12.151)$$

where

$$R_{aa}(\tau, t) = a\left(t - \frac{\tau}{2}\right)a\left(t + \frac{\tau}{2}\right)$$

$$(12.152)$$

$$R_{uu}(\tau) = E\left[u\left(t - \frac{\tau}{2}\right)u\left(t + \frac{\tau}{2}\right)\right]$$

For fixed t, the two-sided frequency–time autospectrum is given by Equation (12.117) as

$$\mathscr{W}_{xx}(f, t) = \int \mathscr{R}_{xx}(\tau, t)e^{-j2\pi f\tau}\, d\tau \quad (12.153)$$

Now let the following terms be defined:

$$S_{aa}(f, t) = \int R_{aa}(\tau, t)e^{-j2\pi f\tau}\, d\tau$$

$$(12.154)$$

$$S_{uu}(f) = \int R_{uu}(\tau)e^{-j2\pi f\tau}\, d\tau$$

Then $\mathscr{W}_{xx}(f, t)$ is the convolution of $S_{aa}(f, t)$ with $S_{uu}(f)$ since $\mathscr{R}_{xx}(\tau, t)$ is the product of $R_{aa}(\tau, t)$ with $R_{uu}(\tau)$, namely,

$$\mathscr{W}_{xx}(f, t) = \int S_{aa}(\alpha, t)S_{uu}(f - \alpha)\, d\alpha \quad (12.155)$$

This is a general relation for arbitrary $a(t)$.

For the special cases where $a(t)$ in Equation (12.150) is at a much lower frequency than $u(t)$, Equation (12.151) is approximated by

$$\mathscr{R}_{xx}(\tau, t) \simeq a^2(t)R_{uu}(\tau) \quad (12.156)$$

The frequency–time autospectrum then becomes

$$\mathscr{W}_{xx}(f, t) \simeq a^2(t)S_{uu}(f) \quad (12.157)$$

where $a^2(t)$ is a slowly varying nonnegative function of t.

Nonstationary random data with autocorrelation and autospectral density functions approximated by Equations (12.156) and (12.157), respectively, are often called *locally stationary* or *uniformly modulated random processes* [Reference 12.4]. For locally stationary random processes, it is convenient to normalize $R_{uu}(\tau)$ in Equation (12.156) so that $R_{uu}(0) = 1$. Then

$$\mathscr{R}_{xx}(0, t) = a^2(t) = \psi_x^2(t) \tag{12.158}$$

and

$$\int_{-\infty}^{\infty} S_{uu}(f)\, df = \int_0^{\infty} G_{uu}(f)\, df = 1 \tag{12.159}$$

Here $G_{uu}(f)$ is the one-sided autospectral density function where $G_{uu}(f) = 2S_{uu}(f)$ for $f \geq 0$ and zero for $f < 0$. This derives the one-sided time varying autospectrum

$$\mathscr{G}_{xx}(f, t) \simeq \psi_x^2(t) G_{uu}(f) \tag{12.160}$$

where

$$\mathscr{G}_{xx}(f, t) = \begin{cases} 2\mathscr{W}_{xx}(f, t) & \text{for } f \geq 0 \\ 0 & \text{for } f < 0 \end{cases} \tag{12.161}$$

Under the assumption that $a(t)$ varies slowly compared to $u(t)$, it follows that

$$G_{uu}(f) \simeq \overline{G}_{xx}(f) \tag{12.162}$$

where $\overline{G}_{xx}(f)$ is the time-averaged autospectrum defined in Equations (12.140) and (12.145).

The result in Equation (12.160) indicates that the time-varying spectrum of a locally stationary random process can be estimated from a single sample record $x(t)$ by two separate operations as follows.

a. Estimate $\psi_x^2(t)$ using short time-averaging or some other appropriate procedure discussed in Section 12.4.3, where the entire data bandwidth is used for the calculations.
b. Estimate $G_{uu}(f)$ by computing $\overline{G}_{xx}(f)$ using the entire available record length to obtain a narrow-band resolution, as would be done with stationary data.

Since each of the above two procedures can be individually accomplished with a relatively large BT product (equivalent number of averages), the desired nonstationary spectrum can be computed with a relatively small random error even though only a single sample record is used for the analysis.

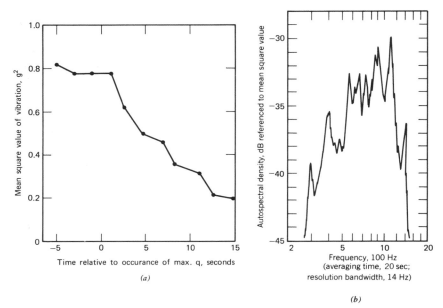

(a)

(b)

Figure 12.9 Time-varying autospectrum of spacecraft launch vibration data. (a) Time-varying mean square value. (b) Average autospectrum. These data resulted from studies funded by the NASA Goddard Space Flight Center, Greenbelt, Maryland, under Contract NAS 5-4590.

***Example 12.11*. Illustration of Time-Averaged Autospectrum Estimate.** Consider the structural vibration of a spacecraft during launch as the spacecraft passes through maximum dynamic pressure ($\frac{1}{2}\rho V^2$, where ρ is the air density and V is velocity). The vibration during this phase of the launch is due primarily to pressure fluctuations in the turbulent boundary layer generated by the spacecraft moving through the air. Since both the velocity and altitude of the spacecraft are changing with time, it would be anticipated that the boundary layer pressures and, hence, structural vibration are nonstationary random processes. Assume that the nonstationary random process representing the vibration at any point on the structure is of the locally stationary form. A time-varying autospectrum of the vibration can then be calculated from a single sample record by separate time-averaging operations, as indicated in Figure 12.9.

Figure 12.9(a) shows the estimate for $\psi^2(t)$ computed with a 2-sec averaging time over a 20-sec interval covering the time of maximum dynamic pressure. Note that the mean square vibration level varies by a factor of 4 to 1 over this interval. Figure 12.9(b) shows the estimate for $G_{uu}(f)$ computed by time averaging over the entire 20-sec interval. If the locally stationary assumption is valid for these data, Figure 12.9 provides an acceptable estimate of the

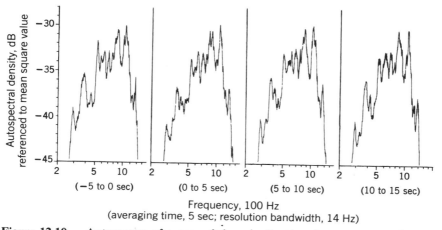

Autospectral density, dB referenced to mean square value

Frequency, 100 Hz
(averaging time, 5 sec; resolution bandwidth, 14 Hz)

Figure 12.10 Autospectra of spacecraft launch vibration data segments. These data resulted from studies funded by the NASA Goddard Space Flight Center, Greenbelt, Maryland, under Contract NAS 5-4590.

time varying autospectrum $\mathcal{G}_{xx}(f, t)$. The B_eT product for each portion of the estimate is well over 200, while the time and frequency resolutions are sufficient to avoid unreasonable bias errors.

The suitability of the locally stationary assumption for this problem can be investigated by comparing independent values of $G_{uu}(f)$ computed from individual 5-sec segments of the 20-sec-long sample record. Such results are presented in Figure 12.10. Remembering that $B_eT \simeq 70$ for each of the estimates ($\varepsilon_r \simeq 0.12$), it is seen that the results of the individual analyses in Figure 12.10 are not significantly different from the overall result for $\hat{G}_{uu}(f)$ in Figure 12.9(b). This concludes Example 12.11.

The product model of Equation (12.150) is simply a generalization of the modulated random process evaluated in Example 12.10. It is seen in that example that the primary influence of the modulation is to spread the frequency range of the time-averaged spectrum $\bar{S}_{xx}(f)$ by $\pm f_0$, the modulating frequency. Hence, as the upper frequency limit of $a(t)$ in Equation (12.150) approaches the frequency range of $u(t)$, the assumption that $\bar{S}_{xx}(f) \simeq S_{uu}(f)$ will no longer be valid, and the previously discussed analysis procedure illustrated in Example 12.11 will produce distorted results. If it can be assumed that $a(t)$ is nonnegative and $\{u(t)\}$ is Gaussian, however, then the product model can be accurately decomposed no matter how rapidly $a(t)$ varies by the procedure detailed in Reference 12.2 and summarized as follows.

Let $a(t)$ in Equation (12.150) be a time-varying standard deviation that can never be negative and $\{u(t)\}$ be a stationary Gaussian random process with a standard deviation of unity and a zero mean value. Given a nonstationary sample record $x(t)$, compute a new stationary sample record $y(t)$ by the

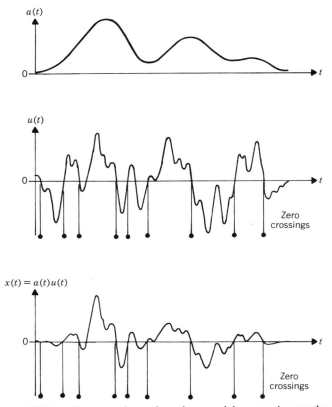

Figure 12.11   Zero crossings of product model nonstationary data.

following nonlinear operation:

$$y(t) = \begin{cases} 1 & \text{if } x(t) \text{ nonnegative} \\ -1 & \text{if } x(t) \text{ negative} \end{cases} \qquad (12.163)$$

The operation in Equation (12.163) is often referred to as *hard clipping* and essentially retains only the zero-crossing information from the original nonstationary record $x(t)$. Under the assumption that $a(t)$ is nonnegative for all t, however, it is clear that $u(t)$ will produce zero-crossings identical to $x(t)$, as illustrated in Figure 12.11. It follows that $y(t)$ represents the hard clipped version of $u(t)$. Then under the assumption that $\{u(t)\}$ is Gaussian, it is shown in Reference 12.5 that the autocorrelation function of $\{u(t)\}$ is given by

$$R_{uu}(\tau) = \sin\left[\frac{\pi}{2} R_{yy}(\tau)\right] \qquad (12.164)$$

where $R_{yy}(\tau)$ is the autocorrelation function of the hard clipped record computed over the entire available record length T using

$$R_{yy}(\tau) = \frac{1}{(T - \tau)} \int_0^{T-\tau} y(t) y(t + \tau)\, dt \qquad (12.165)$$

The autospectrum of $\{u(t)\}$ is finally computed by

$$G_{uu}(f) = 4 \int_0^{\tau_{max}} R_{uu}(\tau) \cos 2\pi f \tau\, d\tau \qquad (12.166)$$

where the resolution of the resulting spectral estimate is given by $\Delta f = 1/\tau_{max}$.

12.7   INPUT/OUTPUT RELATIONS FOR NONSTATIONARY DATA

Consider sample functions from a nonstationary random process $\{x(t)\}$ acting as the input to a time-varying linear system with a weighting function $h(\tau, t)$ and a frequency response function $H(f, t)$, where

$$H(f, t) = \int h(\tau, t) e^{-j2\pi f \tau}\, d\tau \qquad (12.167)$$

For an arbitrary input $x(t)$ belonging to $\{x(t)\}$, the output $y(t)$ belonging to $\{y(t)\}$ is

$$y(t) = \int h(\tau, t) x(t - \tau)\, d\tau \qquad (12.168)$$

It is clear that, in general, $\{y(t)\}$ will be a nonstationary random process since its statistical properties will be a function of t when either $\{x(t)\}$ is nonstationary or $h(\tau, t)$ is a function of t. For constant-parameter linear systems, $h(\tau, t) = h(\tau)$ and $H(f, t) = H(f)$, independent of t. Input/output relations in both the double time correlation domain and the double frequency spectral domain will now be derived for four cases:

1. Nonstationary input and time-varying linear system
2. Nonstationary input and constant-parameter linear system
3. Stationary input and time-varying linear system
4. Stationary input and constant-parameter linear system

The last case reduces to the familiar single time correlation domain and the single frequency spectral domain, as covered in Chapter 6.

12.7.1   *Nonstationary Input and Time-Varying Linear System*

For a pair of times t_1, t_2, the product of $y(t_1)$ with $y(t_2)$ is given by

$$y(t_1)y(t_2) = \int\int h(\alpha, t_1)h(\beta, t_2)x(t_1 - \alpha)x(t_2 - \beta)\,d\alpha\,d\beta$$

Taking expected values produces the nonstationary input/output autocorrelation relation

$$R_{yy}(t_1, t_2) = \int\int h(\alpha, t_1)h(\beta, t_2)R_{xx}(t_1 - \alpha, t_2 - \beta)\,d\alpha\,d\beta \quad (12.169)$$

Similarly, the product of $x(t_1)$ with $y(t_2)$ is given by

$$x(t_1)y(t_2) = \int h(\beta, t_2)x(t_1)x(t_2 - \beta)\,d\beta$$

Again taking expected values yields

$$R_{xy}(t_1, t_2) = \int h(\beta, t_2)R_{xx}(t_1, t_2 - \beta)\,d\beta \qquad (12.170)$$

Equations (12.169) and (12.170) are general results where all operations take place in a real-valued time domain.

To transform to a complex-valued frequency domain, let

$$J(f, g) = \int H(f, t)e^{-j2\pi gt}\,dt \qquad (12.171)$$

Then from Equation (12.167)

$$J(f, g) = \int\int h(\tau, t)e^{-j2\pi(f\tau + gt)}\,d\tau\,dt \qquad (12.172)$$

In words, $J(f, g)$ is the double Fourier transform of $h(\tau, t)$. Also

$$h(\tau, t) = \int\int J(f, g)e^{j2\pi(f\tau + gt)}\,df\,dg \qquad (12.173)$$

Equation (12.168) is thus the same as

$$y(t) = \int\int\int J(f, g)e^{j2\pi(f\tau + gt)}x(t - \tau)\,d\tau\,df\,dg \qquad (12.174)$$

Now let $g = f_1 - f$, $dg = df_1$, $\alpha = t - \tau$, and $d\alpha = -d\tau$. It follows that

$$y(t) = \int\int\int J(f, f_1 - f) e^{j2\pi f_1 t} e^{-j2\pi f\alpha} x(\alpha) \, d\alpha \, df \, df_1$$

$$= \int\int J(f, f_1 - f) X(f) e^{j2\pi f_1 t} \, df \, df_1$$

$$= \int Y(f_1) e^{j2\pi f_1 t} \, df_1 \tag{12.175}$$

where

$$X(f) = \int x(\alpha) e^{-j2\pi f\alpha} \, d\alpha \tag{12.176}$$

$$Y(f_1) = \int J(f, f_1 - f) X(f) \, df \tag{12.177}$$

Equation (12.177) is the key to obtaining the desired nonstationary input/output spectral density relations. Specifically, for a pair of frequencies f_1 and f_2, the product $Y^*(f_1)Y(f_2)$ is given by

$$Y^*(f_1)Y(f_2) = \int\int J^*(\lambda, f_1 - \lambda) J(\eta, f_2 - \eta) X^*(\lambda) X(\eta) \, d\lambda \, d\eta$$

Taking expected values of both sides yields the result

$$S_{yy}(f_1, f_2) = \int\int J^*(\lambda, f_1 - \lambda) J(\eta, f_2 - \eta) S_{xx}(\lambda, \eta) \, d\lambda \, d\eta \tag{12.178}$$

Similarly, the product of $X^*(f_1)$ with $Y(f_2)$ yields

$$X^*(f_1)Y(f_2) = \int J(\eta, f_2 - \eta) X^*(f_1) X(\eta) \, d\eta$$

Taking expected values, one obtains

$$S_{xy}(f_1, f_2) = \int J(\eta, f_2 - \eta) S_{xx}(f_1, \eta) \, d\eta \tag{12.179}$$

Equations (12.178) and (12.179) are general results where all operations take place in a complex-valued frequency domain. These general results are henceforth referred to as the Case 1 results.

12.7.2   *Results for Special Cases*

Further results will now be stated for special cases that follow from the general formulas for Case 1 in Section 12.7.1.

CASE 2.   *Nonstationary Input and Constant-Parameter Linear System*

For the case where the linear system has constant parameters, it follows that

$$h(\tau, t) = h(\tau) \quad J(f, g) = H(f)\delta(g)$$
$$H(f, t) = H(f) \quad Y(f_1) = H(f_1)X(f_1)$$

(12.180)

Then Equations (12.169) and (12.170) become

$$R_{yy}(t_1, t_2) = \int\int h(\alpha)h(\beta)R_{xx}(t_1 - \alpha, t_2 - \beta)\, d\alpha\, d\beta \quad (12.181)$$

$$R_{xy}(t_1, t_2) = \int h(\beta)R_{xx}(t_1, t_2 - \beta)\, d\beta \quad (12.182)$$

and Equations (12.178) and (12.179) become

$$S_{yy}(f_1, f_2) = H^*(f_1)H(f_2)S_{xx}(f_1, f_2) \quad (12.183)$$

$$S_{xy}(f_1, f_2) = H(f_2)S_{xx}(f_1, f_2) \quad (12.184)$$

Note that this last equation involves $H(f_2)$ and *not* $H(f_1)$.

CASE 3.   *Stationary Input and Time-Varying Linear System*

For the case where the input is stationary, it follows that

$$R_{xx}(t_1, t_2) = R_{xx}(t_2 - t_1)$$
$$S_{xx}(f_1, f_2) = S_{xx}(f_1)\delta(f_2 - f_1)$$

(12.185)

Hence, Equations (12.169) and (12.170) become

$$R_{yy}(t_1, t_2) = \int\int h(\alpha, t_1)h(\beta, t_2)R_{xx}(t_2 - t_1 + \alpha - \beta)\, d\alpha\, d\beta \quad (12.186)$$

$$R_{xy}(t_1, t_2) = \int h(\beta, t_2)R_{xx}(t_2 - t_1 - \beta)\, d\beta \quad (12.187)$$

and Equations (12.178) and (12.179) become

$$S_{yy}(f_1, f_2) = \int J^*(f, f_1 - f)J(f, f_2 - f)S_{xx}(f)\, df \quad (12.188)$$

$$S_{xy}(f_1, f_2) = J(f_1, f_2 - f_1)S_{xx}(f_1) \quad (12.189)$$

Note that this last result involves $S_{xx}(f_1)$ and *not* $S_{xx}(f_2)$.

CASE 4. *Stationary Input and Constant-Parameter Linear System*

For the case where the input is stationary and the linear system has constant parameters, all the special relations in Equations (12.180) and (12.185) apply, giving the following well-known results as the simplest form of Equations (12.169), (12.170), (12.178), and (12.179).

$$R_{yy}(\tau) = \int \int h(\alpha)h(\beta)R_{xx}(\tau + \alpha - \beta)\, d\alpha\, d\beta \quad (12.190)$$

$$R_{xy}(\tau) = \int h(\beta)R_{xx}(\tau - \beta)\, d\beta \quad (12.191)$$

$$S_{yy}(f) = |H(f)|^2 S_{xx}(f) \quad (12.192)$$

$$S_{xy}(f) = H(f)S_{xx}(f) \quad (12.193)$$

The results from all four cases are summarized in Table 12.2.

12.7.3 Frequency–Time Spectral Input/Output Relations

Consider Case 2 from the previous section, where nonstationary data passes through a constant-parameter linear system. Frequency–time spectral input/output relations are now derived, starting from the double frequency spectral relations of Equations (12.183) and (12.184).

Let the following transformations be made:

$$f_1 = f - \frac{g}{2} \qquad f_2 = f + \frac{g}{2} \quad (12.194)$$

It then follows that

$$S_{xx}(f_1, f_2) = \mathscr{S}_{xx}(f, g)$$

$$S_{yy}(f_1, f_2) = \mathscr{S}_{yy}(f, g) \quad (12.195)$$

$$S_{xy}(f_1, f_2) = \mathscr{S}_{xy}(f, g) \quad (12.196)$$

Table 12.2

Nonstationary Input/Output Correlation and Spectral Relations

Case	Correlation Relations	Spectral Relations
Nonstationary input, time-varying system	$R_{yy}(t_1, t_2) = \int\int h(\alpha, t_1) h(\beta, t_2) R_{xx}(t_1 - \alpha, t_2 - \beta)\, d\alpha\, d\beta$ $R_{xy}(t_1, t_2) = \int h(\beta, t_2) R_{xx}(t_1, t_2 - \beta)\, d\beta$	$S_{yy}(f_1, f_2) = \int\int J^*(\lambda, f_1 - \lambda) J(\eta, f_2 - \eta) S_{xx}(\lambda, \eta)\, d\lambda\, d\eta$ $S_{xy}(f_1, f_2) = \int J(\eta, f_2 - \eta) S_{xx}(f_1, \eta)\, d\eta$
Nonstationary input, constant-parameter system	$R_{yy}(t_1, t_2) = \int\int h(\alpha) h(\beta) R_{xx}(t_1 - \alpha, t_2 - \beta)\, d\alpha\, d\beta$ $R_{xy}(t_1, t_2) = \int h(\beta) R_{xx}(t_1, t_2 - \beta)\, d\beta$	$S_{yy}(f_1, f_2) = H^*(f_1) H(f_2) S_{xx}(f_1, f_2)$ $S_{xy}(f_1, f_2) = H(f_2) S_{xx}(f_1, f_2)$
Stationary input, time-varying system	$R_{yy}(t_1, t_2) = \int\int h(\alpha, t_1) h(\beta, t_2) R_{xx}(t_2 - t_1 + \alpha - \beta)\, d\alpha\, d\beta$ $R_{xy}(t_1, t_2) = \int h(\beta, t_2) R_{xx}(t_2 - t_1 - \beta)\, d\beta$	$S_{yy}(f_1, f_2) = \int J^*(f, f_1 - f) J(f, f_2 - f) S_{xx}(f)\, df$ $S_{xy}(f_1, f_2) = J(f_1, f_2 - f_1) S_{xx}(f_1)$
Stationary input, constant-parameter system	$R_{yy}(\tau) = \int\int h(\alpha) h(\beta) R_{xx}(\tau + \alpha - \beta)\, d\alpha\, d\beta$ $R_{xy}(\tau) = \int h(\beta) R_{xx}(\tau - \beta)\, d\beta$	$S_{yy}(f) = |H(f)|^2 S_{xx}(f)$ $S_{xy}(f) = H(f) S_{xx}(f)$

This gives in place of Equations (12.183) and (12.184) the spectral input/output results

$$\mathscr{S}_{yy}(f, g) = H^*\left(f - \frac{g}{2}\right)H\left(f + \frac{g}{2}\right)\mathscr{S}_{xx}(f, g) \qquad (12.197)$$

$$\mathscr{S}_{xy}(f, g) = H\left(f + \frac{g}{2}\right)\mathscr{S}_{xx}(f, g) \qquad (12.198)$$

The frequency–time spectra are now calculated from Equations (12.118) and (12.126) by

$$\mathscr{W}_{xx}(f, t) = \int \mathscr{S}_{xx}(f, g)e^{j2\pi gt}\, dg$$

$$\mathscr{W}_{yy}(f, t) = \int \mathscr{S}_{yy}(f, g)e^{j2\pi gt}\, dg \qquad (12.199)$$

$$\mathscr{W}_{xy}(f, t) = \int \mathscr{S}_{xy}(f, g)e^{j2\pi gt}\, dg \qquad (12.200)$$

Let the following functions be defined:

$$\mathscr{S}_{HH}(f, g) = H^*\left(f - \frac{g}{2}\right)H\left(f + \frac{g}{2}\right) \qquad (12.201)$$

$$\mathscr{W}_{HH}(f, t) = \int \mathscr{S}_{HH}(f, g)e^{j2\pi gt}\, dg \qquad (12.202)$$

Then Equation (12.197) becomes

$$\mathscr{S}_{yy}(f, g) = \mathscr{S}_{HH}(f, g)\mathscr{S}_{xx}(f, g) \qquad (12.203)$$

and the frequency–time spectral input/output relation is now

$$\mathscr{W}_{yy}(f, t) = \int \mathscr{W}_{HH}(f, \alpha)\mathscr{W}_{xx}(f, t - \alpha)\, d\alpha \qquad (12.204)$$

showing that $\mathscr{W}_{yy}(f, t)$ is the convolution of $\mathscr{W}_{HH}(f, t)$ with $\mathscr{W}_{xx}(f, t)$. This result in Equation (12.204) can be negative for some values of f and t.

12.7.4 Energy Spectral Input/Output Relations

A special class of nonstationary data are those that physically exist only within a finite, measurable time interval, that is, where the input process $\{x(t)\}$ and the output process $\{y(t)\}$ have nonzero values only for $0 \le t \le T$. Such data are commonly referred to as *transients* and allow for a greatly simplified analysis since for a pair of sample records $x(t)$ and $y(t)$ with zero mean values,

$$X(f) = \int_0^T x(t) e^{-j2\pi ft}\, dt = \int_{-\infty}^{\infty} x(t) e^{-j2\pi ft}\, dt$$

$$Y(f) = \int_0^T y(t) e^{-j2\pi ft}\, dt = \int_{-\infty}^{\infty} y(t) e^{-j2\pi ft}\, dt$$

(12.205)

From Equation (12.135), the expected value of the product of $X^*(f)$ and $Y(f)$ yields the *energy cross-spectral density function* defined by

$$\mathscr{S}_{xy}(f) = E[X^*(f)Y(f)] = \int_0^T \mathscr{W}_{xy}(f, t)\, dt \qquad (12.206)$$

Similarly, the *energy autospectral density functions* are defined by

$$\mathscr{S}_{xx}(f) = E[X^*(f)X(f)] = \int_0^T \mathscr{W}_{xx}(f, t)\, dt$$

$$\mathscr{S}_{yy}(f) = E[Y^*(f)Y(f)] = \int_0^T \mathscr{W}_{yy}(f, t)\, dt$$

(12.207)

It then follows from the input/output relations in Equations (12.197) and (12.198) for $g = 0$ that

$$\mathscr{S}_{yy}(f) = |H(f)|^2 \mathscr{S}_{xx}(f) \qquad (12.208)$$

$$\mathscr{S}_{xy}(f) = H(f)\mathscr{S}_{xx}(f) \qquad (12.209)$$

In terms of one-sided energy spectral density functions that exist only for $f \geq 0$,

$$\mathscr{G}_{yy}(f) = |H(f)|^2 \mathscr{G}_{xx}(f) \qquad (12.210)$$

$$\mathscr{G}_{xy}(f) = H(f)\mathscr{G}_{xx}(f) \qquad (12.211)$$

Note that the input/output relations given above for transients are identical to those developed for stationary random data in Equations (6.6)–(6.9), except "energy" spectra replace "power" spectra. For transient data, the averaging operation needed to compute the energy spectral estimates $\mathscr{G}_{xy}(f)$ and $\mathscr{G}_{xx}(f)$ theoretically requires that the experiment producing the data be repeated many times. In practice, however, transient input/output problems often involve sufficiently high signal-to-noise ratios to allow the calculation of meaningful results from a single experiment, as illustrated in the following example.

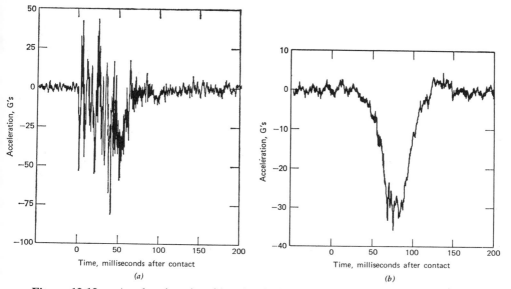

Figure 12.12 Acceleration time histories during automobile collision. (a) Automobile frame. (b) Chest of simulated front seat passenger. These data resulted from studies funded by the FHWA National Highway Safety Bureau, Washington, D.C., under Contract FH-11-7218.

Example 12.12. **Illustration of Energy Spectrum Estimate.** Consider an automobile traveling at a speed of 30 mph that collides head-on with a second automobile traveling at the same speed in the opposite direction. The automobile is carrying a simulated passenger restrained by a conventional seat belt and shoulder harness in the front passenger seat (right side). Figure 12.12 shows the acceleration time histories measured (a) on the vehicle frame just right of the passenger seat and (b) in the chest of the simulated passenger. The energy autospectra for these time histories are presented in Figure 12.13. These spectra were calculated from single records with a frequency resolution of $\Delta f = 1.16$ Hz.

From Equation (12.210), an acceleration gain factor between the vehicle frame and the simulated passenger's chest is given by $|H(f)| = [\mathcal{G}_{yy}(f)/\mathcal{G}_{xx}(f)]^{1/2}$, where $\mathcal{G}_{yy}(f)$ is the energy autospectrum of the passenger acceleration and $\mathcal{G}_{xx}(f)$ is the energy autospectrum of the vehicle acceleration. The results of this calculation are presented in Figure 12.14. A comparison of these results with the transmissibility function illustrated in Figure 2.5 shows the restrained simulated passenger responds to the vehicle impact much like a heavily damped spring supported mass with a natural frequency of about 10 Hz.

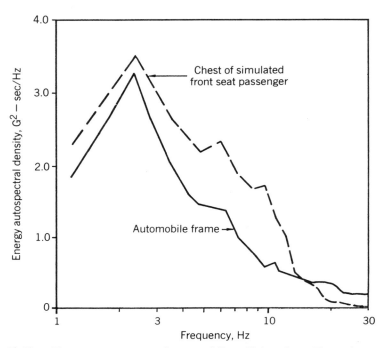

Figure 12.13 Energy autospectra of automobile collision data. These data resulted from studies funded by the FHWA National Highway Safety Bureau, Washington, D.C., under Contract FH-11-7218.

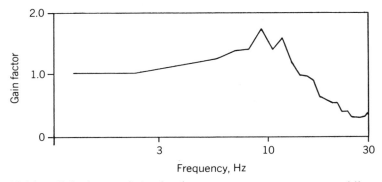

Figure 12.14 Gain factor of simulated passenger response to automobile collision. These data resulted from studies funded by the FHWA National Highway Safety Bureau, Washington, D.C., under Contract FH-11-7218.

PROBLEMS

12.1 Assume a nonstationary random process is defined by

$$\{x(t)\} = a(t) + b(t)\{u(t)\}$$

where $a(t)$ and $b(t)$ are deterministic functions and $\{u(t)\}$ is a stationary random process with zero mean and unity variance. Determine for $\{x(t)\}$ at any time t
(a)   the mean value.
(b)   the variance.
(c)   the mean square value.

12.2 For the nonstationary random process in Problem 12.1, write the equation for the probability density function of $\{x(t)\}$ at any time t assuming $\{u(t)\}$ is Gaussian.

12.3 Consider a nonstationary random process defined by

$$\{x(t)\} = A \sin\frac{\pi t}{T_r} + \{u(t)\} \qquad 0 \le t \le T_r$$

where A is a constant and $\{u(t)\}$ is a stationary random process with zero mean and unity variance. Assume the mean value of $\{x(t)\}$ is estimated by short time-averaging procedures with an averaging time of $T = 0.1T_r$. Determine the bias error in the mean value estimate at $t = 0.5T_r$
(a)   exactly.
(b)   using the approximation in Equation (12.32).

12.4 Assume a nonstationary random process has the form

$$\{x(t)\} = Ae^{-at}\{u(t)\}$$

where A and a are positive constants and $\{u(t)\}$ is a stationary random process with zero mean and unity variance. Determine the double time autocorrelation function of $\{x(t)\}$ in terms of
(a) $R_{xx}(t_1, t_2)$ defined in Equation (12.54).
(b) $\mathscr{R}_{xx}(\tau, t)$ defined in Equation (12.64).

12.5 For the nonstationary random process in Problem 12.4, if $\mathscr{R}_{xx}(\tau, t)$ is an even function of t, determine the double frequency autospectrum in

terms of
(a) $S_{xx}(f_1, f_2)$ defined in Equation (12.77).
(b) $\mathscr{S}_{xx}(f, g)$ defined in Equation (12.93).
Hint: Equations (12.83) and (12.101) will be helpful.

12.6   For the nonstationary random process in Problem 12.4, determine the frequency–time autospectrum $\mathscr{W}_{xx}(f, t)$ defined in Equation (12.123).

12.7   Assume a nonstationary random process has a double frequency autospectrum given by

$$\mathscr{S}_{xx}(f, g) = Ae^{-a(|f|+|g|)}$$

where A and a are positive constants. Determine
(a)   the frequency–time autospectrum $\mathscr{W}_{xx}(f, t)$ for the process.
(b)   the energy autospectral density function for the process.

12.8   For the nonstationary random process in Problem 12.7, determine the double time autocorrelation function $\mathscr{R}_{xx}(\tau, t)$.

12.9   Consider a nonstationary random process described by the product model

$$\{x(t)\} = \cos 2\pi f_0 t \{u(t)\}$$

where f_0 is a constant and $\{u(t)\}$ is a stationary random process with zero mean and unity variance. If the one-sided autospectrum of $\{u(t)\}$ is given by

$$G_{uu}(f) = \begin{cases} \dfrac{1}{B} & f_1 - B/2 \leq f \leq f_1 + B/2 \\ 0 & \text{otherwise} \end{cases}$$

determine the time-averaged autospectrum of $\{x(t)\}$ assuming $f_0 = B/2$ and $f_1 \geq B$.

12.10   Consider a time-varying linear system with a frequency response function defined by

$$H(f, t) = \frac{1}{1 + j2\pi ft}$$

Assume a stationary random input to the system is white noise with an autospectrum of $G_{xx}(f) = G$. Determine the autospectrum of the output of the system in terms of a double frequency spectrum.

REFERENCES

12.1 Claasen, T. A. C. M., and Mecklenbrauker, W. F. G., "The Wigner Distribution—A Tool for Time–Frequency Signal Analysis," *Philips Journal of Research*, Vol. 35, Nos. 3-6, 1980.

12.2 Mark, W. D., and Fischer, R. W., "Investigation of the Effects of Nonhomogeneous (or Nonstationary) Behavior on the Spectra of Atmospheric Turbulence," NASA CR-2745, NASA Langley Research Center, Virginia, February 1976.

12.3 Piersol, A. G., "Power Spectra Measurements for Spacecraft Vibration Data," *J. Spacecraft and Rockets*, Vol. 4, p. 1613, December 1967.

12.4 Silverman, R. A., "Locally Stationary Random Processes," *Transactions of the IRE, Information Theory*, Vol. IT-3, p. 182, September 1957.

12.5 Lawson, J. I., and Uhlenbeck, G. E., *Threshold Signals*, McGraw-Hill, New York, 1950.

CHAPTER 13

THE HILBERT TRANSFORM

The Hilbert transform of a real-valued time domain signal $x(t)$ is another real-valued time domain signal, denoted by $\tilde{x}(t)$, such that $z(t) = x(t) + j\tilde{x}(t)$ is an analytic signal. The Fourier transform of $x(t)$ is a complex-valued frequency domain signal $X(f)$, which is clearly quite different from the Hilbert transform $\tilde{x}(t)$ or the quantity $z(t)$. From $z(t)$, one can define a magnitude function $A(t)$ and a phase function $\theta(t)$, where $A(t)$ describes the envelope of the original function $x(t)$ versus time, and $\theta(t)$ describes the instantaneous phase of $x(t)$ versus time. Section 13.1 gives three equivalent mathematical definitions for Hilbert transforms, followed by examples and basic properties. The intrinsic nature of Hilbert transforms to causal functions and physically realizable systems is also shown. Section 13.2 derives special formulas for Hilbert transforms of correlation functions and their envelopes. Applications are outlined for both nondispersive and dispersive propagation problems. Section 13.3 discusses the computation of two envelope signals followed by correlation of the envelope signals. Further material on Hilbert transforms and its applications appears in References 13.1 to 13.6.

13.1   HILBERT TRANSFORMS FOR GENERAL RECORDS

The Hilbert transform for general records can be defined in three ways as follows, where all integrals shown will exist in practice.

(1)   Definition as Convolution Integral

The *Hilbert transform* of a real-valued function $x(t)$ extending over the range $-\infty < t < \infty$ is a real-valued function $\tilde{x}(t)$ defined by

$$\tilde{x}(t) = \mathscr{H}[x(t)] = \int_{-\infty}^{\infty} \frac{x(u)}{\pi(t-u)}\, du \tag{13.1}$$

Thus $\tilde{x}(t)$ is the convolution integral of $x(t)$ and $(1/\pi t)$, written as

$$\tilde{x}(t) = x(t) * (1/\pi t) \qquad (13.2)$$

Like Fourier transforms, Hilbert transforms are linear operators where

$$\mathcal{H}[a_1 x_1(t) + a_2 x_2(t)] = a_1 \mathcal{H}[x_1(t)] + a_2 \mathcal{H}[x_2(t)] \qquad (13.3)$$

for any constants a_1, a_2 and any functions $x_1(t)$, $x_2(t)$.

(2) Definition as $(\pi/2)$ Phase Shift System

Let $\tilde{X}(f)$ be the Fourier transform of $\tilde{x}(t)$, namely,

$$\tilde{X}(f) = \mathcal{F}[\tilde{x}(t)] = \int_{-\infty}^{\infty} \tilde{x}(t) e^{-j2\pi ft} \, dt \qquad (13.4)$$

Then, from Equation (13.2), it follows that $\tilde{X}(f)$ is the Fourier transform $X(f)$ of $x(t)$, multiplied by the Fourier transform of $(1/\pi t)$. The Fourier transform of $(1/\pi t)$ is given by

$$\mathcal{F}[1/\pi t] = -j \, \mathrm{sgn} \, f = \begin{cases} -j & \text{for } f > 0 \\ j & \text{for } f < 0 \end{cases} \qquad (13.5)$$

At $f = 0$, the function $\mathrm{sgn} \, f = 0$. Hence Equation (13.2) is equivalent to the passage of $x(t)$ through a system defined by $(-j \, \mathrm{sgn} \, f)$ to yield

$$\tilde{X}(f) = (-j \, \mathrm{sgn} \, f) X(f) \qquad (13.6)$$

This complex-valued quantity $\tilde{X}(f)$ is *not* the Hilbert transform of the complex-valued quantity $X(f)$. Its relation to $\tilde{x}(t)$ is

$$\tilde{x}(t) = \int_{-\infty}^{\infty} \tilde{X}(f) e^{j2\pi ft} \, df \qquad (13.7)$$

In words, $\tilde{x}(t)$ is the inverse Fourier transform of $\tilde{X}(f)$.

The Fourier transform $(-j \, \mathrm{sgn} \, f)$ can be represented by

$$B(f) = -j \, \mathrm{sgn} \, f = \begin{cases} e^{-j(\pi/2)} & f > 0 \\ e^{j(\pi/2)} & f < 0 \end{cases} \qquad (13.8)$$

with $B(0) = 0$. Also

$$B(f) = |B(f)| e^{-j\phi_b(f)} \qquad (13.9)$$

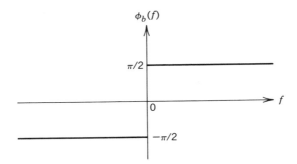

Hence $B(f)$ is a $(\pi/2)$ phase shift system where

$$|B(f)| = 1 \qquad \text{for all } f \neq 0 \tag{13.10}$$

$$\phi_b(f) = \begin{cases} \pi/2 & \text{for } f > 0 \\ -\pi/2 & \text{for } f < 0 \end{cases} \tag{13.11}$$

If one lets

$$X(f) = |X(f)| e^{-j\phi_x(f)} \tag{13.12}$$

it follows that

$$\tilde{X}(f) = \tilde{X}(f) e^{-j\tilde{\phi}_x(f)} = |X(f)| e^{-j[\phi_x(f) + \phi_b(f)]} \tag{13.13}$$

Thus the Hilbert transform consists of passing $x(t)$ through a system which leaves the magnitude of $X(f)$ unchanged, but changes the phase from $\phi_x(f)$ to $\phi_x(f) + \phi_b(f)$ using the $\phi_b(f)$ of Equation (13.11), that is,

$$\phi_x(f) \to \phi_x(f) + (\pi/2) \qquad \text{for } f > 0$$
$$\phi_x(f) \to \phi_x(f) - (\pi/2) \qquad \text{for } f < 0 \tag{13.14}$$

In words, shift $(\pi/2)$ for positive frequencies and shift $(-\pi/2)$ for negative frequencies as sketched above.

(3)   Definition as Imaginary Part of Analytic Signal

A third useful way to understand and to compute the Hilbert transform $\tilde{x}(t)$ of $x(t)$ is via the *analytic signal* $z(t)$ associated with $x(t)$, defined by

$$z(t) = x(t) + j\tilde{x}(t) \tag{13.15}$$

One can also write

$$z(t) = A(t) e^{j\theta(t)} \tag{13.16}$$

where $A(t)$ is called the *envelope signal* of $x(t)$ and $\theta(t)$ is called the *instantaneous phase signal* of $x(t)$. In terms of $x(t)$ and $\tilde{x}(t)$, it is clear that

$$A(t) = \left[x^2(t) + \tilde{x}^2(t)\right]^{1/2} \tag{13.17}$$

$$\theta(t) = \tan^{-1}\left[\frac{\tilde{x}(t)}{x(t)}\right] = 2\pi f_0 t \tag{13.18}$$

The "instantaneous frequency" f_0 is given by

$$f_0 = \left(\frac{1}{2\pi}\right)\frac{d\theta(t)}{dt} \tag{13.19}$$

Let $Z(f)$ be the Fourier transform of $z(t)$, namely,

$$Z(f) = \mathscr{F}[z(t)] = \mathscr{F}[x(t) + j\tilde{x}(t)]$$
$$= \mathscr{F}[x(t)] + j\mathscr{F}[\tilde{x}(t)] = X(f) + j\tilde{X}(f) \tag{13.20}$$

The inverse Fourier transform of $Z(f)$ yields

$$z(t) = \mathscr{F}^{-1}[Z(f)] = x(t) + j\tilde{x}(t) \tag{13.21}$$

where

$$\tilde{x}(t) = \mathscr{H}[x(t)] = \operatorname{Im}[z(t)] \tag{13.22}$$

13.1.1 *Computation of Hilbert Transforms*

To compute $Z(f)$, note from Equation (13.6) that

$$\tilde{X}(f) = (-j\operatorname{sgn} f)X(f)$$

Hence, Equation (13.20) becomes

$$Z(f) = [1 + \operatorname{sgn} f]X(f) = B_1(f)X(f) \tag{13.23}$$

where, as shown in the sketch, $B_1(0) = 1$ and

$$B_1(f) = \begin{cases} 2 & \text{for } f > 0 \\ 0 & \text{for } f < 0 \end{cases} \tag{13.24}$$

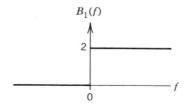

This is a very simple transformation to obtain $Z(f)$ from $X(f)$. One should compute $X(f)$ for all f and then define $Z(f)$ by $Z(0) = X(0)$ and

$$Z(f) = \begin{cases} 2X(f) & \text{for } f > 0 \\ 0 & \text{for } f < 0 \end{cases} \tag{13.25}$$

The inverse Fourier transform of $Z(f)$ then gives $z(t)$ with $\tilde{x}(t) = \text{Im}[z(t)]$. This is the recommended way to compute the Hilbert transform. From Equation (13.25),

$$x(t) = \text{Re}\left[2 \int_0^\infty X(f) e^{j2\pi ft}\, df \right]$$

$$\tilde{x}(t) = \text{Im}\left[2 \int_0^\infty X(f) e^{j2\pi ft}\, df \right] \tag{13.26}$$

Example 13.1. **Digital Formulas for $x(t)$ and $\tilde{x}(t)$.**   For digital computations, from Equations (11.76) and (11.77), one obtains for $n = 0, 1, 2, \ldots,$ $(N - 1)$,

$$x(n\,\Delta t) = 2\,\Delta f\, \text{Re}\left[\sum_{k=0}^{N/2} X(k\,\Delta f) \exp\left(j\frac{2\pi kn}{N} \right) \right]$$

$$\tilde{x}(n\,\Delta t) = 2\,\Delta f\, \text{Im}\left[\sum_{k=0}^{N/2} X(k\,\Delta f) \exp\left(j\frac{2\pi kn}{N} \right) \right]$$

Here, the factor $\Delta f = (1/N\Delta t)$ with

$$X(k\,\Delta f) = \Delta t \sum_{n=0}^{N-1} x(n\,\Delta t) \exp\left(-j\frac{2\pi kn}{N} \right)$$

Note that values of $X(k\,\Delta f)$ are needed only from $k = 0$ up to $k = (N/2)$, where the Nyquist frequency occurs, to obtain the digitized values of $x(n\,\Delta t)$ and its Hilbert transform $\tilde{x}(n\,\Delta t)$. The envelope signal of $x(t)$ is given by

$$A(n\,\Delta t) = \left[x^2(n\,\Delta t) + \tilde{x}^2(n\,\Delta t) \right]^{1/2}$$

13.1.2   Examples of Hilbert Transforms

Table 13.1 gives several $x(t)$ with their associated $\tilde{x}(t)$ and envelopes $A(t)$. Proofs of these results follow directly from previous definitions. These examples are plotted in Figure 13.1. The last result in Table 13.1 is a special case of a general theorem that

$$\mathcal{H}\left[u(t)\cos 2\pi f_0 t \right] = u(t)\sin 2\pi f_0 t \tag{13.27}$$

for any function $u(t)$ that is an even function.

Table 13.1

Examples of Hilbert Transforms

$x(t)$	$\tilde{x}(t)$	$A(t)$
$\cos 2\pi f_0 t$	$\sin 2\pi f_0 t$	1
$\sin 2\pi f_0$	$-\cos 2\pi f_0 t$	1
$\dfrac{\sin t}{t}$	$\dfrac{1 - \cos t}{t}$	$\left\| \dfrac{\sin(t/2)}{(t/2)} \right\|$
$\dfrac{1}{1 + t^2}$	$\dfrac{t}{1 + t^2}$	$\left[\dfrac{1}{1 + t^2} \right]^{1/2}$
$e^{-c|t|}\cos 2\pi f_0 t$	$e^{-c|t|}\sin 2\pi f_0 t$	$e^{-c|t|}$

13.1.3 *Properties of Hilbert Transforms*

A number of properties will now be stated for Hilbert transforms, which follow readily from the basic definition or are proved in the References. Let $\tilde{x}(t) = \mathcal{H}[x(t)]$ and $\tilde{y}(t) = \mathcal{H}[y(t)]$ be Hilbert transforms of $x(t)$ and $y(t)$, with corresponding Fourier transforms $X(f)$ and $Y(f)$.

(a)   *Linear Property*

$$\mathcal{H}\left[ax(t) + by(t) \right] = a\tilde{x}(t) + b\tilde{y}(t) \qquad (13.28)$$

for any functions $x(t)$, $y(t)$ and any constants a, b.

(b)   *Shift Property*

$$\mathcal{H}\left[x(t - a) \right] = \tilde{x}(t - a) \qquad (13.29)$$

(c)   *Hilbert Transform of Hilbert Transform*

$$\mathcal{H}\left[\tilde{x}(t) \right] = -x(t) \qquad (13.30)$$

In words, the application of two successive Hilbert transforms gives the negative of the original function.

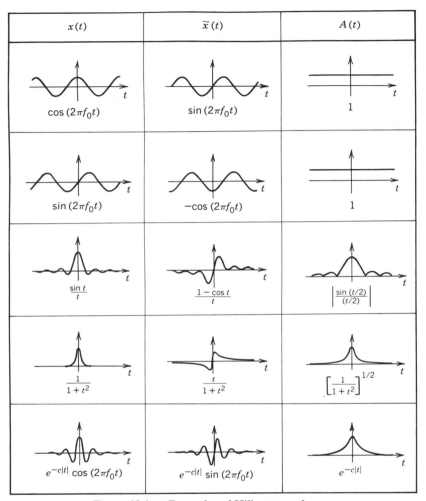

$x(t)$	$\tilde{x}(t)$	$A(t)$
$\cos(2\pi f_0 t)$	$\sin(2\pi f_0 t)$	1
$\sin(2\pi f_0 t)$	$-\cos(2\pi f_0 t)$	1
$\dfrac{\sin t}{t}$	$\dfrac{1-\cos t}{t}$	$\left\|\dfrac{\sin(t/2)}{(t/2)}\right\|$
$\dfrac{1}{1+t^2}$	$\dfrac{t}{1+t^2}$	$\left[\dfrac{1}{1+t^2}\right]^{1/2}$
$e^{-c\|t\|}\cos(2\pi f_0 t)$	$e^{-c\|t\|}\sin(2\pi f_0 t)$	$e^{-c\|t\|}$

Figure 13.1   Examples of Hilbert transforms.

(d)   *Inverse Hilbert Transform of $\tilde{x}(t)$*

$$x(t) = \mathcal{H}^{-1}[\tilde{x}(t)] = -\int_{-\infty}^{\infty} \frac{\tilde{x}(u)}{\pi(t-u)}\, du \qquad (13.31)$$

Thus $x(t)$ is the convolution of $\tilde{x}(t)$ with $(-1/\pi t)$. Alternatively, $x(t)$ can be defined by

$$x(t) = \mathcal{F}^{-1}[(j\,\mathrm{sgn}\,f)\tilde{X}(f)] \qquad (13.32)$$

where

$$\tilde{X}(f) = \mathcal{F}[\tilde{x}(t)] \qquad (13.33)$$

(e)   *Even and Odd Function Properties*

If $x(t)$ is an even (odd) function of t, then $\tilde{x}(t)$ is an odd (even) function of t:

$$x(t) \text{ even} \leftrightarrow \tilde{x}(t) \text{ odd}$$

$$x(t) \text{ odd} \leftrightarrow \tilde{x}(t) \text{ even}$$

(13.34)

(f)   *Similarity Property*

$$\mathscr{H}[x(at)] = \tilde{x}(at)$$

(13.35)

(g)   *Energy Property*

$$\int_{-\infty}^{\infty} x^2(t)\, dt = \int_{-\infty}^{\infty} \tilde{x}^2(t)\, dt$$

(13.36)

This follows from Parseval's theorem since

$$\int_{-\infty}^{\infty} x^2(t)\, dt = \int_{-\infty}^{\infty} |X(f)|^2\, df$$

$$\int_{-\infty}^{\infty} \tilde{x}^2(t)\, dt = \int_{-\infty}^{\infty} |\tilde{X}(f)|^2\, df$$

(13.37)

and the fact that

$$|\tilde{X}(f)|^2 = |X(f)|^2$$

(13.38)

(h)   *Orthogonal Property*

$$\int_{-\infty}^{\infty} x(t)\tilde{x}(t)\, dt = 0$$

(13.39)

This follows from Parseval's theorem since

$$\int_{-\infty}^{\infty} x(t)\tilde{x}(t)\, dt = \int_{-\infty}^{\infty} X^*(f)\tilde{X}(f)\, df$$

(13.40)

and the fact that

$$X^*(f)\tilde{X}(f) = (-j \operatorname{sgn} f)|X(f)|^2$$

(13.41)

is an odd function of f so that the right-hand side of Equation (13.40) is zero.

(i)   *Modulation Property*

$$\mathscr{H}[x(t)\cos 2\pi f_0 t] = x(t)\sin 2\pi f_0 t$$

(13.42)

if $x(t)$ is a signal whose Fourier transform $X(f)$ is bandwidth limited, that is,

$$X(f) = \begin{cases} X(f) & |f| \leq F \\ 0 & \text{otherwise} \end{cases} \tag{13.43}$$

provided f_0 is such that $f_0 > F$. Also

$$\mathcal{H}[x(t)\sin 2\pi f_0 t] = -x(t)\cos 2\pi f_0 t \tag{13.44}$$

(j)   Convolution Property

$$\mathcal{H}[x(t) * y(t)] = \tilde{x}(t) * y(t) = x(t) * \tilde{y}(t) \tag{13.45}$$

This follows from the fact that

$$\mathcal{F}[x(t) * y(t)] = X(f)Y(f) \tag{13.46}$$

and

$$[(-j\,\text{sgn}\,f)\,X(f)]Y(f) = \tilde{X}(f)Y(f)$$

$$= X(f)[(-j\,\text{sgn}\,f)Y(f)] = X(f)\tilde{Y}(f) \tag{13.47}$$

(k)   Lack of Commutation Property

$$\mathcal{F}\{\mathcal{H}[x(t)]\} \neq \mathcal{H}\{\mathcal{F}[x(t)]\} \tag{13.48}$$

In words, Fourier and Hilbert transforms do *not* commute.

13.1.4   *Relation to Physically Realizable Systems*

A *physically realizable constant-parameter linear system* is defined by a weighting function $h(\tau)$ satisfying Equation (2.3), that is,

$$h(\tau) = 0 \quad \text{for } \tau < 0 \tag{13.49}$$

The corresponding frequency response function $H(f)$ of Equation (2.13) is given by

$$H(f) = \mathcal{F}[h(\tau)] = \int_0^\infty h(\tau)e^{-j2\pi f\tau}\,d\tau$$

$$= H_R(f) - jH_I(f) \tag{13.50}$$

Here, $H_R(f)$ equals the real part of $H(f)$ and $H_I(f)$ equals the imaginary part of $H(f)$ as defined by

$$H_R(f) = \int_0^\infty h(\tau)\cos 2\pi f\tau \, d\tau$$

$$H_I(f) = \int_0^\infty h(\tau)\sin 2\pi f\tau \, d\tau \tag{13.51}$$

It will now be proved that for a system to be physically realizable, it is necessary and sufficient that $H_I(f)$ be the Hilbert transform of $H_R(f)$.

This result is a special case of a more general theorem that applies to all causal functions. By definition, a real-valued function $y(t)$ is a *causal function* if

$$y(t) = 0 \qquad \text{for } t < 0 \tag{13.52}$$

Any function $y(t)$ can always be broken down into the sum of an even function $y_e(t)$ plus an odd function $y_o(t)$ by writing

$$y(t) = y_e(t) + y_o(t) \tag{13.53}$$

where

$$y_e(t) = \tfrac{1}{2}[y(t) + y(-t)]$$

$$y_o(t) = \tfrac{1}{2}[y(t) - y(-t)] \tag{13.54}$$

This gives $y_e(-t) = y_e(t)$ and $y_o(-t) = -y_o(t)$. The resolution of any causal function into its even and odd components is illustrated in Figure 13.2.

Assume now that $y(t)$ is a causal function. Then, from Equations (13.52) and (13.54), for $t > 0$,

$$y_e(t) = \tfrac{1}{2}y(t)$$

$$y_o(t) = \tfrac{1}{2}y(t) = y_e(t) \tag{13.55}$$

For $t < 0$, however,

$$y_e(t) = \tfrac{1}{2}y(-t)$$

$$y_o(t) = -\tfrac{1}{2}y(-t) = -y_e(t) \tag{13.56}$$

Hence, for a causal function,

$$y_o(t) = (\text{sgn } t)\,y_e(t) \tag{13.57}$$

where

$$\text{sgn } t = \begin{cases} 1 & t > 0 \\ -1 & t < 0 \end{cases} \tag{13.58}$$

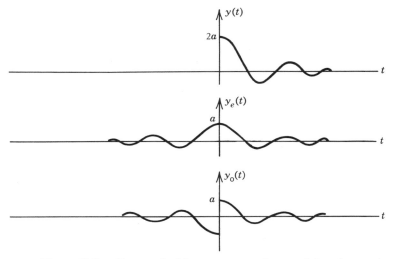

Figure 13.2   Even and odd components of a causal function.

For any causal function $y(t)$, its Fourier transform $Y(f)$ must satisfy

$$Y(f) = \mathscr{F}[y(t)] = \mathscr{F}[y_e(t) + y_o(t)] = Y_R(f) - jY_I(f) \quad (13.59)$$

where

$$\mathscr{F}[y_e(t)] = Y_R(f)$$
$$\mathscr{F}[y_o(t)] = -jY_I(f) \quad (13.60)$$

From Equation (13.57),

$$\mathscr{F}[y_o(t)] \doteq \mathscr{F}[(\operatorname{sgn} t)y_e(t)] = \mathscr{F}[\operatorname{sgn} t] * \mathscr{F}[y_e(t)]$$

where

$$\mathscr{F}[\operatorname{sgn} t] = \frac{-j}{\pi f}$$

Hence

$$\mathscr{F}[y_o(t)] = \int_{-\infty}^{\infty} \frac{-jY_R(u)}{\pi(f-u)} \, du$$

This proves that

$$Y_I(f) = \int_{-\infty}^{\infty} \frac{Y_R(u)}{\pi(f-u)}\, du = \mathcal{H}\left[Y_R(f)\right] \qquad (13.61)$$

In words, $Y_I(f)$ is the Hilbert transform of $Y_R(f)$ when $y(t)$ is a causal function. This completes the proof.

A direct way to determine whether or not a computed $H(f)$ can represent a physically realizable system is now available. It will be physically realizable if $H_I(f)$ is the Hilbert transform of $H_R(f)$ since Equation (13.61) is equivalent to Equation (13.57). Thus Equation (13.61) is both a necessary and sufficient condition for a system to be physically realizable. In equation form,

$$H(f) = H_R(f) - jH_I(f) \qquad \text{with } H_I(f) = \tilde{H}_R(f) \qquad (13.62)$$

It follows that

$$\tilde{H}_I(f) = \tilde{\tilde{H}}_R(f) = -H_R(f) \qquad (13.63)$$

and

$$\tilde{H}(f) = \tilde{H}_R(f) - j\tilde{H}_I(f) = H_I(f) + jH_R(f) = jH(f) \qquad (13.64)$$

Deviations from these results can be used to study nonlinear systems.

Example 13.2. Exponential Causal Function.   Consider the causal function

$$y(t) = \pi e^{-2\pi t} \qquad t > 0 \ (\text{otherwise zero})$$

Here

$$y_e(t) = \frac{\pi}{2} e^{-2\pi|t|} \qquad \text{all } t$$

and

$$y_o(t) = \begin{cases} \dfrac{\pi}{2} e^{-2\pi t} & t > 0 \\[2mm] \dfrac{-\pi}{2} e^{2\pi t} & t < 0 \end{cases}$$

Taking Fourier transforms yields

$$Y_e(f) = Y_R(f) = \frac{1}{2(1+f^2)}$$

and

$$Y_o(f) = Y_I(f) = \frac{f}{2(1 + f^2)}$$

Thus $Y_R(f)$ is an even function of f and $Y_I(f)$ is an odd function of f with

$$Y_I(f) = \mathcal{H}[Y_R(f)]$$

The total Fourier transform of $y(t)$ is then

$$Y(f) = Y_R(f) - jY_I(f) = \frac{1}{2(1 + jf)}$$

Example 13.3. Exponential-Cosine Causal Function.   Consider the causal function

$$y(t) = e^{-at}\cos 2\pi bt \qquad a > 0, \ t > 0 \ (\text{otherwise zero})$$

For this case, the Fourier transform of $y(t)$ is given by

$$Y(f) = \frac{a + j2\pi f}{(a + j2\pi f)^2 + (2\pi b)^2} = Y_R(f) - jY_I(f)$$

where the separate formulas for $Y_R(f)$ and $Y_I(f)$ are quite complicated. Since $y(t)$ is a causal function, however, one will have

$$Y_I(f) = \mathcal{H}[Y_R(f)]$$

with $Y_R(f)$ an even function of f, and $Y_I(f)$ an odd function of f.

13.2   HILBERT TRANSFORMS FOR CORRELATION FUNCTIONS

One of the more important applications for Hilbert transforms involves the calculation of correlation function envelopes to estimate time delays in energy propagation problems. The principles behind such applications are now developed for both nondispersive and dispersive propagation.

13.2.1   *Correlation and Envelope Definitions*

Let $x(t)$ and $y(t)$ represent zero mean value stationary random data with autocorrelation functions $R_{xx}(\tau)$ and $R_{yy}(\tau)$ and cross-correlation function $R_{xy}(\tau)$. Let the associated two-sided autospectral density functions be $S_{xx}(f)$

and $S_{yy}(f)$ and the associated two-sided cross-spectral density function be $S_{xy}(f)$. As defined in Section 5.2, such associated functions are Fourier transform pairs. For each stationary random record $x(t)$ or $y(t)$, let $\tilde{x}(t) = \mathcal{H}[x(t)]$ and $\tilde{y}(t) = \mathcal{H}[y(t)]$ be their Hilbert transforms. Since Hilbert transforms are linear operations, it follows that $\tilde{x}(t)$ and $\tilde{y}(t)$ will also be stationary random data. Various correlation and spectral density functions for $\tilde{x}(t)$ and $\tilde{y}(t)$ can be defined the same as for $x(t)$ and $y(t)$ to yield $R_{\tilde{x}\tilde{x}}(\tau)$, $R_{\tilde{y}\tilde{y}}(\tau)$, $R_{\tilde{x}\tilde{y}}(\tau)$ and $S_{\tilde{x}\tilde{x}}(f)$, $S_{\tilde{y}\tilde{y}}(f)$, $S_{\tilde{x}\tilde{y}}(f)$, where the associated functions are Fourier transform pairs. To be specific, the definitions are

$$R_{\tilde{x}\tilde{x}}(\tau) = E\left[\tilde{x}(t)\tilde{x}(t+\tau)\right]$$

$$R_{\tilde{y}\tilde{y}}(\tau) = E\left[\tilde{y}(t)\tilde{y}(t+\tau)\right] \qquad (13.65)$$

$$R_{\tilde{x}\tilde{y}}(\tau) = E\left[\tilde{x}(t)\tilde{y}(t+\tau)\right]$$

Then their Fourier transforms give

$$S_{\tilde{x}\tilde{x}}(\tau) = \mathcal{F}\left[R_{\tilde{x}\tilde{x}}(\tau)\right]$$

$$S_{\tilde{y}\tilde{y}}(f) = \mathcal{F}\left[R_{\tilde{y}\tilde{y}}(\tau)\right] \qquad (13.66)$$

$$S_{\tilde{x}\tilde{y}}(f) = \mathcal{F}\left[R_{\tilde{x}\tilde{y}}(\tau)\right]$$

Except for a scale factor (which is unimportant for theoretical derivations, since it will appear on both sides of equations and therefore cancel out), one can also use the definitions

$$S_{\tilde{x}\tilde{x}}(f) = E\left[\tilde{X}^*(f)\tilde{X}(f)\right] = E\left[|\tilde{X}(f)|^2\right]$$

$$S_{\tilde{y}\tilde{y}}(f) = E\left[\tilde{Y}^*(f)\tilde{Y}(f)\right] = E\left[|\tilde{Y}(f)|^2\right] \qquad (13.67)$$

$$S_{\tilde{x}\tilde{y}}(f) = E\left[\tilde{X}^*(f)\tilde{Y}(f)\right]$$

Mixing functions and their Hilbert transforms leads to $R_{x\tilde{y}}(\tau)$, $R_{\tilde{x}y}(\tau)$ and the associated $S_{x\tilde{y}}(f)$, $S_{\tilde{x}y}(f)$ defined by

$$R_{x\tilde{y}}(\tau) = E\left[x(t)\tilde{y}(t+\tau)\right]$$

$$R_{\tilde{x}y}(\tau) = E\left[\tilde{x}(t)y(t+\tau)\right] \qquad (13.68)$$

$$S_{x\tilde{y}}(f) = \mathcal{F}\left[R_{x\tilde{y}}(\tau)\right]$$

$$S_{\tilde{x}y}(f) = \mathcal{F}\left[R_{\tilde{x}y}(\tau)\right] \qquad (13.69)$$

Special cases are $R_{x\tilde{x}}(\tau)$, $R_{\tilde{x}x}(\tau)$, $S_{x\tilde{x}}(f)$, and $S_{\tilde{x}x}(f)$. Except for a scale factor, as noted previously, one can also define $S_{x\tilde{y}}(f)$ and $S_{\tilde{x}y}(f)$ directly by

$$S_{x\tilde{y}}(f) = E[X^*(f)\tilde{Y}(f)]$$

$$S_{\tilde{x}y}(f) = E[\tilde{X}^*(f)Y(f)]$$

(13.70)

Envelope functions for $R_{xx}(\tau)$, $R_{yy}(\tau)$, and $R_{xy}(\tau)$ are defined in terms of Hilbert transforms $\tilde{R}_{xx}(\tau)$, $\tilde{R}_{yy}(\tau)$, and $\tilde{R}_{xy}(\tau)$ as follows. The envelope function for $R_{xx}(\tau)$ is

$$A_{xx}(\tau) = \left[R_{xx}^2(\tau) + \tilde{R}_{xx}^2(\tau)\right]^{1/2}$$

(13.71)

The envelope function for $R_{yy}(\tau)$ is

$$A_{yy}(\tau) = \left[R_{yy}^2(\tau) + \tilde{R}_{yy}^2(\tau)\right]^{1/2}$$

(13.72)

The envelope function for $R_{xy}(\tau)$ is

$$A_{xy}(\tau) = \left[R_{xy}^2(\tau) + \tilde{R}_{xy}^2(\tau)\right]^{1/2}$$

(13.73)

Various relationships and properties for all of the above quantities will now be stated.

13.2.2 Hilbert Transform Relations

Table 13.2 gives a number of useful relationships for autocorrelation (auto-spectral density) and cross-correlation (cross-spectral density) functions. Proofs of these results represent straightforward exercises.

Some results in Table 13.2 are worthy of attention. Specifically,

$$\tilde{R}_{xx}(\tau) = \mathscr{H}[R_{xx}(\tau)] = R_{x\tilde{x}}(\tau)$$

(13.74)

In words, the Hilbert transform of $R_{xx}(\tau)$ is the cross-correlation function between $x(t)$ and its Hilbert transform $\tilde{x}(t)$. Also, note that $\tilde{R}_{xx}(\tau)$ is an odd function of τ since $R_{xx}(\tau)$ is an even function of τ. This gives

$$\tilde{R}_{xx}(0) = R_{x\tilde{x}}(0) = 0$$

(13.75)

Thus, a zero crossing of $\tilde{R}_{xx}(\tau)$ occurs at $\tau = 0$, corresponding to a maximum value of $R_{xx}(\tau)$ at $\tau = 0$. This crossing of zero by $\tilde{R}_{xx}(\tau)$ will be with positive slope, that is,

$$\tilde{R}_{xx}(0-) < 0 \quad \text{and} \quad \tilde{R}_{xx}(0+) > 0$$

(13.76)

Table 13.2

Hilbert Transform Relationships

$$R_{\tilde{x}\tilde{x}}(\tau) = R_{xx}(\tau) \qquad S_{\tilde{x}\tilde{x}}(f) = S_{xx}(f)$$

$$R_{\tilde{x}\tilde{x}}(-\tau) = R_{\tilde{x}\tilde{x}}(\tau), \text{ an even function of } \tau$$

$$\sigma_{\tilde{x}}^2 = R_{\tilde{x}\tilde{x}}(0) = R_{xx}(0) = \sigma_x^2$$

$$\tilde{R}_{xx}(\tau) = R_{x\tilde{x}}(\tau) = -R_{\tilde{x}x}(\tau) \qquad \tilde{S}_{xx}(f) = S_{x\tilde{x}}(f) = -S_{\tilde{x}x}(f)$$

$$\tilde{R}_{xx}(-\tau) = -\tilde{R}_{xx}(\tau), \text{ an odd function of } \tau$$

$$\tilde{R}_{xx}(0) = R_{x\tilde{x}}(0) = R_{\tilde{x}x}(0) = 0$$

$$R_{\tilde{x}\tilde{y}}(\tau) = R_{xy}(\tau) \qquad S_{\tilde{x}\tilde{y}}(f) = S_{xy}(f)$$

$$\sigma_{\tilde{x}\tilde{y}} = R_{\tilde{x}\tilde{y}}(0) = R_{xy}(0) = \sigma_{xy}$$

$$\tilde{R}_{xy}(\tau) = R_{x\tilde{y}}(\tau) = -R_{\tilde{x}y}(\tau) \qquad \tilde{S}_{xy}(f) = S_{x\tilde{y}}(f) = -S_{\tilde{x}y}(f)$$

$$\tilde{R}_{xy}(-\tau) = R_{\tilde{y}x}(\tau) \qquad \tilde{S}_{xy}^*(f) = S_{\tilde{y}x}(f)$$

as can be verified for any of the examples in Table 13.1. This property is just the reverse of zero crossings by derivative functions $R'_{xx}(\tau)$ stated in Equation (5.146).

13.2.3 *Analytic Signals for Correlation Functions*

The analytic signal $z_x(t)$ corresponding to a stationary random record $x(t)$ is defined as in Equation (13.15) by

$$z_x(t) = x(t) + j\tilde{x}(t)$$

where $\tilde{x}(t) = \mathcal{H}[x(t)]$. The *complex-valued* autocorrelation function $R_{z_x z_x}(\tau)$ is defined by

$$R_{z_x z_x}(\tau) = E\left[z_x^*(t)z_x(t+\tau)\right] = E\left[\{x(t) - j\tilde{x}(t)\}\{x(t+\tau) + j\tilde{x}(t+\tau)\}\right]$$

$$= R_{xx}(\tau) + R_{\tilde{x}\tilde{x}}(\tau) + j\left[R_{x\tilde{x}}(\tau) - R_{\tilde{x}x}(\tau)\right]$$

But, from Table 13.2,

$$R_{\tilde{x}\tilde{x}}(\tau) = R_{xx}(\tau) \qquad \text{and} \qquad R_{\tilde{x}x}(\tau) = -R_{x\tilde{x}}(\tau)$$

Hence

$$R_{z_x z_x}(\tau) = 2\left[R_{xx}(\tau) + jR_{x\tilde{x}}(\tau)\right] = 2\left[R_{xx}(\tau) + j\tilde{R}_{xx}(\tau)\right]$$

This proves that

$$\frac{R_{z_x z_x}(\tau)}{2} = R_{xx}(\tau) + j\tilde{R}_{xx}(\tau) \tag{13.77}$$

is the analytic signal for $R_{xx}(\tau)$.

The Fourier transform of $R_{z_x z_x}(\tau)$ is $S_{z_x z_x}(f)$, where

$$S_{z_x z_x}(f) = \mathscr{F}\left[R_{z_x z_x}(\tau)\right] = 2\left[S_{xx}(f) + j\tilde{S}_{xx}(f)\right]$$

$$= 2[1 + \text{sgn } f]S_{xx}(f)$$

Hence the autospectrum relation is

$$S_{z_x z_x}(f) = \begin{cases} 4S_{xx}(f) & \text{for } f > 0 \\ 0 & \text{for } f < 0 \end{cases} \tag{13.78}$$

From Equation (13.77), one obtains

$$\mathscr{F}^{-1}\left[S_{z_x z_x}(f)/2\right] = R_{xx}(\tau) + j\tilde{R}_{xx}(\tau) \tag{13.79}$$

Theoretical formulas are now

$$R_{xx}(\tau) = 2\int_0^\infty S_{xx}(f)\cos 2\pi f\tau \, df$$

$$\tilde{R}_{xx}(\tau) = 2\int_0^\infty S_{xx}(f)\sin 2\pi f\tau \, df \tag{13.80}$$

The analytic signal for $R_{xy}(\tau)$ corresponding to stationary random records $x(t)$ and $y(t)$ is defined from

$$z_x(t) = x(t) + j\tilde{x}(t) \qquad z_y(t) = y(t) + j\tilde{y}(t)$$

The *complex-valued* cross-correlation function $R_{z_x z_y}(\tau)$ is defined by

$$R_{z_x z_y}(\tau) = E\left[z_x^*(t)z_y(t + \tau)\right] = E\left[\{x(t) - j\tilde{x}(t)\}\{y(t + \tau) + j\tilde{y}(t + \tau)\}\right]$$

$$= R_{xy}(\tau) + R_{\tilde{x}\tilde{y}}(\tau) + j\left[R_{x\tilde{y}}(\tau) - R_{\tilde{x}y}(\tau)\right]$$

But, from Table 13.2,

$$R_{\tilde{x}\tilde{y}}(\tau) = R_{xy}(\tau) \quad \text{and} \quad R_{\tilde{x}y}(\tau) = -R_{x\tilde{y}}(\tau)$$

Hence

$$R_{z_x z_y}(\tau) = 2\left[R_{xy}(\tau) + jR_{x\tilde{y}}(\tau)\right] = 2\left[R_{xy}(\tau) + j\tilde{R}_{xy}(\tau)\right]$$

This proves that

$$\frac{R_{z_x z_y}(\tau)}{2} = R_{xy}(\tau) + \tilde{R}_{xy}(\tau) \tag{13.81}$$

is the analytic signal for $R_{xy}(\tau)$.

The Fourier transform of $R_{z_x z_y}(\tau)$ is $S_{z_x z_y}(f)$, where

$$S_{z_x z_y}(f) = \mathscr{F}\left[R_{z_x z_y}(\tau)\right] = 2\left[S_{xy}(f) + j\tilde{S}_{xy}(f)\right]$$

$$= 2[1 + \operatorname{sgn} f]S_{xy}(f)$$

Hence the cross-spectrum relation is

$$S_{z_x z_y}(f) = \begin{cases} 4S_{xy}(f) & \text{for } f > 0 \\ 0 & \text{for } f < 0 \end{cases} \tag{13.82}$$

From Equation (13.81), one obtains

$$\mathscr{F}^{-1}\left[\frac{S_{z_x z_y}(f)}{2}\right] = R_{xy}(\tau) + j\tilde{R}_{xy}(\tau) \tag{13.83}$$

This yields the theoretical formulas

$$R_{xy}(\tau) = \operatorname{Re}\left[2\int_0^\infty S_{xy}(f)e^{j2\pi f\tau}\,df\right]$$

$$\tilde{R}_{xy}(\tau) = \operatorname{Im}\left[2\int_0^\infty S_{xy}(f)e^{j2\pi f\tau}\,df\right] \tag{13.84}$$

Example 13.4. Digital Formulas for $R_{xy}(\tau)$ and $\tilde{R}_{xy}(\tau)$. For digital computations, as discussed in Sections 11.4.2 and 11.6.2, augment each of the N data values in $\{x(n\,\Delta t)\}$ and $\{y(n\,\Delta t)\}$ with N zeros so as to obtain new data sequences with $2N$ terms. With $\Delta f = (1/2N\Delta t)$, for $k = 0, 1, 2, \ldots, N$, compute

$$X(k\,\Delta f) = \Delta t \sum_{n=0}^{N-1} x(n\,\Delta t)\exp\left(-j\frac{\pi kn}{N}\right)$$

$$Y(k\,\Delta f) = \Delta t \sum_{n=0}^{N-1} y(n\,\Delta t)\exp\left(-j\frac{\pi kn}{N}\right)$$

$$G_{xy}(k\,\Delta f) = \frac{2}{N\Delta t}\left[X^*(k\,\Delta f)Y(k\,\Delta f)\right]$$

For $r = 0, 1, 2, \ldots, (N - 1)$, one then obtains

$$R_{xy}(r\Delta t) = \left(\frac{N\Delta f}{N - r}\right)\text{Re}\left[\sum_{k=0}^{N} G_{xy}(k\,\Delta f)\exp\left(j\frac{\pi kr}{N}\right)\right]$$

$$\tilde{R}_{xy}(r\Delta t) = \left(\frac{N\Delta f}{N - r}\right)\text{Im}\left[\sum_{k=0}^{N} G_{xy}(k\,\Delta f)\exp\left(j\frac{\pi kr}{N}\right)\right]$$

The squared envelope signal of $R_{xy}(\tau)$ is given by

$$A_{xy}^2(r\Delta t) = R_{xy}^2(r\Delta t) + \tilde{R}_{xy}^2(r\Delta t)$$

Autocorrelation functions are special cases where

$$R_{xx}(r\Delta t) = \left(\frac{N\Delta f}{N - r}\right)\left[\sum_{k=0}^{N} G_{xx}(k\,\Delta f)\cos(\pi kr/N)\right]$$

$$\tilde{R}_{xx}(r\Delta t) = \left(\frac{N\Delta f}{N - r}\right)\left[\sum_{k=0}^{N} G_{xx}(k\,\Delta f)\sin(\pi kr/N)\right]$$

$$A_{xx}^2(r\Delta t) = R_{xx}^2(r\Delta t) + \tilde{R}_{xx}^2(r\Delta t)$$

13.2.4 *Nondispersive Propagation Problems*

Consider the basic nondispersive propagation case previously defined in Equation (5.19). For simplicity, assume that $n(t)$ is essentially zero, so that

$$y(t) = \alpha x(t - \tau_0) \tag{13.85}$$

with

$$R_{yy}(\tau) = \alpha^2 R_{xx}(\tau)$$

$$R_{xy}(\tau) = \alpha R_{xx}(\tau - \tau_0) \tag{13.86}$$

$$\rho_{xy}(\tau) = \frac{R_{xy}(\tau)}{\sqrt{R_{xx}(0)\,R_{yy}(0)}} = \rho_{xx}(\tau - \tau_0)$$

Consider the case where $x(t)$ is bandwidth limited white noise as defined by

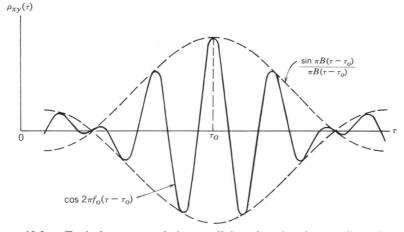

$\rho_{xy}(\tau)$

$\dfrac{\sin \pi B(\tau - \tau_o)}{\pi B(\tau - \tau_o)}$

$\cos 2\pi f_o(\tau - \tau_o)$

0

τ_O

τ

Figure 13.3 Typical cross-correlation coefficient function for nondispersive propagation through a single path.

Equation (5.48), namely,

$$R_{xx}(\tau) = aB \left(\frac{\sin \pi B \tau}{\pi B \tau} \right) \cos 2\pi f_0 \tau \qquad (13.87)$$

Now

$$R_{xx}(0) = aB$$

$$R_{yy}(0) = \alpha^2(aB) \qquad (13.88)$$

$$\rho_{xy}(\tau) = \left[\frac{\sin \pi B(\tau - \tau_0)}{\pi B(\tau - \tau_0)} \right] \cos 2\pi f_0(\tau - \tau_0) \qquad (13.89)$$

Here f_0 is the center of an ideal rectangular filter of bandwidth B, where the autospectral density function $G_{xx}(f)$ is a constant within this band and zero outside. The time delay τ_0 is a constant. Equation (13.89) is plotted in Figure 13.3.

A number of properties of Equation (13.89) for nondispersive propagation are worthy of note.

1. The cosine function $\cos 2\pi f_0 \tau$ at the center frequency f_0 is modulated by the envelope function $[(\sin \pi B \tau)/\pi B \tau]$ that is defined by the bandwidth B.
2. The peak value of the envelope function occurs at the time delay $\tau_0 = (d/c)$, a fixed value independent of frequency.

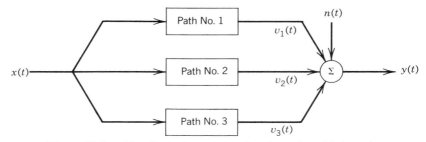

Figure 13.4 Nondispersive propagation through multiple paths.

3. Peak values of the cosine function occur at time delays $\tau_n = \tau_0 \pm (n/f_0)$, where n is any integer. In particular, a peak value of the cosine function coincides with the peak value of the envelope function.

4. The main lobe of the envelope function extends from $[\tau_0 - (1/B)]$ to $[\tau_0 + (1/B)]$, a width of $(2/B)$ sec. The number of oscillations of the cosine function contained in the main lobe of the envelope function is $(2f_0/B)$.

Three important types of nondispersive propagation problems where one desires to measure the envelope of cross-correlation functions are illustrated in the following examples. As in all engineering problems, it is necessary to derive statistical error analysis criteria to help design experiments and to evaluate the estimates of computed results. In particular, formulas are required to determine the statistical errors in (a) magnitude estimates of the envelope peak values, and (b) location estimates of the time delays where the envelope peak values occur. These matters are covered in Reference 13.1 based upon material in Section 8.4.

***Example 13.5.* Nondispersive Propagation Through Multiple Paths.** Figure 13.4 shows a single input passing through multiple paths to produce a single output. For simplicity, only three paths are shown, but the number is arbitrary. The governing equation for this model is

$$y(t) = \alpha_1 x(t - \tau_1) + \alpha_2 x(t - \tau_2) + \alpha_3 x(t - \tau_3) + n(t)$$

where α_i are constant attenuation factors for each path and τ_i are the respective time delays for each path. The noise $n(t)$ is assumed to be uncorrelated with $x(t)$. Simultaneous measurements are made only of $x(t)$ and $y(t)$. It is required to determine the proportion of the output power in $y(t)$ that goes via each path and the respective time delays for passage of $x(t)$ through each path. Cross-correlation measurements are made between $x(t)$ and $y(t)$, followed by Hilbert transform computations for its envelope. Here

$$R_{xy}(\tau) = \alpha_1 R_{xx}(\tau - \tau_1) + \alpha_2 R_{xx}(\tau - \tau_2) + \alpha_3 R_{xx}(\tau - \tau_3)$$

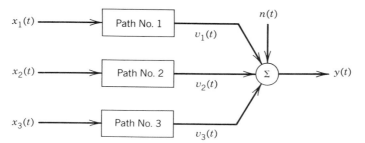

Figure 13.5   Nondispersive propagation from multiple uncorrelated sources.

Example 13.6. **Nondispersive Propagation from Multiple Uncorrelated Sources.**   Figure 13.5 shows separate inputs with different paths to produce a single output. Simultaneous measurements are made of all inputs and the output. For simplicity, only three sources and associated paths are shown, but the number of uncorrelated sources can be arbitrary. It is required to determine the proportion of the output power in $y(t)$ that comes from each source and the respective time delays for passage of each source through its particular path. The output noise $n(t)$ is assumed to be uncorrelated with the inputs $x_i(t)$, and the three sources are also assumed to be uncorrelated with each other. The governing equation for this model is

$$y(t) = \alpha_1 x_1(t - \tau_1) + \alpha_2 x_2(t - \tau_2) + \alpha_3 x_3(t - \tau_3) + n(t)$$

where α_i are constant attenuation factors for each path and τ_i are the respective time delays for each path. Cross-correlation measurements are made between each $x_i(t)$ and $y(t)$, followed by Hilbert transform computations to obtain their envelopes. Here

$$R_{x_i y}(\tau) = \alpha_i R_{xx}(\tau - \tau_i)$$

Example 13.7. **Nondispersive Propagation from an Unmeasured Single Source to Measured Multiple Outputs.**   Figure 13.6 shows separate outputs due to an unmeasured single input. Simultaneous measurements are made of all output signals. For simplicity, only three outputs are shown, but the number of different outputs can be arbitrary. It is required to determine the relative time delay between all pairs of output signals without knowing the actual time delays from the input $x(t)$ to any of the outputs, assuming $x(t)$ is not or cannot be measured. The extraneous noise terms $n_i(t)$ are all assumed to be uncorrelated with each other and with $x(t)$. The governing equations for this model are that each output for $i = 1, 2, 3$ is of the form

$$y_i(t) = v_i(t) + n_i(t) = \alpha_i x(t - \tau_i) + n_i(t)$$

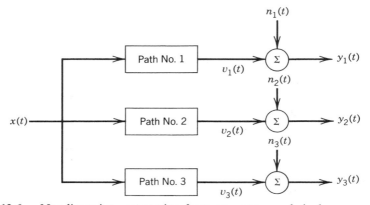

Figure 13.6   Nondispersive propagation from an unmeasured single source to measured multiple outputs.

Cross-correlation measurements are made between pairs of output records, followed by Hilbert transform computations of their envelopes. Here

$$R_{y_i y_j}(\tau) = \alpha_i \alpha_j R_{xx}(\tau + \tau_i - \tau_j)$$

13.2.5   *Dispersive Propagation Problems*

Material in Section 13.2.4 applies to nondispersive propagation problems where the velocity of propagation is a constant independent of frequency. Consider other situations where the propagation paths are frequency dispersive as discussed in References 13.7 and 13.8. In particular for flexural waves in structures, the "apparent" propagation speed of the waves at a given frequency is called the *group velocity* c_g. This c_g is related but not equal to the *phase velocity* c_p. It is known that the group velocity of flexural waves in thin beams satisfies

$$c_g = 2c_p \sim \sqrt{f} \tag{13.90}$$

In words, c_g is twice c_p and both are proportional to the square root of frequency.

For dispersive propagation problems governed by Equation (13.90), as a first order of approximation, Reference 13.8 proves that the cross-correlation coefficient function corresponding to Equation (13.89) now takes the form

$$\rho_{xy}(\tau) \approx \left[\frac{\sin \pi B_0(\tau - \tau_2)}{\pi B_0(\tau - \tau_2)} \right] \cos 2\pi f_0(\tau - \tau_1) \tag{13.91}$$

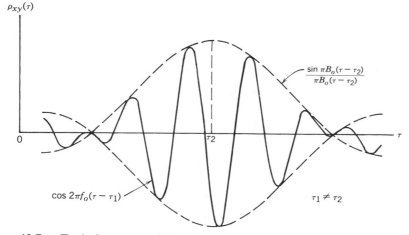

Figure 13.7 Typical cross-correlation coefficient function for dispersive propagation through a single path.

where

$$\tau_1 = \left(d/c_p\right) \sim f_0^{-1/2} \tag{13.92}$$

$$\tau_2 = \left(d/c_g\right) \sim f_0^{-1/2} \tag{13.93}$$

with $\tau_1 = 2\tau_2$ since $c_g = 2c_p$. Equation (13.91) is plotted in Figure 13.7.

Observe that Equation (13.91) is similar in nature to Equation (13.89), but has two important differences.

1. The peak value of the envelope function occurs at the time delay $\tau_2 = (d/c_g)$, which is now a function of frequency since $c_g \sim \sqrt{f_0}$.
2. Peak values of the cosine function occur at time delays $\tau_n = \tau_1 \pm (n/f_0)$, where n is any integer. In general, a peak value of the cosine function does *not* coincide with the peak value of the envelope function.

Equation (13.91) shows that the "apparent" propagation speed for the envelope function at a given frequency is determined by the group velocity c_g rather than by the phase velocity c_p. For such dispersive propagation problems, the peak value of $\rho_{xy}(\tau)$ from its fine structure due to the cosine function in Equation (13.91) may *not* coincide with the envelope peak value that occurs at the time delay τ_2. To find τ_2, one must compute the envelope function of $\rho_{xy}(\tau)$, as can be done using Hilbert transform techniques.

The derivation of Equation (13.91) is based on the following ideas. Start with the general relation

$$R_{xy}(\tau) = \int_{-\infty}^{\infty} S_{xy}(f)e^{j2\pi f\tau}\, df \tag{13.94}$$

For dispersive cases leading to Equation (13.91), the corresponding $S_{xy}(f)$ is

$$S_{xy}(f) = \alpha S_{xx}(f) e^{-j2\pi f \tau_p} \tag{13.95}$$

where the delay $\tau_p = (d/c_p)$ is a function of frequency since the phase velocity $c_p \sim \sqrt{f}$. Here $\tau_p = \tau_p(f)$ can be expressed at frequency f as

$$\tau_p = \frac{d}{c_p} = \frac{a}{\sqrt{2\pi f}} \tag{13.96}$$

with a as a suitable proportionality constant. Substitution of Equation (13.95) into Equation (13.94) gives

$$R_{xy}(\tau) = \alpha \int_{-\infty}^{\infty} S_{xx}(f) e^{j2\pi f(\tau - \tau_p)} \, df$$

$$= \alpha \int_{0}^{\infty} G_{xx}(f) \cos 2\pi f(\tau - \tau_p) \, df \tag{13.97}$$

where the one-sided autospectrum $G_{xx}(f) = 2S_{xx}(f)$ for $f \geq 0$ and is otherwise zero. For bandwidth limited white noise,

$$G_{xx}(f) = \begin{cases} K & 0 \leq f_0 - (B_0/2) \leq f \leq f_0 + (B_0/2) \\ 0 & \text{otherwise} \end{cases} \tag{13.98}$$

Thus

$$R_{xy}(\tau) = \alpha K \int_{f_0 - (B_0/2)}^{f_0 + (B_0/2)} \cos 2\pi f(\tau - \tau_p) \, df \tag{13.99}$$

Also

$$R_{xx}(0) = K \int_{f_0 - (B_0/2)}^{f_0 + (B_0/2)} df = K B_0$$

$$R_{yy}(0) = \alpha^2 R_{xx}(0) = \alpha^2 K B_0 \tag{13.100}$$

Hence, the cross-correlation coefficient function becomes

$$\rho_{xy}(\tau) = \frac{1}{B_0} \int_{f_0 - (B_0/2)}^{f_0 + (B_0/2)} \cos 2\pi f(\tau - \tau_p) \, df \tag{13.101}$$

Consider situations where the frequency bandwidth B_0 is less than one octave, so that for any $\varepsilon < 1$, f can be replaced by

$$f = f_0(1 + \varepsilon) \qquad df = f_0 \, d\varepsilon \tag{13.102}$$

This change of variable gives

$$\rho_{xy}(\tau) = \frac{f_0}{B_0} \int_{-B_0/2f_0}^{B_0/2f_0} \cos 2\pi f_0(1 + \varepsilon)(\tau - \tau_p) \, d\varepsilon \qquad (13.103)$$

and the τ_p of Equation (13.96) becomes

$$\tau_p = \frac{a}{\sqrt{2\pi f}} = \frac{a}{\sqrt{2\pi f_0(1 + \varepsilon)}} \qquad (13.104)$$

For small ε, neglecting terms of order ε^2,

$$2\pi f_0(1 + \varepsilon)(\tau - \tau_p) = 2\pi f_0(1 + \varepsilon)\tau - a\sqrt{2\pi f_0(1 + \varepsilon)}$$

$$\approx 2\pi f_0(1 + \varepsilon)\tau - a\sqrt{2\pi f_0}\left(1 + \frac{\varepsilon}{2}\right)$$

$$\approx 2\pi f_0\left(\tau - \frac{a}{\sqrt{2\pi f_0}}\right) + 2\pi f_0\left(\tau - \frac{a}{2\sqrt{2\pi f_0}}\right)\varepsilon \quad (13.105)$$

Then, neglecting sine terms compared to cosine terms,

$$\cos 2\pi f_0(1 + \varepsilon)(\tau - \tau_1) \approx \cos[2\pi f_0(\tau - \tau_1) + 2\pi f_0(\tau - \tau_2)\varepsilon]$$

$$\approx [\cos 2\pi f_0(\tau - \tau_1)][\cos 2\pi f_0(\tau - \tau_2)\varepsilon] \quad (13.106)$$

where

$$\tau_1 = \frac{a}{\sqrt{2\pi f_0}} = \frac{d}{c_p} \qquad (13.107)$$

$$\tau_2 = \frac{a}{2\sqrt{2\pi f_0}} = \frac{d}{c_g} \qquad (13.108)$$

Here c_p and c_g represent the phase velocity at frequency f_0 and the group velocity at frequency f_0 with $c_g = 2c_p$. Finally, one should substitute Equation (13.106) into Equation (13.103) and integrate to obtain

$$\rho_{xy}(\tau) \approx \left(\frac{f_0}{B_0}\right) \cos 2\pi f_0(\tau - \tau_1) \int_{-B_0/2f_0}^{B_0/2f_0} \cos 2\pi f_0(\tau - \tau_2)\varepsilon \, d\varepsilon$$

$$\approx \left[\frac{\sin \pi B_0(\tau - \tau_2)}{\pi B_0(\tau - \tau_2)}\right] \cos 2\pi f_0(\tau - \tau_1) \qquad (13.109)$$

This is the stated result of Equation (13.91).

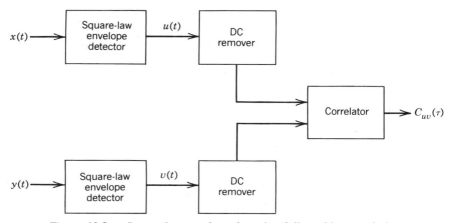

Figure 13.8   Square-law envelope detection followed by correlation.

13.3   ENVELOPE DETECTION FOLLOWED BY CORRELATION

Consider Figure 13.8, where

$$u(t) = x^2(t) + \tilde{x}^2(t) = \text{squared envelope of } x(t)$$
$$v(t) = y^2(t) + \tilde{y}^2(t) = \text{squared envelope of } y(t) \tag{13.110}$$

The cross-correlation function $R_{uv}(\tau)$ of these squared envelope signals is given by

$$
\begin{aligned}
R_{uv}(\tau) &= E[u(t)v(t+\tau)] = E\big[\{x^2(t) + \tilde{x}^2(t)\}\{y^2(t+\tau) + \tilde{y}^2(t+\tau)\}\big] \\
&= E\big[x^2(t)y^2(t+\tau)\big] + E\big[x^2(t)\tilde{y}^2(t+\tau)\big] \\
&\quad + E\big[\tilde{x}^2(t)y^2(t+\tau)\big] + E\big[\tilde{x}^2(t)\tilde{y}^2(t+\tau)\big]
\end{aligned} \tag{13.111}
$$

Assume that $x(t)$, $\tilde{x}(t)$, $y(t)$, and $\tilde{y}(t)$ are jointly normally distributed with zero mean values. Then

$$
\begin{aligned}
E\big[x^2(t)y^2(t+\tau)\big] &= \sigma_x^2\sigma_y^2 + 2R_{xy}^2(\tau) \\
E\big[x^2(t)\tilde{y}^2(t+\tau)\big] &= \sigma_x^2\sigma_{\tilde{y}}^2 + 2R_{x\tilde{y}}^2(\tau) \\
E\big[\tilde{x}^2(t)y^2(t+\tau)\big] &= \sigma_{\tilde{x}}^2\sigma_y^2 + 2R_{\tilde{x}y}^2(\tau) \\
E\big[\tilde{x}^2(t)\tilde{y}^2(t+\tau)\big] &= \sigma_{\tilde{x}}^2\sigma_{\tilde{y}}^2 + 2R_{\tilde{x}\tilde{y}}^2(\tau)
\end{aligned} \tag{13.112}
$$

Substituting Equation (13.112) into (13.111) and using results from Table 13.2 yields

$$R_{uv}(\tau) = 4\sigma_x^2\sigma_y^2 + 4\left[R_{xy}^2(\tau) + \tilde{R}_{xy}^2(\tau)\right] \tag{13.113}$$

Also

$$\bar{u} = E[u(t)] = 2\sigma_x^2$$
$$\bar{v} = E[v(t)] = 2\sigma_y^2 \tag{13.114}$$

Hence

$$R_{uv}(\tau) - (\bar{u})(\bar{v}) = 4\left[R_{xy}^2(\tau) + \tilde{R}_{xy}^2(\tau)\right] = 4A_{xy}^2(\tau) \tag{13.115}$$

where, as shown in Equation (13.73), the quantity $A_{xy}^2(\tau)$ is the squared envelope of $R_{xy}(\tau)$. Now

$$R_{uu}(0) = E[u^2(t)] = E\left[\left\{x^2(t) + \tilde{x}^2(t)\right\}^2\right]$$
$$= E[x^4(t)] + 2E[x^2(t)\tilde{x}^2(t)] + E[\tilde{x}^4(t)]$$
$$= 3\sigma_x^4 + 2\sigma_x^2\sigma_{\tilde{x}}^2 + 3\sigma_{\tilde{x}}^4 = 8\sigma_x^4 \tag{13.116}$$

Thus

$$R_{uu}(0) - (\bar{u})^2 = 4\sigma_x^4 \tag{13.117}$$

Similarly,

$$R_{vv}(0) - (\bar{v})^2 = 4\sigma_y^4 \tag{13.118}$$

The preceding relations prove that the correlation coefficient function $\rho_{uv}(\tau)$ for the squared envelope signals $u(t)$ and $v(t)$ is given by

$$\rho_{uv}(\tau) = \frac{R_{uv}(\tau) - (\bar{u})(\bar{v})}{\sqrt{\left[R_{uu}(0) - (\bar{u})^2\right]\left[R_{vv}(0) - (\bar{v})^2\right]}} = \frac{A_{xy}^2(\tau)}{\sigma_x^2\sigma_y^2} \tag{13.119}$$

This particular result in Equation (13.119) should be compared with the usual correlation coefficient function $\rho_{xy}(\tau)$ for the original signals $x(t)$ and $y(t)$ as given by Equation (5.16), namely,

$$\rho_{xy}(\tau) = \frac{R_{xy}(\tau) - (\bar{x})(\bar{y})}{\sqrt{\left[R_{xx}(0) - (\bar{x})^2\right]\left[R_{yy}(0) - (\bar{y})^2\right]}} = \frac{R_{xy}(\tau)}{\sigma_x\sigma_y} \tag{13.120}$$

Upon taking Hilbert transforms

$$\tilde{\rho}_{xy}(\tau) = \mathscr{H}\left[\rho_{xy}(\tau)\right] = \frac{\tilde{R}_{xy}(\tau)}{\sigma_x \sigma_y} = \frac{R_{x\tilde{y}}(\tau)}{\sigma_x \sigma_y} = \rho_{x\tilde{y}}(\tau) \qquad (13.121)$$

It follows now using Equation (13.73) that

$$\rho_{uv}(\tau) = \rho_{xy}^2(\tau) + \tilde{\rho}_{xy}^2(\tau) \qquad (13.122)$$

Thus the function $\rho_{uv}(\tau)$, with mean values removed prior to correlation, measures the squared envelope value of $\rho_{xy}(\tau)$. The quantity $\rho_{uv}(\tau)$ is the correlation coefficient function for

$$C_{uv}(\tau) = R_{uv}(\tau) - (\bar{u})(\bar{v}) = E\left[\{u(t) - \bar{u}\}\{v(t+\tau) - \bar{v}\}\right] \quad (13.123)$$

where $u(t)$, $v(t)$, $\bar{u}$ and $\bar{v}$ satisfy Equations (13.110) and (13.114). The computation of $C_{uv}(\tau)$ is sketched in Figure 13.8.

Three special points should be noted from Equation (13.122) regarding the nature of the envelope correlation coefficient $\rho_{uv}(\tau)$ compared to its underlying $R_{xy}(\tau)$ and its envelope $A_{xy}(\tau)$.

1. The quantity $\rho_{uv}(\tau)$, like $A_{xy}(\tau)$, will be independent of the fine structure in $R_{xy}(\tau)$.
2. The quantity $\rho_{uv}(\tau)$ will sharpen the correlation function of $R_{xy}(\tau)$ and of $\rho_{xy}(\tau)$ in the vicinity of τ where the peak value occurs.
3. The quantity $\rho_{uv}(\tau)$ will also sharpen the correlation function of $A_{xy}(\tau)$ in the vicinity of τ where the peak value occurs.

Thus the result $\rho_{uv}(\tau)$ is *superior* to both $\rho_{xy}(\tau)$ and to $A_{xy}(\tau)$ in locating where peak values occur.

***Example 13.8.* Exponential-Cosine Cross-Correlation Function.**   Consider a cross-correlation coefficient function of the form

$$\rho_{xy}(\tau) = \frac{R_{xy}(\tau)}{\sigma_x \sigma_y} = e^{-b|\tau|}\cos 2\pi f_0 \tau$$

The Hilbert transform is then

$$\tilde{\rho}_{xy}(\tau) = \frac{\tilde{R}_{xy}(\tau)}{\sigma_x \sigma_y} = e^{-b|\tau|}\sin 2\pi f_0 \tau$$

It follows that

$$A_{xy}(\tau) = \sigma_x \sigma_y e^{-b|\tau|}$$

and

$$\rho_{uv}(\tau) = \frac{A_{xy}^2(\tau)}{\sigma_x^2 \sigma_y^2} = e^{-2b|\tau|}$$

Observe that $\rho_{uv}(\tau)$ and $A_{xy}(\tau)$ are both independent of the modulating frequency f_0. Also, near $\tau = 0$, where the peak value of $\rho_{xy}(\tau)$ occurs, both $\rho_{xy}(\tau)$ and $A_{xy}(\tau)$ behave like $e^{-b|\tau|}$ but the associated $\rho_{uv}(\tau)$ behaves like $e^{-2b|\tau|}$. Clearly, $\rho_{uv}(\tau)$ will have a sharper peak at $\tau = 0$ than $A_{xy}(\tau)$. This concludes the example.

Consider next the cross-spectral density functions and the autospectral density functions for the squared envelope signals $u(t)$ and $v(t)$ given in Equation (13.110). From Equation (13.115), the envelope cross-spectral density function

$$S_{uv}(f) = \mathscr{F}\left[R_{uv}(\tau) - \bar{u}\bar{v}\right] = 4\mathscr{F}\left[A_{xy}^2(\tau)\right] \tag{13.124}$$

The squared envelope of $R_{xy}(\tau)$, namely,

$$A_{xy}^2(\tau) = R_{xy}^2(\tau) + \tilde{R}_{xy}^2(\tau) \tag{13.125}$$

can be computed by the procedure outlined in Example 13.4. Now

$$\mathscr{F}\left[R_{xy}^2(\tau)\right] = \int_{-\infty}^{\infty} S_{xy}(\alpha) S_{xy}(f - \alpha)\, d\alpha \tag{13.126}$$

$$\mathscr{F}\left[\tilde{R}_{xy}^2(\tau)\right] = \int_{-\infty}^{\infty} \tilde{S}_{xy}(\alpha) \tilde{S}_{xy}(f - \alpha)\, d\alpha$$

$$= \int_{-\infty}^{\infty} B(\alpha) B(f - \alpha) S_{xy}(\alpha) S_{xy}(f - \alpha)\, d\alpha \tag{13.127}$$

where $B(f) = (-j\,\mathrm{sgn}\,f)$ as in Equation (13.8). Hence

$$S_{uv}(f) = 4\int_{-\infty}^{\infty} \left[1 + B(\alpha) B(f - \alpha)\right] S_{xy}(\alpha) S_{xy}(f - \alpha)\, d\alpha \tag{13.128}$$

For any $f > 0$, the quantity

$$B(\alpha) B(f - \alpha) = \begin{cases} 1 & \alpha < 0 \\ -1 & 0 < \alpha < f \\ 1 & \alpha > f \end{cases} \tag{13.129}$$

Similarly, the envelope autospectral density functions

$$S_{uu}(f) = 4\int_{-\infty}^{\infty} [1 + B(\alpha)B(f - \alpha)] S_{xx}(\alpha)S_{xx}(f - \alpha)\, d\alpha$$

$$\tag{13.130}$$

$$S_{vv}(f) = 4\int_{-\infty}^{\infty} [1 + B(\alpha)B(f - \alpha)] S_{yy}(\alpha)S_{yy}(f - \alpha)\, d\alpha$$

Thus, knowledge of the basic spectral density functions for the signals $x(t)$ and $y(t)$ enables one to compute the associated spectral density functions for the squared envelope signals $u(t)$ and $v(t)$.

PROBLEMS

13.1   Which of the following statements are correct?
  (a)   The Hilbert transform is a linear operator.
  (b)   The Hilbert transform of a time dependent function is also a time dependent function.
  (c)   Given $x(t)$ and its Hilbert transform $\tilde{x}(t)$ with Fourier transforms $X(f)$ and $\tilde{X}(f)$, respectively, $\tilde{X}(f)$ equals the Hilbert transform of $X(f)$.
  (d)   Given $x(t)$ and its Hilbert transform $\tilde{x}(t)$, the magnitude of the Fourier transforms of $x(t)$ and $\tilde{x}(t)$ are equal.
  (e)   If the Fourier transform of $x(t)$ is real valued, the Fourier transform of $\tilde{x}(t)$ will also be real valued.

13.2   Determine the Hilbert transform of the function

$$x(t) = ae^{-b|t|} \qquad \text{where } b > 0.$$

13.3   Determine the Hilbert transform of the function

$$x(t) = \frac{t - a}{1 + (t - a)^2}$$

13.4   Consider a function $x(t)$ with a Fourier transform given by

$$X(f) = \frac{1}{1 + j2\pi f}$$

Determine the Fourier transform of the Hilbert transform $\tilde{x}(t)$.

13.5   Given a complex-valued function $z(t) = x(t) + jy(t)$, under what circumstances will the Hilbert transform of $z(t)$, denoted by $\tilde{z}(t)$, be equal to $jz(t)$?

13.6   The real part of the frequency response function for a physically realizable constant parameter linear system is given by

$$H_R(f) = \frac{1 - f^2}{\left(1 - f^2\right)^2 + \left(af\right)^2}$$

Determine the imaginary part of the frequency response function.

13.7   Assume the cross-correlation function between the excitation and response of a physical system has the form

$$R_{xy}(\tau) = aB\left(\frac{\sin \pi B|\tau - \tau_2|}{\pi B|\tau - \tau_2|}\right)\cos 2\pi f_0|\tau - \tau_1|$$

where $B < f_0$. If $\tau_1 = \tau_2$, what is the propagation time through the system at frequency $f = f_0 + B$?

13.8   In Problem 13.7, if $\tau_1 \neq \tau_2$, what is the propagation time through the system at frequency $f = f_0 + B$ assuming the group velocity of propagating waves is $c_g \sim \sqrt{f}$?

13.9   Given a modulated signal $y(t) = x(t)\cos 2\pi f_0 t$ where the spectrum of $x(t)$ includes no frequency components above $F < f_0$, determine the following:
   (a)   The Hilbert transform of $y(t)$, denoted by $\tilde{y}(t)$.
   (b)   The correlation function of $y(t)$, denoted by $R_{yy}(\tau)$, in terms of $R_{xx}(\tau)$.
   (c)   The Hilbert transform of the correlation function $R_{yy}(\tau)$, denoted by $\tilde{R}_{yy}(\tau)$.
   (d)   The envelope of the correlation function $R_{yy}(\tau)$.

13.10   Consider an analytic function $z(t) = x(t) + j\tilde{x}(t)$ where the two-sided autospectrum of $x(t)$ is given by

$$S_{xx}(f) = \frac{2a}{a^2 + 4\pi^2 f^2}$$

Determine the autospectra of (a) $\tilde{x}(t)$ and (b) $z(t)$.

REFERENCES

13.1   Bendat, J. S., "The Hilbert Transform and Applications to Correlation Measurements," *Bruel & Kjaer*, Denmark, 1985.

13.2   Bracewell, R., *The Fourier Transform and Its Applications*, McGraw-Hill, New York, 1965.

13.3 Oppenheim, A. V. and Shafer, R. W., *Digital Signal Processing*, Prentice-Hall, New Jersey, 1975.

13.4 Thrane, N., "The Hilbert Transform," *Bruel & Kjaer Technical Review*, No. 3, 1984.

13.5 Herlufsen, H., "Duel Channel FFT Analysis (Parts 1 and 11)," *Bruel & Kjaer Technical Review*, Nos. 1 and 2, 1984.

13.6 Dugundji, J., "Envelopes and Pre-Envelopes of Real Waveforms," *IRE Transactions on Information Theory*, Vol. IT-4, p. 53, March 1958.

13.7 Cremer, L., Heckl, M., and Unger, E. E., *Structure-Borne Sound*, Springer-Verlag, New York, 1973.

13.8 White, P. H., "Cross-Correlation in Structural Systems: Dispersive and Nondispersive Waves," *Journal of Acoustical Society of America*, Vol. 45, p. 1118, May 1969.

REFERENCES

Adams, E. P., *Smithsonian Mathematical Formulae and Tables of Elliptic Functions*, Smithsonian Institution, Washington, D.C., 1947.

Beauchamp, K. G., and Yuen, C. K., *Data Acquisition for Signal Analysis*, Allan & Unwin, London, 1980.

Bendat, J. S., *Principles and Applications of Random Noise Theory*, Wiley, New York, 1958. Reprinted by Krieger, Melbourne, Florida, 1977.

Bendat, J. S., and Piersol. A. G., *Engineering Applications of Correlation and Spectral Analysis*, Wiley-Interscience, New York, 1980.

Bendat, J. S., "Statistical Errors in Measurement of Coherence Functions and Input/Output Quantities," *Journal of Sound and Vibration*, Vol. 59, p. 405, 1978.

Bendat, J. S., "Modern Analysis Procedures for Multiple Input/Output Problems," *Journal of the Acoustical Society of America*, Vol. 68, p. 498, 1980.

Bendat, J. S. "The Hilbert Transform and Applications to Correlation Measurements," *Bruel and Kjaer*, Denmark, 1985.

Bracewell, R., *The Fourier Transform and Its Applications*, 2nd ed., McGraw-Hill, New York, 1978.

Brigham, E. O., *The Fast Fourier Transform*, Prentice-Hall, Englewood Cliffs, New Jersey, 1974.

Brownlee, K. A., *Statistical Theory and Methodology in Science and Engineering*, 2nd ed., Wiley, New York, 1965.

Carter, G. C., Knapp, C. H., and Nuttall, A. H., "Estimation of the Magnitude-Squared Coherence via Overlapped Fast Fourier Transform Processing," *IEEE Transactions on Audio and Electroacoustics*, Vol. AU-21, p. 337, August 1973.

Childers, D. G., *Modern Spectral Analysis*, IEEE Press, New York, 1978.

Chung, J. Y, "Rejection of Flow Noise Using a Coherence Function Method," *Journal of the Acoustical Society of America*, Vol. 62, p. 388, 1977.

Claasen, T. A. C. M., and Mecklenbrauker, W. F. G., "The Wigner Distribution—A Tool for Time–Frequency Signal Analysis," *Philips Journal of Research*, Vol. 35, Nos. 3-6, 1980.

Cooley, J. W., and Tukey, J. W., "An Algorithm for the Machine Calculation of Complex Fourier Series," *Mathematics of Computation*, Vol. 19, p. 297, April 1965.

Crandall, S. H., and Mark, W. D., *Random Vibration in Mechanical Systems*, Academic, Orlando, Florida, 1963.

Cremer, L., Heckl, M., and Ungar, E. E., *Structure-Borne Sound*, Springer-Verlag, New York, 1973.

517

Davies, G. L., *Magnetic Tape Instrumentation*, McGraw-Hill, New York, 1961.

Dixon, W. J. and Massey, F. J., Jr., *Introduction to Statistical Analysis*, 3rd ed., McGraw-Hill, New York, 1969.

Dodds, C. J., and Robson, J. D., "Partial Coherence in Multivariate Random Processes," *Journal of Sound and Vibration*, Vol. 42, p. 243, 1975.

Doeblin, E. O., *Measurement Systems: Application and Design*, McGraw-Hill, New York, 1966.

Doob, J. L., *Stochastic Processes*, Wiley, New York, 1953.

Dugundji, J., "Envelopes and Pre-Envelopes of Real Waveforms," *IRE Transactions on Information Theory*, Vol. IT-4, p. 53, March 1958.

Fitzgerald, J., and Eason, T. S., *Fundamentals of Data Communication*, Wiley, New York, 1978.

Goodman, N. R., "Measurement of Matrix Frequency Response Functions and Multiple Coherence Functions," AFFDL TR 65-56, Air Force Flight Dynamics Laboratory, Wright-Patterson AFB, Ohio, February 1965.

Gregg, W. D., *Analog and Digital Communication*, Wiley, New York, 1977.

Guttman, I., Wilks, S. S., and Hunter, J. S., *Introductory Engineering Statistics*, 3rd ed., Wiley, New York, 1982.

Hamming, R. W., *Digital Filters*, 2nd ed., Prentice-Hall, Englewood Cliffs, New Jersey, 1983.

Harris, F. J., "On the Use of Windows for Harmonic Analysis with the Discrete Fourier Transform," *Proceedings of the IEEE*, Vol. 66, p. 51 January 1978.

Herlufsen, H., "Dual Channel FFT Analysis (Parts I and II)," *Bruel & Kjaer Technical Review*, Nos. 1 and 2, 1984.

Hurty, W. G., and Rubinstein, M. F., *Dynamics of Structures*, Prentice-Hall, Englewood Cliffs, New Jersey, 1964.

Jenkins, G. M., and Watts, D. G., *Spectral Analysis and Its Applications*, Holden-Day, San Francisco, 1968.

Johnson, N. L., and Leone, F. C., *Statistics and Experimental Design in Engineering and the Physical Sciences*, 2nd ed., Wiley, New York, 1977.

Kendall, M. G., and Stuart, A., *The Advanced Theory of Statistics*, Vol. 2, "Inference and Relationship," Hafner, New York, 1961.

Laha, R. G., and Rohatgi, V. K., *Probability Theory*, Wiley, New York, 1979.

Laning, J. H., Jr., and Battin, R. H., *Random Processes in Automatic Control*, McGraw-Hill, New York, 1956.

Lawson, J. I., and Uhlenbeck, G. E., *Threshold Signals*, McGraw-Hill, New York, 1950.

Liebeck, H., *Algebra for Scientists and Engineers*, Wiley, New York, 1969.

Loeve, M. M., *Probability Theory*, 4th ed., Springer-Verlag, New York, 1977.

Mark, W. D., and Fischer, R. W., "Investigation of the Effects of Nonhomogeneous (or Nonstationary) Behavior on the Spectra of Atmospheric Turbulence," NASA CR-2745, NASA Langley Research Center, Virginia, February 1976.

Nuttall, A. H., "Some Windows with Very Good Side Lobe Behavior," *IEEE Transactions on Acoustics, Speech, and Signal Processing*, Vol. ASSP-29, p. 84, February 1981.

Nuttall, A. H., and Carter, G. C., "Spectral Estimation Using Combined Time and Lag Weighting," *Proceedings of the IEEE*, Vol. 70, p. 1115, September 1982.

Oden, J. I., and Reddy, J. N., *An Introduction to the Mathematical Theory of Finite Elements*, Wiley-Interscience, New York, 1976.

Oppenheim, A. V., and Schafer, R. W., *Digital Signal Processing*, Prentice-Hall, Englewood Cliffs, New Jersey, 1975.

Otnes, R. K., and Enochson, L., *Applied Time Series Analysis*, Vol. 1, Wiley, New York, 1978.

Papoulis, A., *Probability, Random Variables, and Stochastic Processes*, McGraw-Hill, New York, 1965.

Patel, J. K., and Read, C. B., *Handbook of the Normal Distribution*, Dekker, New York, 1982.

Piersol, A. G., "Power Spectra Measurements for Spacecraft Vibration Data," *Journal of Spacecraft and Rockets*, Vol. 4, p. 1613, December 1967.

Rabiner, L. R., and Gold, B., *Theory and Applications of Digital Signal Processing*, Prentice-Hall, Englewood Cliffs, New Jersey, 1975.

Rice, S. O., "Mathematical Analysis of Random Noise," *Selected Papers on Noise and Stochastic Processes* (N. Wax, Ed.) Dover, New York, 1954.

Robinson, E. A., and Treitel, S., *Geophysical Signal Analysis*, Prentice-Hall, Englewood Cliffs, New Jersey, 1980.

Schmidt, H., "Resolution Bias Errors in Spectral Density, Frequency Response and Coherence Function Measurements," *Journal of Sound and Vibration*, Vol. 101, p. 347, 1985.

Seybert, A. F., and Hamilton, J. F., "Time Delay Bias Errors in Estimating Frequency Response and Coherence Functions," *Journal of Sound and Vibration*, Vol. 60, p. 1, 1978.

Silverman, R. A., "Locally Stationary Random Processes," *Transactions of the IRE, Information Theory*, Vol. IT-3, p. 182, September 1957.

Stokey, W. F., "Vibration of Systems Having Distributed Mass and Elasticity," Chapter 7 in *Shock and Vibration Handbook*, 2nd ed. (C. M. Harris and C. E. Crede, Eds.), McGraw-Hill, New York, 1976.

Telemetry Standards IRIG Document 106-66. Secretariat Range Commanders Council, White Sands Missile Range, New Mexico, March 1966.

Thrane, N., "The Hilbert Transform," *Bruel & Kjaer Technical Review*, No. 3, 1984.

Upton, R., "Innovative Functions for Two-Channel FFT Analyzers, *Sound and Vibration*, Vol. 18, p. 18, March 1984.

White, P. H., "Cross-Correlation in Structural Systems: Dispersive and Nondispersive Waves," *Journal of the Acoustical Society of America*, Vol. 45, p. 1118, May 1969.

Winograd, S., "On Computing the Discrete Fourier Transform," *Mathematics of Computation*, Vol. 32, p. 175, January 1978.

APPENDIX A

STATISTICAL TABLES

Table A.1

Ordinates of the Standardized Normal Density Function

$$p(z) = \frac{1}{\sqrt{2\pi}}\, e^{-z^2/2}$$

z	0.00	0.01	0.02	0.03	0.04	0.05	0.06	0.07	0.08	0.09
0.0	0.3989	0.3989	0.3989	0.3988	0.3986	0.3986	0.3982	0.3980	0.3977	0.3973
0.1	0.3970	0.3966	0.3961	0.3956	0.3951	0.3945	0.3939	0.3932	0.3925	0.3918
0.2	0.3910	0.3902	0.3894	0.3884	0.3876	0.3867	0.3857	0.3847	0.3836	0.3825
0.3	0.3814	0.3802	0.3790	0.3778	0.3765	0.3752	0.3739	0.3725	0.3712	0.3697
0.4	0.3683	0.3668	0.3653	0.3637	0.3621	0.3605	0.3589	0.3572	0.3555	0.3538
0.5	0.3521	0.3503	0.3485	0.3467	0.3448	0.3429	0.3410	0.3391	0.3372	0.3352
0.6	0.3332	0.3312	0.3292	0.3271	0.3251	0.3230	0.3209	0.3187	0.3166	0.3144
0.7	0.3123	0.3101	0.3079	0.3056	0.3034	0.3011	0.2989	0.2966	0.2943	0.2920
0.8	0.2897	0.2874	0.2850	0.2827	0.2803	0.2780	0.2756	0.2732	0.2709	0.2685
0.9	0.2661	0.2637	0.2613	0.2589	0.2565	0.2541	0.2516	0.2492	0.2468	0.2444
1.0	0.2420	0.2396	0.2371	0.2347	0.2323	0.2299	0.2275	0.2251	0.2227	0.2203
1.1	0.2179	0.2155	0.2131	0.2107	0.2083	0.2059	0.2036	0.2012	0.1989	0.1965
1.2	0.1942	0.1919	0.1895	0.1872	0.1849	0.1826	0.1804	0.1781	0.1758	0.1736
1.3	0.1714	0.1691	0.1669	0.1647	0.1626	0.1605	0.1582	0.1561	0.1539	0.1518
1.4	0.1497	0.1476	0.1456	0.1435	0.1415	0.1394	0.1374	0.1354	0.1334	0.1315
1.5	0.1295	0.1276	0.1257	0.1238	0.1219	0.1200	0.1282	0.1163	0.1145	0.1127
1.6	0.1109	0.1092	0.1074	0.1057	0.1040	0.1023	0.1006	0.0989	0.0973	0.0957
1.7	0.0940	0.0925	0.0909	0.0893	0.0878	0.0863	0.0848	0.0833	0.0818	0.0804
1.8	0.0790	0.0775	0.0761	0.0748	0.0734	0.0721	0.0707	0.0694	0.0681	0.0669
1.9	0.0656	0.0644	0.0632	0.0620	0.0608	0.0596	0.0584	0.0573	0.0562	0.0051
2.0	0.0540	0.0529	0.0519	0.0508	0.0498	0.0488	0.0478	0.0468	0.0459	0.0449
2.1	0.0440	0.0431	0.0422	0.0413	0.0404	0.0396	0.0387	0.0379	0.0371	0.0363
2.2	0.0355	0.0347	0.0339	0.0332	0.0325	0.0317	0.0310	0.0303	0.0297	0.0290
2.3	0.0283	0.0277	0.0270	0.0264	0.0258	0.0252	0.0246	0.0241	0.0235	0.0229
2.4	0.0224	0.0219	0.0213	0.0208	0.0203	0.0198	0.0194	0.0189	0.0184	0.0180

z	0.00	0.01	0.02	0.03	0.04	0.05	0.06	0.07	0.08	0.09
2.5	0.0175	0.0171	0.0167	0.0163	0.0158	0.0154	0.0151	0.0147	0.0143	0.0139
2.6	0.0136	0.0132	0.0129	0.0126	0.0122	0.0119	0.0116	0.0113	0.0110	0.0107
2.7	0.0104	0.0101	0.0099	0.0096	0.0093	0.0091	0.0088	0.0086	0.0084	0.0081
2.8	0.0079	0.0077	0.0075	0.0073	0.0071	0.0069	0.0067	0.0065	0.0063	0.0061
2.9	0.0060	0.0058	0.0056	0.0055	0.0053	0.0051	0.0050	0.0048	0.0047	0.0046
3.0	0.0044	0.0043	0.0042	0.0040	0.0039	0.0038	0.0037	0.0036	0.0035	0.0034
3.1	0.0033	0.0032	0.0031	0.0030	0.0029	0.0028	0.0027	0.0026	0.0025	0.0025
3.2	0.0024	0.0023	0.0022	0.0022	0.0021	0.0020	0.0020	0.0019	0.0018	0.0018
3.3	0.0017	0.0017	0.0016	0.0016	0.0015	0.0015	0.0014	0.0014	0.0013	0.0013
3.4	0.0012	0.0012	0.0012	0.0011	0.0011	0.0010	0.0010	0.0010	0.0009	0.0009
3.5	0.0009	0.0008	0.0008	0.0008	0.0008	0.0007	0.0007	0.0007	0.0007	0.0006
3.6	0.0006	0.0006	0.0006	0.0005	0.0005	0.0005	0.0005	0.0005	0.0005	0.0004
3.7	0.0004	0.0004	0.0004	0.0004	0.0004	0.0004	0.0003	0.0003	0.0003	0.0003
3.8	0.0003	0.0003	0.0003	0.0003	0.0003	0.0002	0.0002	0.0002	0.0002	0.0002
3.9	0.0002	0.0002	0.0002	0.0002	0.0002	0.0002	0.0002	0.0002	0.0001	0.0001

Areas under Standardized Normal Density Function

$$\text{Value of } \alpha = \int_{z_a}^{\infty} \frac{1}{\sqrt{2\pi}} \, e^{-z^2/2} \, dz = \text{Prob}\,[z > z_a]$$

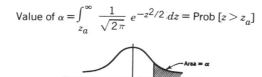

z_α	0.00	0.01	0.02	0.03	0.04	0.05	0.06	0.07	0.08	0.09
0.0	0.5000	0.4960	0.4920	0.4880	0.4840	0.4801	0.4761	0.4721	0.4681	0.4641
0.1	0.4602	0.4562	0.4522	0.4483	0.4443	0.4404	0.4364	0.4325	0.4286	0.4247
0.2	0.4207	0.4168	0.4129	0.4090	0.4052	0.4013	0.3974	0.3936	0.3897	0.3859
0.3	0.3821	0.3783	0.3745	0.3707	0.3669	0.3632	0.3594	0.3557	0.3520	0.3483
0.4	0.3446	0.3409	0.3372	0.3336	0.3300	0.3264	0.3228	0.3192	0.3156	0.3121
0.5	0.3085	0.3050	0.3015	0.2981	0.2946	0.2912	0.2877	0.2843	0.2810	0.2776
0.6	0.2743	0.2709	0.2676	0.2643	0.2611	0.2578	0.2546	0.2514	0.2483	0.2451
0.7	0.2420	0.2389	0.2358	0.2327	0.2296	0.2266	0.2236	0.2206	0.2177	0.2148
0.8	0.2119	0.2090	0.2061	0.2033	0.2005	0.1977	0.1949	0.1922	0.1894	0.1867
0.9	0.1841	0.1814	0.1788	0.1762	0.1736	0.1711	0.1685	0.1660	0.1635	0.1611
1.0	0.1587	0.1562	0.1539	0.1515	0.1492	0.1469	0.1446	0.1423	0.1401	0.1379
1.1	0.1357	0.1335	0.1314	0.1292	0.1271	0.1251	0.1230	0.1210	0.1190	0.1170
1.2	0.1151	0.1131	0.1112	0.1093	0.1075	0.1056	0.1038	0.1020	0.1003	0.0985
1.3	0.0968	0.0951	0.0934	0.0918	0.0901	0.0885	0.0869	0.0853	0.0838	0.0823
1.4	0.0808	0.0793	0.0778	0.0764	0.0749	0.0735	0.0721	0.0708	0.0694	0.0681
1.5	0.0668	0.0655	0.0643	0.0630	0.0618	0.0606	0.0594	0.0582	0.0571	0.0559
1.6	0.0548	0.0537	0.0526	0.0516	0.0505	0.0495	0.0485	0.0475	0.0465	0.0455
1.7	0.0446	0.0436	0.0427	0.0418	0.0409	0.0401	0.0392	0.0384	0.0375	0.0367
1.8	0.0359	0.0351	0.0344	0.0336	0.0329	0.0322	0.0314	0.0307	0.0301	0.0294
1.9	0.0287	0.0281	0.0274	0.0268	0.0262	0.0256	0.0250	0.0244	0.0239	0.0233
2.0	0.0228	0.0222	0.0217	0.0212	0.0207	0.0202	0.0197	0.0192	0.0188	0.0183
2.1	0.0179	0.0174	0.0170	0.0166	0.0162	0.0158	0.0154	0.0150	0.0146	0.0143
2.2	0.0139	0.0136	0.0132	0.0129	0.0125	0.0122	0.0119	0.0116	0.0113	0.0110
2.3	0.0107	0.0104	0.0102	0.00990	0.00964	0.00939	0.00914	0.00889	0.00866	0.00842
2.4	0.00820	0.00798	0.00776	0.00755	0.00734	0.00714	0.00695	0.00676	0.00657	0.00639
2.5	0.00621	0.00604	0.00587	0.00570	0.00554	0.00539	0.00523	0.00508	0.00494	0.00480
2.6	0.00466	0.00453	0.00440	0.00427	0.00415	0.00402	0.00391	0.00379	0.00368	0.00357
2.7	0.00347	0.00336	0.00326	0.00317	0.00307	0.00298	0.00289	0.00280	0.00272	0.00264
2.8	0.00256	0.00248	0.00240	0.00233	0.00226	0.00219	0.00212	0.00205	0.00199	0.00193
2.9	0.00187	0.00181	0.00175	0.00169	0.00164	0.00159	0.00154	0.00149	0.00144	0.00139

The data in Tables A.2 thru A.6 are extracted from *Handbook of Statistical Tables* by Donald B. Owen with the permisson of the publisher, Addison-Wesley Publishing Company, Reading, Mass.

Percentage Points of Chi-Square Distribution

Value of $\chi^2_{n;\alpha}$ such that $\text{Prob}[\chi_n^2 > \chi^2_{n;\alpha}] = \alpha$

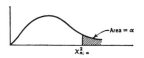

n	0.995	0.990	0.975	0.950	0.900	0.10	0.05	0.025	0.010	0.005
1	0.000039	0.00016	0.00098	0.0039	0.0158	2.71	3.84	5.02	6.63	7.88
2	0.0100	0.0201	0.0506	0.103	0.211	4.61	5.99	7.38	9.21	10.60
3	0.0717	0.115	0.216	0.352	0.584	6.25	7.81	9.35	11.34	12.84
4	0.207	0.297	0.484	0.711	1.06	7.78	9.49	11.14	13.28	14.86
5	0.412	0.554	0.831	1.15	1.61	9.24	11.07	12.83	15.09	16.75
6	0.676	0.872	1.24	1.64	2.20	10.64	12.59	14.45	16.81	18.55
7	0.989	1.24	1.69	2.17	2.83	12.02	14.07	16.01	18.48	20.28
8	1.34	1.65	2.18	2.73	3.49	13.36	15.51	17.53	20.09	21.96
9	1.73	2.09	2.70	3.33	4.17	14.68	16.92	19.02	21.67	23.59
10	2.16	2.56	3.25	3.94	4.87	15.99	18.31	20.48	23.21	25.19
11	2.60	3.05	3.82	4.57	5.58	17.28	19.68	21.92	24.73	26.76
12	3.07	3.57	4.40	5.23	6.30	18.55	21.03	23.34	26.22	28.30
13	3.57	4.11	5.01	5.89	7.04	19.81	22.36	24.74	27.69	29.82
14	4.07	4.66	5.63	6.57	7.79	21.06	23.68	26.12	29.14	31.32
15	4.60	5.23	6.26	7.26	8.55	22.31	25.00	27.49	30.58	32.80
16	5.14	5.81	6.91	7.96	9.31	23.54	26.30	28.85	32.00	34.27
17	5.70	6.41	7.56	8.67	10.08	24.77	27.59	30.19	33.41	35.72
18	6.26	7.01	8.23	9.39	10.86	25.99	28.87	31.53	34.81	37.16
19	6.84	7.63	8.91	10.12	11.65	27.20	30.14	32.85	36.19	38.58
20	7.43	8.26	9.59	10.85	12.44	28.41	31.41	34.17	37.57	40.00
21	8.03	8.90	10.28	11.59	13.24	29.62	32.67	35.48	38.93	41.40
22	8.64	9.54	10.98	12.34	14.04	30.81	33.92	36.78	40.29	42.80
23	9.26	10.20	11.69	13.09	14.85	32.01	35.17	38.08	41.64	44.18
24	9.89	10.86	12.40	13.85	15.66	33.20	36.42	39.36	42.98	45.56
25	10.52	11.52	13.12	14.61	16.47	34.38	37.65	40.65	44.31	46.93
26	11.16	12.20	13.84	15.38	17.29	35.56	38.88	41.92	45.64	48.29
27	11.81	12.88	14.57	16.15	18.11	36.74	40.11	43.19	46.96	49.64
28	12.46	13.56	15.31	16.93	18.94	37.92	41.34	44.46	48.28	50.99
29	13.12	14.26	16.05	17.71	19.77	39.09	42.56	45.72	49.59	52.34
30	13.79	14.95	16.79	18.49	20.60	40.26	43.77	46.98	50.89	53.67
40	20.71	22.16	24.43	26.51	29.05	51.81	55.76	59.34	63.69	66.77
60	35.53	37.48	40.48	43.19	46.46	74.40	79.08	83.30	88.38	91.95
120	83.85	86.92	91.58	95.70	100.62	140.23	146.57	152.21	158.95	163.65

For $n > 120$, $\chi^2_{n;\alpha} \approx n\left[1 - \dfrac{2}{9n} + z_\alpha\sqrt{\dfrac{2}{9n}}\right]^3$ where z_α is the desired percentage point for a standardized normal distribution.

Table A.4

Percentage Points of Student t Distribution

Value of $t_{n;\alpha}$ such that $\text{Prob}[t_n > t_{n;\alpha}] = \alpha$

n	α				
	0.10	0.050	0.025	0.010	0.005
1	3.078	6.314	12.706	31.821	63.657
2	1.886	2.920	4.303	6.965	9.925
3	1.638	2.353	3.182	4.541	5.841
4	1.533	2.132	2.776	3.747	4.604
5	1.476	2.015	2.571	3.365	4.032
6	1.440	1.943	2.447	3.143	3.707
7	1.415	1.895	2.365	2.998	3.499
8	1.397	1.860	2.306	2.896	3.355
9	1.383	1.833	2.262	2.821	3.250
10	1.372	1.812	2.228	2.764	3.169
11	1.363	1.796	2.201	2.718	3.106
12	1.356	1.782	2.179	2.681	3.055
13	1.350	1.771	2.160	2.650	3.012
14	1.345	1.761	2.145	2.624	2.977
15	1.341	1.753	2.131	2.602	2.947
16	1.337	1.746	2.120	2.583	2.921
17	1.333	1.740	2.110	2.567	2.898
18	1.330	1.734	2.101	2.552	2.878
19	1.328	1.729	2.093	2.539	2.861
20	1.325	1.725	2.086	2.528	2.845
21	1.323	1.721	2.080	2.518	2.831
22	1.321	1.717	2.074	2.508	2.819
23	1.319	1.714	2.069	2.500	2.807
24	1.318	1.711	2.064	2.492	2.797
25	1.316	1.708	2.060	2.485	2.787
26	1.315	1.706	2.056	2.479	2.779
27	1.314	1.703	2.052	2.473	2.771
28	1.313	1.701	2.048	2.467	2.763
29	1.311	1.699	2.045	2.462	2.756
30	1.310	1.697	2.042	2.457	2.750
40	1.303	1.684	2.021	2.423	2.704
60	1.296	1.671	2.000	2.390	2.660
120	1.289	1.658	1.980	2.358	2.617

$\alpha = 0.995, 0.990, 0.975, 0.950,$ and 0.900 follow from $t_{n;1-\alpha} = -t_{n;\alpha}$

Table A.5(a)

Percentage Points of F Distribution

Values of $F_{n_1,n_2;0.05}$ such that $\text{Prob}[F_{n_1,n_2} > F_{n_1,n_2;0.05}] = 0.05$

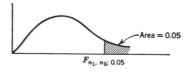

$F_{n_1,\,n_2;\,0.05}$

n_2 \ n_1	1	2	3	4	5	6	7	8	9	10	11	12	13	14	16
1	161	200	216	225	230	234	237	239	241	242	243	244	245	245	246
2	18.5	19.0	19.2	19.2	19.3	19.3	19.4	19.4	19.4	19.4	19.4	19.4	19.4	19.4	19.4
3	10.1	9.55	9.28	9.12	9.01	8.94	8.89	8.85	8.81	8.79	8.76	8.74	8.73	8.71	8.69
4	7.71	6.94	6.59	6.39	6.26	6.16	6.09	6.04	6.00	5.96	5.94	5.91	5.89	5.87	5.84
5	6.61	5.79	5.41	5.19	5.05	4.95	4.88	4.82	4.77	4.73	4.70	4.68	4.66	4.64	4.60
6	5.99	5.14	4.76	4.53	4.39	4.28	4.21	4.15	4.10	4.06	4.03	4.00	3.98	3.96	3.92
7	5.59	4.74	4.35	4.12	3.97	3.87	3.79	3.73	3.68	3.64	3.60	3.57	3.55	3.53	3.49
8	5.32	4.46	4.07	3.84	3.69	3.58	3.50	3.44	3.39	3.35	3.31	3.28	3.26	3.24	3.20
9	5.12	4.26	3.86	3.63	3.48	3.37	3.29	3.23	3.18	3.14	3.10	3.07	3.05	3.03	2.99
10	4.96	4.10	3.71	3.48	3.33	3.22	3.14	3.07	3.02	2.98	2.94	2.91	2.89	2.86	2.83
11	4.84	3.98	3.59	3.36	3.20	3.09	3.01	2.95	2.90	2.85	2.82	2.79	2.76	2.74	2.70
12	4.75	3.89	3.49	3.25	3.11	3.00	2.91	2.85	2.80	2.75	2.72	2.69	2.66	2.64	2.60
13	4.67	3.81	3.41	3.18	3.03	2.92	2.83	2.77	2.71	2.67	2.63	2.60	2.58	2.55	2.51
14	4.60	3.74	3.35	3.11	2.96	2.85	2.76	2.70	2.65	2.60	2.57	2.53	2.51	2.48	2.44
16	4.49	3.63	3.24	3.01	2.85	2.74	2.66	2.59	2.54	2.49	2.46	2.42	2.40	2.37	2.33
18	4.41	3.55	3.16	2.93	2.77	2.66	2.58	2.51	2.46	2.41	2.37	2.34	2.31	2.29	2.25
20	4.35	3.49	3.10	2.87	2.71	2.60	2.51	2.45	2.39	2.35	2.31	2.28	2.25	2.22	2.18
22	4.30	3.44	3.05	2.82	2.66	2.55	2.46	2.40	2.34	2.30	2.26	2.23	2.20	2.17	2.13
24	4.26	3.40	3.01	2.78	2.62	2.51	2.42	2.36	2.30	2.25	2.21	2.18	2.15	2.13	2.09
26	4.23	3.37	2.98	2.74	2.59	2.47	2.39	2.32	2.27	2.22	2.18	2.15	2.12	2.09	2.05
28	4.20	3.34	2.95	2.71	2.56	2.45	2.36	2.29	2.24	2.19	2.15	2.12	2.09	2.06	2.02
30	4.17	3.32	2.92	2.69	2.53	2.42	2.33	2.27	2.21	2.16	2.13	2.09	2.06	2.04	1.99
40	4.08	3.23	2.84	2.61	2.45	2.34	2.25	2.18	2.12	2.08	2.04	2.00	1.97	1.95	1.90
50	4.03	3.18	2.79	2.56	2.40	2.29	2.20	2.13	2.07	2.03	1.99	1.95	1.92	1.89	1.85
60	4.00	3.15	2.76	2.53	2.37	2.25	2.17	2.10	2.04	1.99	1.95	1.92	1.89	1.86	1.82
80	3.96	3.11	2.72	2.49	2.33	2.21	2.13	2.06	2.00	1.95	1.91	1.88	1.84	1.82	1.77
100	3.94	3.09	2.70	2.46	2.31	2.19	2.10	2.03	1.97	1.93	1.89	1.85	1.82	1.79	1.75
200	3.89	3.04	2.65	2.42	2.26	2.14	2.06	1.98	1.93	1.88	1.84	1.80	1.77	1.74	1.69
500	3.86	3.01	2.62	2.39	2.23	2.12	2.03	1.96	1.90	1.85	1.81	1.77	1.74	1.71	1.66
∞	3.84	3.00	2.60	2.37	2.21	2.10	2.01	1.94	1.88	1.83	1.79	1.75	1.72	1.69	1.64

Table A.5(a) (*continued*)

18	20	22	24	26	28	30	40	50	60	80	100	200	500	∞	n_1 / n_2
247	248	249	249	249	250	250	251	252	252	252	253	254	254	254	1
19.4	19.5	19.5	19.5	19.5	19.5	19.5	19.5	19.5	19.5	19.5	19.5	19.5	19.5	19.5	2
8.67	8.66	8.65	8.64	8.63	8.62	8.62	8.59	8.59	8.57	8.56	8.55	8.54	8.53	8.53	3
5.82	5.80	5.79	5.77	5.76	5.75	5.75	5.72	5.70	5.69	5.67	5.66	5.65	5.64	5.63	4
4.58	3.56	4.54	4.53	4.52	4.50	4.50	4.46	4.44	4.43	4.41	4.41	4.39	4.37	4.37	5
3.90	3.87	3.86	3.84	3.83	3.82	3.81	3.77	3.75	3.74	3.72	3.71	3.69	3.68	3.67	6
3.47	3.44	3.43	3.41	3.40	3.39	3.38	3.34	3.32	3.30	3.29	3.27	3.25	3.24	3.23	7
3.17	3.15	3.13	3.12	3.10	3.09	3.08	3.04	3.02	3.01	2.99	2.97	2.95	2.94	2.93	8
2.96	2.94	2.92	2.90	2.89	2.87	2.86	2.83	2.80	2.79	2.77	2.76	2.73	2.72	2.71	9
2.80	2.77	2.75	2.74	2.72	2.71	2.70	2.66	2.64	2.62	2.60	2.59	2.56	2.55	2.54	10
2.67	2.65	2.63	2.61	2.59	2.58	2.57	2.53	2.51	2.49	2.47	2.46	2.43	2.42	2.40	11
2.57	2.54	2.52	2.51	2.49	2.48	2.47	2.43	2.40	2.38	2.36	2.35	2.32	2.31	2.30	12
2.48	2.46	2.44	2.42	2.41	2.39	2.38	2.34	2.31	2.30	2.27	2.26	2.23	2.22	2.21	13
2.41	2.38	2.37	2.35	2.33	2.32	2.31	2.27	2.24	2.22	2.20	2.19	2.16	2.14	2.13	14
2.30	2.28	2.25	2.24	2.22	2.21	2.19	2.15	2.12	2.11	2.08	2.07	2.04	2.02	2.01	16
2.22	2.19	2.17	2.15	2.13	2.12	2.11	2.06	2.04	2.02	1.99	1.98	1.95	1.93	1.92	18
2.15	2.12	2.10	2.08	2.07	2.05	2.04	1.99	1.97	1.95	1.92	1.91	1.88	1.86	1.84	20
2.10	2.07	2.05	2.03	2.01	2.00	1.98	1.94	1.91	1.89	1.86	1.85	1.82	1.80	1.78	22
2.05	2.03	2.00	1.98	1.97	1.95	1.94	1.89	1.86	1.84	1.82	1.80	1.77	1.75	1.73	24
2.02	1.99	1.97	1.95	1.93	1.91	1.90	1.84	1.82	1.80	1.78	1.76	1.73	1.71	1.69	26
1.99	1.96	1.93	1.91	1.90	1.88	1.87	1.82	1.79	1.77	1.74	1.73	1.69	1.67	1.65	28
1.96	1.93	1.91	1.89	1.87	1.85	1.84	1.79	1.76	1.74	1.71	1.70	1.66	1.64	1.62	30
1.87	1.84	1.81	1.79	1.77	1.76	1.74	1.69	1.66	1.64	1.61	1.59	1.55	1.53	1.51	40
1.81	1.78	1.76	1.74	1.72	1.70	1.69	1.63	1.60	1.58	1.54	1.52	1.48	1.46	1.44	50
1.78	1.75	1.72	1.70	1.68	1.66	1.65	1.59	1.56	1.53	1.50	1.48	1.44	1.41	1.39	60
1.73	1.70	1.68	1.65	1.63	1.62	1.60	1.54	1.51	1.48	1.45	1.43	1.38	1.35	1.32	80
1.71	1.68	1.65	1.63	1.61	1.59	1.57	1.52	1.48	1.45	1.41	1.39	1.34	1.31	1.28	100
1.66	1.62	1.60	1.57	1.55	1.53	1.52	1.46	1.41	1.39	1.35	1.32	1.26	1.22	1.19	200
1.62	1.59	1.56	1.54	1.52	1.50	1.48	1.42	1.38	1.34	1.30	1.28	1.21	1.16	1.11	500
1.60	1.57	1.54	1.52	1.50	1.48	1.46	1.39	1.35	1.32	1.27	1.24	1.17	1.11	1.00	∞

Table A.5(b)

Percentage Points of F Distribution

Values of $F_{n_1,n_2;0.025}$ such that Prob $[F_{n_1,n_2} > F_{n_1,n_2;0.025}] = 0.025$

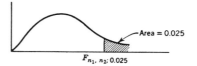

n_2 \ n_1	1	2	3	4	5	6	7	8	9	10	11	12	13	14	16
1	648	800	864	900	922	937	948	957	963	969	973	977	980	983	987
2	38.5	39.0	39.2	39.2	39.3	39.3	39.4	39.4	39.4	39.4	39.4	39.4	39.4	39.4	39.4
3	17.4	16.0	15.4	15.1	14.9	14.7	14.6	14.5	14.5	14.4	14.4	14.3	14.3	14.3	14.2
4	12.2	10.6	9.98	9.60	9.36	9.20	9.07	8.98	8.90	8.84	8.79	8.75	8.72	8.69	8.64
5	10.0	8.43	7.76	7.39	7.15	6.98	6.85	6.76	6.68	6.62	6.57	6.52	6.49	6.46	6.41
6	8.81	7.26	6.60	6.23	5.99	5.82	5.70	5.60	5.52	5.46	5.41	5.37	5.33	5.30	5.25
7	8.07	6.54	5.89	5.52	5.29	5.12	4.99	4.90	4.82	4.76	4.71	4.67	4.63	4.60	4.54
8	7.57	6.06	5.42	5.05	4.82	4.65	4.53	4.43	4.36	4.30	4.24	4.20	4.16	4.13	4.08
9	7.21	5.71	5.08	4.72	4.48	4.32	4.20	4.10	4.03	3.96	3.91	3.87	3.83	3.80	3.74
10	6.94	5.46	4.83	4.47	4.24	4.07	3.95	3.85	3.78	3.72	3.66	3.62	3.58	3.55	3.50
11	6.72	5.26	4.63	4.28	4.04	3.88	3.76	3.66	3.59	3.53	3.47	3.43	3.39	3.36	3.30
12	6.55	5.10	4.47	4.12	3.89	3.73	3.61	3.51	3.44	3.37	3.32	3.28	3.24	3.21	3.15
13	6.41	4.97	4.35	4.00	3.77	3.60	3.48	3.39	3.31	3.25	3.20	3.15	3.12	3.08	3.03
14	6.30	4.86	4.24	3.89	3.66	3.50	3.38	3.29	3.21	3.15	3.09	3.05	3.01	2.98	2.92
16	6.12	4.69	4.08	3.73	3.50	3.34	3.22	3.12	3.05	2.99	2.93	2.89	2.85	2.82	2.76
18	5.98	4.56	3.95	3.61	3.38	3.22	3.10	3.01	2.93	2.87	2.81	2.77	2.73	2.70	2.64
20	5.87	4.46	3.86	3.51	3.29	3.13	3.01	2.91	2.84	2.77	2.72	2.68	2.64	2.60	2.55
22	5.79	4.38	3.78	3.44	3.22	3.05	2.93	2.84	2.76	2.70	2.65	2.60	2.56	2.53	2.47
24	5.72	4.32	3.72	3.38	3.15	2.99	2.87	2.78	2.70	2.64	2.59	2.54	2.50	2.47	2.41
26	5.66	4.27	3.67	3.33	3.10	2.94	2.82	2.73	2.65	2.59	2.54	2.49	2.45	2.42	2.36
28	5.61	4.22	3.63	3.29	3.06	2.90	2.78	2.69	2.61	2.55	2.49	2.45	2.41	2.37	2.32
30	5.57	4.18	3.59	3.25	3.03	2.87	2.75	2.65	2.57	2.51	2.46	2.41	2.37	2.34	2.28
40	5.42	4.05	3.46	3.13	2.90	2.74	2.62	2.53	2.45	2.39	2.33	2.29	2.25	2.21	2.15
50	5.34	3.98	3.39	3.06	2.83	2.67	2.55	2.46	2.38	2.32	2.26	2.22	2.18	2.14	2.08
60	5.29	3.93	3.34	3.01	2.79	2.63	2.51	2.41	2.33	2.27	2.22	2.17	2.13	2.09	2.03
80	5.22	3.86	3.28	2.95	2.73	2.57	2.45	2.36	2.38	2.21	2.16	2.11	2.07	2.03	1.97
100	5.18	3.83	3.25	2.92	2.70	2.54	2.42	2.32	2.24	2.18	2.12	2.08	2.04	2.00	1.94
200	5.10	3.76	3.18	2.85	2.63	2.47	2.35	2.26	2.18	2.11	2.06	2.01	1.97	1.93	1.87
500	5.05	3.72	3.14	2.81	2.59	2.43	2.31	2.22	2.14	2.07	2.02	1.97	1.93	1.89	1.83
∞	5.02	3.69	3.12	2.79	2.57	2.41	2.29	2.19	2.11	2.05	1.99	1.94	1.90	1.87	1.80

18	20	22	24	26	28	30	40	50	60	80	100	200	500	∞	n_1 / n_2
990	993	995	997	999	1000	1001	1006	1008	1010	1012	1013	1016	1017	1018	1
39.4	39.4	39.5	39.5	39.5	39.5	39.5	39.5	39.5	39.5	39.5	39.5	39.5	39.5	39.5	2
14.2	14.2	14.1	14.1	14.1	14.1	14.1	14.0	14.0	14.0	14.0	14.0	13.9	13.9	13.9	3
8.60	8.56	8.53	8.51	8.49	8.48	8.46	8.41	8.38	8.36	8.33	8.32	8.29	8.27	8.26	4
6.37	6.33	6.30	6.28	6.26	6.24	6.23	6.18	6.14	6.12	6.10	6.08	6.05	6.03	6.01	5
5.21	5.17	5.14	5.12	5.10	5.08	5.07	5.01	4.98	4.96	4.93	4.92	4.88	4.86	4.85	6
4.50	4.47	4.44	4.42	4.39	4.38	4.36	4.31	4.28	4.25	4.23	4.21	4.18	4.16	4.14	7
4.03	4.00	3.97	3.95	3.93	3.91	3.89	3.84	3.81	3.78	3.76	3.74	3.70	3.68	3.67	8
3.70	3.67	3.64	3.61	3.59	3.58	3.56	3.51	3.47	3.45	3.42	3.40	3.37	3.35	3.33	9
3.45	3.42	3.39	3.37	3.34	3.33	3.31	3.26	3.22	3.20	3.17	3.15	3.12	3.09	3.08	10
3.26	3.23	3.20	3.17	3.15	3.13	3.12	3.06	3.03	3.00	2.97	2.96	2.92	2.90	2.88	11
3.11	3.07	3.04	3.02	3.00	2.98	2.96	2.91	2.87	2.85	2.82	2.80	2.76	2.74	2.72	12
2.98	2.95	2.92	2.89	2.87	2.85	2.84	2.78	2.74	2.72	2.69	2.67	2.63	2.61	2.60	13
2.88	2.84	2.81	2.79	2.77	2.75	2.73	2.67	2.64	2.61	2.58	2.56	2.53	2.50	2.49	14
2.72	2.68	2.65	2.63	2.60	2.58	2.57	2.51	2.47	2.45	2.42	2.40	2.36	2.33	2.32	16
2.60	2.56	2.53	2.50	2.48	2.46	2.44	2.38	2.35	2.32	2.29	2.27	2.23	2.20	2.19	18
2.50	2.46	2.43	2.41	2.39	2.37	2.35	2.29	2.25	2.22	2.19	2.17	2.13	2.10	2.09	20
2.43	2.39	2.36	2.33	2.31	2.29	2.27	2.21	2.17	2.14	2.11	2.09	2.05	2.02	2.00	22
2.36	2.33	2.30	2.27	2.25	2.23	2.21	2.15	2.11	2.08	2.05	2.02	1.98	1.95	1.94	24
2.31	2.28	2.24	2.22	2.19	2.17	2.16	2.09	2.05	2.03	1.99	1.97	1.92	1.90	1.88	26
2.27	2.23	2.20	2.17	2.15	2.13	2.11	2.05	2.01	1.98	1.94	1.92	1.88	1.85	1.83	28
2.23	2.20	2.16	2.14	2.11	2.09	2.07	2.01	1.97	1.94	1.90	1.88	1.84	1.81	1.79	30
2.11	2.07	2.03	2.01	1.98	1.96	1.94	1.88	1.83	1.80	1.76	1.74	1.69	1.66	1.64	40
2.03	1.99	1.96	1.93	1.91	1.88	1.87	1.80	1.75	1.72	1.68	1.66	1.60	1.57	1.55	50
1.98	1.94	1.91	1.88	1.86	1.83	1.82	1.74	1.70	1.67	1.62	1.60	1.54	1.51	1.48	60
1.93	1.88	1.85	1.82	1.79	1.77	1.75	1.68	1.63	1.60	1.55	1.53	1.47	1.43	1.40	80
1.89	1.85	1.81	1.78	1.76	1.74	1.71	1.64	1.59	1.56	1.51	1.48	1.42	1.38	1.35	100
1.82	1.78	1.74	1.71	1.68	1.66	1.64	1.56	1.51	1.47	1.42	1.39	1.32	1.27	1.23	200
1.78	1.74	1.70	1.67	1.64	1.62	1.60	1.51	1.46	1.42	1.37	1.34	1.25	1.19	1.14	500
1.75	1.71	1.67	1.64	1.61	1.59	1.57	1.48	1.43	1.39	1.33	1.30	1.21	1.13	1.00	∞

Table A.5(c)

Percentage Points of F Distribution

Values of $F_{n_1,n_2;0.01}$ such that $\text{Prob}[F_{n_1,n_2} > F_{n_1,n_2;0.01}] = 0.01$

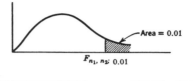

$F_{n_1,n_2;0.01}$

n_2\\n_1	1	2	3	4	5	6	7	8	9	10	11	12	13	14	16
*1	405	500	540	563	576	586	593	598	602	606	608	611	613	614	617
2	98.5	99.0	99.2	99.2	99.3	99.3	99.4	99.4	99.4	99.4	99.4	99.4	99.4	99.4	99.4
3	34.1	30.8	29.5	28.7	28.2	27.9	27.7	27.5	27.3	27.2	27.1	27.1	27.0	26.9	26.8
4	21.2	18.0	16.7	16.0	15.5	15.2	15.0	14.8	14.7	14.5	14.4	14.4	14.3	14.2	14.2
5	16.3	13.3	12.1	11.4	11.0	10.7	10.5	10.3	10.2	10.1	9.96	9.89	9.82	9.77	9.68
6	13.7	10.9	9.78	9.15	8.75	8.47	8.26	8.10	7.98	7.87	7.79	7.72	7.66	7.60	7.52
7	12.2	9.55	8.45	7.85	7.46	7.19	6.99	6.84	6.72	6.62	6.54	6.47	6.41	6.36	6.27
8	11.3	8.65	7.59	7.01	6.63	6.37	6.18	6.03	5.91	5.81	5.73	5.67	5.61	5.56	5.48
9	10.6	8.02	6.99	6.42	6.06	5.80	5.61	5.47	5.35	5.26	5.18	5.11	5.05	5.00	4.92
10	10.0	7.56	6.55	5.99	5.64	5.39	5.20	5.06	4.94	4.85	4.77	4.71	4.65	4.60	4.52
11	9.65	7.21	6.22	5.67	5.32	5.07	4.89	4.74	4.63	4.54	4.46	4.40	4.34	4.29	4.21
12	9.33	6.93	5.95	5.41	5.06	4.82	4.64	4.50	4.39	4.30	4.22	4.16	4.10	4.05	3.97
13	9.07	6.70	5.74	5.21	4.86	4.62	4.44	4.30	4.19	4.10	4.02	3.96	3.91	3.86	3.78
14	8.86	6.51	5.56	5.04	4.70	4.46	4.28	4.14	4.03	3.94	3.86	3.80	3.75	3.70	3.62
16	8.53	6.23	5.29	4.77	4.44	4.20	4.03	3.89	3.78	3.69	3.62	3.55	3.50	3.45	3.37
18	8.29	6.01	5.09	4.58	4.25	4.01	3.84	3.71	3.60	3.51	3.43	3.37	3.32	3.27	3.19
20	8.10	5.85	4.94	4.43	4.10	3.87	3.70	3.56	3.46	3.37	3.29	3.23	3.18	3.13	3.05
22	7.95	5.72	4.82	4.31	3.99	3.76	3.59	3.45	3.35	3.26	3.18	3.12	3.07	3.02	2.94
24	7.82	5.61	4.72	4.22	3.90	3.67	3.50	3.36	3.26	3.17	3.09	3.03	2.98	2.93	2.85
26	7.72	5.53	4.64	4.14	3.82	3.59	3.42	3.29	3.18	3.09	3.02	2.96	2.90	2.86	2.78
28	7.64	5.45	4.57	4.07	3.75	3.53	3.36	3.23	3.12	3.03	2.96	2.90	2·84	2.79	2.72
30	7.56	5.39	4.51	4.02	3.70	3.47	3.30	3.17	3.07	2.98	2.91	2.84	2.79	2.74	7.66
40	7.31	5.18	4.31	3.83	3.51	3.29	3.12	2.99	2.89	2.80	2.73	2.66	2.61	2.56	2.48
50	7.17	5.06	4.20	3.72	3.41	3.19	3.02	2.89	2.79	2.70	2.63	2.56	2.51	2.46	2.38
60	7.08	4.98	4.13	3.65	3.34	3.12	2.95	2.82	2.72	2.63	2.56	2.50	2.44	2.39	2.31
80	6.96	4.88	4.04	3.56	3,26	3.04	2.87	2.74	2.64	2.55	2.48	2.42	2.36	2.31	2.23
100	6.90	4.82	3.98	3.51	3.21	2.99	2.82	2.69	2.59	2.50	2.43	2.37	2.31	2.26	2.19
200	6.76	4.71	3.88	3.41	3.11	2.89	2.73	2.60	2.50	2.41	2 41	2.34	2.27	2.22	2.09
500	6.69	4.65	3.82	3.36	3.05	2.84	2.68	2.55	2.44	2.36	2.28	2.22	2.17	2.12	2.04
∞	6.63	4.61	3.78	3.32	3.02	2.80	2.64	2.51	2.41	2.32	2.25	2.18	2.13	2.08	2.00

Table A.5(c) (*continued*)

18	20	22	24	26	28	30	40	50	60	80	100	200	500	∞	n_1/n_2
619	621	622	623	624	625	626	629	630	631	633	633	635	636	637	1
99.4	99.4	99.5	99.5	99.5	99.5	99.5	99.5	99.5	99.5	99.5	99.5	99.5	99.5	99.5	2
26.8	26.7	26.6	26.6	26.6	26.5	26.5	26.4	26.4	26.3	26.3	26.2	26.2	26.1	26.1	3
14.1	14.0	14.0	13.9	13.9	13.9	13.8	13.7	13.7	13.7	13.6	13.6	13.5	13.5	13.5	4
9.61	9.55	9.51	9.47	9.43	9.40	9.38	9.29	9.24	9.20	9.16	9.13	9.08	9.04	9.02	5
7.45	7.40	7.35	7.31	7.28	7.25	7.23	7.14	7.09	7.06	7.01	6.99	6.93	6.90	6.88	6
6.21	6.16	6.11	6.07	6.04	6.02	5.99	5.91	5.86	5.82	5.78	5.75	5.70	5.67	5.65	7
5.41	5.36	5.32	5.28	5.25	5.22	5.20	5.12	5.07	5.03	4.99	4.96	4.91	4.88	4.85	8
4.86	4.81	4.77	4.73	4.70	4.67	4.65	4.57	4.52	4.48	4.44	4.42	4.36	4.33	4.31	9
4.46	4.41	4.36	4.33	4.30	4.27	4.25	4.17	4.12	4.08	4.04	4.01	3.96	3.93	3.91	10
4.15	4.10	4.06	4.02	3.99	3.96	3.94	3.86	3.81	3.78	3.73	3.71	3.66	3.62	3.60	11
3.91	3.86	3.82	3.78	3.75	3.72	3.70	3.62	3.57	3.54	3.49	3.47	3.41	3.38	3.36	12
3.72	3.66	3.62	3.59	3.56	3.53	3.51	3.43	3.38	3.34	3.30	3.27	3.22	3.19	3.16	13
3.56	3.51	3.46	3.43	3.40	3.37	3.35	3.27	3.22	3.18	3.14	3.11	3.06	3.03	3.00	14
3.31	3.26	3.22	3.18	3.15	3.12	3.10	3.02	2.97	2.93	2.89	2.86	2.81	2.78	2.75	16
3.13	3.08	3.03	3.00	2.97	2.94	2.92	2.84	2.78	2.75	2.70	2.68	2.62	2.59	2.57	18
2.99	2.94	2.90	2.86	2.83	2.80	2.78	2.69	2.64	2.61	2.56	2.54	2.48	2.44	2.42	20
2.88	2.83	2.78	2.75	2.72	2.69	2.67	2.58	2.53	2.50	2.45	2.42	2.36	2.33	2.31	22
2.79	2.74	2.70	2.66	2.63	2.60	2.58	2.49	2.44	2.40	2.36	2.33	2.27	2.24	2.21	24
2.72	2.66	2.62	2.58	2.55	2.53	2.50	2.42	2.36	2.33	2.28	2.25	2.19	2.16	2.13	26
2.65	2.60	2.56	2.52	2.49	2.46	2.44	2.35	2.30	2.26	2.22	2.19	2.13	2.09	2.06	28
2.60	2.55	2.51	2.47	2.44	2.41	2.39	2.30	2.25	2.21	2.16	2.13	2.07	2.03	2.01	30
2.42	2.37	2.33	2.29	2.26	2.23	2.20	2.11	2.06	2.02	1.97	1.94	1.87	1.83	1.80	40
2.32	2.27	2.22	2.18	2.15	2.12	2.10	2.01	1.95	1.91	1.86	1.82	1.76	1.71	1.68	50
2.25	2.20	2.15	2.12	2.08	2.05	2.03	1.94	1.88	1.84	1.78	1.75	1.68	1.63	1.60	60
2.17	2.12	2.07	2.03	2.00	1.97	1.94	1.85	1.79	1.75	1.69	1.66	1.58	1.53	1.49	80
2.12	2.07	2.02	1.98	1.94	1.92	1.89	1.80	1.73	1.69	1.63	1.60	1.52	1.47	1.43	100
2.02	1.97	1.93	1.89	1.85	1.82	1.79	1.69	1.63	1.58	1.52	1.48	1.39	1.33	1.28	200
1.97	1.92	1.87	1.83	1.79	1.76	1.74	1.63	1.56	1.52	1.45	1.41	1.31	1.23	1.16	500
1.93	1.88	1.83	1.79	1.76	1.72	1.70	1.59	1.52	1.47	1.40	1.36	1.25	1.15	1.00	∞

* Multiply the number of the first row ($n_2 = 1$) by 10.

Table A.6

Percentage Points of Run Distribution

Values of $r_{n;\alpha}$ such that Prob $[r_n > r_{n;\alpha}] = \alpha$, where $n = N_1 = N_2 = N/2$

$n = N/2$	α					
	0.99	0.975	0.95	0.05	0.025	0.01
5	2	2	3	8	9	9
6	2	3	3	10	10	11
7	3	3	4	11	12	12
8	4	4	5	12	13	13
9	4	5	6	13	14	15
10	5	6	6	15	15	16
11	6	7	7	16	16	17
12	7	7	8	17	18	18
13	7	8	9	18	19	20
14	8	9	10	19	20	21
15	9	10	11	20	21	22
16	10	11	11	22	22	23
18	11	12	13	24	25	26
20	13	14	15	26	27	28
25	17	18	19	32	33	34
30	21	22	24	37	39	40
35	25	27	28	43	44	46
40	30	31	33	48	50	51
45	34	36	37	54	55	57
50	38	40	42	59	61	63
55	43	45	46	65	66	68
60	47	49	51	70	72	74
65	52	54	56	75	77	79
70	56	58	60	81	83	85
75	61	63	65	86	88	90
80	65	68	70	91	93	96
85	70	72	74	97	99	101
90	74	77	79	102	104	107
95	79	82	84	107	109	112
100	84	86	88	113	115	117

Table A.7

Percentage Points of Reverse Arrangement Distribution

Values of $A_{N;\alpha}$ such that Prob $[A_N > A_{N;\alpha}] = \alpha$ where N = total number of measurements

N	α					
	0.99	0.975	0.95	0.05	0.025	0.01
10	9	11	13	31	33	35
12	16	18	21	44	47	49
14	24	27	30	60	63	66
16	34	38	41	78	81	85
18	45	50	54	98	102	107
20	59	64	69	120	125	130
30	152	162	171	263	272	282
40	290	305	319	460	474	489
50	473	495	514	710	729	751
60	702	731	756	1013	1038	1067
70	977	1014	1045	1369	1400	1437
80	1299	1344	1382	1777	1815	1860
90	1668	1721	1766	2238	2283	2336
100	2083	2145	2198	2751	2804	2866

APPENDIX B

DEFINITIONS FOR RANDOM DATA ANALYSIS

Autocorrelation Function

The autocorrelation function $R_{xx}(\tau)$ of a quantity $x(t)$ is the average of the product of the quantity at time t with the quantity at time $(t + \tau)$ for an appropriate averaging time T:

$$R_{xx}(\tau) = \frac{1}{T} \int_0^T x(t) x(t + \tau) \, dt$$

The delay τ can be either positive or negative.

For an ergodic process, T should approach infinity; but, in practice, T must be finite.

The total mean square value $\overline{x^2}$ can be estimated by

$$\overline{x^2} = R_{xx}(0) = \frac{1}{T} \int_0^T x^2(t) \, dt$$

Autospectral Density Function

By finite Fourier transform techniques, the autospectral (also called power spectral) density function $G_{xx}(f)$ is defined for $0 \leq f < \infty$ by

$$G_{xx}(f) = \frac{2}{T} E\left[|X_T(f)|^2 \right]$$

where $E[\ ]$ is an ensemble average, for fixed f, over n_d available sample records of $X_T(f)^2$. The quantity $X_T(f)$ is a finite Fourier transform of $x(t)$ of length T. The quantity $G_{xx}(f) = 0$ for $f < 0$ and is physically measurable.

For theoretical studies, a two-sided autospectral density function $S_{xx}(f)$ can be defined for $-\infty < f < \infty$ by setting

$$S_{xx}(f) = \tfrac{1}{2}G_{xx}(f) \qquad \text{when } f \geqq 0$$

$$S_{xx}(-f) = S_{xx}(f)$$

For stationary random data, the autospectral density function $G_{xx}(f)$ is twice the Fourier transform of the autocorrelation function $R_{xx}(\tau)$.

The total mean square value $\overline{x^2}$ can be obtained by integrating $G_{xx}(f)$ or $S_{xx}(f)$ as follows:

$$\overline{x^2} = \int_0^\infty G_{xx}(f)\,df = \int_{-\infty}^\infty S_{xx}(f)\,df$$

Coherence Function

The coherence function $\gamma_{xy}^2(f)$ of two quantities $x(t)$ and $y(t)$ is the ratio of the square of the absolute value of the cross-spectral density function to the product of the autospectral density functions of the two quantities:

$$\gamma_{xy}^2(f) = \frac{\left|G_{xy}(f)\right|^2}{G_{xx}(f)G_{yy}(f)}$$

For all f, the quantity $\gamma_{xy}^2(f)$ satisfies $0 \leqq \gamma_{xy}^2(f) \leqq 1$.

This ordinary coherence function measures the extent to which $y(t)$ may be predicted from $x(t)$ by an optimum linear least squares relationship.

Coherent Output Spectrum

The coherent output spectrum $G_{vv}(f)$ for a single-input/single-output problem is the product of the coherence function between the input signal $x(t)$ and the output signal $y(t)$, multiplied by the output autospectral density function:

$$G_{vv}(f) = \gamma_{xy}^2(f)G_{yy}(f)$$

The associated noise output spectrum $G_{nn}(f)$ is given by

$$G_{nn}(f) = \left[1 - \gamma_{xy}^2(f)\right]G_{yy}(f)$$

Cross-Correlation Coefficient Function

The cross-correlation coefficient function $\rho_{xy}(\tau)$ of two quantities $x(t)$ and $y(t)$ is the ratio of the cross-correlation function $R_{xy}(\tau)$ to the square root of the product of the autocorrelation functions of the two quantities at $\tau = 0$:

$$\rho_{xy}(\tau) = \frac{R_{xy}(\tau)}{\sqrt{R_{xx}(0)R_{yy}(0)}}$$

For all τ, the quantity $\rho_{xy}(\tau)$ satisfies $-1 \le \rho_{xy}(\tau) \le 1$.

This cross-correlation coefficient function and the ordinary coherence function are *not* Fourier transforms of each other.

Cross-Correlation Function

The cross-correlation function $R_{xy}(\tau)$ of two quantities $x(t)$ and $y(t)$ is the average of the product of $x(t)$ at time t with $y(t)$ at time $(t + \tau)$ for an appropriate averaging time T:

$$R_{xy}(\tau) = \frac{1}{T}\int_0^T x(t)y(t + \tau)\,dt$$

For a pair of ergodic processes, T should approach infinity; but, in practice, T must be finite.

The autocorrelation function $R_{xx}(\tau)$ is a special case of $R_{xy}(\tau)$ when $x(t) = y(t)$.

Cross-Spectral Density Function

By finite Fourier transform techniques, the cross-spectral density function is defined for $0 \le f < \infty$ by

$$G_{xy}(f) = \frac{2}{T}E\left[X_T^*(f)Y_T(f)\right]$$

where $E[\ ]$ is an ensemble average, for fixed f, over n_d available associated pairs of $X_T^*(f)$ and $Y_T(f)$ computed from sample records of $x(t)$ and $y(t)$, each of length T. The quantity X_T^* is the complex conjugate of the finite Fourier transform $X_T(f)$ of $x(t)$, while $Y_T(f)$ is the finite Fourier transform of $y(t)$. The quantity $G_{xy}(f) = 0$ for $f < 0$ and is physically measurable.

For theoretical studies, a two-sided cross-spectral density function $S_{xy}(f)$ can be defined for $-\infty < f < \infty$ by setting

$$S_{xy}(f) = \tfrac{1}{2}G_{xy}(f) \qquad \text{when } f \ge 0$$

$$S_{xy}(-f) = S_{xy}^*(f)$$

For stationary random data, the cross-spectral density function $G_{xy}(f)$ is twice the Fourier transform of the cross-correlation function $R_{xy}(\tau)$.

Energy Autospectral Density Function

By finite Fourier transform techniques, the energy autospectral density function is defined for $0 \leq f < \infty$ by

$$\mathscr{G}_{xx}(f) = 2E\left[\left|X_T(f)\right|^2\right]$$

where $E[\ ]$ is an ensemble average, for fixed f, over n_d available sample (transient) records of $x(t)$, each of finite length T. This quantity $\mathscr{G}_{xx}(f) = 0$ for $f < 0$ and is physically measurable.

The energy autospectral density function for transient random data is related to the "power" autospectral density function for the same transient data by

$$\mathscr{G}_{xx}(f) = TG_{xx}(f)$$

where T is the length of the transient records. Observe that $G_{xx}(f)$ for transient data must approach zero as T approaches infinity.

For theoretical studies, a two-sided energy autospectral density function can be defined for $-\infty < f < \infty$ by setting

$$\mathscr{S}_{xx}(f) = \tfrac{1}{2}\mathscr{G}_{xx}(f) \qquad \text{when } f \geq 0$$

$$\mathscr{S}_{xx}(-f) = \mathscr{S}_{xx}(f)$$

Energy Cross-Spectral Density Function

By finite Fourier transform techniques, the energy cross-spectral density function is defined for $0 \leq f < \infty$ by

$$\mathscr{G}_{xy}(f) = 2E\left[X_T^*(f)Y_T(f)\right]$$

where $E[\ ]$ is an ensemble average, for fixed f, over n_d available associated pairs of sample (transient) records of $x(t)$ and $y(t)$, each of finite length T. This quantity $\mathscr{G}_{xy}(f) = 0$ for $f < 0$ and is physically measurable.

The energy cross-spectral density function for transient random data is related to the usual cross-spectral density function for the same transient data by

$$\mathscr{G}_{xy}(f) = TG_{xy}(f)$$

where T is the length of the transient records.

For theoretical studies, a two-sided energy cross-spectral density function can be defined for $-\infty < f < \infty$ by setting

$$\mathscr{S}_{xy}(f) = \tfrac{1}{2}\mathscr{G}_{xy}(f) \qquad \text{when } f \geq 0$$

$$\mathscr{S}_{xy}(-f) = \mathscr{S}_{xy}^*(f)$$

Ergodic Process

An ergodic process is a stationary random process involving a collection of time-history records where time-averaged results are the same for every record. It follows that these time-averaged results from any single record will then be equal to corresponding ensemble averaged results over the collection of records.

To help explain this definition, consider a stationay random process $\{x_k(t)\}$ where $k = 1, 2, 3, \ldots$, represents the different records. For any particular record $x_k(t)$, a time-averaged result such as the mean square value is given theoretically by

$$\overline{x_k^2} = \lim_{T \to \infty} \frac{1}{T} \int_0^T x_k^2(t)\, dt$$

This result must be independent of k and the same for all records if the process is ergodic. The corresponding ensemble averaged result over the collection of records is given theoretically by

$$E[x^2] = \lim_{K \to \infty} \frac{1}{K} \sum_{k=1}^{K} x_k^2(t)$$

and is independent of t for stationary processes. For an ergodic process, the two types of averages given above will be equal.

Fourier Series

A Fourier series expresses a periodic quantity $x(t)$ in terms of its individual frequency components. If $x(t)$ is periodic of period T where $x(t) = x(t + T)$, then

$$x(t) = \frac{a_0}{2} + \sum_{n=1}^{\infty} a_n \cos 2\pi nft + \sum_{n=1}^{\infty} b_n \sin 2\pi nft$$

The frequency $f = (1/T)$ is the fundamental frequency. The coefficients

$$a_n = \frac{2}{T} \int_0^T x(u) \cos 2\pi nfu\, du$$

$$b_n = \frac{2}{T} \int_0^T x(u) \sin 2\pi nfu\, du$$

$$n = 0, 1, 2, \ldots$$

where u is a dummy variable of integration.

Fourier Transform

The Fourier transform, also called the Fourier spectrum, $X(f)$ of a quantity $x(t)$, is a complex-valued function of frequency f defined by

$$X(f) = \int_{-\infty}^{\infty} x(t) e^{-j2\pi ft} \, dt \qquad -\infty < f < \infty$$

assuming $x(t)$ is such that $X(f)$ exists. The time function $x(t)$ is obtained from $X(f)$ by

$$x(t) = \int_{-\infty}^{\infty} X(f) e^{j2\pi ft} \, df \qquad -\infty < t < \infty$$

$X(f)$ and $x(t)$ are known as the direct and inverse Fourier transforms, respectively.

In terms of real and imaginary parts,

$$X(f) = \text{Re}[X(f)] - j \, \text{Im}[X(f)]$$

where

$$\text{Re}[X(f)] = \int_{-\infty}^{\infty} x(t) \cos 2\pi ft \, dt$$

$$\text{Im}[X(f)] = \int_{-\infty}^{\infty} x(t) \sin 2\pi ft \, dt$$

In actual practice, $x(t)$ will be of finite length T, so that $X(f)$ is estimated by computing the finite Fourier transform

$$X_T(f) = X(f, T) = \int_{0}^{T} x(t) e^{-j2\pi ft} \, dt$$

Such finite integrals always exist.

Frequency Response Function

The frequency response function $H(f)$ for a constant-parameter linear system is the Fourier transform of the unit impulse response function $h(\tau)$ which describes this system. In equation form,

$$H(f) = \int_{-\infty}^{\infty} h(\tau) e^{-j2\pi f\tau} \, d\tau$$

The quantity $H(f)$ is often called the transfer function by engineers, although the term transfer function should be reserved for the Laplace transform of the unit impulse response function.

In complex polar notation,

$$H(f) = |H(f)|e^{-j\phi(f)}$$

where

$$|H(f)| = \text{gain factor of system}$$

$$\phi(f) = \text{phase factor of system}$$

For linear systems, $H(f)$ can be estimated using deterministic data, transient data, or stationary random data since its properties are independent of the nature of data passing through the system.

Gain Factor

See frequency response function.

Gaussian Process

A Gaussian process is a stationary random process $x(t)$ whose instantaneous values at any time t have the following probability density function:

$$p(x) = \frac{1}{\sigma_x\sqrt{2\pi}} \exp\left[-(x - \mu_x)^2/2\sigma_x^2\right] \qquad -\infty < x < \infty$$

where μ_x is the true mean value of $x(t)$ and σ_x^2 is the true variance of $x(t)$. This probability density function defines the Gaussian distribution.

Histogram

A histogram is a plot of the number of observed values of a quantity $x(t)$ that falls within various specified magnitude ranges, called class intervals.

Hilbert Transform

The Hilbert transform of a real-valued function $x(t)$ extending over the range $-\infty < t < \infty$ is a real-valued function $\tilde{x}(t)$ defined by

$$\tilde{x}(t) = \mathcal{H}[x(t)] = \int_{-\infty}^{\infty} \frac{x(u)}{\pi(t - u)} du$$

Thus the Hilbert transform $\tilde{x}(t)$ is the original function $x(t)$ convolved with $(1/\pi t)$.

Line Spectrum

A line spectrum is a spectrum whose components occur at a number of discrete frequencies, as in a Fourier series representation.

Linear System

A linear system is one that is additive and homogeneous. Given two inputs x_1 and x_2 that individually produce outputs y_1 and y_2, the system is additive if the input $x_1 + x_2$ produces the output $y_1 + y_2$, and homogeneous if the input cx_1 produces the output cy_1, where c is an arbitrary constant.

Mean Value

The mean value $\bar{x}$ of a quantity $x(t)$ is the time average of the quantity for an appropriate averaging time T:

$$\bar{x} = \frac{1}{T} \int_0^T x(t)\, dt$$

For an ergodic process, the true mean value μ_x can be obtained by letting T approach infinity.

Mean Square Value

The mean square value $\overline{x^2}$ of a quantity $x(t)$ is the time average of the square of the quantity for an appropriate averaging time T:

$$\overline{x^2} = \frac{1}{T} \int_0^T x^2(t)\, dt$$

For an ergodic process, the true mean square value can be obtained by letting T approach infinity. See also notes under autocorrelation and autospectral density functions.

Multiple Coherence Function

The multiple coherence function $\gamma^2_{y:x}(f)$ between a quantity $y(t)$ and a set of other quantities $x_i(t)$, $i = 1, 2, \ldots, q$, measures the extent to which $y(t)$ may be predicted from these various $x_i(t)$ by optimum linear least squares relationships. For all f, the multiple coherence function satisfies $0 \leq \gamma^2_{y:x}(f) \leq 1$. The multiple coherence function includes the ordinary coherence function as a special case.

Narrow-Band Random Data

Narrow-band random data are data whose spectral values are significant only within a narrow bandwidth around a center frequency. If such data have instantaneous values that follow a Gaussian distribution, then their peak values approximate a Rayleigh distribution.

Nonlinear System

A nonlinear system is one that is either not additive or not homogeneous, as defined under linear system.

Nonstationary Process

A nonstationary process is any process that is not a stationary process (refer to definition of stationary process). Statistical averages computed over an ensemble of time-history records are not invariant with respect to translations in time but are a function of the times being analyzed.

In general, time-averaged results from any single record do not represent this record or any other record, since information about essential time-varying properties are lost by the averaging operation.

Partial Coherence Function

The partial coherence function between a quantity $y(t)$ and any subset of a larger set of known quantities $x_i(t)$, $i = 1, 2, \ldots, q$, measures the extent to which $y(t)$ may be predicted from this subset by optimum linear least squares relationships. Such partial coherence functions will be bounded between zero and unity.

Peak Value

A peak value of a quantity $x(t)$ is the value of $x(t)$ at a maximum or minimum value. Peak counts per unit time can indicate the number of maxima only, the number of minima only, or both.

Phase Factor

See frequency response function.

Power Spectral Density Function

See autospectral density function.

Probability Density Function

The probability density function $p(x)$ of a quantity $x(t)$ is the probability that $x(t)$ at any value of t will assume the value x per unit magnitude window for an appropriate magnitude window W:

$$p(x) = \frac{P(x, W)}{W}$$

where $P(x, W)$ is the probability that $x(t)$ falls in the magnitude window W centered at x. In words, $p(x)$ is an estimate of the rate of change of probability with magnitude.

For stationary random data, W should approach zero; but, in practice, W must be greater than zero. For all values of x, $p(x) \geq 0$, with the area under the probability density curve equal to unity.

Probability Distribution Function

The probability distribution function $P(x)$ defines the probability that $x(t) \leq x$ at any value of t. In terms of the probability density function $p(x)$,

$$P(x) = \int_{-\infty}^{x} p(u) \, du$$

where u is a dummy variable of integration. Observe that $P(-\infty) = 0$ and $P(\infty) = 1$.

Random Process

A random process is a collection of time-history records that can be described by appropriate statistical parameters, such as averaged properties of these records at a number of fixed times.

Root Mean Square Value

The root mean square (rms) value is the positive square root of the mean square value. The rms value is equal to the standard deviation if the mean value is zero.

Spectral Density

See autospectral and cross-spectral density function.

Spectrum

A spectrum is a description of a quantity in terms of any function of frequency. The spectrum may be either a line spectrum or a continuous spectrum.

Standard Deviation

The standard deviation is the positive square root of the variance. The standard deviation is equal to the rms value if the mean value is zero.

Stationary Process

A stationary random process is a collection of time-history records having statistical properties that are invariant with respect to translations in time. Stationary processes may be either ergodic or nonergodic.

Time-History Record

A time-history record is the waveform of any quantity expressed as a function of time, where the data may be either deterministic or random. The reciprocal of the period for a record that is periodic in time is the frequency of the record. Any other independent variable may replace time in time-history records, provided a corresponding change is made in the interpretation of frequency.

Transfer Function

The transfer function for a constant-parameter linear system is the Laplace transform of the unit impulse response function that describes this system. Along the imaginary axis, this quantity becomes the frequency response function of the system. See frequency response function.

Transient Data

Transient data are data of limited duration, which may be either deterministic or random.

Unit Impulse Response (Weighting) Function

The unit impulse response function, also called the weighting function, $h(\tau)$ of a constant-parameter linear system describes the response of this system to a unit impulse (delta) function. It is the inverse Fourier transform of the frequency response function of the system.

Variance

The variance s_x^2 of a quantity $x(t)$ is the time average of the square of the deviation from the mean value $\bar{x}$ for an appropriate averaging time T:

$$s_x^2 = \frac{1}{T} \int_0^T [x(t) - \bar{x}]^2 \, dt$$

For an ergodic process, the true variance σ_x^2 can be obtained by letting T approach infinity. The quantity $s_x^2 = \overline{x^2}$ if $\bar{x} = 0$.

Wide-Band Random Data

Wide-band random data are data whose spectral values are significant over a wide frequency range relative to the center of this band.

LIST OF FIGURES

LIST OF TABLES

LIST OF EXAMPLES

INDEX